This book is in the
ADDISON-WESLEY SERIES IN ELECTRICAL ENGINEERING

Consulting Editors:
David K. Cheng
Leonard A. Gould
Fred K. Manasse

Portions of this manuscript originally appeared in *Introduction to Semiconductor Circuit Design* by David J. Comer. Copyright © 1968 by Addison-Wesley Publishing Company, Inc.

Copyright © 1976 by Addison-Wesley Publishing Company, Inc. Philippines copyright 1976 by Addison-Wesley Publishing Company, Inc.

All rights reserved. No part of this publication may be reproduced, stored in a retrieval system, or transmitted, in any form or by any means, electronic, mechanical, photocopying, recording, or otherwise, without the prior written permission of the publisher. Printed in the United States of America. Published simultaneously in Canada. Library of Congress Catalog Card No. 75-9008.

ISBN 0-201-01008-9
ABCDEFGHIJ-MA-798765

PREFACE

This book is intended for the electronics sequence of courses in an undergraduate electrical engineering curriculum. There is sufficient material for a three-semester sequence beginning with introductory electronics and finishing with advanced electronic circuits.

Since there are several good electronics textbooks available today, one is prompted to question the necessity of yet another. This question is answered, however, by listing the unique features of this text.

1. Extraneous detail is avoided. A great deal of editing has been done to rid the text of material that is too complex or too specialized for the undergraduate student. This should allow the instructor to proceed smoothly without the necessity of skipping sections at regular intervals. As in my first book *Introduction to Semiconductor Circuit Design*, sections appearing with an asterisk can be used or avoided at the instructor's discretion.

2. After the first two chapters the primary emphasis is on design of circuits rather than analysis. It is hoped that the student will come to appreciate the fact that engineers must often make decisions based on broad constraints rather than specific formulas.

3. A modest emphasis is placed on digital system design in addition to circuit design. Today's young engineer can scarcely avoid the digital field, and much of the design in this field is based on integrated circuit systems.

4. Each new electronics textbook must update the older ones as a result of the unbelievable speed with which the field is developing. Since this text covers several important new topics it is as modern as its title indicates.

5. The homework problems are subdivided by section and an asterisk is placed next to each problem that has the answer listed in the back of the book. Hopefully, this will aid the instructor in assigning meaningful exercises.

I am grateful to those around me who have encouraged this project. My family, in particular, serves as an inspiration in my writing efforts. I am grateful to my parents for instilling in me the confidence necessary to embark on such projects. The several typists who have assisted with the manuscript also have my gratitude, especially Mrs. Bonnie Smith whose expert skill in producing the final manuscript is greatly appreciated.

Chico, California D. J. C.
October 1975

CONTENTS

Chapter 1 Semiconductor processes

    1.1   Introduction 1
    1.2   Charge carriers 1
    1.3   Conduction processes in semiconductors 11

Chapter 2 Discrete devices and integrated circuits

    2.1   pn-Junction diodes 26
    2.2   The bipolar transistor 49
    2.3   The field-effect transistor 60
    2.4   Construction of transistors and integrated circuits 71

Chapter 3 Transistor biasing

    3.1   Purpose of biasing 82
    3.2   Base current bias 85
    3.3   Temperature effects 87
    3.4   Temperature stability of base-current bias stage 89
    3.5   Emitter bias 90
    3.6   Collector-base bias 94
    3.7   Stability with respect to changes in β_0 95
    3.8   Temperature compensation techniques 99
    3.9   Variation of I_{co} with temperature 102
    3.10   Variation of V_{BE} with temperature 104
    3.11   Bias design of FET amplifiers 105
    3.12   Integrated circuit biasing 109

Chapter 4 Equivalent circuits and modeling

    4.1   The diode equivalent circuit 117
    4.2   Bipolar transistor equivalent circuits 122
    4.3   FET equivalent circuits 138
    4.4   Computer-aided design models 144

Chapter 5 Low-frequency amplifiers

5.1	Voltage amplifiers	161
5.2	Low frequency falloff due to coupling capacitor	169
5.3	Frequency response	173
5.4	Low frequency falloff due to emitter-bypass capacitor	178
5.5	The combination of a coupling capacitor and an emitter-bypass capacitor	181
5.6	Low frequency reactive effects in the FET	182
5.7	Gain stability	184
5.8	Common-base amplifier	186
5.9	Emitter follower	187
5.10	Design examples	191
5.11	DC amplification	202
5.12	The differential amplifier	205

Chapter 6 Feedback amplifiers

6.1	The ideal feedback amplifier	220
6.2	Practical feedback amplifiers	234
6.3	Stability of feedback systems	246
6.4	Operational amplifiers	248

Chapter 7 High-frequency amplifiers

7.1	The transistor at high frequencies	271
7.2	Single-stage broadbanding techniques	282
7.3	Multistage amplifiers	294
7.4	Tuned stages	319

Chapter 8 Large-signal circuits

8.1	Low-speed transistor switching	341
8.2	High-speed transistor switching	347
8.3	Tunnel diode switching	365
8.4	The silicon-controlled rectifier	366

Chapter 9 Logic components and families

9.1	Logic gates	380
9.2	Logic families	385
9.3	Multivibrators	395

Chapter 10 Digital systems

10.1	The digital computer	423
10.2	Boolean algebra	428
10.3	Encoding and decoding	439
10.4	Registers and timing circuits	448
10.5	Arithmetic circuits	464
10.6	Memories	469

Chapter 11 Waveform generation

11.1	Sinusoidal oscillators	487
11.2	Waveform generation	506

Chapter 12 Power amplifiers

12.1	Classification of power stages	519
12.2	Allowable dissipation	522
12.3	Class-A stages	528
12.4	Class-B power output stages	536
12.5	Class-C amplifiers	548
12.6	Pulse-width modulated audio power amplifiers	554
12.7	Miscellaneous aspects of power amplification	558

Chapter 13 Applications of the op amp

13.1	Digital-to-analog conversion	566
13.2	Integrators and sweep circuits	569
13.3	Active filters	581

Chapter 14 Modulators, receivers, and communication systems

14.1	Communications	596
14.2	Modulation	598

Chapter 15 Semiconductor power supplies

15.1	Rectifiers	644
15.2	Filters	648
15.3	Regulators	661
15.4	High-current, variable voltage supply	674

Answers to selected problems	681
Index	687

1
SEMICONDUCTOR PROCESSES

1.1 INTRODUCTION

The semiconductor field has been rather successful in the continual development of new devices and has not yet approached the bounds of operation. Junction transistors appear to have a theoretical frequency limit of operation, but this limit depends to some extent on the manufacturing process. In the past when the frequency limits were approached, new processes were developed to extend these limits. Furthermore, as these limits were increased for the transistor, new semiconductor devices were developed which exceeded the frequency performance of transistors. Alloy transistors were followed by grown junction units and eventually mesa or planar transistors. The tunnel diode, avalanche diode, and varactor diode extended semiconductor frequency capability into the GHz range. So too has the power capability of semiconductor devices increased, and the improvement in transistor performance has been accompanied by the development of such new power devices as the silicon-controlled rectifier. Other important developments in the last two decades include semiconductor lasers, field-effect transistors, and integrated circuits. The bipolar junction transistor remains one of the more important devices as a result of its widespread usage both in discrete circuits and as a component part of the integrated circuit. While several semiconductor devices are treated in succeeding chapters, the bipolar transistor is most heavily emphasized. This chapter reviews those processes that are basic to the operation of transistors and other semiconductor devices.

1.2 CHARGE CARRIERS

A semiconductor is a material possessing an electrical conductivity between that of an insulator and that of a conductor. A good conductor might have a conductivity of 6×10^5 ℧/cm while the conductivity of an insulator may be 6×10^{-18} ℧/cm. On a macroscopic level it is difficult to visualize a property varying from one material to another by a factor of 10^{20} or more. Only at the atomic level can variations of this order of magnitude be explained. The effective conduction of electrical current is not, however, explainable simply in terms of total numbers of charged particles present in a material. Certainly insulators such as wood, glass, or rubber contain many atoms with accompanying protons and electrons. These charged particles are bound very closely to the parent atoms in an insulator. On the other hand, a conductor contains many electrons that are only weakly bound to the parent atom. At normal temperatures these electrons possess enough thermal energy to leave the parent atom and move randomly throughout the lattice. A slight electric field applied to the conductor imparts a net movement to these free electrons. Current is then conducted by means of the movement of charge resulting from this free electron motion. Of course, there is a tremendous number of charged particles, electrons and protons, that remain tightly bound to the lattice atoms and do not contribute to the conduction process. The basic difference between an insulator and conductor then is not in the total number of charged particles, but in the number of free charges or electrons available in the material.

At room temperature semiconducting materials contain a density of free electrons that is intermediate to that of a conductor and that of an insulator. The resulting conductivity also falls somewhere between the conductivities of an insulator and a metal, hence the name semiconductor. Germanium has a conductivity of approximately 2.2×10^{-2} ℧/cm, while pure silicon may approach values as low as 4×10^{-6} ℧/cm.

One point that must be remembered is that while metals and semiconductors have free electrons or mobile charge carriers, the net charge over a metal or semiconducting bar will be zero. Each immobile parent atom that contributes a free electron has a net charge of $+q$ on the departure of the electron.

We will now consider the presence of free charges in metals and semiconductors. We will see that while thermal energy plays an important part in the generation of free electrons, the total concentration can be controlled in semiconductors by the addition of certain impurity atoms.

A. Free Electrons in Metals and Semiconductors

Although we are primarily concerned with semiconductors, a brief discussion of the origin of free electrons in metal provides useful background information. It has already been noted that metals contain many more free or conduction electrons than do semiconductor materials. To provide the complete picture of conduction electrons in metals requires a knowledge of quantum mechanics and Fermi–Dirac statistics. It is considered beyond the scope and intent of this book to provide a detailed treatment of these topics. Thus, only the high points of established theory in these areas will be covered.

Basic chemistry tells us that an atom of any material includes a nucleus and any number of electrons that are considered to be in "orbit" about the nucleus. These orbits are not perfectly circular paths in which the electrons rotate, but are more appropriately regarded as modes of motion of the electrons. A rather important parameter in specifying these orbits is the angular momentum of the electron. No one fully understands why certain atomic rules or laws are never violated; however, certain principles have been developed based on observation that tend to support these rules. It is known, for example, that an electron associates with a nucleus only in discrete states. That is, each electron has a particular value of orbital angular momentum, orbital and electron spin angular momenta, and total energy. The principle describing the fact that no two electrons in a system can occupy the same state is called the Pauli exclusion principle.

If the discrete values of allowable energy are associated with electron orbits, the Pauli exclusion principle can be used to predict the number of electrons required to completely fill each orbit. At lower energies fewer values of angular momenta are permitted and fewer electrons are required to fill the shell. More electrons can be accommodated in higher energy orbits as a result of more values of permitted angular momenta. Not all possible states are occupied by electrons, although in general the lower energy orbits tend to be full. Each chemical element has a particular configuration of electrons.

It is possible to modify the electronic configuration of an atom. Of particular importance to our discussion is the change in configuration due to ionization. As energy is imparted to the electrons of the outermost or highest energy shell, an electron can break its association with the nucleus and become a free electron. The atom is said to be ionized when one or more electrons leaves the atom, which then has a net charge equal in magnitude but opposite in sign to the electronic charge of the free electrons. There are various means of ionizing atoms. Thermal energy is an important ionizing agent in semiconductors. Optical energy is used in photoconductive devices. Electrical energy is used to ionize gas atoms in various devices including gas-filled tubes such as neon tubes.

In metals and semiconductors we are dealing with large numbers of atoms that generally fit into a well-defined structure termed a lattice. It is useful to consider the electron energy levels that exist when the atoms form this lattice. The Pauli exclusion principle tells us that no two electrons of the system can occupy the same state, yet there are literally millions of electrons in any given volume, each of which must occupy some discrete state. If we postulate a model consisting of many noninteracting atoms (separate systems) that can be brought nearer together to form a single system, an energy diagram such as that shown in Fig. 1.1 would result [2]. As the atoms approach each other, the interatomic forces interact with neighboring atoms and electrons and cause the allowable energy levels to spread into bands. Electrons in states nearer their nucleus are influenced to a lesser extent than more distant electrons. The allowable energy states can now be defined in terms of bands rather than the discrete energy levels of the single atom. It can be noted that some bands overlap, and in fact this phenomenon results in unique conductivity properties for some metallic materials. The regions between the allowed energy bands are referred to as forbidden regions or energy gaps. Electrons can be excited to higher states, but to do so must experience an increase in energy sufficient to cross the adjacent forbidden region.

The allowable energy bands of greatest importance in the semiconductor are the valence band and the conduction band. Schematically, we might represent these energy levels as shown in Fig. 1.2.

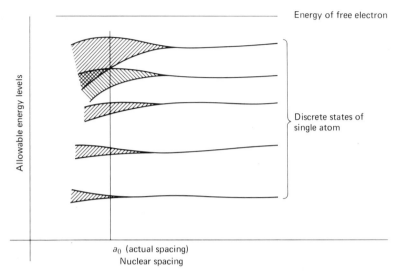

Fig. 1.1 Allowable electronic energy levels as a function of interatomic spacing.

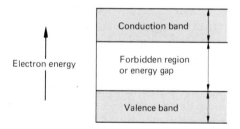

Fig. 1.2 Schematic representation of valence and conduction bands.

The conduction band represents permissible energy levels of free electrons while the valence band represents energy levels corresponding to the outer orbits of the electrons. In silicon or germanium at very low temperatures, the valence band is completely filled with electrons and the conduction band is void of electrons. The semiconductor at very low temperatures is a nonconductor of electricity.

At temperatures approaching room temperature enough thermal energy is imparted to the electrons to cause a small fraction to leave the valence band and become free carriers. There are two results of this modification of electron energy distribution that contribute to electrical conduction. As mentioned previously, there are now free charge carriers in the conduction band. A second result is the creation of empty states in the valence band. At first thought one might be tempted to dismiss the importance of these empty states. However, these empty states almost equal the importance of the free electrons in the conduction process. When the valence band was completely filled there could be no net movement of valence electrons. All electrons are, of course, in continuous motion about the parent atom and can possibly leave the parent atom to become bound to a new parent atom. For this to occur, however, an electron from the second parent atom must simultaneously become bound to the first parent atom since there are only two available states for these two electrons. Thus, even if collisions of valence electrons cause exchange of electrons between parent atoms or energy levels, no net movement of valence electrons occurs. When empty states exist in the valence band, a given electron may move to a neighboring parent atom that contains an available state without requiring that an electron immediately fill the vacated state. Thus, relocation of valence electrons can cause local variations of net charge. Later it will be shown that the movement of valence electrons when empty states exist can be influenced by an externally applied electric field. It might be noted that certain types of metals that are excellent conductors of electricity depend on the movement of electrons within a partially filled valence band. This process is called conduction by holes and will shortly be considered in more detail since it is so important to semiconductor operation.

B. Intrinsic Semiconductors

It has been noted that conduction is dependent on the number of free electrons (conduction electrons) and on the number of holes in the valence band. It is possible to control either of these quantities through a process known as doping. By controlling the electron or hole concentration, the overall electrical conductivity of the semiconductor is determined. It is important to investigate the effect of doping on the number of free carriers within the semiconductor. A pure semiconductor material having an equal number of holes and free electrons is called intrinsic.

Before considering doping, we must take a more detailed look at intrinsic semiconductor material. We assume that it is possible to obtain pure germanium or silicon with a perfect lattice structure. Practically this assumption is invalid; however, the matter of impurities will be related to doping theory later and included in conductivity calculations as appropriate.

There are several types of atomic bonding that occur in the different solids. The four major crystal types are metal, ionic, molecular, and covalent. There are crystals that exhibit combinations of these types of bonding, but silicon and germanium both clearly demonstrate covalent bonding wherein valence band electrons are shared in electron pairs with other atoms. The diamond structure is exemplified by the semiconductors silicon and germanium. In this arrangement each atom is surrounded by four nearest-neighbor atoms. While each atom has several orbits or shells of electrons, the inner shells are unimportant to the conduction process as a result of the tight binding of these electrons to their respective nuclei. The outer shell contains four valence electrons, yet this shell has eight available states. In the diamond structure each nearest-neighbor atom shares one of its outer orbit electrons with the central atom. To better visualize this arrangement a two-dimensional schematic is shown in Fig. 1.3.

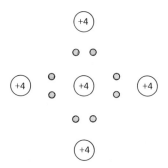

Fig. 1.3 Schematic representation of covalent bonding.

When each atom is viewed as a central atom, there will be eight electrons associated with this atom. This electron-sharing arrangement is called covalent bonding. It can be noted that the entire material will be space-charge neutral since the net charge of $+4q$ is balanced by the negative charge of the four electrons.

Each electron that is covalently shared has an energy that lies in the valence band of energy levels. At very low temperatures a negligible number of electrons possess enough thermal energy to break the bond. Alternatively, we could say that the electrons do not possess enough energy to cross the forbidden region or energy gap to become free carriers. This explains why the semiconductor is such a poor conductor at lower temperatures. Even at higher temperatures (for example, $T = 300°K$) the conductivity is poor although several electrons now have gained enough energy to break the bond and become free carriers.

As an electron crosses the energy gap to generate a free carrier, a vacancy or hole is left in the valence band. At all temperatures in pure germanium or silicon the density of free electrons equals the density of free holes. The density of free electrons and holes can be calculated at any temperature for a given material. It should be emphasized that the calculation is based on statistics. If the result of the calculation tells us that there are 10^8 free electrons per cubic centimeter, we must recognize that there is a continuous exchange of electrons between conduction and valence bands with the average concentration of conduction electrons equaling $10^8/\text{cm}^3$.

To calculate the density of free electrons or holes in intrinsic material requires the use of Fermi–Dirac or Boltzmann statistics and some quantum-mechanical considerations. Rather than attempt such a development, we will use only the result. It has been found theoretically and verified experimentally that the concentration of free electrons (n_i) and free holes (p_i) in pure germanium or silicon is given by

$$n_i(T) = p_i(T) = AT^{3/2}e^{-E_g/2kT}, \tag{1.1}$$

where E_g is the energy gap of the material, T is the absolute temperature, A is a constant, and k is Boltzmann's constant. Near room temperature (300°K), the intrinsic concentrations for silicon and germanium are (in number/cm³)

	Si	Ge
n_i	1.4×10^{10}	2.5×10^{13}

The difference in concentrations for the two materials is primarily a result of energy gap differences. The band gap for germanium is approximately 0.68 eV while for silicon it is 1.1 eV. The factor of less than 2 in energy gaps for the two materials results in a factor of almost 2×10^3 in free carrier concentration.

C. Creation of Free Electrons or Holes in Semiconductors (Doping)

It is possible to sprinkle the semiconductor lattice with certain types of atoms that can contribute free electrons or free holes. Since the number of free carriers is controlled by the addition of these impurity atoms, so also is the conductivity of the semiconductor material.

Acceptor impurities. In order to create additional holes in the valence band, electrons must be removed from this band. This can be accomplished by introducing acceptor atoms such as boron, gallium, or indium which contain only three electrons in their valence band. When acceptor atoms become an integral part of the semiconductor lattice, covalent bonding requirements are no longer satisfied for each atom. The acceptor atoms tend to have only seven surrounding valence electrons whereas the covalent bonding arrangement requires eight valence electrons. An electron from some neighboring atom can easily fall into this vacancy resulting in the creation of a hole.

From an energy standpoint, the acceptor atoms provide available electron states at energy levels very near the valence band. The thermal energy of the valence electrons can easily excite electrons to these new states. Thus, at room temperature one hole is created for almost every acceptor atom that is present in the lattice. The schematic representation of an acceptor atom surrounded by semiconductor atoms is given in Fig. 1.4(a). Figure 1.4(b) shows the energy levels of the electron states of a semiconductor material doped with acceptor atoms.

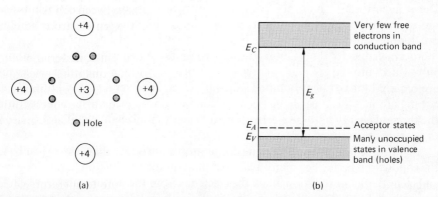

Fig. 1.4 (a) Acceptor atom surrounded by semiconductor atoms. (b) Energy diagram for semiconductor with acceptor impurities.

Impurity atoms can be added to a semiconductor in several ways. Vapors of acceptor atoms can be diffused into the lattice, impurities can be added to molten semiconductor material before the lattice is grown, or the doped material can be produced by a rate-grown process. These and other manufacturing techniques will be considered later.

Before adding the acceptor atoms the only available states to which the bound valence electrons could move were in the conduction band. Since a relatively high energy is required to excite electrons to the conduction band, very few electrons or holes could act as free carriers. Addition of the acceptor atoms creates available states to which the bound valence electrons can easily move. For the popular acceptor impurities, all available acceptor states are occupied at room temperature so long as the concentration of doping atoms is not excessive. At a concentration of 10^{16} atoms/cm^3 approximately 99% of the acceptor states are occupied. In modern transistors higher concentrations of impurities are encountered and some modification of theory is in order. We will not consider the case of degenerate or near-degenerate doping, noting that even transistors with heavily doped emitters approach a more typical impurity concentration near the base-emitter junction. For fabrication of most semiconductor devices the concentration of doping atoms is greater by several orders of magnitude than is the intrinsic carrier concentration. Since $N_A \gg p_i$, then

$$p = N_A. \tag{1.2}$$

Material doped with an acceptor impurity is called p-type.

Even with doping the net charge is zero throughout the semiconductor. An isolated acceptor atom is electrically neutral as are all lattice atoms of the material. When a valence electron leaves its parent atom, the vacancy or hole results in a charge of $+q$ for the atom. The acceptor atom now has a charge of $-q$ as a result of the additional electron. Obviously the net charge will be zero.

There are impurities called double acceptors that create two holes per atom, but these impurities are not so easy to control as the popular ones mentioned earlier.

Donor impurities. An atom having five electrons in its outer orbit (such as antimony, arsenic, or phosphorus) can act as a donor impurity. Again it is possible to use double donor impurities with six valence electrons, but single donor atoms are more amenable to the manufacturing processes. When a pentavalent atom is substituted into the germanium or silicon lattice, only four of the valence electrons are required for covalent bonding. The fifth electron is very weakly bound to the doping atom and can easily be freed for conduction. In fact, at room temperature the thermal energy of these electrons is sufficient to ionize almost all donor atoms of the types mentioned above.

From an energy standpoint, the donor atoms introduce filled states just below the conduction band. The many available states in the conduction band along with sufficient thermal excitation allow the electrons from the donor states to reach the conduction band. The schematic representation of a donor atom surrounded by

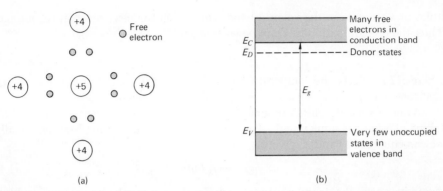

Fig. 1.5 (a) Donor atom surrounded by semiconductor atoms. (b) Energy diagram for semiconductor with donor impurities.

semiconductor atoms is given in Fig. 1.5(a). Figure 1.5(b) shows the energy level diagram for a semiconductor material doped with donor atoms.

For fabrication of most devices the concentration of doping atoms is greater by several orders of magnitude than the intrinsic carrier concentration. Since $N_D \gg n_i$, then

$$n = N_D. \tag{1.3}$$

Material doped with a donor impurity is called n-type.

D. Generation and Recombination of Carriers

We have noted that the thermal energy possessed by the semiconductor material is sufficient to break several bonds and free valence electrons at room temperature. The generation of free electrons is a rather complex process involving electron-electron collisions and electron-lattice interactions. Fortunately, the generation process can be treated statistically to derive some rather interesting results. By physical reasoning we recognize that the thermal generation rate $g(T)$ must depend in a given material only on the temperature of the material, since thermal energy is the only available means of breaking a bond and creating an electron-hole pair. Of course, if a higher density of lattice atoms were present, all other factors being equal, the generation rate would be higher. In considering a specific material, however, we must conclude that the generation rate is governed only by the temperature T.

Just as electrons can gain energy by lattice and electron interactions, they can also lose energy by these same means. A conduction electron can give up energy and fall back into the valence band by recombining with an available hole. This process also takes place continuously within the material at a rate dependent on three key quantities. First of all, the recombination rate must depend on the temperature since the energy of the lattice and the conduction electrons are determined by T. This rate must also depend on the number of conduction electrons available to recombine, and it must depend on the number of holes or vacancies in the valence band with which the electrons can recombine. Thus, the recombination rate can be written as

$$R = r(T)np, \qquad (1.4)$$

where $r(T)$ reflects the temperature dependence of R, n is the density of free electrons, and p is the density of holes in the valence band.

As electron-hole pairs are generated at a rate $g(T)$, recombination must be taking place at the same rate if the system is to be in equilibrium. Mathematically we can equate $g(T)$ and R, giving

$$g(T) = r(T)np. \qquad (1.5)$$

Again by physical reasoning we can visualize the self-regulating mechanism built into Eq. (1.5). If $g(T)$ were assumed larger than R, the values of n and p would continue to increase. The result of increased n and p would be to increase R until the two rates were equal. If the generation rate is assumed smaller than the recombination rate, the values of n and p would decrease continually, leading to a decreasing value of R, until the two rates were equal.

For an intrinsic material where $p_i = n_i$ we can use Eq. (1.5) to write

$$n_i p_i = n_i^2 = g(T)/r(T). \qquad (1.6)$$

Can we extend the results of the above physical arguments to the case of a doped semiconductor material? The answer to this question is yes if we can demonstrate that $g(T)$ and $r(T)$ remain unchanged after doping. The generation rate depends on temperature and on the density of valence electrons. Under normal doping conditions, N_D or N_A will be several orders of magnitude smaller than the density of the lattice atoms; hence, $g(T)$ will be unaffected by doping. If $g(T)$ remains constant then R must also remain constant with doping. It is apparent, though, that R could remain constant while $r(T)$, p, and n change. The quantity $r(T)$ should be unaffected by the addition of donor or acceptor atoms so long as the density of lattice atoms remains much greater than that of the doping atoms. Here again, we conclude that $r(T)$ remains constant with doping, at least under practical doping conditions. Equation (1.5) allows us to write for a doped material

$$pn = g(T)/r(T),$$

and Eq. (1.6) leads to

$$pn = n_i^2. \qquad (1.7)$$

Equation (1.7) is a very important relationship that can also be proven by Fermi–Dirac statistics. We know that when donors are added in normal amounts to the semiconductor, the number of free electrons increases to $n = N_D$; therefore, the number of free holes in this case must reduce to

$$p = n_i^2/N_D. \qquad (1.8)$$

As one might conclude from physical reasoning, adding many free electrons reduces the number of holes in the material. This is often referred to as hole suppression. Since n_i is a constant at a given temperature and since N_D is controllable in the manufacturing process, p can be controlled if necessary.

When a concentration of N_A acceptor atoms is added to intrinsic material, the number of holes created is $p = N_A$ and electrons are suppressed as calculated from the equation

$$n = n_i^2/N_A. \tag{1.9}$$

Equations (1.8) and (1.9) are very important in deriving the theoretical V–I characteristics of a semiconductor diode or transistor.

1.3 CONDUCTION PROCESSES IN SEMICONDUCTORS

The conduction of electrical current in a solid is related to the number of free carriers in the solid and the ease or speed with which these carriers can move through the lattice. When an electric field exists in free space, the motion of an electron placed in that field can easily be predicted by classical methods based on Newton's laws of motion. The motion of a conduction electron through the semiconductor lattice cannot be described directly in classical terms. To accurately treat the conduction process in semiconductors it is necessary to apply quantum mechanics. In free space the particles do not continually collide with accompanying quantum transfers of momenta. As an electron travels in the semiconductor lattice, electron-electron and electron-lattice interactions cause several changes of state to occur.

Fortunately, it is possible to analyze the movement of electrons or holes through a lattice with periodically varying potential and represent the essential results in a classical way. A one-dimensional model such as the Kronig–Penny model is often used to represent the semiconductor lattice. In this model a periodic potential having a very large value but small width at the lattice atom positions, is assumed. Between atoms the potential is taken as zero. Once the analysis has been carried out, it is found that similar conduction properties result if one assumes effective masses for the electrons and holes and applies classical analysis methods [4]. Surprisingly, the effective mass of an electron is less than the rest mass in some materials and greater than the rest mass in others. It should be remembered, however, that the concept of effective mass is merely a tool to bridge the gap between quantum particle behavior and classical analysis methods.

A. Conduction by Electrons

Electrical current requires a net movement of charge to take place. In metals or n-type semiconductors, groups of electrons are free to move when acted on by an external field. The current that flows can be found by measuring the time rate of flow of charges passing a specified point in the circuit. Application of an electric field to the semiconductor material tends to accelerate the free electrons in the appropriate direction. Lattice collisions limit the velocity of any given electron and we speak of an average velocity of the electrons for the given applied field. A measure of the ease with which the particles can move through the lattice is the mobility of the particles. The mobility is given by

$$\mu = |\mathbf{v}/\mathscr{E}|, \qquad (1.10)$$

where $\mathbf{v}$ is the drift velocity of the carriers and $\mathscr{E}$ is the applied electric field. The mobility is a function of the lattice parameters and the lattice vibrations, among other things. The thermal vibrations of the lattice depend on the lattice temperature; thus, mobility for a given material is dependent on temperature. For example, the variation of electron mobility in silicon is often approximated by $= KT^{-5/2}$. At room temperature typical mobilities of electrons in intrinsic silicon and germanium are (in cm^2/volt-sec):

	Si	Ge
μ_n	1300	3800

These values decrease with increasing doping levels.

For conduction in a material dominated by free electrons, the conductivity can be calculated rather easily. If n is the concentration of electrons and $\mathbf{v}_n$ is the average or drift velocity, the time rate of charge flow across a fictitious plane perpendicular to the electron flow is equal to the current density $\mathbf{J}$. The plane is taken to have unit cross-sectional area. This quantity is given by

$$\mathbf{J} = -nq\mathbf{v}_n, \qquad (1.11)$$

where the negative sign results from the negative charge on an electron. The velocity of the electrons can be written in terms of mobility as

$$\mathbf{v}_n = -\mu_n \mathscr{E}. \qquad (1.12)$$

Conductivity due to electrons is now found to be

$$\sigma_n = \left|\frac{\mathbf{J}}{\mathscr{E}}\right| = qn\mu_n. \qquad (1.13)$$

If q is expressed in units of coulombs, n in number per cubic centimeter, and μ_n in cm^2/volt-sec, the conductivity will have units of ℧/cm. The conductivity is a function of the density of free carriers (which can be determined during the doping process) and the temperature. The conductivity is temperature-dependent to the extent that μ_n is a function of temperature. This fact allows one to construct temperature-sensitive resistors.

B. Conduction by Holes

The concept of effective mass has allowed us to base calculations on classical principles and simplify this description of the conduction process. We must now investigate the movement of valence electrons in the presence of holes. If some of the atoms of the material are lacking valence electrons, there are holes near these atoms. These holes can be created as previously found by doping a semiconductor with acceptor impurities. Any hole can be filled if a valence electron from a neighboring atom happens to fall into the vacancy, creating a new vacancy at the atom from which the electron came. All valence electrons move over a relatively large range at room temperature and a particular hole will move at random throughout the conductor when no field is applied. Application of an electric field tends to influence the movement of any valence electrons that happen to break their thermal bonds. A given vacancy or hole tends to move in the direction opposite to that of the electrons. It is easy to see from Fig. 1.6 that the movement of the holes from left to right results from a flow of electrons from right to left.

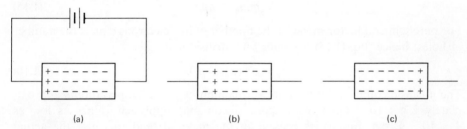

Fig. 1.6 Idealized movement of holes under the influence of an electric field: (a) original assumed distribution; (b) distribution after short period of time; (c) distribution after holes have reached the right edge of material.

A very useful representation for the hole is that of a particle of electronic charge $+q$. From quantum mechanics an effective mass can be associated with the hole. Thus, although the actual current is carried by electrons, we can analyze the process classically by considering the movement of a positively charged hole. This simplifies the analysis and allows a parallel treatment of holes and electrons in the conduction process.

We must note that the mechanism of current flow through hole conduction is quite different than that of free electron conduction. For the case of free electrons, the particles move through the lattice participating in various interactions, but remain as free particles (statistically speaking). In hole conduction, electrons not only interact with the lattice and other electrons, but also fall into the vacancies made available by doping. The latter process is somewhat more involved and it is not surprising that the mobility of the hole is less than that of the free electron. At room temperature the mobilities of holes in intrinsic silicon and germanium are (in cm^2/volt-sec):

	Si	Ge
μ_p	480	1800

Hole mobility theoretically exhibits the same temperature dependence as electron mobility.

The conductivity of p-type material, wherein negligible current is carried by free electrons, is

$$\sigma_p = qp\mu_p, \tag{1.14}$$

where p is the concentration of holes.

C. Intrinsic Conductivity

When a valence electron of an intrinsic semiconductor atom gains enough energy to break its bond, it contributes to the conduction process in two ways. Current will flow as a result of the movement of the free electron and as a result of the newly created hole. Both means of conduction are carried out independently of the other, except that the concentrations of each type of carrier will be related. The general expression for conductivity must reflect the effects of both carrier types and can be written as

$$\sigma = qn\mu_n + qp\mu_p. \tag{1.15}$$

For pure semiconductor material the number of free electrons equals the number of holes; hence, Eq. (1.15) becomes for intrinsic material

$$\sigma_i = qn_i(\mu_n + \mu_p). \tag{1.16}$$

The intrinsic conductivity for pure germanium at room temperature is approximately 2.2×10^{-2} ℧/cm while pure silicon may approach values as low as 4×10^{-6} ℧/cm. Impurities present in the material tend to cause the actual value of conductivity to differ from the theoretical value.

If impurities in the semiconductor material cannot be decreased to the desired point, it is often possible to use compensation to achieve lower values of conductivity. When the residual impurities are predominantly of one type (that is, donor or acceptor), the controlled addition of the opposite type impurity can reduce the number of free carriers in the semiconductor. If a concentration of N_D free electrons are present, doping the specimen with an equal concentration of acceptor atoms ($N_A = N_D$) will reduce the number of free electrons by providing a number of holes with which the electrons can recombine. In the case in which both N_D and N_A are much greater than n_i, the concentration of electrons after doping with both impurities is $N_D - N_A$.

D. Current Flow by Diffusion

Current flow in a conductor or a semiconductor with a uniform distribution of charge carriers is the result of an applied electric field. Under the influence of the field a net drift velocity of electrons or holes results in a charge flow that is referred to as drift current.

There is a different conduction process that requires no electric field to be present. This process is called diffusion and is the result of a nonuniform carrier distribution or an excess number of free carriers in a particular region. An understanding of diffusion current is very important in the development of semiconductor diode and transistor theory; thus, a detailed look at this phenomenon is now in order.

Consider a sample of materials such as those shown in Fig. 1.7. If there is no electric field present in either of the samples, no drift current will flow. While no net electron flow occurs for the uniform distribution, an electron flow can occur in the material having a nonuniform distribution even in the absence of an electric

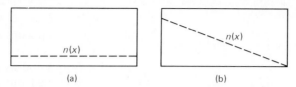

Fig. 1.7 (a) Material with uniform free electron distribution. (b) Material with nonuniform distribution.

field. The tendency of the electrons will be to redistribute so that a uniform distribution results. If the concentrations at both ends of the material can be forced to remain at their initial levels (and they can), electrons will continue to flow. To keep the concentration at the left side constant requires that a source of electrons be present to replenish the diffusing electrons. At the right side of the material, a sink for electrons must absorb all the diffusing electrons to avoid an accumulation of charge at this point. In the presence of a source and sink, a continuous electron flow will take place as the carriers attempt, but fail, to reach a uniform distribution. We will see later how the magnitude of diffusion current is controlled in diodes or transistors by controlling the concentrations at the end points of a region.

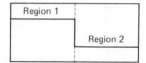

Fig. 1.8 Nonuniform distribution of free carriers.

While the distribution of electrons in Fig. 1.7(a) does not tend to change, each individual particle is continuously moving in a random manner. The probability of finding a given particle at one point in the material must equal that of finding that particle at any other point. Let us assume that we could obtain a distribution of free carriers as shown in Fig. 1.8, where each region occupies an equal volume. In this figure there are more free carriers in Region 1 than in Region 2. As a result of the random motion of all carriers more particles will move from Region 1 to Region 2 than will move from Region 2 to Region 1. If the regions are very small, the particle flow rate from Region 1 to Region 2 will be proportional to the number of particles or concentration in Region 1. The particle flow rate from Region 2 to Region 1 will be proportional to the number of particles or concentration in Region 2. The net flow rate of particles is (from Region 1 to Region 2)

$$F = KN_1 - KN_2 = K(N_1 - N_2),$$

where N_1 and N_2 are the total numbers of particles in the corresponding regions and K is a constant that involves the rate of movement of the particles and the average distance a particle must travel to cross the boundary.

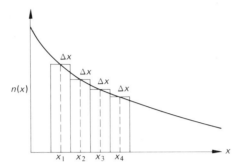

Fig. 1.9 Free carrier distribution.

Any general distribution can be considered on a differential basis by applying the same line of reasoning as above. Given the free carrier distribution of Fig. 1.9, differential volume elements can be taken as shown. The particle flow rate between differential elements at x_1 and x_2 is

$$F_{12} = H[n(x_1)\Delta x A - n(x_2)\Delta x A],$$

where H is a constant for a given material and a fixed value of Δx, $n(x)$ is the density of carriers at x, and A is the cross-sectional area of the material. If $n(x)$ is a slowly varying function of x, we can use the first two terms of a Taylor series expansion about x_1 to find $n(x_2)$. Thus

$$n(x_2) = n(x_1) + \left.\frac{dn}{dx}\right|_{x_1} (\Delta x)$$

if

$$\left.\frac{dn}{dx}\right|_{x_1} \gg \left.\frac{d^2 n}{dx^2}\right|_{x_1}.$$

It follows that

$$F_{12} = HA\left[n(x_1) - n(x_1) - \left.\frac{dn}{dx}\right|_{x_1}\Delta x\right]\Delta x$$

$$= -HA(\Delta x)^2 \left.\frac{dn}{dx}\right|_{x_1}.$$

For a given value of Δx (made small enough to approach a differential quantity), $H(\Delta x)^2$ will be constant and the flow rate of particles between the differential elements at x_1 and x_2 is

$$F_{12} = -DA \left.\frac{dn}{dx}\right|_{x_1}. \tag{1.17}$$

The constant D is called the diffusion constant and depends on the type of carrier involved, doping levels, the mobility of the carriers, and the temperature of the lattice. Equation (1.17) reflects a rather important result, namely, that the particle flow rate at any point is a function of the concentration gradient at that point. We should also note that the negative sign indicates a particle flow away from the area of highest concentration.

The current flow carried by the particles depends on the polarity of the charge of the particles. For electrons each particle carries a charge of $-q$. The electron diffusion current flow at a point x is

$$I_n = qD_n \frac{dn}{dx} A \quad \text{(diffusion equation for electrons)}. \tag{1.18}$$

For holes the diffusion current at x is

$$I_p = -qD_p \frac{dp}{dx} A \quad \text{(diffusion equation for holes)}. \tag{1.19}$$

Often these equations are expressed in terms of current densities by dividing both sides by the area A or by taking A as unit area. Typical values of diffusion constants for electrons and holes in intrinsic silicon and germanium at room temperature are (in cm^2/sec):

	Si	Ge
D_n	34	98
D_p	13	46

We note from Eqs. (1.18) and (1.19) that if the carrier concentration is constant as in Fig. 1.7(a), no diffusion current flow will result. If the slope of the carrier concentration is constant as in Fig. 1.7(b), the diffusion current is constant at each point in the material. This situation is typical of the base region in a uniformly doped transistor. The current is determined by the concentration of carriers injected from some source (emitter) and the concentration of carriers absorbed at the side nearest the sink (collector). For the distribution of Fig. 1.9, the diffusion current decreases with x. This occurs only when some other component of current such as drift current increases with x to result in constant total current flow at all values of x.

E. Recombination of Charge Carriers

In Section 1.2D the processes of carrier generation and recombination were discussed. It was noted that under equilibrium conditions, the recombination and generation rates are equal. For an intrinsic semiconductor the concentrations of holes and electrons are equal to n_i and, although recombination and generation are both taking place continuously, these concentrations remain statistically constant. For a material doped with donor impurities the situation is similar except that $n \gg p$. The recombination and generation rates are again equal and both n and p remain constant at a given temperature. The net change in free carrier concentration is zero.

Very often in the operation of semiconductor devices, a region will have excess carriers injected into it, upsetting the equilibrium conditions. The concentration of excess carriers will generally be much smaller than the concentration of lattice atoms and, thus, the generation rate will be unaffected. If excess holes are injected into an n-type material, we expect the recombination rate to increase from Eq. (1.4):

$$R = r(T)np.$$

Before injection, the generation rate equalled the recombination rate, giving

$$g(T) = r(T)n_0 p_0,$$

where n_0 and p_0 are the equilibrium concentrations.

Assuming the concentration of holes after injection to be p, the recombination rate increases to

$$R = r(T)n_0 p.$$

Since R is now greater than $g(T)$, there will be a net change (reduction) in both n_0 and p. However, n_0 is much greater than p, so changes in electron concentration can be neglected. The net rate of change in hole concentration is found by subtracting R from $g(T)$, giving

$$\frac{dp}{dt} = r(T)n_0 p_0 - r(T)n_0 p = -r(T)n_0(p - p_0).$$

If we define the difference $(p - p_0)$ as the excess hole concentration p_e, and let $r(T)n_0$ be represented by a constant we can write

$$\frac{dp_e}{dt} = -\frac{p_e}{\tau_p}, \qquad (1.20)$$

where τ_p is called the average recombination time or lifetime of a hole in the material considered. Equation (1.20) indicates that if an excess density of holes is present in n-type material, the holes will decay to the equilibrium value over a period of time. The recombination time τ_p is obviously a function of temperature and doping.

A similar result can be derived for the decay of excess electrons in p-type material, that is,

$$\frac{dn_e}{dt} = -\frac{n_e}{\tau_n}. \qquad (1.21)$$

The practical importance of recombination is exhibited in the diffusion process for a wide base transistor or diode. Diffusion current in a region is established by injecting excess carriers at one edge and absorbing all carriers at the other end. When negligible recombination takes place, the current calculation is very simple, since the slope of excess carriers will remain constant over the region. The following discussion indicates the approach to the case in which recombination is not negligible and the slope of excess carriers varies with distance from the point of injection.

F. Diffusion Current in the Presence of Recombination

Many transistors contain a base region that is extremely narrow. In this case, recombination of excess carriers is small enough to be negligible and Fig. 1.7(b) accurately describes the carrier distribution within the base region. Knowing this distribution allows one to calculate the total current flow which, in this case, would be entirely diffusion current. Let us consider the case of n-type material into which holes are injected at one end and absorbed at the other end, as indicated in Fig. 1.10. Recombination is assumed to take place throughout the region. We further assume that no external field is applied to the material. In order to calculate diffusion current at any point we must know the slope of the hole distribution. The first problem then is to solve for the hole distribution.

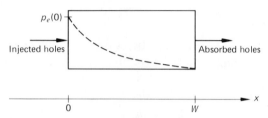

Fig. 1.10 Hole distribution in the presence of recombination.

To make the problem take on more practical significance, we note that the total current flow must be constant at any cross-section in the region, since the current entering the material must equal the current leaving. We also observe that the electron flow in the external circuit has not been specified. We shall now assume that electrons can move freely into the material from the right side as necessary, but do not enter from the left side. It will later be seen that this assumption is realistic in that it corresponds directly to conditions in a pn-junction.

The holes injected at the left edge of the material result in a current I. Since there is recombination throughout the region, far fewer holes reach the right edge than were injected. The current due to hole diffusion is maximum at the point of injection and decreases throughout the region. The total current must remain constant and equal to I; therefore, we conclude that electrons flowing into the right edge of the material must account for the difference between I and the hole diffusion current. While it can be seen that this electron flow is necessary, we should examine the origin of the flow in order to obtain a better understanding of it. Initially the material is space-charge neutral. When holes are injected, an equal number of electrons must also enter the semiconductor material to preserve charge neutrality. In our idealized case, the electrons will enter the right edge of the material. These electrons take on a spatial distribution very nearly the same as the hole distribution to preserve neutrality throughout the region.

At this point we note that there can be four distinct current components in the material. Diffusion current will be carried by both holes and electrons and a very slight electric field will lead to hole and electron drift current. Fortunately, the hole drift current is negligible when compared to the electron drift current since these components are directly proportional to carrier densities. The electron drift and diffusion currents can be combined as an electron recombination current. The total current I will then consist of hole-diffusion current and electron current feeding the recombination process. The hole-diffusion current at any point is given by Eq. (1.19) as

$$I_p = -qD_p \frac{dp}{dx} A.$$

The slope of the total hole distribution is equal to the slope of the excess hole distribution and we can write

$$I_p = -qD_p \frac{dp_e}{dx} A.$$

The electron current due to recombination can be found by noting that holes and electrons must recombine in pairs. Therefore from Eqs. (1.20) and (1.21) it follows that

$$\frac{dn_e}{dt} = \frac{dp_e}{dt} = -\frac{p_e}{\tau_p}.$$

The preceding expression indicates the net rate of change of excess carrier concentration at any point. To find the recombination current we must calculate the total rate of annihilation of electrons over the entire volume and multiply the result by the charge of an electron. Integrating the expression for dn_e/dt and multiplying by the appropriate constant give for electron current,

$$I_n = -qA \int_0^x \frac{dn_e}{dt} dx = qA \int_0^x \frac{p_e}{\tau_p} dx. \tag{1.22}$$

The total current is then

$$I = I_p + I_n = -qD_p \frac{dp_e}{dx} A + \frac{qA}{\tau_p} \int_0^x p_e \, dx.$$

Differentiating this equation gives a second-order differential equation in p_e, namely,

$$\frac{d^2 p_e}{dx^2} - \frac{p_e}{D_p \tau_p} = 0. \tag{1.23}$$

Solving this differential equation for the excess hole distribution results in

$$p_e = p_e(0) e^{-x/L_p}, \tag{1.24}$$

where $L_p = \sqrt{D_p \tau_p}$ and is called the mean recombination or diffusion length, and $p_e(0)$ is the excess concentration of holes at $x = 0$. We can now return to our expression for hole-diffusion current and use Eq. (1.24) to yield

$$I_p = \frac{qD_pAp_e(0)}{L_p} e^{-x/L_p} \quad \text{(hole-diffusion current)}. \tag{1.25}$$

The electron recombination current from Eq. (1.22) is

$$I_n = \frac{qD_pAp_e(0)}{L_p} (1 - e^{-x/L_p}). \tag{1.26}$$

Total current is the sum of the two components and is

$$I = I_p + I_n = \frac{qD_pAp_e(0)}{L_p}. \tag{1.27}$$

Plots of the exponential hole distribution and the current components are shown in Fig. 1.11.

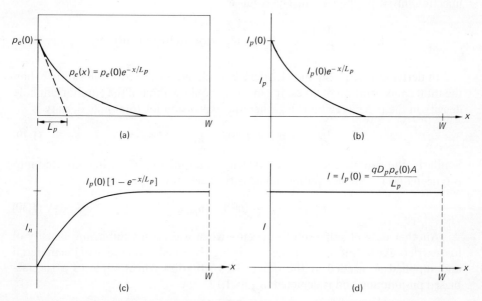

Fig. 1.11 (a) Excess hole distribution. (b) Hole-diffusion current. (c) Electron current. (d) Total current.

Several points that are applicable to device operation become apparent from the previous discussion. First of all, we see that the injection of excess holes at one edge of a doped material results in an exponentially decaying distribution as a result of recombination. Second, the total current is carried by hole diffusion at the point of injection, assuming no electrons are exiting through the left edge of the region. Thus, rather than carry out complex calculations for total current, we could apply Eq. (1.19) at $x = 0$. This gives

$$I = I_p(0) = -qD_pA \left.\frac{dp_e}{dx}\right|_{x=0}.$$

The slope dp_e/dx at $x = 0$ is simply

$$\left.\frac{dp_e}{dx}\right|_{x=0} = -\frac{p_e(0)}{L_p};$$

therefore,

$$I = \frac{qD_pAp_e(0)}{L_p}.$$

If $p_e(0)$ is known along with A, D_p, and L_p, then I can be calculated without reference to I_n. This fact will be used later in the analysis of pn-junctions.

An equation similar to Eq. (1.25) can be developed for the case of electron injection into a p-type region. The result is

$$I_n = \frac{qD_nAn_e(0)}{L_n} e^{-x/L_n} \qquad \text{(electron-diffusion current)}. \tag{1.28}$$

In deriving Eqs. (1.24) and (1.25) we considered only excess holes rather than the total concentration of holes. It can easily be shown that if the equilibrium hole density in the n-type region is p_{n0}, then the expression for total hole density is

$$p(x) = p_{n0} + (p(0) - p_{n0})e^{-x/L_p}. \tag{1.29}$$

Similarly, the equation for electron density as a function of x when injected into p-type material with an equilibrium density of electrons equal to n_{p0} is

$$n(x) = n_{p0} + (n(0) - n_{p0})e^{-x/L_n}. \tag{1.30}$$

Another case of some interest occurs when a certain equilibrium density of free carriers exists, but these carriers are being absorbed at one edge of the material and supplied as needed at the other. This situation corresponds to a reverse-biased pn-junction and is depicted in Fig. 1.12.

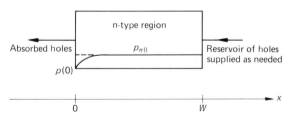

Fig. 1.12 Hole distribution when holes are absorbed at left edge.

A consideration of generation and recombination rates leads to the result that the hole concentration will experience a net generation rate at any point x of

$$\frac{dp(x)}{dt} = \frac{p_{n0}}{\tau_p} - \frac{p(x)}{\tau_p}. \tag{1.31}$$

Proceeding as before we can calculate a hole distribution given by Eq. (1.29). A hole-diffusion component of current and an electron component could again be found, but total current can be related more easily to the diffusion current at $x = 0$. This quantity is

$$I = I_p(0) = -\frac{qD_p A}{L_p}(p_{n0} - p(0)). \tag{1.32}$$

The corresponding equation for electrons in p-type material is

$$I = I_n(0) = \frac{qD_n A}{L_n}(n_{p0} - n(0)). \tag{1.33}$$

G. Majority Carriers and Minority Carriers

If a bar of doped semiconductor material is subjected to an electric field, current is carried by both holes and electrons. In n-type material the normal concentration of electrons is many orders of magnitude greater than the concentration of holes. This can be seen from an examination of Eq. (1.8). For typical values of $n_i = 1.4 \times 10^{10}/\text{cm}^3$ and $N_D = 1 \times 10^{15}/\text{cm}^3$, a hole concentration of $p = n_i^2/N_D = 1.96 \times 10^5/\text{cm}^3$ results. The electron concentration is $1 \times 10^{15}/\text{cm}^3$, and consequently virtually all current is conducted by electrons. Electrons are called majority carriers and holes are minority carriers for n-type material. In p-type material the situation is reversed. Holes are present in far greater numbers than are free electrons and therefore carry most of the current. In this material holes are the majority carriers and electrons are minority carriers. Minority current is important in the operation of diodes and bipolar transistors whereas the operation of the field-effect transistor is dependent entirely on majority current carriers.

REFERENCES AND SUGGESTED READING

1. R. B. Adler, A. C. Smith, and R. L. Longini, *Introduction to Semiconductor Physics*, SEEC Volume 1. New York: Wiley, 1964.
2. J. Allison, *Electronic Engineering Materials and Devices*. London: McGraw-Hill, 1971.
3. D. J. Comer, *Introduction to Semiconductor Circuit Design*. Reading, Mass.: Addison-Wesley, 1968.
4. D. K. Ferry and D. R. Fannin, *Physical Electronics*. Reading, Mass.: Addison-Wesley, 1971.
5. M. S. Ghausi, *Electronic Circuits*. New York: D. Van Nostrand, 1971.

PROBLEMS

Sections 1.2A–1.2B

* **1.1** If $n_i = 2.5 \times 10^{13}/\text{cm}^3$ at $T = 300°\text{K}$ for germanium, find n_i when $T = 350°\text{K}$. Assume a value of 8.65×10^{-5} eV/°K for Boltzmann's constant.

1.2 Repeat Problem 1.1 by assuming that the change in n_i with temperature results only from the exponential term; that is, assume $AT^{3/2}$ remains constant with T.

1.3 If $n_i = 1.4 \times 10^{10}/\text{cm}^3$ at $T = 300°\text{K}$ for silicon, find n_i when $T = 270°\text{K}$. Assume a value of 8.65×10^{-5} eV/°K for Boltzmann's constant.

* **1.4** Repeat Problem 1.3 by assuming that the change in n_i with temperature results only from the exponential term; that is, assume $AT^{3/2}$ remains constant with T.

Sections 1.2C–1.2D

1.5 If $n_i = 2.5 \times 10^{13}/\text{cm}^3$ at $T = 300°\text{K}$ for germanium and a germanium sample is doped with $N_D = 1 \times 10^{17}/\text{cm}^3$, determine the concentrations of free electrons and holes in the sample at $T = 300°\text{K}$ assuming all donor atoms are ionized.

* **1.6** Calculate the concentrations of electrons and holes in the germanium sample of Problem 1.5 if the temperature is raised to $T = 330°\text{K}$.

* **1.7** If $n_i = 1.4 \times 10^{10}/\text{cm}^3$ at $T = 300°\text{K}$ for silicon and a silicon sample is doped with $N_A = 2 \times 10^{14}/\text{cm}^3$, determine the concentrations of free electrons and holes in the sample at $T = 300°\text{K}$.

1.8 Calculate the concentrations of electrons and holes in the silicon sample of Problem 1.7 if the temperature is lowered to $T = 270°\text{K}$.

Sections 1.3A–1.3C

1.9 Calculate the conductivity of pure germanium at a temperature of $340°\text{K}$. Assume that $\mu_n = 2900$ and $\mu_p = 1350$ cm^2/volt-sec at this temperature.

*
1.10 Calculate the conductivity of pure silicon at a temperature of $270°\text{K}$. Assume that $\mu_n = 1700$ and $\mu_p = 520$ cm^2/volt-sec at this temperature.

1.11 It is desired to create a conductivity of $\sigma = 30$ ℧/cm in a germanium bar at room temperature. Calculate the concentration of impurity atoms required to satisfy this conductivity for n-type material.

*
1.12 Repeat Problem 1.11 for p-type material.

1.13 Calculate the conductivity at room temperature of a germanium wafer that is doped with donor atoms of concentration $N_D = 1 \times 10^{17}/\text{cm}^3$ and compare to the intrinsic case.

Section 1.3D

1.14 The concentration of holes at the right edge of the silicon bar of unit area shown in the figure is maintained at zero. Given that the hole-diffusion current flowing through the bar is 1 mA, find the concentration of holes at the left edge of the bar.

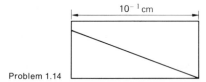

Problem 1.14

*1.15 If the concentration of holes at the left edge of the bar of Problem 1.14 is $2 \times 10^{14}/cm^3$ and drops with constant slope to $4 \times 10^{13}/cm^3$ at the right edge, calculate the diffusion current flow.

*1.16 Repeat Problem 1.14 for a germanium bar.

1.17 Repeat Problem 1.15 for a germanium bar.

Sections 1.3E–1.3G

1.18 Explain why n can be considered constant while p changes when excess holes are added to an n-type material in connection with the derivation of Eq. (1.20).

*1.19 Solve Eq. (1.20) for p_e as a function of time.

1.20 Derive Eq. (1.21) from a consideration of generation and recombination rates.

1.21 The concentration of electrons at the left edge of the silicon bar of unit area shown in the figure is maintained at $1 \times 10^{15}/cm^3$. If $L_n = 1 \times 10^{-2}$ cm, calculate the electron-diffusion current flowing through the bar, assuming the equilibrium concentration of electrons is negligible. In which direction does conventional current flow?

*1.22 If 10 µA of electron-diffusion current flows at $x = 0$ in the bar of Problem 1.21, calculate the concentration of electrons that must be maintained at the left edge of the bar. Assume that the equilibrium concentration of electrons is negligible.

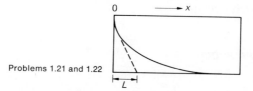

Problems 1.21 and 1.22

---

*The answers to these problems are provided at the end of the book.

DISCRETE DEVICES AND INTEGRATED CIRCUITS

The operation of several important semiconductor devices will be outlined in this chapter. These discrete devices have played a major role in the history of electronics and, while the integrated circuit is continuing to grow in importance, the discrete semiconductor device continues to contribute to the electronic field. An integrated circuit is a collection of discrete elements (diodes, resistors, capacitors, and transistors) created by means of a single construction process in which all elements are formed. An understanding of discrete element operation is prerequisite to the study of integrated circuits.

2.1 pn-JUNCTION DIODES

The two types of rectifying junction of major import in electronics are the pn-junction and the metal-semiconductor junction. Since the rectifying properties of both junctions are somewhat similar to those of a vacuum tube diode, the name diode has been applied to these rectifiers.

The metal-semiconductor junction has great historical importance since it led to the development of the point-contact transistor one or two years prior to the development of the junction transistor. Surface barrier transistors which are constructed as a metal-semiconductor-metal sandwich also achieved some importance in early transistor circuit work. More recently a metal-oxide-semiconductor combination is used in constructing silicon MOS field-effect transistors.

The pn-junction is presently the most widely used, especially when serving as the basis for the bipolar junction transistor. This junction is made up of two regions of semiconducting material (usually silicon or germanium) with each region containing a different type of impurity. One region is doped to create free electrons; the other is doped to have free holes. In this chapter we will consider first the pn-junction. While the rectifying properties of both types of diode are similar (exponential), the prominent role played by the pn-junction in transistor operation justifies maximum coverage for this device. The metal-oxide-semiconductor arrangement will be considered in connection with the field-effect transistor which can be constructed in a pn-junction or MOS configuration.

A. Junction Formation

Modern fabrication methods generally apply vapor diffusion techniques to create pn-junctions within a single semiconductor mass. This leads to impurity profiles within the two regions that are not constant. To simplify the discussion of diode operation and highlight the more important mechanisms, we will assume that the junction is created by bringing together two separate, uniformly doped regions. Practical departures from the behavior of this idealized junction will be treated later in this chapter.

The original p- and n-type materials are shown in Fig. 2.1. The majority-carrier concentrations can be controlled during doping and will be given by

$$p_p = N_A \quad \text{and} \quad n_n = N_D,$$

where N_A is the concentration of acceptor atoms and N_D is the concentration of donor atoms.

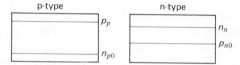

Fig. 2.1 p- and n-type materials showing majority- and minority-carrier concentrations.

The minority-carrier concentrations can be calculated from Eqs. (1.8) and (1.9),

$$p_{n0} = \frac{n_i^2}{n_n} = \frac{n_i^2}{N_D} \quad \text{and} \quad n_{p0} = \frac{n_i^2}{p_p} = \frac{n_i^2}{N_A}.$$

Note that there is a very sharp gradient of free carriers at the end points of each region. The concentration of p_p, for example, appears as in Fig. 2.2. There is a very strong diffusive tendency for these holes, but they cannot leave the p-type region, since they are bound by the extremes of the material. The same reasoning applies to the n-type material also; that is, the electrons have a strong diffusive tendency due to the high concentration gradient, but are constrained from leaving the material by the extremes of the material. If the p- and n-type materials are now joined together, a pn-junction is formed.

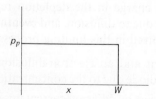

Fig. 2.2 Majority-carrier distribution in p-type material.

The holes that are near the edge of the p-type material can now diffuse. They will tend to move into the n-type material. The electrons near the edge of the n-type material will flow into the p-type region. As these carriers move, the concentration gradients decrease. If there were no electrostatic forces acting on the particles, that is, if they behaved like gas molecules, the diffusion would proceed until no gradient existed. The carriers would then be uniformly distributed throughout both regions. Fortunately, there is an electrostatic effect which limits this diffusion current and ultimately allows external control of the junction current flow.

B. The Depletion Region

It is important to recognize that, while free holes exist in one region and free electrons in the other, prior to joining, both regions are space-charge neutral. The negative charge of each free electron is offset by the positive charge of an immobile parent atom in the n-type region. The positive charge of each free hole is offset by the negative charge of an immobile parent atom in the p-type region. After joining the regions, the diffusion of free carriers causes a charge to build up near the junction. When a hole leaves the p-region, it uncovers a parent acceptor atom which now has a negative charge. As an electron leaves the n-region, it uncovers a parent donor atom which has a positive charge. The doping atoms are bound to the lattice and hence are immobile. The resulting junction appears in Fig. 2.3.

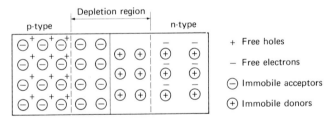

Fig. 2.3 Schematic representation of carriers and doping atoms in a pn-junction.

There is a region near the junction where few free carriers are present due to the fact that they have diffused across the junction and recombined with the opposite-type carriers. This region is called the depletion region or transition region. The voltage appearing across the depletion region is called the barrier voltage. Its presence is due to the net charge in the depletion region. The barrier voltage opposes the flow of carriers due to diffusion, and eventually limits the depletion of further area. The steps involved in this limiting process might well be listed.

1. A hole-diffusion current and an electron-diffusion current exist across the depletion region of the diode due to the concentration gradients of the carriers.
2. The removal of free carriers by diffusion current upsets space-charge neutrality; hence a voltage is developed across this area.

3. This barrier voltage causes drift currents for both electrons and holes which cancel the diffusion components exactly.
4. The cancellation of the drift and diffusion currents is a "built-in" feature of the junction, as demonstrated by the following reasoning:

   Assume that the diffusion components are larger than the drift components of current. This means that more free carriers will leave their associated impurity atoms, thereby widening the depletion region. The barrier voltage will then increase, causing the drift current to increase. It is obvious that this process will result in an equilibrium situation wherein the width of the depletion region is constant in the absence of an applied voltage.

The depletion region is generally two to three orders of magnitude narrower than the p- or n-type regions of a diode. In Figs. 2.4, 2.5, and 2.6 the extent of this region is greatly exaggerated to demonstrate the carrier concentrations in the depletion region.

C. Current Flow in a pn-Junction

In order to evaluate the current through a diode for different applied voltages, the free carrier distributions must be found. Three assumptions are used to solve the diode problem. The validity of the resulting solution has been verified experimentally for germanium diodes, and certain modifications to the theory make it applicable to silicon diodes. The assumptions to be used are as follows.

1. Maxwell–Boltzmann statistics are in effect throughout the depletion region; that is, the number of free carriers at any point x is given by

$$n(x) = n_0 \exp\left[\frac{-q}{kT}(V(x) - V_0)\right] \quad (2.1)$$

and

$$p(x) = p_0 \exp\left[\frac{q}{kT}(V(x) - V_0)\right], \quad (2.2)$$

where p_0 and n_0 refer to carrier concentrations at $x = 0$, V_0 is the potential at $x = 0$, and $V(x)$ is the potential at x. Points other than $x = 0$ may serve as the reference point for Eqs. (2.1) and (2.2), so long as the corresponding carrier concentrations and potential are used.

2. The densities of free carriers in the depletion region are small in comparison to the densities of doping atoms and can be neglected in finding the total charge.
3. The electron and hole currents are constant throughout the depletion region; that is, no recombination takes place in this region. This assumption is not accurate for silicon diodes, but is quite good for germanium.

The second approximation results in a charge distribution, as shown in Fig. 2.4(a).

Zero bias. Figure 2.4(b) shows the free-carrier distribution throughout the diode. It is to be noted that majority and minority concentrations are constant outside the

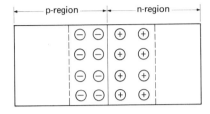
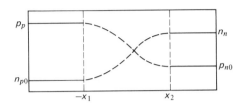
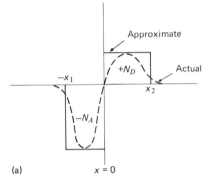
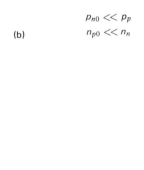

Fig. 2.4 Charge distributions in a pn-junction: (a) net charge distribution in a pn-junction with no applied bias; (b) free carrier distribution with no applied bias.

depletion region. Therefore, with no applied bias, the only currents present are the diffusion currents within the depletion region and the offsetting drift currents due to the barrier voltage. If the voltage is taken as zero at $x = -x_1$, the barrier voltage can be found from either Eq. (2.1) or (2.2):

$$\phi = V(x_2) = -\frac{kT}{q} \ln \frac{p_p}{p_{n0}} = -\frac{kT}{q} \ln \frac{n_n}{n_{p0}}. \tag{2.3}$$

In terms of the densities of doping atoms, ϕ becomes

$$\phi = -\frac{kT}{q} \ln \frac{N_D N_A}{n_i^2}. \tag{2.4}$$

The barrier voltage depends on the doping densities N_D and N_A. For a germanium junction a typical value of ϕ is 0.3 V; for silicon this value might be 0.7 V.

Forward bias. If a small forward bias ($V < |\phi|$) is applied to the diode, the results will be as shown in Fig. 2.5.

The applied voltage opposes the barrier voltage and causes a net voltage across the junction of $|\phi| - V$. The width of the depletion region decreases with forward bias. If the left edge of the depletion region is again taken as the point of zero potential, the potential at x_2' is $V_j = V - |\phi|$. The density of holes at x_2' can be found by writing Eq. (2.2) as

$$p_n(x_2') = p_p \exp\left[\frac{-q}{kT}(|\phi| - V)\right]. \tag{2.5}$$

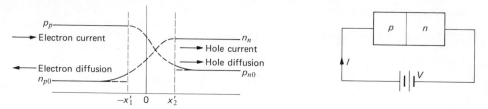

Fig. 2.5 Carrier distribution in a forward-biased junction.

Using Eq. (2.3), we find that this is simply

$$p_n(x_2') = p_{n0} e^{qV/kT}. \tag{2.6}$$

In a similar manner the number of electrons at $-x_1'$ can be found as

$$n_p(-x_1') = n_{p0} e^{qV/kT}. \tag{2.7}$$

The densities of minority carriers at the junction edges are strong functions of the applied voltage. A forward voltage as shown in Fig. 2.5 increases these densities exponentially with applied voltage. While the majority-carrier densities must increase to preserve neutrality in each region, this increase is small compared to the total density under low-level injection conditions. Thus, the bias voltage has negligible effect on the majority-carrier numbers.

The reduction of junction voltage due to the applied voltage upsets the balance between the drift and diffusion currents across the junction, allowing a net current to flow. This imbalance results in the injection of holes at x_2' and the injection of electrons at $-x_1'$. The nonuniform distributions of carriers give rise to a current flow that can be related to the applied voltage. Fortunately, we can calculate the diode current by considering the hole-diffusion current at $x = x_2'$ and the electron-diffusion current at $x = -x_1'$. Before performing this calculation we will discuss briefly the other components of current flow and show why it is necessary only to consider diffusion currents at the junctions.

The total current through any cross-section throughout the diode is constant and can be expressed in terms of four components:

$$I = I_p(x)\text{drift} + I_p(x)\text{diffusion} + I_n(x)\text{drift} + I_n(x)\text{diffusion}.$$

Since drift components are directly proportional to densities of free carriers, the minority drift current will always be negligible with respect to majority drift current. If we now apply this equation at $x = x_2'$ we have

$$I = I_p(x_2')\text{diffusion} + I_n(x_2')\text{diffusion} + I_n(x_2')\text{drift}.$$

Applying the equation at $x = -x_1'$ leads to

$$I = I_n(-x_1')\text{diffusion} + I_p(-x_1')\text{drift} + I_p(-x_1')\text{diffusion}.$$

Since we have assumed that no recombination occurs in the depletion region, both hole and electron components are constant in the region from $-x_1'$ to x_2'. For

32 DISCRETE DEVICES AND INTEGRATED CIRCUITS

hole current we can then write

$$I_p(-x_1')\text{drift} + I_p(-x_1')\text{diffusion} = I_p(x_2')\text{diffusion}.$$

Using this relationship we can now write

$$I = I_n(-x_1')\text{diffusion} + I_p(x_2')\text{diffusion},$$

leading to a relatively simple expression for the total current.

Although other current components exist, the preceding equation shows that only diffusion components at the depletion region edges need be considered in calculating the total diode current.

The hole-diffusion current is found by using Eq. (1.25) and noting that the x-variable in Eq. (1.25) is referenced to the point of injection. Therefore, $x = 0$ at the point of injection (x_2') and

$$I_p = \frac{AqD_p p_e(0)}{L_p} e^{-x/L_p}\bigg|_{x=0} = \frac{AqD_p}{L_p} p_{no}(e^{qV/kT} - 1), \quad (2.8)$$

where A is the cross-sectional area. Note that the excess number of holes at $x = x_2'$ is given by

$$p_e(x_2') = [p_n(x_2') - p_{no}] = p_{no}(e^{qV/kT} - 1).$$

The electron-diffusion current at $-x_1'$ is

$$I_n = \frac{AqD_n}{L_n} n_{po}(e^{qV/kT} - 1) \quad (2.9)$$

and

$$n_e(-x_1') = [n_p(-x_1') - n_{po}].$$

The total current must be the sum of these two currents, or

$$I = A\left(\frac{qD_p p_{no}}{L_p} + \frac{qD_n n_{po}}{L_n}\right)(e^{qV/kT} - 1). \quad (2.10)$$

This is referred to as the diode equation and can also be written in the following two forms:

$$I = Aqn_i^2 \left(\frac{D_p}{L_p N_D} + \frac{D_n}{L_n N_A}\right)(e^{qV/kT} - 1), \quad (2.11)$$

or

$$I = I_S(e^{qV/kT} - 1), \quad (2.12)$$

where

$$I_S = Aqn_i^2 \left(\frac{D_p}{L_p N_D} + \frac{D_n}{L_n N_A}\right).$$

Dependent on temperature but not on voltage, I_S is called the reverse saturation current for reasons that will become clear in the next section.

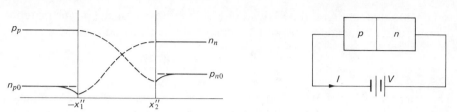

Fig. 2.6 Carrier distribution in a reverse-biased junction.

Reverse bias. When a reverse bias is applied to the diode, the voltage across the junction is $V_j = -(|\phi| + V)$. The increased voltage widens the depletion region and again upsets the balance of drift and diffusion currents that occurs when no bias is present. In this case the drift current exceeds the diffusion current in the depletion region and the resulting carrier distribution appears as in Fig. 2.6.

The reverse-bias current can be calculated by again considering the hole-diffusion current at x_2'' and the electron-diffusion current at $-x_1''$. Using Eqs. (1.32), (1.33), (2.6), and (2.7), we obtain

$$I = A\left(\frac{qD_p p_{n0}}{L_p} + \frac{qD_n n_{p0}}{L_n}\right)(e^{-qV/kT} - 1),$$

which takes the same form as Eq. (2.10). However, the voltage is negative in this case and the resulting current will be negative. Equation (2.12) then is valid for both positive and negative voltages. The graph of this equation is shown in Fig. 2.7.

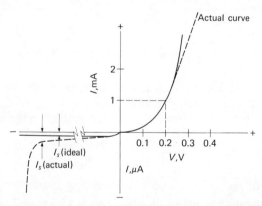

Fig. 2.7 V–I characteristics of a germanium diode. (Note the change in current scale for positive and negative currents.)

In the reverse direction, the current rapidly reaches a value of I_S, and then remains essentially constant for all reverse voltages.

Throughout the above derivation, it was assumed that the voltage drop across the bulk material was negligible. This assumption can now be checked. From the graph of Fig. 2.7, an applied voltage of 200 mV causes 1 mA of current to flow. Assuming a typical resistivity of 2×10^{-3} Ω-cm for the p-type material, a length of 0.1 cm, and an area of 10^{-2} cm², we can calculate the voltage drop as

$$V_p = \frac{\rho L}{A} I = \frac{(2 \times 10^{-3})(10^{-1})}{10^{-2}} \times 1 \times 10^{-3} = 2 \times 10^{-5} \text{ V}.$$

Thus there is a drop of 0.02 mV out of an applied voltage of 200 mV. Assuming a drop of the same order of magnitude across the n-type material shows that it was a good approximation indeed to assume that none of the applied voltage drops across the bulk material.

D. Physical Description of Diode Operation

If the last section were reviewed, it would be found that a quantitative knowledge of carrier distributions within the depletion region was not required. In fact, a knowledge of the carrier densities and gradients at the edges of the depletion region is all that is used in the calculation of the V–I relationship. Because of this, it is convenient to neglect the depletion region entirely in the dc analysis of the diode. The diode can be pictured as shown in Fig. 2.8(a).

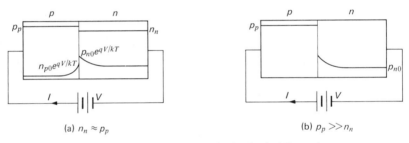

Fig. 2.8 Carrier distributions neglecting the depletion region.

Normally, one region of a practical diode is doped much more heavily than the other region. If N_A is much larger than N_D, then $p_p \gg n_n$, and the picture of Fig. 2.8(b) results. The diffusion of electrons in the p-region is negligible and I_S becomes

$$I_S = \frac{Aqn_i^2 D_p}{L_p N_D} = \frac{AqD_p p_{n0}}{L_p}. \tag{2.13}$$

A forward-bias voltage causes the injection of holes from the p-region into the n-region, increasing p_n over the equilibrium value. This causes a gradient of holes in the n-region, giving rise to a diffusion current. This is the basic principle underlying semiconductor diode behavior. The applied voltage controls the gradient of minority carriers, and hence controls the current flow. Since the gradient or slope at the point of injection increases exponentially with applied voltage, the current also increases exponentially with applied voltage. The derivation of the diode equation for the diode of Fig. 2.8(b) consists of two steps: (1) relating the gradient of carriers at the point of injection to the applied voltage, and (2) relating this gradient to the current flow by using the diffusion equation. This situation corresponds to the idealized example considered in Section 1.3F.

E. Current Components in the Diode

Since we have derived the diode equation and have discussed the basic mechanism underlying its behavior, we could continue to other topics without considering the various current components throughout the diode. For the sake of completeness, we have shown these components in Fig. 2.9. The distributions of carriers correspond to those cases considered in Section 1.3. The holes in the n-region and the electrons in the p-region recombine, causing an exponential variation of minority carriers. Equations (1.29) and (1.30) can be used to calculate these distributions, taking the respective points of injection as $x = 0$. While both diffusion and drift currents are present, only minority-carrier diffusion at the points of injection need be considered. The diode current is found by evaluating these current components; there is no necessity to consider drift current.

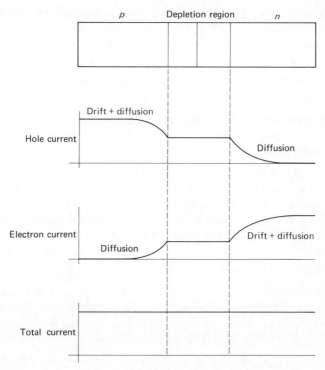

Fig. 2.9 Current components in a pn-junction.

F. Silicon Diodes

It is found that the above theory predicts quite accurately the experimental V–I curve of a germanium diode. It is not as accurate for the silicon diode, especially at lower current levels. Shockley [4] has shown that as the energy gap of a semiconductor material becomes larger, recombination of current carriers in the depletion region becomes more important. This recombination is a function of "traps" brought about by irregularities of the lattice or impurities. The current at low levels approaches the value

$$I = K_1 n_i e^{qV/2kT}, \tag{2.14}$$

where K_1 is a constant. At higher current levels the standard diode equation can again be used.

G. Departure from Ideal Behavior

There are three major reasons why the actual diode characteristics do not correspond exactly to those given in Fig. 2.7. These are listed below.

1. Ohmic resistance and contact resistance in series with the diode causes the V–I characteristics to become linear at high forward currents.
2. Avalanche or Zener breakdown takes place at high reverse voltages, causing an abrupt increase in current. This characteristic can be controlled and used to advantage in the Zener diode, which is discussed in Section 2.2.
3. Surface contaminants cause an ohmic layer to form across the junction. As reverse voltage is increased, the reverse current increases slightly, instead of remaining constant at I_S.

The dashed curve of Fig. 2.7 demonstrates these effects.

H. Low-Frequency Analysis

When a dc voltage is applied to the diode, the resulting current can be found from the V–I curve or from Eq. (2.12) if I_S is known. For transient or repetitive applied signals, the calculation of the resulting current waveform as a function of time is more involved. It is convenient to classify each applied signal as a large signal or a small signal to consider the effects of time-dependent waveforms. Generally, if the transient current is small compared to the dc current through the diode, the small-signal case pertains.

Large-signal operation. While it is often tedious, large-signal operation of a diode can be handled by graphical methods. Fortunately, there is an approximate analytical method that is simpler to use while satisfying the accuracy requirements of most practical circuits. We will first discuss the more general graphical method.

Load-line construction. In the circuit of Fig. 2.10 we are to find the voltage appearing between points a and b or across the device. We shall make the reasonable assumption that there is a nonlinear relationship between device current and voltage expressible in terms of a V–I curve. We might ask whether it is the device or the external circuit consisting of E_{CC} and R_L that determines the value of current flow in the circuit. Obviously, both device and external circuit will affect the current flow; therefore, we must consider both to find this current value. One method of handling this problem is to consider two separate sets of characteristics. The graphical V–I characteristic of the external circuit is called the load-line. The device characteristics are usually given graphically by the manufacturer or can be generated from the nonlinear equation for the device used. The load-line relationship can be expressed as

$$V_{ab} = E_{CC} - R_L I. \qquad (2.15)$$

It is important to note that the load-line equation is independent of the device characteristics and depends only on the external circuit. Equation (2.15) has two unknown quantities and more information is required to solve our problem. An analytic equation for the device is often unknown or too complex to be of value; thus we resort to a graphical solution of two simultaneous equations. The V–I characteristics representing the device are shown in Fig. 2.11.

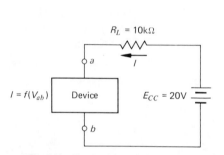

Fig. 2.10 Circuit with nonlinear device.

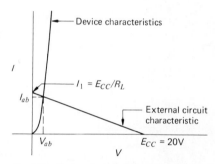

Fig. 2.11 Device characteristics with load-line.

The plot of the external circuit behavior or load-line is superimposed on the device characteristics. The external circuit requires a current that lies somewhere along the load-line while the device requires a current that lies on the characteristic curve. Since equal currents must flow through the device and through the external circuit, the actual operating point must lie at the intersection of the load-line and the device characteristics. The device current and voltage can then be read from the graph.

For given values of R_L and E_{CC} it is very easy to locate the load-line on the device curves. When I is assumed zero in Eq. (2.15), $V_{ab} = E_{CC}$. When V_{ab} is assumed zero, $I = E_{CC}/R_L$. Since Eq. (2.15) is linear, a straight line intersecting the current axis at E_{CC}/R_L and the voltage axis at E_{CC} represents the load-line.

If the applied voltage changes over a range of values, device variables can be found corresponding to several specific values of applied voltage. Interpolation can then be used to find the variables between these calculated points. For example, if the applied voltage changes from -1 V to $+4$ V, one might calculate device voltages for -1 V, 0 V, 2 V, and 4 V. As we shall see in Example 2.1, the device voltages can easily be interpolated between known values.

Piecewise-linear approximation. It is possible to approximate a nonlinear device V–I curve with a relatively small number of straight line segments. For many large-signal diode circuits, sufficient accuracy is obtained with just two line segments as shown in Fig. 2.12. In this approximation the diode appears as a small constant resistance of value R_1 when forward-biased and as a large constant resistance of value R_2 when reverse-biased. This approximation is especially appropriate if the external circuit resistance is much larger than R_1 and much smaller than R_2. Typical values of R_1 and R_2 for a low-power silicon diode are 10 Ω and 10 MΩ; thus, an external resistance from several hundred ohms to several hundred kilohms allows this diode approximation to be used with good accuracy. An example will demonstrate the use of both preceding methods.

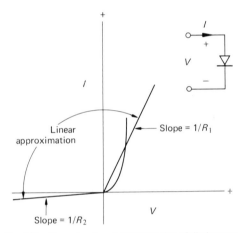

Fig. 2.12 Piecewise linear approximation of diode curve.

Example 2.1. The applied voltage of the circuit of Fig. 2.13 changes linearly from -1 V to 20 V. If the diode characteristics are as shown in Fig. 2.14 and high-frequency effects can be neglected, show the output voltage.

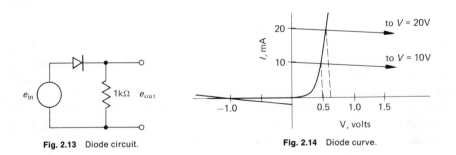

Fig. 2.13 Diode circuit.　　　　　　　　**Fig. 2.14** Diode curve.

Solution. Drawing a load-line for applied voltages of -1 V, 10 V, and 20 V results in device voltages of -1 V, 0.5 V, and 0.6 V respectively. The values of e_{out} for these three input voltages are found by subtracting the corresponding diode voltage values from e_{in}. Thus $e_{out} = 0$ V, 9.5 V, and 19.4 V. Noting that the output voltage will always be zero for a negative input voltage allows the output waveform of Fig. 2.15 to be plotted.

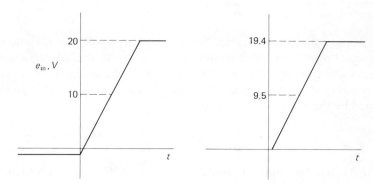

Fig. 2.15 Waveforms for the circuit of Fig. 2.13.

The second method of solution begins by approximating the forward portion of the diode curve by a straight-line segment of 20 mA/V slope which corresponds to a 50-Ω resistance. The reverse portion of the curve can be approximated by zero slope or infinite resistance. When e_{in} is negative, the diode is represented by an open circuit. For positive values of e_{in}, the diode is replaced by a 50-Ω resistance and the output voltages for $e_{in} = 10$ V and 20 V are calculated in the simple resistive circuit of Fig. 2.16 to be 9.5 V and 19.0 V respectively.

Note that the two methods disagree slightly. We will discuss more accurate diode equivalent circuits in Chapter 4. The graphical method is often avoided since the approximate circuit method is more straightforward for large-signal analysis. We might note that for this problem and many others the diode could be approximated by a short circuit in the forward direction and an open circuit in the reverse direction. This allows a great number of practical diode problems to be worked with no knowledge of the diode characteristics beyond the maximum reverse voltage and forward current ratings.

Small-signal operation. When the diode is used in an application where the forward bias is considerably greater than the ac fluctuations, small-signal operation is in effect. Although the V–I curve for the diode is quite nonlinear, any region of this curve small enough to be considered linear can be represented by a resistance. For example, let us assume that a diode with the characteristics represented in Fig. 2.7 is used in the circuit of Fig. 2.17.

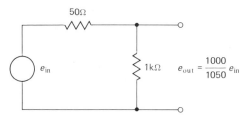
Fig. 2.16 Equivalent forward-biased diode circuit.

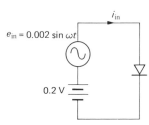
Fig. 2.17 Small-signal diode circuit.

We want to find the ac current flowing into the diode. We may now define a small signal resistance which relates the ac current to the ac voltage. If the ac voltage is small compared to the dc voltage across the diode, the V–I curve will have a constant slope throughout the cycle. In this case, the current and voltage are related by the slope of the V–I curve as

$$\frac{\Delta I}{\Delta V} = \text{slope} = \frac{dI}{dV}. \tag{2.16}$$

Since $\Delta I/\Delta V$ has the units of conductance, Eq. (2.16) can be inverted to give units of resistance. Therefore

$$\frac{dV}{dI} = \frac{1}{dI/dV} \equiv r_d, \tag{2.17}$$

where r_d is defined as the small-signal resistance and relates small changes of current to small changes in voltage. We can easily find r_d by considering the diode equation. Since

$$I = I_s(e^{qV/kT} - 1) \simeq I_s e^{qV/kT},$$

then

$$\frac{dI}{dV} = \left(\frac{q}{kT}\right) I_s e^{qV/kT} = \frac{q}{kT} I.$$

Now r_d equals the reciprocal of dI/dV; thus we can write

$$r_d = \frac{kT}{qI} = \frac{0.026}{I} \quad \text{(at room temperature)}. \tag{2.18}$$

The dynamic resistance depends on the dc bias current of the diode. For the circuit of Fig. 2.17 the dc bias current will be 1 mA (from Fig. 2.7); therefore

$$r_d = \frac{0.026}{1 \times 10^{-3}} = 26 \, \Omega.$$

The input ac current can be found as

$$i_{in} = \frac{e_{in}}{r_d} = \frac{0.002 \sin \omega t}{26} = 0.077 \sin \omega t \text{ mA}.$$

Note that this value of ac current can be changed, even when e_{in} is kept constant, if the applied bias is changed. We must always be careful to use the concept of dynamic resistance only when the ac current flow is small compared to the dc bias current.

I. Depletion-Region Capacitance

It was previously mentioned that the width of the depletion region varies with voltage applied to the diode. With no bias the width of the depletion region depends on the doping levels of the p-type and n-type materials making up the diode. It follows that the total immobile charge contained in the depletion region is also a function of doping in an unbiased diode. When a voltage source is used to apply a forward bias, the width of the depletion layer is decreased as charge is supplied by the source. A reverse bias results in a wider depletion layer and charge is removed by the source. Thus, any change in applied voltage results in a change in the net charge included in the depletion region. The change in net charge resulting from a given change of voltage can be found; however, this calculation is laborious. Rather than repeat this calculation each time the applied voltage is modified, the effect can be represented by a capacitance for small voltage changes. If calculated correctly, the capacitor will yield a change in charge equal to that of the depletion region for a given applied voltage change. As we shall presently see, the charge-voltage relationship is a nonlinear one and consequently the value of equivalent capacitance is a function of applied voltage. For many applications, the applied voltage change is small and a single value of equivalent capacitance can accurately be used. This value is referred to as the small-signal capacitance. For switching circuits where large voltage changes are ordinarily encountered, an average value of capacitance that is related to the small-signal value can be found. This average value will be discussed in detail in Chapter 8 relative to high-speed switching considerations.

To find the small-signal depletion-region capacitance, it is helpful to assume that the number of free carriers in the depletion region is negligible compared to the doping densities. We also make use of the fact that the practical case will normally find one region of the diode doped much more heavily than the other; for example, N_A may be 100 to 1000 times N_D. This results in a situation such as that pictured in Fig. 2.18. There must be an equal amount of uncovered charge on each side of the origin, or $N_A x_1 = N_D x_2$. Since $N_A \gg N_D$, then $x_1 \ll x_2$; that is, the depletion region must extend much further into the n-type region than into the p-type. In this case almost the entire barrier voltage appears across the n-type region from $x = 0$ to $x = x_2$. Poisson's equation can be applied to find the voltage across this region,

$$\frac{d^2 V_j}{dx^2} = -\frac{qN_D}{\varepsilon}, \tag{2.19}$$

where ε is the dielectric constant of the material. If V_j is used to represent the voltage within the depletion region, and if we assume that $V_j = 0$ and $dV_j/dx = 0$

at $x = 0$, we obtain
$$V_j = \frac{-qN_D x^2}{2\varepsilon}.$$

At $x = x_2$, the junction voltage equals $V - |\phi|$, where V is the applied voltage. Thus,
$$V - |\phi| = \frac{-qN_D x_2^2}{2\varepsilon}. \tag{2.20}$$

The charge that is uncovered over this region is
$$Q = qN_D x_2 A. \tag{2.21}$$

Since the voltage as a function of charge is desired, Eq. (2.20) is written as
$$V - |\phi| = \frac{-Q^2}{2qN_D \varepsilon A^2}. \tag{2.22}$$

The small-signal capacitance can be defined as
$$C = \left|\frac{dQ}{dV}\right|.$$

Differentiating Eq. (2.22) gives
$$\frac{dV}{dQ} = \frac{-Q}{qN_D \varepsilon A^2}.$$

Thus, for small voltage changes, the depletion-layer capacitance is
$$C_{dl} = \left|\frac{dQ}{dV}\right| = \frac{qN_D \varepsilon A^2}{Q} = \frac{\varepsilon A}{x_2}. \tag{2.23}$$

Equation (2.23) shows that the formula for depletion-layer capacitance is the same as that for a parallel-plate capacitor of area A, with a plate separation x_2 and a dielectric constant of ε. However, this equation is not very useful in its present form, since x_2 is usually not known. It is more useful when x_2 is replaced by the appropriate term involving the applied voltage from Eq. (2.20):
$$C_{dl} = \frac{\varepsilon A}{[2\varepsilon/qN_D]^{1/2}\sqrt{(|\phi| - V)}} = \frac{k}{\sqrt{(|\phi| - V)}}, \tag{2.24}$$

where $k = [qN_D \varepsilon/2]^{1/2} A$. Quite often k can be found simply by measurement, so that C_{dl} can be evaluated at any value of applied voltage. Since V will be negative for reverse-bias voltages, C_{dl} will become less for large reverse voltages. It should be noted that for forward-bias voltages, V must always be smaller than the barrier voltage for normal operation. Most manufacturers will specify the depletion-layer capacitance at a given reverse-bias voltage for diodes that are used in high frequency applications.

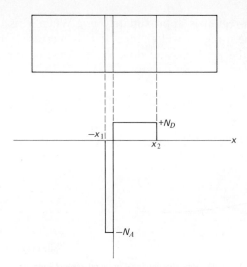

Fig. 2.18 Net charge density vs. distance.

Generally, this capacitance is of more interest for the reverse bias case since another capacitance called the diffusion capacitance becomes the dominant reactive effect for forward-biased diodes. The use of the equivalent depletion-layer capacitance will be treated in a later chapter in connection with diode circuit models.

For the planar construction process used for many diodes, there is a non-uniform impurity profile. In this case, the variation of small-signal capacitance is more nearly approximated by

$$C_{dl} = \frac{k_1}{(|\phi| - V)^{1/3}}. \tag{2.25}$$

J. Diffusion Capacitance

This effect is due to the change in the minority-carrier distribution as a result of change in applied voltage.

Consider the diode of Fig. 2.19 which is forward-biased and has a heavily doped p-region. Again it will be assumed that the concentration of holes at the

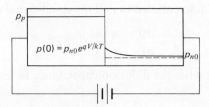

Fig. 2.19 Minority-carrier distribution in forward-biased diode.

right edge of the n-region is approximately p_{no}. When V is changed, the hole concentration in the n-region will change. If V is increased, more holes will be required to set up the gradient, and thus charge must be supplied, as shown in Fig. 2.20.

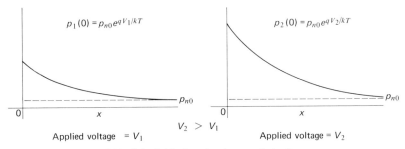

Fig. 2.20 Hole distributions for given applied voltages.

To find the diffusion capacitance, the change in excess charge in the n-region for a given change in voltage must be found. From Eq. (1.24) we can determine the total excess charge to be

$$Q = qA \int_0^W p_e(x)\, dx = qA \int_0^W p_e(0) e^{-x/L_p}\, dx.$$

Therefore

$$Q = qA p_e(0) L_p (1 - e^{-W/L_p}). \tag{2.26}$$

The number of excess holes at $x = 0$ is a function of applied voltage and can be expressed as

$$p_e(0) = p(0) - p_{no} = p_{no}(e^{qV/kT} - 1).$$

The total excess charge then is

$$Q = qAL_p p_{no}(1 - e^{-W/L_p})(e^{qV/kT} - 1),$$

and, since $W \gg L_p$, this equation reduces to

$$Q = qAL_p p_{no}(e^{qV/kT} - 1). \tag{2.27}$$

Equation (2.27) can be differentiated to find the small-signal diffusion capacitance,

$$C_d = \frac{dQ}{dV} = \frac{q^2 AL_p p_{no}}{kT} e^{qV/kT}. \tag{2.28}$$

Using Eq. (2.8), we can write the diffusion capacitance in terms of the dc current:

$$C_d = \frac{L_p^2 qI}{D_p kT}. \tag{2.29}$$

The diffusion capacitance increases directly with forward current.

When a diode is reverse-biased, the current is very small and the diffusion capacitance is quite small. For this case the depletion-layer capacitance dominates the reactive behavior. It is only for significant forward current flow that the diffusion capacitance exceeds the depletion-layer capacitance. When transistors are used as linear amplifiers, the diffusion capacitance of the collector-base junction is neglected since this junction is reverse-biased. At the same time, the diffusion capacitance of the forward-biased emitter-base junction is the larger of the two capacitances associated with this junction.

It is significant to note that the product of C_d and r_d for a forward-biased junction is constant with bias current, that is,

$$\tau = r_d C_d = \frac{kT}{qI} \times \frac{L_p^2 qI}{D_p kT} = \frac{L_p^2}{D_p}.$$

This is an important result and will be discussed in more detail in later chapters.

K. The Zener Diode

The diode characteristics of Fig. 2.7 indicate that a large reverse voltage can cause the normally small reverse current to increase extremely rapidly. This point is referred to as breakdown, but is nondestructive if the power dissipation of the diode is limited to reasonable values.

In many applications the diode must block current for all values of reverse voltage. For these situations, the breakdown voltage of the diode must exceed the maximum reverse voltage that will be applied. Other applications such as power supply regulators utilize the region beyond breakdown where a large current change is accompanied by a very small diode voltage change. Diodes designed for this purpose are called Zener diodes.

The magnitude of the breakdown voltage can be controlled by doping to fall in a range from about 2.6 V to 200 V. The actual mechanism of breakdown depends on the electric field intensity of the reverse-biased junction.

Diodes with lower doping levels lead to smaller electric field intensities and larger breakdown voltages. The governing mechanism in this case is avalanche breakdown. At higher levels of doping, large electric field intensities result from small reverse voltages and Zener breakdown takes place.

There are not enough minority current carriers crossing the depletion region of a reverse-biased junction to support the large breakdown currents of the Zener diode. It is then obvious that there must be a source of charge carriers within the depletion region. The valence electrons associated with the crystal structure supply these necessary charge carriers for the breakdown current of the Zener diode. Below breakdown, these electrons are bound to the atoms of the lattice. For very heavily doped junctions, the electric field can become so great that valence electrons are dislodged from the lattice atoms and become free carriers. This process is called Zener breakdown. In lightly doped junctions, avalanche breakdown occurs with the free carriers being accelerated to the point that ionizing collisions occur in the depletion region. The newly created free electrons can then accelerate and collide with the lattice producing even more hole-electron pairs. This results in avalanche multiplication and leads to the high breakdown currents of the Zener diode. Both breakdown mechanisms are nondestructive so long as the external circuit limits the current flow to an acceptable level. The symbol for the Zener diode is shown in Fig. 2.21. Applications will be considered in connection with power supply regulators in a later chapter.

Fig. 2.21 Zener diode symbol.

L. The Tunnel Diode

Shortly after the first published work on tunneling in heavily doped pn-junctions appeared in 1958, those in the electronics field became very interested in the tunnel diode. This diode promised higher switching speeds, amplification with greater bandwidths, less temperature dependence, and lower noise figures than transistors. The enthusiasm with which the tunnel diode was met was unwarranted due to the rather serious shortcomings of the device. It is a two-terminal device resulting in no isolation between input and output. The negative resistance region extends over only a few tenths of a volt; thus, the output signal is quite small. Negative resistance also leads to stability problems and tunnel diode circuits require precautions in this regard. Earlier tunnel diodes exhibited an aging problem, but this no longer appears to be a problem. An interesting sidelight to this aging problem relates to a firm that manufactures digital computers. After an extensive effort to build an advanced computer in the early 1960s with a design based on the gallium arsenide tunnel diode, it was discovered that the V–I characteristics of the device changed with age to the point that the computer was no longer reliable. Thus, the entire design effort was wasted.

While the expected takeover of the electronics field by the tunnel diode never developed, the device remains useful in several applications. Its operating frequency extends into the microwave region and it is a useful component in microwave communications. It is also well used in low-level switching circuits and low-noise amplifiers.

Theory of operation. The distinguishing characteristic of the tunnel diode is that it possesses a negative resistance region. In this region, a voltage increase causes an incremental decrease in current flow.

In a normal pn-junction the barrier voltage presents an obstacle to the flow of carriers from one region to another. Only a very few carriers have sufficient energy to surmount the barrier and enter the other region. A forward-bias voltage opposes the barrier voltage, reducing the net junction voltage. The number of carriers with sufficient energy to surmount the net junction voltage increases exponentially with applied voltage. When the diode is reverse-biased, the carriers flow in the opposite direction. In this case, the junction voltage aids the carrier flow; however, the total current is limited to a very small value (saturation current), due to the small number of available minority carriers.

The tunnel diode is doped very heavily, causing an extremely narrow depletion region. The barrier-voltage gradient then extends over a very small distance. The transition region might be 5×10^{-7} cm in a tunnel diode, compared to 10^{-4} cm for a conventional diode. The fact that the barrier voltage is very narrow allows carriers with insufficient energy to surmount and "tunnel" through the barrier. The probability of a carrier penetrating the barrier decreases exponentially with increasing width of the barrier. An additional requirement for tunneling to occur is that there be available states at the proper energy level on the other side of the barrier.

With very heavy (degenerate) doping, the energy bands of the diode appear schematically as shown in Fig. 2.22. In part (a), showing the no bias condition, the tunneling of electrons from both sides of the barrier is equal and opposite; thus the net current is zero. For a small forward bias, the electrons in the n-region possess a higher energy than those in the p-region. More electrons then flow from n- to p-region than from p- to n-region. As the forward applied voltage is increased, the current continues to increase. As the energy level of the electrons in the n-region is raised, the level of the higher-energy electrons corresponds to the energy of the forbidden gap in the p-region. Since these electrons have no available states in the p-region, they cannot tunnel through the barrier. Further increase in voltage increases the number of n-region electrons that have no corresponding available states in the p-region. The current then decreases as the applied voltage is raised. Figure 2.22(d) shows the condition for zero tunnel current, because no available states exist in the p-region that can accept n-region electrons. The current flow does not drop to zero at this point, since the normal diode current will flow in

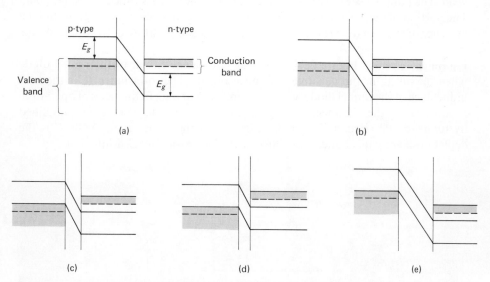

Fig. 2.22 Schematic energy diagrams of the tunnel diode for various bias conditions: (a) no bias; (b) small forward bias; (c) bias for maximum tunneling current; (d) bias for zero tunneling current; and (e) reverse bias.

addition to tunnel current. The superposition of these currents gives the total current flow. The characteristics are shown in Fig. 2.23. In the reverse direction, more electrons in the p-region see available states in the n-region and a high reverse conductance results.

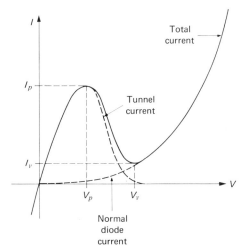

Fig. 2.23 V–I characteristics of the tunnel diode.

The point corresponding to maximum tunneling current is called the *peak point*. The point where the tunnel current approaches zero is called the *valley point*. The portion of the curve between V_p and V_v is a negative resistance region, and the presence of this negative resistance is the distinguishing feature of the tunnel diode.

Equivalent circuit. Figure 2.24 shows the equivalent circuit of the tunnel diode when operating in the negative resistance region. The inductance L_t is the lead inductance of the tunnel diode, which is significant at high frequencies of operation. Frequencies in the kilo-megahertz (gigahertz) range can be generated or amplified by tunnel-diode circuits. The bulk resistance of the device is R_t, while C is the depletion-layer capacitance associated with the forward-biased junction.

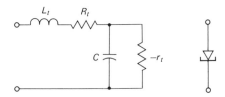

Fig. 2.24 (a) Small-signal equivalent circuit of the tunnel diode. (b) Symbol for the tunnel diode.

2.2 THE BIPOLAR TRANSISTOR

The transistor is a device consisting of three regions of doped semiconductor materials. The three regions are connected as shown in Fig. 2.25(a) to form two junctions. The base region is very narrow compared to the emitter and collector regions; it is often in the range of 0.00001 to 0.0001 inch in thickness. The processes governing the transistor's operation take place in the base region, and therefore a magnified base region is usually drawn as shown in Fig. 2.25(b). The bias batteries shown in part (b) cause the transistor to be in its active region; that is, the emitter-base junction is forward-biased and the collector-base junction is reverse-biased.

There are many different construction processes that can be used for transistor fabrication. Each process results in a different set of device parameters. Basically, the operation of the different types of transistors is similar and, fortunately, a "textbook" transistor can be used to analyze this operation. In the textbook transistor, abrupt junctions are assumed to exist, a unit cross-sectional area is assumed, and the area is assumed constant throughout the various regions.

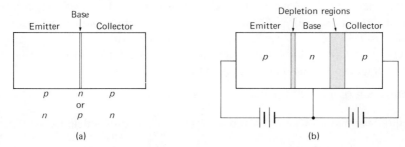

Fig. 2.25 The transistor: (a) physical picture; (b) base region magnified.

A. Depletion Regions

A depletion region will exist at each junction. The effective base region is the region between the two depletion layers. It was shown in Section 2.1 that a knowledge of the depletion region was unimportant in the analysis of the diode. This is true to a first-order approximation in the case of the transistor. However, since the depletion-region thicknesses depend on the applied voltages, so also does the effective base width. This dependence of base width on applied voltage gives rise to a "second-order" effect, which in some cases can be neglected. Aside from this so-called base-width modulation, the depletion regions can again be neglected in calculating the carrier distributions in the transistor.

B. Minority-Carrier Distribution

For a pnp-transistor the hole distribution for the condition of no bias is as shown in Fig. 2.26. The hole distribution is shown because holes are minority carriers in the base region, which is a region of great importance in the transistor. The number of

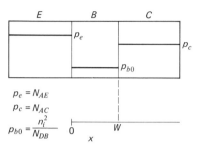

Fig. 2.26 Hole distribution in unbiased transistor.

minority electrons in the emitter will be negligible compared to p_{b0} due to the doping requirement that N_{AE} be much greater than N_{DB}.

$$n_{e0} = \frac{n_i^2}{N_{AE}} \ll p_{b0} = \frac{n_i^2}{N_{DB}}, \quad \text{if} \quad N_{AE} \gg N_{DB}.$$

It will be recalled that when a typical voltage is applied to a diode, the voltage is dropped across the junction. The majority-carrier distribution on either side of the junction is negligibly changed, but the minority-carrier distribution depends strongly on the applied voltage. The general equation for minority carriers at a junction edge can be found [Eqs. (2.6) and (2.7)] to be

concentration of minority carriers at junction edge

$$= \text{equilibrium minority concentration} \times \exp\left[\frac{q}{kT} \times (\text{applied junction voltage})\right]$$

or

$$p_{\text{junction edge}} = p_{b0} e^{qV/kT}. \tag{2.30}$$

If the collector-base junction is reverse-biased, the concentration of holes at this junction becomes zero for any reasonable bias voltage:

$$p(W) = p_{b0} e^{-V_{CB}/0.026} \to 0 \tag{2.31}$$

($kT/q = 0.026$ at room temperature). The emitter-base junction, when forward-biased, will have a concentration of

$$p(0) = p_{b0} e^{V_{EB}/0.026}. \tag{2.32}$$

The above equations indicate that the collector-base junction is a sink for holes which will absorb all holes near this junction, and that the emitter-base junction is a source that supplies the necessary holes to satisfy Eq. (2.32). It is obvious that there will be some sort of nonuniform distribution of holes across the base region. The problem is to find this distribution, so that we can evaluate the current flow through the base. Since we assume that the applied voltages drop across the junctions and not across the base, we see that only diffusion current will flow in this region if negligible recombination takes place.

C. Base Diffusion Current

It is easy to find the hole distribution in the base region. The average distance that a hole travels before recombining is given by L_p. A typical value of L_p is 0.005 in. Therefore, since the base region is usually much narrower than this value, very little recombination is expected to take place there. If there is negligible carrier recombination, then the diffusion current must be constant at all points throughout the base. This means that the gradient of carriers must also be constant in the base region (as noted in Section 1.3). The only possible distribution is as shown in Fig. 2.27. The diffusion current can easily be calculated from the diffusion equation,

$$I_p = -qD_p \frac{dp}{dx} A = +qD_p \frac{[p(0) - p(W)]}{W} A$$

$$= \frac{qD_p A}{W} p_{b0} e^{V_{EB}/0.026} \quad \text{(at room temperature).} \quad (2.33)$$

Equation (2.33) shows that the current that travels from emitter to collector across the base region depends exponentially on the emitter-base voltage.

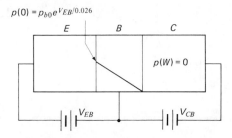

Fig. 2.27 Minority distribution in base region.

This simple discussion of diffusion current through the base region forms the complete basis of operation of the transistor. Current through the transistor is proportional to the negative of the gradient of minority carriers in the base region. When the collector is reverse-biased, this gradient is controlled by the emitter-base voltage. Therefore, the emitter-base voltage controls the collector and emitter currents by controlling the gradient of minority carriers.

We have neglected recombination in the base region. Actually, a very small fraction of the injected holes recombine with electrons in the base region. In order to replace these electrons so that equilibrium is still in effect, electrons flow in from the base lead. This amounts to a conventional current flow out of the base lead. The base current is only about $\frac{1}{50}$ or $\frac{1}{100}$ of the emitter current, but it is very important to the operation of the transistor.

The parameter α is generally used to show the various current relationships; α is defined as

$$\alpha = \frac{\Delta I_C}{\Delta I_E}\bigg|_{V_C = \text{const}}. \quad (2.34)$$

A dc α is sometimes defined as

$$\alpha_{dc} = \left.\frac{I_C}{I_E}\right|_{V_C = \text{const}}.$$

Usually α_{dc} and α differ by less than one or two percent and can often be assumed to have the same value. The parameter α tells the portion of current injected into the emitter that reaches the collector. Typically, α ranges from 0.97 to 0.999.

D. Relation of the Characteristic Curves to the Minority-Carrier Distribution

Common-base configuration. The characteristic curves for a transistor are plots of the output current vs. output voltage for various values of input current. In the common-base configuration the collector-to-base voltage is the output voltage and the emitter current is the input current. This configuration is shown in Fig. 2.28. Note that the base terminal is the common or reference to both input and output voltages.

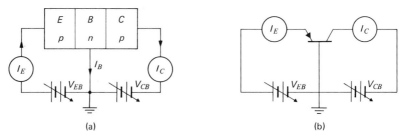

Fig. 2.28 Common-base measurements: (a) common-base configuration; (b) alternate form.

CASE 1. Zero I_E–nonzero V_{CB}.

Figure 2.29 shows this case. The concentration of holes at the CB-junction is equal to zero for appreciable values of V_{CB}. The concentration at the EB-junction is nonzero, since there is no sink for holes in this vicinity. The collector current that flows is

$$I_C = -qAD_p \frac{dp}{dx} = \frac{qAD_p p(0)}{W} = I_{co}. \tag{2.35}$$

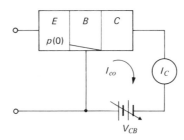

Fig. 2.29 Hole distribution for $I_E = 0$.

We have neglected the component of leakage current due to minority electrons in the collector region in Fig. 2.29 and Eq. (2.35). In planar transistors it is likely that this component will be comparable to the hole component. In this case the collector current will be

$$I_C = -qA\left(\frac{D_p p(0)}{W} + \frac{D_n n_{co}}{L_n}\right). \tag{2.36}$$

This current that flows with the emitter open is called the CB-leakage current and is usually denoted by I_{co} or I_{cbo}. It is very small, typically 1 to 10 μA for low-power germanium transistors and 0.001 to 0.1 μA for low-power silicon units. Ideally, I_{co} would not vary with collector voltage. However, since larger V_{CB} values cause the base width to decrease (due to increased depletion-layer thickness), I_{co} increases slightly with increased V_{CB}. This curve is plotted in Fig. 2.30.

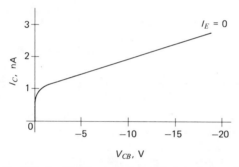

Fig. 2.30 Common-base output characteristics for $I_E = 0$.

CASE 2. Nonzero I_E–nonzero V_{CB}.

The collector current in this case will consist of I_{co} plus the contribution from the emitter current. This is shown in Fig. 2.31. The collector current in this case is again weakly dependent on the collector voltage due to the base-width modulation effect. The base current can also be written in terms of the emitter current, since $I_B = I_E - I_C$:

$$I_C = \alpha I_E + I_{co}, \tag{2.37}$$
$$I_B = (1 - \alpha)I_E - I_{co}. \tag{2.38}$$

The characteristic curves are shown in Fig. 2.32 for several values of I_E. The slight upward slope is due to the base-width modulation effect. It is noted that for finite emitter currents, I_C flows even when $V_{CB} = 0$. Since I_E is present, a certain amount of holes will be injected into the base across the EB-junction. A concentration of $p(0)$ will exist. Unless this same concentration occurs at the CB-junction, a gradient will exist and collector current will flow. A slight forward bias at the CB-junction causes $p(W) = p(0)$, and hence $I_C = 0$. This situation is shown in Fig. 2.33, neglecting minority-carrier concentrations in both emitter and collector.

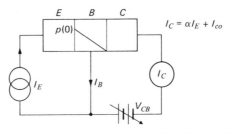

Fig. 2.31 Hole distribution in the base when it is biased in the active region.

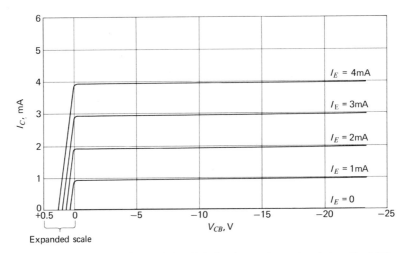

Fig. 2.32 Common-base characteristics. (Note the change of current scale from Fig. 2.30.)

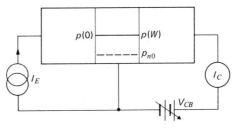

Fig. 2.33 Conditions necessary for $I_C = 0$.

The region of the characteristic curves where both *CB*- and *EB*-junctions are forward-biased is called the saturation region.

Common-emitter configuration. The common-emitter configuration is the most commonly used configuration. It appears in Fig. 2.34. In this case the two quantities, I_C and V_{CE}, are the output quantities. The base current is the input parameter and the emitter terminal is common to both input and output. The voltage V_{BE} still appears across the *EB*-junction as it previously did, but the *CB*-junction has a voltage of $V_{CE} - V_{BE}$ across it. Since we want to find I_C as a function of I_B, we shall

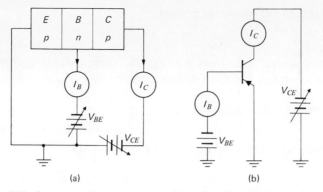

Fig. 2.34 Common-emitter measurements: (a) configuration; (b) alternate form.

use Eq. (2.37). The following results are obtained:

$$I_C = \alpha I_E + I_{co}, \quad (2.37)$$
$$I_B = I_E - I_C,$$

$$I_C = \frac{\alpha}{1-\alpha} I_B + \frac{1}{1-\alpha} I_{co}, \quad (2.39a)$$

$$I_C = \beta I_B + (\beta + 1) I_{co}, \quad (2.39b)$$

where $\beta = \alpha/(1 - \alpha)$. These latter two equations hold for all configurations in the active mode, but are used only when I_B is the input current, that is, for the common-emitter configuration. Note that β can be very large, typically 50 to 200. This quantity is defined as the base-to-collector, short-circuit current gain,

$$\beta = \frac{\Delta I_C}{\Delta I_B}\bigg|_{V_{CE} = \text{const}}; \quad (2.40)$$

β and β_{dc} can differ, but for now will be assumed equal. The characteristic curves are shown in Fig. 2.35.

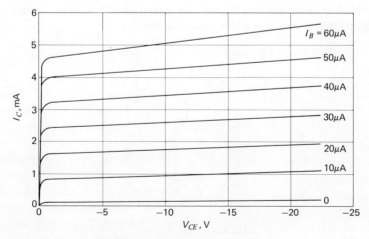

Fig. 2.35 Common-emitter characteristics.

There are several differences between the common-base and the common-emitter characteristics. For the common emitter:

1. I_C is greater for zero-input current,
2. I_C increases more with output voltage for a given input current, and
3. The slope of the curves increases at higher values of I_C.

The first point results from the fact that when $I_B = 0$, I_E is nonzero; that is, current is being injected into the emitter even when $I_B = 0$. From Eq. (2.39b), with $I_B = 0$,

$$I_C = (\beta + 1)I_{co}. \tag{2.41}$$

Here I_C is $(\beta + 1)$ times as large for the common-emitter configuration than for the common-base configuration when no input current is present. The slopes of the lines are steeper in this case because α changes slightly as V_{CB} increases. This change is negligible in the common-base configuration; however, a very slight change in α causes a large change in β. A change of one percent in α from 0.98 to 0.99 results in a 100 percent change in β. It is this change in β that causes points 2 and 3 to occur; β usually increases with higher V_{CB}, and with I_C, up to a point. At very high I_C, β starts dropping again. It should be noted that there is quite a high current gain for the common-emitter configuration. For low-power silicon transistors, I_{co} is so small that collector current is zero when base current is zero. For high-power units Eq. (2.41) may become important.

Common-collector configuration (emitter follower). The emitter follower is depicted in Fig. 2.36. Characteristics for the emitter follower are not usually plotted, since the common-emitter characteristics can be used. This circuit will be analyzed in some detail in later chapters because its importance is intermediate to that of the common emitter and that of the common base.

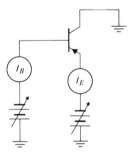

Fig. 2.36 Emitter follower.

E. Practical Aspects of Transistor Amplification

We are now in a position to qualitatively discuss the mechanisms involved in current and voltage amplification by transistors. Consider the common-base amplifier of Fig. 2.37. Assuming that the sources V_{EB} and V_{CC} will bias the transistor in its active region, we can note the effects of e_{in}. The small ac voltage will modulate the gradient of minority carriers in the base region. Since the density of carriers at the emitter-base junction will vary exponentially, this modulation will be appreciable even for very small applied voltages. The change of the gradient also changes the current proportionately. Therefore large emitter current changes are produced by the small input voltage changes. This ac current travels through the base region with very small loss due to recombination, since

$$i_c = \alpha i_e \quad \text{(incremental equation)}. \tag{2.42}$$

The collector current develops a voltage across R_L, which is the output voltage and can be much larger than e_{in}. The operation can be summarized as follows:

1. e_{in} produces large emitter current changes (exponential); and
2. the current travels through the transistor to R_L, developing a large output voltage.

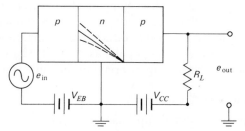

Fig. 2.37 Common-base amplifier.

With the configuration shown, voltage gains as high as 100 to 200 are obtained. The current gain is quite low, however, since the source current is i_e and the output current is αi_e. This gain is equal to α, which is slightly less than unity. We should also note that the phase shift between input and output voltages is zero.

The operation of the common-emitter amplifier shown in Fig. 2.38 is quite similar to that of the common base. The input source is again applied across the emitter-base junction; therefore the same magnitude of voltage gain is expected. There are two differences between the common-emitter and common-base configurations. The first is that when the source goes positive, it is actually lowering the emitter-base voltage and hence the emitter current. The output voltage is then 180° out of phase with the input voltage. The other difference is that the source current is base current in the common-emitter configuration. This current is much smaller than either emitter or collector current, meaning there is a high current gain in this configuration. Since there is both current and voltage gain for a

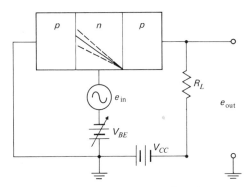

Fig. 2.38 Common-emitter configuration.

common-emitter stage, the power gain is quite high. It is this fact that makes this stage the most widely used.

Up to this point we have discussed only transistors with uniformly doped regions. Actually, most mass production processes now lead to devices with graded doping profiles in all regions. Transport of charge carriers across the base region takes place with a large drift component of current for the graded base transistor. While there are differences in the two devices, similar equivalent circuits apply, differing only in element value rather than configuration. Thus, although the uniform base transistor is not as popular as the graded base device, a discussion of the operation of the former is appropriate to the understanding of both types.

F. Graphical Analysis of the Transistor

In general, an equivalent circuit approach to transistor design is much more effective than graphical analysis, but it is instructive to investigate the load-line analysis for the purpose of obtaining more information about the transistor.

Consider the circuit of Fig. 2.39(a) with the corresponding V–I characteristics of Fig. 2.39(b). The load-line is sketched on the V–I characteristics of the device with the slope $= -(1/R_L) = -(1/5 \text{ k}\Omega)$. Actually, only one set of device characteristics is shown. An infinite number of curves could be drawn corresponding to other I_B values than those given. In effect, there is an infinite number of possible operating points. On the other hand, once I_B is specified, the operating point is uniquely determined. If I_B is 20 μA, the operating point is given by point a, and the output quantities are then found to be $V_{CE} = 7.5$ V and $I_C = 2.5$ mA. Now let us assume that I_B changes slowly to zero current. In this case, the operating point will move slowly to the right along the load-line, reaching point b when $I_B = 0$. This point corresponds to the transistor being cut off, and V_{CE} is almost equal to the power supply voltage. The collector current is made up of leakage current only; thus, there will be a very small drop of voltage across the load. For all silicon transistors and for many germanium transistors the leakage is negligible, and this fact results in $V_{CE} = V_{CC} = 20$ V.

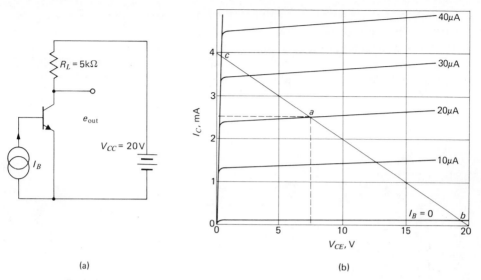

Fig. 2.39 Common-emitter amplifier: (a) common-emitter circuit; (b) characteristics.

If I_B is now slowly increased to 40 μA, point c on the load-line will be reached. This corresponds to saturation and the transistor will have only a small voltage drop from collector to emitter. A typical value might be 0.1 V. The collector current is equal to approximately 4 mA, that is,

$$V_{CE} = 0.1 \simeq 0 \text{ V},$$

so

$$I_C = \frac{V_{CC} - V_{CE}}{5 \text{ k}\Omega} \simeq \frac{20 \text{ V}}{5 \text{ k}\Omega} = 4 \text{ mA}.$$

Now I_B can be increased beyond this value of 40 μA. If 70 μA of base current is used, the operating point does not change, since the $I_B = 70$ μA curve also passes through point c. In most transistors, the base current can become 10 to 20 times the minimum current required for saturation without destroying the transistor.

The common-emitter amplifier can then take on any output voltage ranging from approximately zero volts to the power supply voltage, depending on the input current. If the input consists of a dc current plus an ac component, then the output voltage will also consist of these two components. For example, if

$$I_B = (20 + 10 \sin \omega t) \text{ μA},$$

then the output voltage will be $(7.5 - 5 \sin \omega t)$ V. The quiescent level (dc level with no ac input) will be $V_{CEQ} = 7.5$ V. As I_B swings to 30 μA (peak value), V_{CE} moves to 2.5 V; as I_B swings to 10 μA, V_{CE} swings to 12.5 V. An ac current of 10 μA peak enters the base of the transistor and causes a 5 V peak voltage output (inverted in phase). The problem now is to find the voltage gain of the stage. To do this the input characteristics would be required. Figure 2.40 shows a set of input characteristics. The input-current swing is seen to correspond to an input-voltage swing from 0.54 V to 0.66 V, or a peak input voltage of 0.06 V. The voltage gain of the stage is then $-(5/0.06) = -83$. The above method can be used to solve transistor amplifier problems, but as mentioned before, it is not a very effective method.

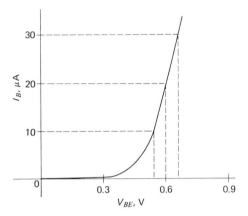

Fig. 2.40 Input characteristics of a transistor.

The load-line approach has demonstrated a very important point that can be applied later; that is, the active region of a common-emitter amplifier with no emitter resistance is essentially equal to the power supply voltage.

2.3 THE FIELD-EFFECT TRANSISTOR

The field-effect transistor or FET was proposed prior to the development of the bipolar transistor; however, it has been more recently that this device has achieved real significance in the electronics field. While it is true that a number of special-purpose applications have used the discrete FET to advantage, the integrated circuit industry now fabricates many types of linear and digital circuits based on the FET. The fabrication process can consist of fewer diffusive steps and the packing density can be greater for many FET circuits than for bipolar counterparts. Thus, some yield and cost advantages are exhibited for FET integrated circuits.

There are basically two different types of FET. The junction FET was the first type to become practically significant near 1960. Ten years later the metal-oxide-silicon FET or MOSFET became more prominent. Both devices have similar operating characteristics and both can exhibit very high values of input impedance. Bipolar transistors are more useful in high-frequency amplifier design than are junction FETs, but maximum oscillation frequencies of 30 GHz have been reported for some configurations of the FET.

This section will develop the basic theory of operation of the junction FET and extend this theory to the MOSFET.

A. The Junction FET

The actual configuration of the junction FET may be quite different from the simple geometry to be discussed here. It is rather instructive though to consider the simpler device in developing the basic principles governing the operation of the FET.

Voltage controlled resistance. A doped silicon bar exhibits a resistance from one end to the other that depends on length, cross-sectional area, and free carrier concentration. It is possible to control the effective area of a bar by forming junctions along the length of the bar on opposite sides as shown in Fig. 2.41. The heavily doped p-regions cause the depletion region to extend far into the n-region for appreciable values of reverse-bias voltage. These depleted areas have no free carriers and consequently the conductivity of the material in the depletion regions is extremely low. The electrical cross-sectional area is reduced under the p-regions and the resistance of the bar is increased.

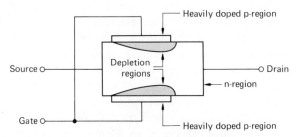

Fig. 2.41 Voltage-controlled resistor (VCR).

The depletion-region widths are functions of the voltage applied between the source and gate terminals and the voltage drop from drain to source. When no current flows from drain to source resulting in no voltage drop between these terminals, the depletion-region widths are controlled only by the gate-to-source voltage as shown in Fig. 2.42. Obviously, V_{GS} effectively controls the resistance of the device if there is no drain-to-source voltage difference. Larger reverse-bias

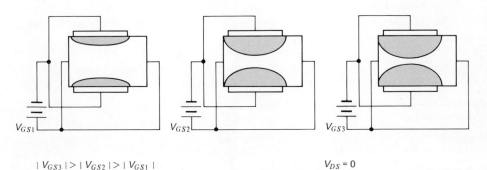

$|V_{GS3}| > |V_{GS2}| > |V_{GS1}|$ $V_{DS} = 0$

Fig. 2.42 Effect of V_{GS} on effective area of n-channel VCR.

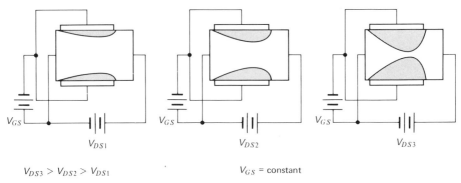

$V_{DS3} > V_{DS2} > V_{DS1}$ V_{GS} = constant

Fig. 2.43 Effect of V_{DS} on effective area of n-channel VCR.

voltages lead to increased resistance of the channel. Unfortunately, the practical resistor is always used to conduct a certain amount of current, establishing a voltage difference from one end of the resistor to the other. If a positive drain-to-source voltage is applied to the resistance, the reverse bias of the junction will be greater near the drain than near the source. The widths of the depletion regions will then be greater near the drain, as shown in Fig. 2.43.

When V_{DS} is small and the depletion regions are narrow, small changes in V_{DS} result in almost negligible changes in effective area of the bar. Resistance is only slightly affected by V_{DS} for small values of this voltage. As V_{DS} reaches higher values, the depletion regions extend over almost the entire width of the bar. When $V_{DS} = V_{DS3}$, as shown in Fig. 2.43, a small change in depletion-region width may change the effective area by a large factor. The effective resistance for larger values of V_{DS} is a strong function of this voltage. Figure 2.44 shows a plot of drain current as a function of V_{DS} for various constant values of V_{GS}. We note that if V_{GS} has a more negative value, the depletion regions can be made to extend completely across the width of the bar at lower values of V_{DS}.

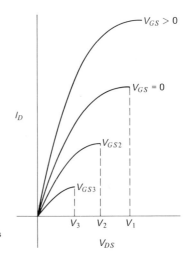

Fig. 2.44 V–I characteristics of the VCR.

The resistance of the n-channel is fairly constant over the initial portion of the curves, as indicated by the constant slope. The linear range of resistance is extended for values of V_{GS} that are near zero or slightly positive. Of course, V_{GS} should not exceed the positive value required to forward bias the gate-source junction. If this value exceeds a few tenths of a volt, injection from gate to channel takes place, leading to a different behavior than indicated thus far. As the depletion regions extend over almost the entire width of the bar, the increase in resistance is reflected by the flattened portion of the curves.

Often the FET is used as a VCR and the linear operating region will be limited to a few volts. The linearity of the device can be improved by referencing the gate voltage to the drain rather than the source. As drain voltage increases, the maximum gate-to-channel voltage remains more nearly constant and the variation of effective area is minimized.

Quantitative relations can be developed for the VCR, but since the succeeding section on FET theory will consider these relations, no further theory will be developed here.

Operation of a simple FET. A practical configuration for the junction FET is shown in Fig. 2.45.

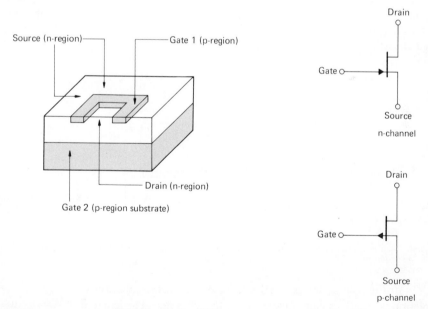

Fig. 2.45 Junction FET geometry and symbols.

Gate 1 is doped very heavily while the p-type substrate making up Gate 2 is low conductivity. The depletion region extends from Gate 1 into the channel with almost no extension from Gate 2 into the channel. Figure 2.46 indicates the idealized device that we will use to develop the governing equations.

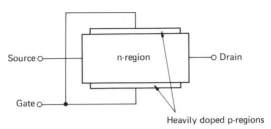

Fig. 2.46 Idealized n-channel FET.

The idealized FET assumes that both p-regions are doped equally and much more heavily than the n-channel. This allows a simpler derivation of characteristics. Since the p-regions are heavily doped, the depletion region extends into the n-region as shown in Fig. 2.47. The current through the n-region depends on the resistance of this region, but because the effective area of the n-region varies, the resistance is nonuniform. The resistance of any incremental element dx is

$$dR(x) = \rho \frac{dx}{A(x)}, \tag{2.43}$$

where ρ is the resistivity of the n-type material and $A(x)$ is the effective cross-sectional area of the nondepleted region. The incremental resistance obviously increases with distance from the source. The voltage variation is definitely nonlinear from source to drain, due to the nonuniform resistance.

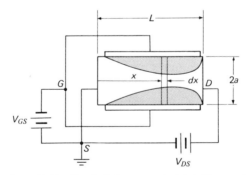

Fig. 2.47 Field-effect transistor with bias. The shaded areas are depleted of free carriers due to the reverse bias applied to pn-junctions.

As drain voltage increases the reverse bias of the junction increases and the depletion regions extend further into the channel. The current increases with voltage, but the magnitude of this increase depends on the effective channel width. As in the case of the VCR, the overall resistance of the channel depends on drain-to-gate voltage. At low values of drain voltage, the depletion region widths are very small compared to the channel width. If drain voltage increases slightly, the channel width and thus channel resistance change very little. The current varies linearly with voltage when drain voltage is small.

2.3 THE FIELD-EFFECT TRANSISTOR

Higher values of drain voltage decrease the effective channel width. When the channel width is narrow a slight change in drain voltage will change the channel width and resistance markedly. Figure 2.48 illustrates the effects on absolute channel width as drain-to-gate voltage modifies depletion-region widths. In part (a) the effective channel width changes only minutely with depletion-region width change. In Fig. 2.48(b) the channel width may change by 50% or more as depletion region widths vary. At higher voltages then, the resistance increases considerably with small changes in drain-to-gate voltage.

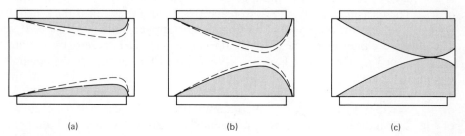

Fig. 2.48 Demonstration of effect of channel width change on effective channel area: (a) wide channel; (b) narrow channel; (c) pinchoff region.

Figure 2.48(c) shows the pinchoff condition that occurs when the effective channel width equals zero at one point along the channel. At this point the maximum depletion region widths are equal to half the channel width. When pinchoff occurs the current flowing prior to pinchoff continues to flow. Free carriers from the nondepleted area of the channel are swept across the depleted region of the channel by the electric field. Further increase of drain voltage affects the drain current very little since most of the additional voltage drops across the high resistance region of the channel that is depleted of free carriers. The drain characteristics of the FET are shown in Fig. 2.49.

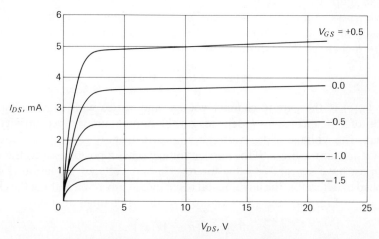

Fig. 2.49 Typical V–I characteristics of a junction FET.

For a given value of gate-to-source voltage the drain current variation shown by the characteristics can be easily related to the preceding discussion. When V_{DS} is small, the current variation with this voltage is approximately linear. As the pinchoff region is approached, the current variation becomes highly nonlinear varying much more slowly with V_{DS}. When pinchoff is reached, the current limits at a constant value with only slight increases as V_{DS} reaches higher values.

Control of the effective channel area is exerted by both the gate voltage and the drain voltage. If V_{GS} approaches some large negative value, the entire channel is depleted of free carriers and drain current is approximately zero for all values of V_{DS}. The measured value of V_{GS} to reach this condition is called the pinchoff voltage V_P. Near the drain end of the channel the applied voltage is $V_{GS} - V_{DS}$. The effective channel area near the drain can be reduced to zero by varying either V_{GS} or V_{DS}. Thus, when $V_{GS} - V_{DS} = V_P$, the FET has entered the pinchoff region. If the measured pinchoff voltage is -3.0 V, pinchoff is reached for the combinations $V_{GS} = -2.0$ V and $V_{DS} = 1.0$ V, $V_{GS} = 0$ V and $V_{DS} = 3.0$ V, or $V_{GS} = -3.0$ V and $V_{DS} = 0$ V. Examination of the characteristic curves of Fig. 2.49 indicates that V_P is approximately -2.0 V for this FET and the pinchoff region for each curve is entered at a value of $V_{DS} = V_{GS} - V_P$. For example when $V_{GS} = -1.5$ V, the pinchoff region is entered near $V_{DS} = 0.5$ V.

While it is possible to develop an analytical description of the curves in the area below pinchoff it is quite difficult to describe the region above pinchoff analytically. In active circuits the FET is used primarily in the pinchoff region. By noting that the drain current limits at the edge of the pinchoff region, we can use the theory developed for the region below pinchoff and easily approximate the extension of the curves into the operating region.

Since the construction of the FET generally involves the diffusion process, the density profiles of the gate and channel are nonuniform. It is instructive to consider an idealized device with uniform doping of both regions. Procedurally, the derivations of characteristics are similar, but the uniformly doped device results in simpler expressions.

For an n-channel FET, the p-region is doped much more heavily than the n-region and the depletion-region width can be approximated by its extension into the n-channel. This width is given by

$$W(x) = k\sqrt{\phi - V_J(x)}, \qquad (2.44)$$

where $V_J(x)$ is the gate-to-channel voltage and varies with x since $V_J(x)$ is a function of x. The constant k depends on the doping and the permittivity of the material. The voltage $V_J(x)$ varies nonlinearly from V_{GS} near the source to $V_{GS} - V_{DS}$ near the drain. If the barrier voltage ϕ has a positive value, the voltage $V_J(x)$ must have a negative value for reverse-bias voltages. Equation (2.43) can be applied to determine the incremental resistance at any point x along the channel. Referring to Fig. 2.47, the effective cross-sectional channel area is

$$A(x) = 2[a - W(x)]b,$$

where b is the channel depth. This allows us to write for incremental resistance along the channel

$$dR(x) = \frac{\rho \, dx}{2b[a - W(x)]} = \frac{\rho \, dx}{2b\{a - k[\phi - V_J(x)]^{1/2}\}}. \quad (2.45)$$

We can write the voltage $V_J(x)$ in terms of the applied gate-to-source voltage and the voltage drop along the channel. Noting that $V_J(x)$ is to be negative for reverse bias and positive for forward bias we can write

$$V_J(x) = V_{GS} - I_{DS} \int_0^x dR(x) = V_{GS} - I_{DS} \int_0^x \frac{\rho \, dx}{2b\{a - k[\phi - V_J(x)]^{1/2}\}}. \quad (2.46)$$

Since $V_J(x)$ appears on both sides of the integral equation it is helpful to differentiate with respect to x. After differentiation and some manipulation we can write

$$\{a - k[\phi - V_J(x)]^{1/2}\} dV_J(x) = -\frac{\rho I_{DS}}{2b} dx.$$

Integrating both sides of this equation with the limits on $V_J(x)$ of V_{GS} at $x = 0$ and $V_{GS} - V_{DS}$ at $x = L$, we obtain

$$aV_{DS} - \tfrac{2}{3}k[(\phi - V_{GS} + V_{DS})^{3/2} - (\phi - V_{GS})^{3/2}] = \frac{\rho I_{DS} L}{2b}. \quad (2.47)$$

Equation (2.47) relates I_{DS} to values of V_{GS} and V_{DS} in the region below or just including pinchoff. We are interested in the region where the current reaches its limit and this occurs when $V_{GS} - V_{DS} = V_P$. The drain current corresponding to pinchoff can be expressed as

$$I_{DSP} = -\frac{2ba}{\rho L} V_P \left(1 - \frac{V_{GS}}{V_P}\right) - \frac{4bk}{3\rho L} (\phi - V_P)^{3/2} \left[1 - \left(\frac{\phi - V_{GS}}{\phi - V_P}\right)^{3/2}\right]. \quad (2.48)$$

When the junction voltage near the drain reaches the pinchoff value, the entire channel must be depleted. From Eq. (2.44) the width at $x = L$ must equal a or

$$W(L) = a = k(\phi - V_P)^{1/2}$$

and

$$\phi - V_P = \left(\frac{a}{k}\right)^2.$$

Using this substitution where appropriate, the limiting value of drain current becomes

$$I_{DSP} = -\frac{2ba}{3\rho L} (V_P - \phi) \left[1 - 3\left(\frac{V_{GS} - \phi}{V_P - \phi}\right) + 2\left(\frac{V_{GS} - \phi}{V_P - \phi}\right)^{3/2}\right]. \quad (2.49)$$

If we define V_0 to be

$$V_0 = V_P - \phi$$

and the constant I_{DS0} to be

$$I_{DS0} = -\frac{2baV_0}{3\rho L}, \qquad (2.50)$$

the current can be written as

$$I_{DSP} = I_{DS0}\left[1 - 3\left(\frac{V_{GS} - \phi}{V_0}\right) + 2\left(\frac{V_{GS} - \phi}{V_0}\right)^{3/2}\right]. \qquad (2.51)$$

The voltage V_{DS} does not appear in Eq. (2.51), but for each value of V_{GS} used a specific value of V_{DS} is implied from the expression $V_P = V_{GS} - V_{DS}$. Equation (2.51) defines a single point on the characteristics of the FET for each specified value of V_{GS}. Figure 2.50 indicates the points calculated from Eq. (2.51).

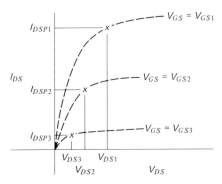

Fig. 2.50 FET characteristics showing points calculated from Eq. (2.51).

The transconductance at the edge of the pinchoff region can be found by differentiating I_{DSP} with respect to V_{GS}. This gives

$$g_m = \frac{\partial I_{DSP}}{\partial V_{GS}} = -\frac{3I_{DS0}}{V_0}\left[1 - \left(\frac{V_{GS} - \phi}{V_0}\right)^{1/2}\right]. \qquad (2.52)$$

This equation is strictly true at the edge of the pinchoff region, but since the current saturates at this point, it is approximately true in the region of operation.

The FET can easily be constructed with a p-channel and the resulting characteristics show the polarities of V_{GS} and V_{DS} reversed along with the opposite direction for I_{DS}.

B. The MOSFET

The junction FET is based on the principle that the gate-to-source or control voltage depletes the conducting channel of free carriers. This results in a reduced channel conductance until pinchoff is reached and then the conductance decreases quite abruptly to the pinched-off value. The metal-oxide-silicon FET or MOSFET can be constructed to operate as an enhancement device, or the construction may lead to a device that operates in the enhancement mode for a given polarity gate voltage and the depletion mode for the opposite polarity gate voltage.

Figure 2.51 shows an enhancement mode, n-channel MOSFET. The p-type substrate is of very high resistivity while the n-type source and drain are heavily doped. A thin oxide layer is formed over the substrate to protect the surface and to insulate the gate from the channel. Silicon nitride is used to shield the oxide layer from contamination. The metal gate is deposited so that it covers the entire region between the two n-regions.

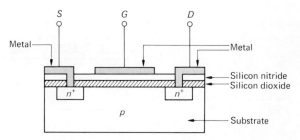

Fig. 2.51 Enhancement mode, n-channel-FET.

The metal gate and substrate act as the plates of a capacitance with the nitride and oxide acting as the dielectric. The substrate is often tied internally to the source terminal. A positive gate-to-source voltage results in positive charge collecting on the metal gate and an equal amount of negative charge induced near the surface of the substrate. This negative charge is present in sufficient quantity for relatively small gate voltages to convert the lightly doped p-material to n-type material. These free carriers induced near the substrate surface form an effective n-channel which conducts current from drain to source. The channel resistivity is controlled by the gate-to-source voltage and generates the characteristics shown in Fig. 2.52. The current limits near the drain end of the channel as the drain-to-gate voltage becomes positive enough so that negative charge is no longer induced in the channel. The governing equations are somewhat similar to those of the junction FET.

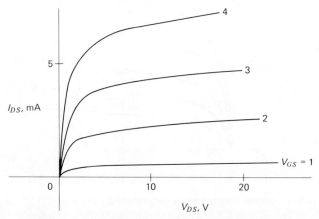

Fig. 2.52 V–I characteristics of enhancement MOSFET (n-channel).

The polarity of the gate voltage and drain voltage is the same for the enhancement mode MOSFET. A p-channel enhancement mode MOSFET can also be constructed that operates with negative values of V_{GS} and V_{DS}. For either of these MOSFET units, a zero value of gate voltage results in very small values of drain current. The junction FET conducts considerable current for the zero gate voltage condition and generally operates at lower current levels in practical applications.

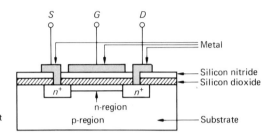

Fig. 2.53 A depletion/enhancement MOSFET (n-channel).

A second type of MOSFET can operate for both polarities of gate-to-source voltage. Figure 2.53 shows the depletion/enhancement MOSFET. This device is similar to the enhancement mode unit except that an n-region is added between the two heavily doped n-regions. A reasonable conductance now exists between source and drain even when the gate-to-source voltage is zero. A finite current flows from drain to source for $V_{GS} = 0$. If V_{GS} becomes negative, the n-channel contains fewer free carriers as the depletion region extends into the channel. The device behaves much like the junction FET for negative values of V_{GS}. When V_{GS} becomes positive, negative charge is induced in the n-channel and the conductivity is enhanced. If this MOSFET is properly designed, the characteristics for negative and positive values of V_{GS} are symmetrical enough to allow amplification with

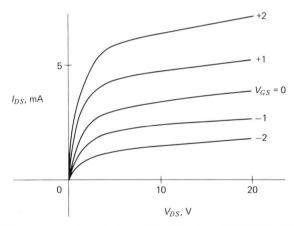

Fig. 2.54 V–I characteristics for the depletion/enhancement MOSFET (n-channel).

Fig. 2.55 Symbols for n-channel MOSFETs: (a) enhancement mode; (b) depletion/enhancement MOSFET.

no bias. A typical set of characteristics is shown in Fig. 2.54. The symbols for the two types of MOSFET are shown in Fig. 2.55.

The analytic equations relating I_{DS} to V_{GS} and V_{DS} will not be considered here. While these equations are quite similar to those developed for the junction FET, they are not of great value in design situations.

2.4 CONSTRUCTION OF TRANSISTORS AND INTEGRATED CIRCUITS

Early transistor circuit design was done almost exclusively with germanium transistors. The relatively unsophisticated fabrication methods were easier to control for germanium than for silicon. The higher yields of acceptable germanium transistors from the various construction processes led to considerably lower prices for these devices compared to silicon.

In 1960, the planar process was introduced and silicon devices experienced a phenomenal reduction in cost. Good silicon transistors that previously sold for five to ten dollars could now be obtained for 25 cents in bulk quantities. Furthermore, as the planar process was improved, the silicon integrated circuit became a reality, diminishing the importance of germanium to an almost negligible level in transistor or integrated circuit design.

The improving technology of fabrication has led to a rapid growth of the integrated circuit industry. The three outstanding advantages of these devices over conventional, discrete-element circuits are (1) the large number of networks that can be contained in a small volume, (2) low cost, and (3) increased reliability.

The first practical integrated circuits appearing on the market were digital circuits and this area continues to lead all others in application of integrated circuits. Less stringent demands on drift and component tolerances made the digital field attractive to the small-scale integration capabilities of the mid-1960s. Typically, two or three gates or a flip-flop were included on a single chip. Toward the end of the decade medium-scale and large-scale integrated circuits became available. Medium-scale integrated circuits may include the equivalent of from 10 to 100 logic gates on the chip, while large-scale integration may use a single silicon wafer to synthesize an entire control unit for a modern electronic calculator.

The linear integrated circuit field now offers extensive lines of operational and differential amplifiers. Much less discrete element amplifier design is now required, although some special purpose amplifiers must be constructed from discrete elements.

A. Construction Methods

The most important construction method at the present time is the planar process which can be used for producing batch transistors or integrated circuits. All components of the integrated circuit are formed in or on a single silicon chip. These elements may include resistors, capacitors, diodes, bipolar or field-effect transistors.

There are five basic steps [3] in the planar fabrication process: epitaxy, surface passivation, photolithography, diffusion, and deposition. Prior to application of any of these steps the silicon material must be refined and formed or grown with the appropriate crystal structure. Generally, zone refining is used to reduce the concentration of residual impurities to less than one impurity atom in 10^9 silicon atoms.

To achieve the appropriate crystal structure, the refined silicon may be melted and a seed crystal dipped into the molten material and slowly withdrawn. A continuous bar grows from the melted silicon with a uniform crystal structure that will now exhibit homogeneous electrical properties. Doping can be achieved during the growing process by adding impurities to the molten silicon. This material is then used as the substrate or starting material in the planar process.

It is sometimes advantageous to deposit very thin layers of silicon having a different impurity concentration than the substrate material. This epitaxial region is generally a low-conductivity layer that is formed by vapor-phase deposition of silicon at elevated temperatures (1000–1200°C). The deposited atoms form a continuation of the original crystal structure.

The diffusion process allows impurity atoms to diffuse into the semiconductor material at elevated temperatures to form a pn-junction. Boron is the most popular acceptor impurity while phosphorus, arsenic, or antimony can be used for donor impurities. The starting material, either the substrate or an epitaxial layer on the substrate, is uniformly doped with a relatively low concentration of impurity atoms. To form a junction, this material is either exposed to a vapor of the proper dopant (constant source) or the dopant is deposited directly onto the semiconductor surface prior to the diffusion process (limited source). At elevated temperatures, the impurity atoms diffuse into the semiconductor material away from the region of highest concentration. The profile of the impurity atoms takes on a complementary-error function distribution for constant source diffusion and a Gaussian distribution for the limited source diffusion. The depth of the diffusion is controlled by temperature and time.

Figure 2.56(a) shows the impurity profile after a single diffusion. The junctions are formed at the points where $N_D - N_A$ changes algebraic sign. When N_D is greater than N_A, the material is n-type; when N_A is greater than N_D, the material is p-type. After a single diffusion the p-type substrate of Fig. 2.56 becomes a pn-junction. The second diffusion creates a pnp transistor. The accuracy of junction formation is excellent in the diffusion process with base regions of less than one micron being typical in the fabrication of planar transistors or integrated circuits.

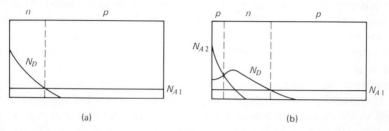

Fig. 2.56 Impurity profiles for a planar transistor: (a) impurity profile after single diffusion; (b) impurity profile after double diffusion.

Passivation, the process of forming a thin layer of oxide on the surface of the semiconductor material, forms the basis for batch construction of planar transistors and integrated circuits. This layer of silicon dioxide is thermally grown during exposure of the material at an elevated temperature (900–1200°C) to an oxidizing atmosphere. Silicon nitride is also used for passivation in some cases as it is more resistant to ionic contamination than silicon dioxide.

The passivation layer prevents impurities and contaminants from reaching the semiconductor material. Obviously, this has the long-term advantage of stabilizing the device characteristics against moisture or impurities in the atmosphere. A more immediate advantage, however, is the shielding of the semiconductor material from impurity diffusion. The diffusion process is unable to take place if the material is completely covered with the oxide layer. It is possible to remove the oxide coating at any desired location by photographic etching techniques.

When a batch of transistors or integrated circuits is to be created within a silicon wafer the areas of exposure to the diffusion process must first be determined. A layout representing the first diffusion is then produced at a scale many hundred times larger than the finished chip. The mask is produced by photographically reducing this pattern.

The oxide-coated wafer to be used is covered with a material known as photoresist. This material, after exposure to ultraviolet light, becomes resistant to the etching solution used to remove the silicon dioxide layer from the wafer. If the entire wafer were exposed to the ultraviolet light, no diffusion of impurities into the silicon could take place. However, it is possible to remove the photoresist material at selected locations using the photographic mask created earlier. The mask is placed against the wafer and exposed to an ultraviolet light source. Only those areas where no diffusion is desired are allowed by the mask to receive the ultraviolet light rays. As a result, the photoresist polymerizes or hardens below the transparent portions of the mask. The unexposed areas on the wafer surface can then have the photoresist material washed from them by washing the entire wafer with the proper solution. The wafer is then etched to remove the silicon dioxide coating, but the hardened photoresist does not allow the etchant to remove the oxide coating from the areas covered by this material. After etching, a special solution is used to remove the hardened photoresist from the wafer. At this point the silicon wafer is covered with a silicon dioxide layer that contains windows at all points where diffusion of impurities into the wafer is to take place.

The diffusion step is now performed, creating junctions beneath each window in the oxide layer. Successive diffusion steps are done to complete the entire process, with each diffusion preceded by the masking-photoresist-etching procedure. Hundreds or thousands of components can be created simultaneously on a single wafer by these means. A single planar transistor might appear as shown in the side view of Fig. 2.57.

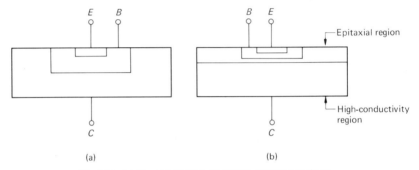

Fig. 2.57 (a) Planar transistor. (b) Planar epitaxial transistor.

The starting material must be a weakly doped, low-conductivity material since the doping of the base must exceed the collector level by several orders of magnitude. The emitter is doped even more heavily than the base in order to create the emitter-base junction. Because the collector region is lightly doped, the bulk resistance is somewhat high. The epitaxial layer shown in Fig. 2.57(b) can be used to overcome this problem. The thin, low-conductance epitaxial layer is deposited on a high conductance wafer. The collector region now has the necessary mechanical strength while providing a low bulk resistance for this region. Formation of the junctions takes place within the epitaxial region as shown.

The interconnection of circuits or components on a chip is accomplished by depositing the proper pattern of thin conductive films on the wafer surface. Resistors can also be formed this way by choosing the correct geometry of the thin film. Aluminum is the most commonly used material for interconnection purposes. For resistors materials such as tantalum, nickel-chromium alloys, or tin oxide are useful [3]. Vacuum deposition or other similar methods are used in this step.

We have considered the formation of single junctions by means of diffusion and selective etching techniques. The preciseness of the photographic etching and diffusion methods is so great that several thousands of junctions or transistors can be created within a five-mil wafer of one inch diameter. Either a very complex network with thousands of components or several less complex networks can be formed within the wafer. The latter choice is the logical one in terms of the present state of the art.

Since all components are created within a single wafer, the problem of isolating each component from all others must be considered. If not isolated, each component would be connected to others through the resistance of the silicon substrate, which would be quite low. Isolation of components is effectively achieved by the use of reverse-biased pn-junctions. It is instructive to consider the step-by-step procedure involved in a typical integrated circuit fabrication [2].

The starting material might be a p-type silicon wafer, one inch in diameter and a few mils thick. This material is quite heavily doped to form low-resistivity silicon. A thin epitaxial layer of n-type silicon of high resistivity is grown onto the starting wafer. An oxide layer is then grown on the surface of the epitaxial layer. Impurity diffusion cannot take place through the oxide layer, as explained previously. Using photoresist material and photographic masking and etching techniques, the oxide coating can be removed corresponding to any desired pattern. The pattern that is chosen allows a p-type diffusion that effectively isolates many islands of n-type material. Figure 2.58 shows the wafer after the isolation diffusion.

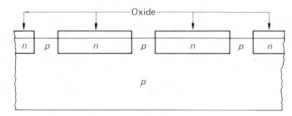

Fig. 2.58 Silicon wafer after isolation diffusion.

A second oxide coating is formed to cover the windows that allowed the isolation diffusion to take place. The etching process is repeated except that a different pattern is used to prepare for the next diffusion. This second diffusion again creates p-type material, but is a shallow diffusion taking place within the n-type islands. Transistor bases, p-type resistances, or anodes of diodes are formed during this step. The resulting wafer is indicated in Fig. 2.59.

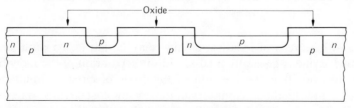

Fig. 2.59 Silicon wafer after second diffusion.

The oxide layer is reformed after this step and selectively etched to allow an n-type diffusion that forms emitters, cathodes of diodes, and contact regions for the collectors, as shown in Fig. 2.60.

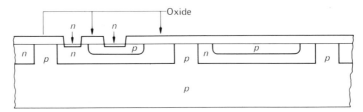

Fig. 2.60 Silicon wafer after final diffusion.

Again the diffusion windows are closed by formation of an oxide layer. Selective etching now is done to open areas in the oxide for deposition of metal contacts and interconnection of components. A metal film such as aluminum is then evaporated onto the surface, as shown in Fig. 2.61. The metal that is not needed for interconnections is etched away leaving the finished product, except for external connections and mounting as shown in Fig. 2.62.

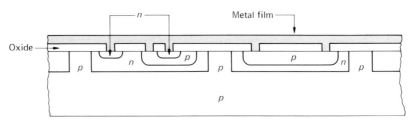

Fig. 2.61 Silicon wafer after metal evaporation step.

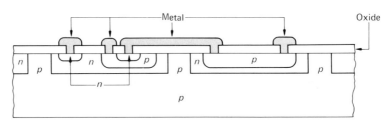

Fig. 2.62 Silicon slice ready for packaging.

Several hundreds of identical circuits may be fabricated on a single wafer. A diamond scribe is generally used to aid the separation of the individual chips.

It is obvious that the creation of hundreds of circuits simultaneously by seemingly simple methods should result in very low cost circuits. While integrated circuits for the most part are far less expensive than their discrete-element counterparts, the ultimate prices will be even less than present prices. There are several factors that tend to slow the price decrease in integrated circuits. One factor is related to the testing of all circuits fabricated. Since faults can arise in single circuits due to contamination, crystal imperfections, or breakage, the yield is never 100 percent. All circuits must then be tested and the acceptable ones must be packaged. These steps, although largely automated, add to the final price of each circuit. The fact that some defective elements are to be expected also explains why very large complex circuits result in low yields of workable circuits, since one bad element may render the entire circuit inoperative. As yields, testing methods, and packaging methods improve, prices will continue to drop.

REFERENCES AND SUGGESTED READING

1. J. Allison, *Electronic Engineering Materials and Devices.* London: McGraw-Hill, 1971, Chapter 13.
2. H. R. Camenzind, *Circuit Design for Integrated Electronics.* Reading, Mass.: Addison-Wesley, 1968, Chapters 1 and 3.
3. A. B. Grebene, *Analog Integrated Circuit Design.* Cincinnati, Ohio: Van Nostrand-Reinhold, 1972.
4. W. Shockley, C. T. Sah, and R. N. Noyce, "Carrier generation and recombination in pn-junctions and pn-junction characteristics." *Proc. IRE* **45**, 9 (September, 1957).

PROBLEMS

Sections 2.1A–2.1C

* **2.1** Calculate the barrier voltage of a silicon diode at room temperature (300°K) given that $N_D = 10^{15}/cm^3$ and $N_A = 10^{18}/cm^3$.

2.2 Calculate ϕ for the diode of Problem 2.1 at $T = 320°K$.

2.3 Plot Eq. (2.12) for $I_S = 10^{-7}$ A and $T = 300°K$. Use different scales for negative and positive currents.

* **2.4** Calculate the current flowing through an ideal germanium diode at room temperature if $I_S = 1$ μA and a forward bias of 0.15 V is applied.

Sections 2.1D–2.1G

2.5 The saturation current for a diode is given by Eq. (2.13). What quantities in this equation depend on temperature? Assuming that only n_i varies with temperature, derive an expression for the variation of I_S with temperature, dI_S/dT in terms of I_S. Is this a reasonable assumption? Explain.

* **2.6** Given that $I_S = 0.01$ μA for a silicon diode at $T = 300°K$.
  a) Calculate I_S at $T = 350°K$.
  b) Find the temperature at which $I_S = 0.04$ μA.

2.7 Given that $I_S = 4$ μA for a germanium diode at $T = 300°K$.
  a) Calculate I_S at $T = 350°K$.
  b) Find the temperature at which $I_S = 10$ μA.

Section 2.1H

2.8 Using the diode characteristics of Fig. 2.14, sketch the output voltage as a function of time. (See the figure.)

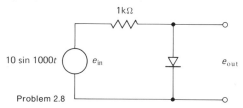

Problem 2.8

2.9 Repeat Problem 2.8 using a piecewise-linear model of the diode. Assume an infinite back resistance and a 50-Ω forward resistance for the diode.

2.10 Calculate the small-signal dynamic resistance of the diode at room temperature with a forward dc current of (a) 0.5 mA; (b) 2 mA. (c) Repeat the calculations if T changes from 300°K to 340°K.

*2.11 In the circuit of Fig. 2.17 the dc voltage is increased to a value that results in 1.4 mA of dc current. If e_{in} remains constant at 0.002 sin 1000t, find the ac component of diode current.

2.12 If the forward-biased diode drop is 0.5 V (dc), plot the output voltage as a function of time. (See the figure.)

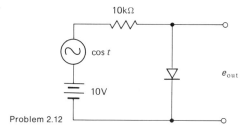

Problem 2.12

Sections 2.1I–2.1J

*2.13 If the small-signal depletion-layer capacitance for a reverse-biased, abrupt junction is 20 pF (1 pF = 10^{-12} F) at $V = -8$ V and $\phi = -1$ V, what is the capacitance of the junction when (a) $V = -3$ V; (b) $V = -16$ V.

2.14 Repeat Problem 2.13 for a forward-bias voltage of $V = 0.5$ V.

2.15 If the diffusion capacitance of a forward-biased diode is $C_d = 500$ pF when the diode current is 2 mA, calculate C_d when the current is (a) 1 mA; (b) 3 mA.

Sections 2.1K–2.1L

*2.16 The regulator diode has a reverse breakdown voltage of 9 V. Calculate the output voltage and the power dissipation of the diode. (See the figure.)

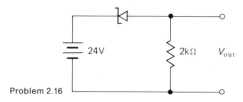

Problem 2.16

2.17 Repeat Problem 2.16 if the reverse breakdown voltage of the diode is 16 V.

2.18 The regulator diode shown in the figure has a reverse breakdown voltage of 6 V. Sketch a complete cycle of the output voltage.

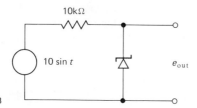

Problem 2.18

Section 2.2

*__2.19__ Given that the transistor in the figure has the characteristics shown in Fig. 2.32, find the output voltage when
a) $I_E = (1.0 + 0.5 \sin \omega t)$ mA,
b) $I_E = (2.0 + 0.5 \sin \omega t)$ mA.
Compare the ac output voltages in (a) and (b) and explain any differences.

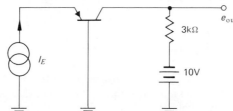

Problem 2.19

2.20 Repeat Problem 2.19 for $I_E = (2.0 + 2.0 \sin \omega t)$ mA. Sketch the output voltage waveform.

*__2.21__ Repeat Problem 2.19 for $I_E = (3.0 + 2.0 \sin \omega t)$ mA.

2.22 Given that the transistor in the figure has the characteristics shown in Fig. 2.35, find the output voltage when
a) $I_B = (20 + 10 \sin \omega t)$ μA,
b) $I_B = (40 + 10 \sin \omega t)$ μA.
Compare the ac output voltages in (a) and (b), and explain any differences.

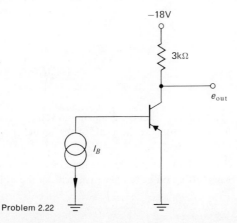

Problem 2.22

2.23 Repeat Problem 2.22 for $I_B = (25 + 25 \sin \omega t)\ \mu A$.

*2.24 Repeat Problem 2.22 for $I_B = (40 + 25 \sin \omega t)\ \mu A$.

2.25 Assuming that the transistor in Problem 2.22 has the output characteristics shown in Fig. 2.35 and the input characteristics shown in Fig. 2.40 with negative values of V_{BE}, find the output voltage, the input voltage, and the ac voltage gain of the stage for $I_B = (20 + 5 \sin \omega t)\ \mu A$. Sketch both input voltage and output voltage waveforms.

*2.26 Repeat Problem 2.25, changing the load resistor from 3 kΩ to 4.5 kΩ. How does ac voltage gain vary with R_L?

2.27 Repeat Problem 2.25, changing the load resistor from 3 kΩ to 0.5 kΩ. How does ac voltage gain vary with R_L?

2.28 The short-circuit ac current gain from base to collector is defined as

$$\beta = \left.\frac{\Delta I_C}{\Delta I_B}\right|_{V_{CE} = \text{constant}}.$$

Taking ΔI_B as 10 μA, use the characteristics of Fig. 2.35 to evaluate β at $V_{CE} = -10$ V and
a) $I_C = 1$ mA, b) $I_C = 2$ mA,
c) $I_C = 3$ mA, d) $I_C = 4$ mA,
e) $I_C = 5$ mA.

*2.29 Work Problem 2.28 given that I_C is 5 mA and
a) $V_{CE} = -5$ V, b) $V_{CE} = -10$ V,
c) $V_{CE} = -15$ V, d) $V_{CE} = -20$ V.

2.30 In the circuit shown in the figure, β is assumed to be 100. Find the output ac voltage given that $I_B = (20 + 5 \sin \omega t)\ \mu A$. If the dc β is assumed to equal 100 also, what is the quiescent (dc) output voltage?

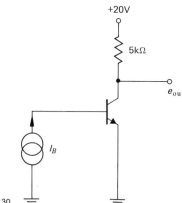

Problem 2.30

*2.31 If I_B in Problem 2.30 is $(36 + A \sin \omega t)\ \mu A$, what is the maximum value that A can have and still keep the transistor from saturating at the peak of the input current swing? Assume that $\beta = \beta_{dc} = 100$.

2.32 If I_B in Problem 2.30 is $(6 + B \sin \omega t)\ \mu A$, what is the maximum value that B can have and still keep the transistor from cutting off at the negative peak of the input current swing?

2.33 In Section 2.2D it was explained why I_C is nonzero when V_{CB} is equal to zero for the common-base configuration. In the common-emitter configuration $I_C = 0$, when $V_{CE} = 0$. Explain why, in terms of minority-carrier concentration in the base region.

Section 2.3

*__2.34__ Explain why Eq. (2.51), the expression for drain current, does not involve the drain-to-source voltage V_{DS}. Can this equation be used to predict I_{DS} as a function of V_{DS} for fixed values of V_{GS}?

2.35 Sketch a set of V–I characteristics for the FET based on Eq. (2.51). Assume that $I_{DS0} = 4$ mA, $V_P = -3.0$ V, and $\phi = 0.9$ V. Show curves for $V_{GS} = +0.5, 0, -0.5, -1.0,$ and -2.0 V.

*__2.36__ Evaluate the transconductance of the FET of Problem 2.35 at a gate-to-source voltage of -1.0 V.

2.37 Evaluate the output voltage of the circuit, assuming the characteristics of Fig. 2.54 apply to the MOSFET used in the figure shown.

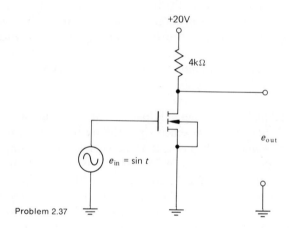

Problem 2.37

TRANSISTOR BIASING

3.1 PURPOSE OF BIASING

Biasing is a very important aspect of linear amplifier design. In Chapter 2 it was found that the output voltage swing of the transistor is limited at one extreme by cutoff and at the other extreme by saturation. Cutoff occurs when the emitter-base junction is no longer forward-biased. Further reduction in base-emitter voltage causes no corresponding change in collector current when the transistor is in cutoff; hence the input signal is not able to control the output signal. Saturation occurs when the collector-base junction becomes forward-biased. A further increase of input current will not affect the output voltage for this case. Again the control of output signal by the input signal is lost. In a linear circuit, it is important that neither of these boundaries be reached since the output signal must be proportional to the input signal at all times. If an ac signal is to be amplified, the transistor must allow the output voltage to swing in both directions from the quiescent or no-signal value. Cutoff and saturation must be avoided, even at the extreme values of output voltage swing. The process by which the quiescent output voltage is caused to fall somewhere between the cutoff and saturated values is referred to as biasing. Figure 3.1 demonstrates the points of cutoff and saturation for a common-emitter stage. Point c corresponds to saturation. A further increase of input current above 40 μA does not change the output voltage or current at this point. Point a corresponds to cutoff occurring for zero or negative input currents, where again the collector voltage or current is no longer controlled by the input current. A logical choice for the quiescent output voltage would be the center of the active region, corresponding to point b on the characteristics. The quiescent output voltage would then be 10 V, with the possibility of a positive swing to approximately 20 V (at cutoff) and a negative swing to 0 V (at saturation). In practice, distortion will take place for large swings, due to the unequal spacing of the curves, but this will be neglected for now. Since this quiescent level is exactly at the center of the active region, the maximum possible symmetrical output swing is allowed. Occasionally, for reasons to be considered in following chapters, the quiescent point will be offset from the midpoint. However, for many practical cases the midpoint is a reasonable choice for the quiescent output voltage.

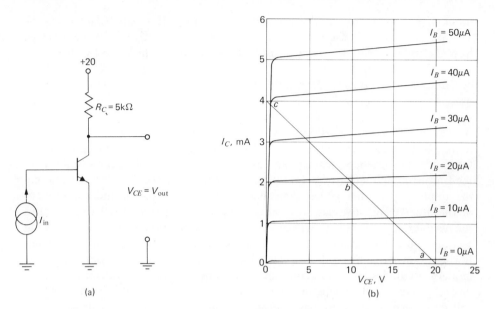

Fig. 3.1 Common-emitter stage: (a) common-emitter configuration; (b) characteristics.

After the choice of quiescent point has been made, the method of obtaining that point must be considered. Biasing could be accomplished by using a separate source, such as shown in Figs. 3.2(a) and (b). From the characteristic curves the quiescent base current I_{BQ}, corresponding to the desired V_{CEQ}, can be selected. From Fig. 3.1(b), this corresponds to 18 μA. Then I_{bb} in Fig. 3.2(b) would be set at 18 μA, and the correct bias would result. Alternatively, E_{bb} can be chosen from the

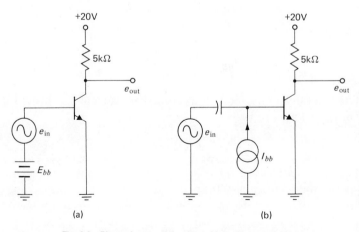

Fig. 3.2 Bias schemes: (a) voltage bias; (b) current bias.

input characteristics to give this particular value of I_{BQ}. It is usually far too expensive and inefficient to use a separate bias source. Normally, a resistive network from the collector supply voltage is used instead to supply the correct value of I_{BQ}.

In the electronics industry where an engineer does a great deal of amplifier design, it is impractical to determine the characteristic curves for each transistor used. As will be discussed in Chapter 4, the characteristics of two transistors of the same type can be very different; hence with transistors there is no such thing as typical characteristics. This presents two important problems: (1) it is difficult to design an amplifier with little knowledge of the characteristics, and (2) the performance of several identical circuits can be very different due to the variability of the transistor characteristics. Both of these problems can be overcome by making the amplifier performance less dependent on the transistor and more dependent on the circuit element values, such as resistors and voltage sources. This eliminates the need for an accurate knowledge of the transistor parameters. In the following sections, we will assume that I_{co} and β_0 are known; however, later in the chapter, it will be shown that an accurate knowledge of these values is unnecessary. Actually, modern circuit design uses silicon transistors almost exclusively and for low power units I_{co} is generally negligible. The characteristic curves for silicon show no collector current for zero base current, regardless of the collector-to-emitter voltage applied. The following sections will carry the terms involving I_{co} throughout the derivations for completeness. The I_{co} terms will be dropped to give the design equation used for low-power silicon transistor circuits. However, higher-power silicon transistors, germanium transistors, or special applications may require these terms to be considered.

Returning to Fig. 3.1(b), we see that the active region extends from a collector voltage of approximately zero volts to $+V_{CC}$ volts. As explained before, a reasonable choice of bias voltage is $V_{CEQ} = V_{CC}/2$. Once R_C is known, the quiescent collector current that will result in $V_{CEQ} = V_{CC}/2$ can be found, since

$$V_{CC} - V_{CEQ} = I_{CQ} R_C.$$

This gives a value of

$$I_{CQ} = \frac{V_{CC} - V_{CEQ}}{R_C}. \tag{3.1}$$

Establishing this value of I_{CQ} requires a certain base current I_{BQ}. From Eq. (2.39b), the value of I_{BQ} is found to be

$$I_{BQ} = \frac{I_{CQ}}{\beta_0} - \frac{(\beta_0 + 1)}{\beta_0} I_{co}. \tag{3.2a}$$

For silicon, the value of I_{BQ} becomes

$$I_{BQ} = \frac{I_{CQ}}{\beta_0}, \tag{3.2b}$$

since I_{co} can be neglected. The next step is to select an appropriate bias network to supply the value of I_{BQ}, as calculated above, to the transistor base. This procedure for biasing is summarized below.

1. Select the desired quiescent collector voltage. Often V_{CEQ} will be chosen to be $V_{CC}/2$.
2. Find the required value of I_{CQ} from Eq. (3.1).
3. Using Eq. (3.2) find the necessary I_{BQ} that will cause the required value of I_{CQ} to flow.
4. Select the bias network to supply the proper value of I_{BQ}.

3.2 BASE CURRENT BIAS

The circuit of Fig. 3.3 demonstrates one method of obtaining bias current. This method of bias is seldom used in practice because it allows the quiescent level to shift drastically with changes in temperature. However, it serves to familiarize one with the techniques involved in biasing a stage. The coupling capacitor isolates the source from bias considerations, since it appears as an open circuit for dc signals.

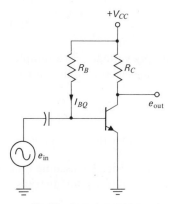

Fig. 3.3 Base current bias.

We can choose I_{BQ} if the voltage at the base of the transistor is known:

$$I_{BQ} = \frac{V_{CC} - V_{BE}}{R_B}. \tag{3.3}$$

Usually, V_{BE} is not known accurately, but for germanium transistors it will be approximately 0.2 V and for silicon transistors, 0.5 V. A particular value of V_{BE} can be used for all cases because of the sharpness of the input characteristics. Figures 3.4(a) and (b) show the rapid increase of base current with applied voltage for both germanium and silicon transistors. The curves break so sharply that it

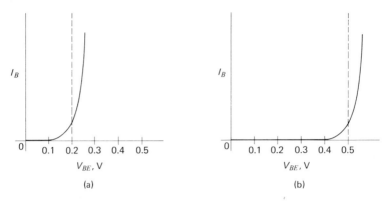

Fig. 3.4 Input characteristics: (a) germanium; (b) silicon.

leads to little error to assume that the dotted lines represent the input characteristics. Equation (3.3) reduces to

$$I_{BQ} = \frac{V_{CC} - 0.2 \text{ V}}{R_B} \quad \text{(Ge)} \quad (3.4a)$$

and

$$I_{BQ} = \frac{V_{CC} - 0.5 \text{ V}}{R_B} \quad \text{(Si)}. \quad (3.4b)$$

We note that for typical supply voltages, for example 10 V or greater, the voltage from base to emitter can be neglected with little error. Using the simpler equation

$$I_{BQ} = \frac{V_{CC}}{R_B}$$

generally results in an error smaller than five percent.

Example 3.1. For the circuit of Fig. 3.3, assume that $R_C = 5 \text{ k}\Omega$, $\beta_0 = 120$, and $V_{CC} = 20 \text{ V}$. Select a value of R_B to bias a silicon transistor to the center of its active region.

Solution.

1. $V_{CEQ} = \dfrac{V_{CC}}{2} = 10 \text{ V}.$

2. $I_{CQ} = \dfrac{20 - 10}{5 \text{ k}\Omega} = 2 \text{ mA}.$

3. $I_{BQ} = \dfrac{2}{120} = 16.7 \text{ }\mu\text{A}$

4. $R_B = \dfrac{20 - 0.5}{16.7 \text{ }\mu\text{A}} = 1.17 \text{ M}\Omega.$

For the case where V_{BE} is assumed to be zero, a value of
$$R_B = 2\beta_0 R_C = 1.2 \text{ M}\Omega$$
will set V_{CEQ} at $V_{CC}/2$. This can be seen by considering the value of I_{BQ} resulting from using this value of R_B. The base current is then approximately
$$I_{BQ} = \frac{V_{CC}}{2\beta_0 R_C}.$$
The collector current is
$$I_{CQ} = \beta_0 I_{BQ} = \frac{V_{CC}}{2R_C},$$
giving a quiescent collector voltage of
$$V_{CEQ} = V_{CC} - I_{CQ}R_C = V_{CC} - \frac{V_{CC}}{2} = \frac{V_{CC}}{2}.$$

A. Problems of the Base-Current Bias Stage

Consider the circuit of Fig. 3.3 with the values of R_C and R_B as calculated in Example 3.1. If $\beta_0 = 120$, the quiescent collector voltage is 10 V. What happens if we replace the transistor with one having a β_0 of 60? The quiescent voltage is then 15 V. If β_0 is 200, the quiescent voltage becomes 3.3 V. The variation in quiescent voltage with change in β_0 is too great to allow the base-current bias stage to be of much practical value. Furthermore, the stage allows an excessive change in V_{CEQ} with temperature change, as will be seen in the following sections.

3.3 TEMPERATURE EFFECTS

Achieving the proper bias point for a given stage is quite a simple problem. Unfortunately, additional constraints on the bias network will be imposed, due to temperature effects. Certain types of bias network allow a large change in the quiescent output voltage as the temperature of the surroundings changes. Others will minimize the change in quiescent point with temperature. The origin and treatment of the more important temperature effects will now be considered.

In both germanium and silicon transistors, the leakage current I_{co} is a very strong function of temperature. For germanium, I_{co} is approximately doubled for each 9°C increase in temperature (Section 3.9). The fractional change is

$$\frac{1}{I_{co}} \frac{dI_{co}}{dT} = 0.09 \qquad \text{(Ge, near room temperature).} \qquad (3.5a)$$

For silicon the fractional change in I_{co} is about 0.15, but I_{co} changes are much less important for low-power silicon transistors than for germanium. The absolute value of I_{co} for silicon is a factor of 100 to 1000 times smaller than for germanium, giving a negligible value for dI_{co}/dT. It must be pointed out, however, that for higher-power silicon transistors changes in I_{co} must often be considered. In the following discussion I_{co} changes are neglected for silicon. If such changes are important, the treatment can easily be extended, as will be indicated shortly. The equation for fractional change in leakage current is

$$\frac{1}{I_{co}}\frac{dI_{co}}{dT} = 0.15 \quad \text{(Si)}. \tag{3.5b}$$

Another factor that changes with temperature is the emitter-base voltage drop. This value decreases by about 1.6 mV/°C for germanium and 2.0 mV/°C for silicon (Section 3.10).

The third temperature effect is the change in β_0 that occurs as the temperature is changed. This effect is much more marked for silicon than for germanium; β_0 can double over its room temperature value when the temperature increases by 50–60°C for silicon transistors.

In considering the temperature effects, we will initially consider only the variations in I_{co} and V_{BE}. After developing suitable circuits that minimize quiescent-point variations with respect to changes in I_{co} and V_{BE}, we will consider the effect of changes in β_0. Actually, whether the change in β_0 arises from a temperature change or a replacement of one transistor by another is unimportant. The circuit must be designed to accommodate these changes without an excessive shift in operating point.

Since the collector current or voltage is usually the output quantity of interest, it is reasonable to relate the temperature effects to one or both of these quantities. A voltage stability factor, related directly to temperature, is defined as

$$S_V = \frac{dV_{CQ}}{dT}, \tag{3.6}$$

where V_{CQ} depends on the variables V_{BE} and I_{co}, as shown by Eqs. (3.2) and (3.3). Therefore, S_V can be written as

$$S_V = \frac{\partial V_{CQ}}{\partial V_{BE}}\frac{\partial V_{BE}}{\partial T} + \frac{\partial V_{CQ}}{\partial I_{co}}\frac{\partial I_{co}}{\partial T}. \tag{3.7}$$

For germanium Eq. (3.7) becomes

$$S_V = -0.0016\frac{\partial V_{CQ}}{\partial V_{BE}} + 0.09I_{co}\frac{\partial V_{CQ}}{\partial I_{co}} \quad \text{(Ge)}. \tag{3.8a}$$

For silicon it becomes

$$S_V = -0.002\frac{\partial V_{CQ}}{\partial V_{BE}} \quad \text{(Si)}. \tag{3.8b}$$

For medium- or high-power silicon transistors I_{co} changes can be included, giving a voltage stability factor of

$$S_V = -0.002\frac{\partial V_{CQ}}{\partial V_{BE}} + 0.15I_{co}\frac{\partial V_{CQ}}{\partial I_{co}}. \quad \text{(high power Si)} \tag{3.8c}$$

It should be emphasized that the stability factor has little value in calculations involving sizable temperature changes since S_V also depends on temperature. This factor is important as an index of the stability of one amplifier configuration compared to another at the same temperature. It also allows a comparison of the effects of I_{co} changes to V_{BE} changes within a given stage. We can now return to the case of the base-current bias circuit to examine the effect of temperature on V_{CEQ}.

3.4 TEMPERATURE STABILITY OF BASE-CURRENT BIAS STAGE

Before evaluating the stability factor for this circuit, let us examine the reasons for change in output voltage with temperature change. Equation (3.3) shows that the base current is a function of V_{BE}, or

$$I_{BQ} = \frac{V_{CC} - V_{BE}}{R_B},$$

where R_B is chosen to set I_{BQ} at the proper value. It follows that I_{CQ} and hence V_{CEQ} are then fixed so long as the temperature remains constant. As the temperature rises, however, V_{BE} decreases, causing I_{BQ} and I_{CQ} to increase. Since

$$V_{CEQ} = V_{CC} - I_{CQ}R_C,$$

the quiescent voltage decreases with temperature. Although I_{BQ} does not change rapidly with temperature, I_{CQ} changes can be appreciable. From Eq. (2.39b), I_{CQ} is seen to increase β_0 times as fast as I_{BQ}.

Equation (2.39b) also explains why I_{co} changes affect the output voltage. As I_{co} increases with temperature, I_{CQ} increases $(\beta_0 + 1)$ times as fast, again causing V_{CEQ} to decrease.

The expression for the stability factor for a silicon common-emitter stage is

$$S_V = -0.002 \frac{\partial V_{CQ}}{\partial V_{BE}}.$$

From Eq. (3.3) and Eq. (2.39b) we see that the collector current is

$$I_{CQ} = \beta_0 \frac{(V_{CC} - V_{BE})}{R_B}.$$

The quiescent collector voltage is then

$$V_{CQ} = V_{CEQ} = V_{CC} - \frac{\beta_0(V_{CC} - V_{BE})R_C}{R_B}. \tag{3.9}$$

From Eq. (3.9), $\partial V_{CQ}/\partial V_{BE}$ is found to be

$$\frac{\partial V_{CQ}}{\partial V_{BE}} = \frac{\beta_0 R_C}{R_B}.$$

Thus S_V now becomes

$$S_V = -0.002 \frac{\beta_0 R_C}{R_B} \quad \text{(Si)}. \tag{3.10}$$

For a given β_0 and R_C, the stability factor S_V becomes smaller with larger values of R_B. However, there is an upper limit on R_B, since I_{CQ} becomes smaller as R_B increases. There will be some minimum allowed value for I_{CQ}, due to either circuit requirements or the fact that β_0 falls sharply at low values of I_{CQ} for silicon transistors. Typical values of S_V for this case range from 1 mV/°C to 10 mV/°C.

Example 3.2. Using the same silicon transistor and the results of Example 3.1, evaluate the voltage stability factor.

Solution. By direct substitution into Eq. (3.10), we obtain

$$S_V = \frac{-0.002 \times 120 \times 5 \times 10^3}{1.17 \times 10^6} = 1 \text{ mV/°C}.$$

One might be tempted to minimize the voltage drift with temperature by reducing R_C. This is not always practical, since either gain considerations, output impedance requirements, or current drain requirements fix the value of R_C.

The value of S_V for the silicon transistor is quite reasonable for many applications. We have seen in Section 3.2A that the circuit is very unstable with respect to changes in β_0. For this reason, the circuit is seldom used in practice.

3.5 EMITTER BIAS

Emitter bias is a popular method of bias because it leads to fairly low stability factors and good stability with respect to changes in β_0. The circuit of Fig. 3.5 shows the emitter-bias scheme. This circuit employs current feedback to stabilize the temperature variations.

Taking a Thévenin equivalent circuit of R_1, R_2, and V_{CC}, we obtain the circuit shown in Fig. 3.5(b). To find I_B, we write the loop equation

$$E_{th} = R_{th}I_B + V_{BE} + R_E I_E.$$

From Eq. (2.39b),

$$I_E = I_B + I_C = (\beta_0 + 1)I_B + (\beta_0 + 1)I_{co}. \tag{3.11}$$

Solving these two equations for I_{BQ} gives

$$I_{BQ} = \frac{E_{th} - V_{BE} - (\beta_0 + 1)I_{co}R_E}{R_{th} + (\beta_0 + 1)R_E}. \tag{3.12}$$

Again neglecting I_{co} for silicon transistors leads to

$$I_{BQ} = \frac{E_{th} - V_{BE}}{R_{th} + (\beta_0 + 1)R_E}. \tag{3.13}$$

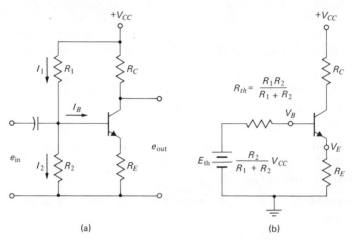

Fig. 3.5 Emitter bias: (a) actual circuit; (b) equivalent dc circuit.

There are two methods of selecting R_1 and R_2 for the proper bias. Since Eq. (3.13) relates I_{BQ} to E_{th} and R_{th}, once I_{BQ} is known, E_{th} and R_{th} can be chosen to satisfy this equation. Then R_2 and R_1 can be found (subject to stability constraints) from the equations

$$\frac{R_2}{R_1 + R_2} V_{CC} = E_{th} \quad \text{and} \quad \frac{R_1 R_2}{R_1 + R_2} = R_{th}.$$

A second method, which is simpler for design purposes, is to first select R_2 on the basis of the stability requirements and then to select R_1 to give the proper value of I_{BQ}.

We should note that the addition of R_E to the circuit decreases the size of the active region. The edge of the active region defined by cutoff remains the same. When the transistor turns off, the collector voltage becomes $V_C = +V_{CC}$. However, the edge of the region defined by saturation is no longer approximately zero volts. When the transistor saturates, the transistor can be approximated by a short circuit and V_C is obviously

$$V_C = \frac{R_E}{R_E + R_C} V_{CC}.$$

That is, the entire power-supply voltage drops across R_E and R_C when the transistor is saturated. This value might be 1 or 2 V or even greater in some applications. In order to allow an equal swing on both sides of the quiescent point, the quiescent voltage must be set at the midpoint of the two collector voltage extremes. This point is

$$V_{CQ} = \frac{V_{C\,max} + V_{C\,min}}{2} = \frac{V_{CC}}{2}\left(1 + \frac{R_E}{R_E + R_C}\right)$$

$$= \text{midpoint of active region.} \qquad (3.14)$$

When this condition is met, half of the power-supply voltage drops across the transistor and half drops across the sum of R_E and R_C. An example will demonstrate the second method of selecting R_2 and R_1 for a midpoint-bias voltage.

Example 3.3. In the emitter-bias circuit of Fig. 3.5 $R_C = 10$ kΩ, $R_E = 1$ kΩ, and $V_{CC} = 20$ V. Given that $\beta_0 = 80$ and $V_{BE} = 0.6$ V, find R_1 and R_2 to set V_{CQ} at the center of the active region.

Solution. We will first assume that $R_2 = 10R_E$ is a condition imposed on the circuit by stability requirements. It will be shown later that this is a reasonable choice of R_2. This gives a value of $R_2 = 10$ kΩ for this example. We now must find the desired value of I_{BQ}.

1. The center of the active region must be located. Using Eq. (3.14), we find that this value is

$$V_{CQ} = 10(1 + \tfrac{1}{11}) = 10.9 \text{ V}.$$

2. The required collector current and base current can now be found to give this value of V_{CQ}:

$$I_{CQ} = \frac{V_{CC} - V_{CQ}}{R_C} = \frac{20 - 10.9}{10 \text{ k}\Omega} = 0.91 \text{ mA},$$

$$I_{BQ} = \frac{I_{CQ}}{\beta_0} = \frac{0.91}{80} = 0.0114 \text{ mA}.$$

3. We now select R_1 to give the calculated value of I_{BQ}. By noting in Fig. 3.5(a) that $I_1 = I_{BQ} + I_2$, we can find the necessary value of R_1:

$$I_2 = \frac{V_{BQ}}{R_2} = \frac{I_{EQ}R_E + V_{BE}}{R_2} = \frac{(I_{BQ} + I_{CQ})R_E + 0.6}{R_2}$$

$$= \frac{(0.92)1 \text{ k}\Omega + 0.6}{10 \text{ k}\Omega} = 0.152 \text{ mA};$$

$$I_1 = I_{BQ} + I_2 = 0.011 + 0.152 = 0.163 \text{ mA}.$$

Since the voltage across R_1 is $V_{CC} - V_{BQ}$, then

$$R_1 = \frac{V_{CC} - V_{BQ}}{I_1} = \frac{20 - 1.52}{0.163} = 113 \text{ k}\Omega.$$

Generally, for design purposes the second method of selecting R_1 and R_2 is preferable to taking a Thévenin equivalent of the bias network. For analysis of a circuit having all values specified, the calculation is simplified by using the Thévenin equivalent.

The temperature stability of the emitter-bias circuit using a silicon transistor is not greatly improved over that of the base-bias case, when changes in β_0 are neglected. This circuit is quite well used, however, because it has a very stable operating point with changes in β_0. The voltage stability is found by solving for

I_{CQ} from Eq. (3.13) giving

$$I_{CQ} = \beta_0 I_{BQ} = \frac{\beta_0[E_{th} - V_{BE}]}{R_{th} + (\beta_0 + 1)R_E}.$$

The quiescent voltage is then

$$V_{CQ} = V_{CC} - I_{CQ}R_C = V_{CC} - \frac{\beta_0 R_C[E_{th} - V_{BE}]}{R_{th} + (\beta_0 + 1)R_E}.$$

It follows that S_V is given by

$$S_V = \frac{dV_{CQ}}{dT} = \frac{dV_{CQ}}{dV_{BE}} \frac{\partial V_{BE}}{\partial T} = \frac{-0.002\beta_0 R_C}{R_{th} + (\beta_0 + 1)R_E} \quad \text{(Si)}. \quad (3.15)$$

Equation (3.15) is very similar to Eq. (3.10) (base bias), with R_B replaced by $R_{th} + (\beta_0 + 1)R_E$.

The voltage stability factor for a germanium emitter-bias circuit can be found to be

$$S_V = \frac{\partial V_{CQ}}{\partial I_{co}} \frac{\partial I_{co}}{\partial T} + \frac{\partial V_{CQ}}{\partial V_{BE}} \frac{\partial V_{BE}}{\partial T}$$

$$= -\left[\frac{0.09(\beta_0 + 1)I_{co}R_C[R_{th} + R_E] + 0.0016\beta_0 R_C}{R_{th} + (\beta_0 + 1)R_E}\right] \quad \text{(Ge)}. \quad (3.16)$$

Examining Eq. (3.16), we see that for S_V to be minimized $R_{th} + (\beta_0 + 1)R_E$ is required to be much larger than $R_{th} + R_E$. This will be true only if $(\beta_0 + 1)R_E$ is much larger than R_{th}. A normal choice for R_{th} is one that satisfies the inequality

$$R_{th} \leq 0.1(\beta_0 + 1)R_E \quad (3.17)$$

As pointed out in connection with Example 3.3, the starting point of the bias circuit design is often the choice of R_2 to equal 10 R_E. If β_0 is near 100, then Eq. (3.17) leads to this value. After R_2 is selected, R_1 is calculated to give proper bias. Since $R_2 = 10 R_E$, then R_{th} will always be smaller than 10 R_E.

For silicon, the stability factor is minimized when $R_{th} + (\beta_0 + 1)R_E$ is maximized. Thus, one might conclude that a large value of R_{th} should be used. Section 3.7 will show that this conclusion is valid. For good β_0 stability, however, the same condition on R_{th} is imposed. That is,

$$R_{th} \leq 0.1(\beta_0 + 1)R_E$$

and if β_0 is near 100, $R_{th} = 10 R_E$.

Example 3.4. Calculate S_V for the circuit of Example 3.3.

Solution. Using Eq. (3.15) and the element values found in Example 3.3, we obtain

$$S_V = \frac{-0.002 \times 80 \times 10^4}{9.2 \times 10^3 + 81 \times 10^3}$$

$$= -18 \text{ mV/}^\circ\text{C}.$$

3.6 COLLECTOR-BASE BIAS

This method uses voltage feedback to offset the increase in collector current with increasing temperatures. In Fig. 3.6, the current through R_C is equal to the sum of I_C and I_B, which also equals I_E. The procedure for biasing this circuit is outlined below.

1. Select the proper quiescent output voltage. Often this will be the center of the active region, which is

$$V_{CQ} = \frac{V_{CC}}{2}\left(1 + \frac{R_E}{R_E + R_C}\right).$$

2. Calculate the required current through R_C to set the quiescent voltage.
3. Calculate the voltage at the base of the transistor:

$$V_{BQ} = V_{EQ} + V_{BE} = I_{EQ}R_E + V_{BE}.$$

4. Find the base current required.
5. Select R_F such that

$$R_F = \frac{V_{CQ} - V_{BQ}}{I_{BQ}}. \tag{3.18}$$

The constraints on selecting R_F are discussed in the following section.

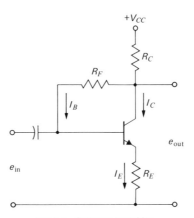

Fig. 3.6 Collector-base bias.

If I_{CQ} increases due to temperature effects, the quiescent collector voltage will decrease. The decrease in V_{CQ} will lower I_{BQ} which, from Eq. (2.39b), tends to offset the original increase in I_{CQ}.

The base current through R_F is

$$I_{BQ} = \frac{V_{CQ} - V_{EQ} - V_{BE}}{R_F}. \tag{3.19}$$

The current through R_C is $I_{CQ} + I_{BQ} = I_{EQ}$, which, upon substitution, can be written as

$$I_{EQ} = (\beta_0 + 1)I_{BQ} = \frac{(\beta_0 + 1)}{R_F}[V_{CQ} - V_{EQ} - V_{BE}].$$

This gives a value for $V_{CQ} = V_{CC} - I_{EQ}R_C$ of

$$V_{CQ} = V_{CC} - \frac{(\beta_0 + 1)R_C}{R_F}[V_{CQ} - V_{EQ} - V_{BE}].$$

Note that V_{CQ} appears on both sides of the above equation and also that V_{EQ} is related to V_{CQ} by

$$V_{EQ} = \frac{R_E}{R_C}[V_{CC} - V_{CQ}].$$

Solving for V_{CQ}, we get

$$V_{CQ} = \frac{V_{CC} + (\beta_0 + 1)[R_E V_{CC} + R_C V_{BE}]/R_F}{1 + (\beta_0 + 1)R_E/R_F + (\beta_0 + 1)R_C/R_F}. \tag{3.20}$$

The stability factor for silicon is

$$S_V = \frac{dV_{CQ}}{dV_{BE}}\frac{\partial V_{BE}}{\partial T} = -0.002\frac{(\beta_0 + 1)R_C}{R_F + (\beta_0 + 1)(R_C + R_E)} \quad \text{(Si)}. \tag{3.21}$$

The upper limit of S_V from Eq. (3.21) is -2 mV/°C. This value occurs when R_F and R_E are small; S_V becomes smaller as R_E and R_F become larger.

The stability factor for germanium is found to be

$$S_V = \frac{-0.0016(\beta_0 + 1)R_C}{R_F + (\beta_0 + 1)(R_E + R_C)} - \frac{0.09I_{co}R_C(\beta_0 + 1)R_F}{R_F + (\beta_0 + 1)(R_E + R_C)} \quad \text{(Ge)}. \tag{3.22}$$

Generally, the largest contribution to drift comes from the I_{co}-term. If

$$(\beta_0 + 1)(R_C + R_E)$$

is fairly large compared to R_F, this term will be minimized and the stability will be quite good. A typical value for S_V is -6 mV/°C for germanium transistors. Note that the stability can be quite good even when $R_E = 0$.

3.7 STABILITY WITH RESPECT TO CHANGES IN β_0

In the previous discussions of temperature stability the variation of β_0 with temperature was not considered. While this effect can be treated analytically, it leads to quite involved design equations. Furthermore, β_0 variations will occur when transistors are interchanged, resulting in the same effect as temperature variations of β_0. A figure for $\partial V_{CQ}/\partial T$, related to β_0, would not be very meaningful because it would not predict quiescent-point variations for interchanged transistors. Usually, values of $\beta_{0(min)}$ and $\beta_{0(max)}$ are given by the manufacturer and all transistors of the same type will have a β_0 within this range. If the circuit is designed to operate properly when β_0 assumes either of these two extreme values, the circuit is stable with respect to β_0 changes. In Section 3.2, we found that the base-current bias circuit is very unstable for β_0 changes, even though S_V can have a reasonable value for a silicon transistor. We can now consider the stages for the emitter bias and the collector-base bias in this regard.

A. Emitter-Bias Stage

The circuit of Fig. 3.5(b) is an emitter-bias stage, but it can also represent a base-bias circuit if R_E is allowed to become zero, $R_{th} = R_B$, and $E_{th} = V_{CC}$. For a silicon transistor the collector voltage is given by

$$V_{CQ} = V_{CC} - \frac{\beta_0 R_C [E_{th} - V_{BE}]}{R_{th} + (\beta_0 + 1) R_E}. \tag{3.23}$$

If R_{th} is much greater than $(\beta_0 + 1) R_E$, the equation reduces to

$$V_{CQ} = V_{CC} - \frac{\beta_0 R_C}{R_{th}} [E_{th} - V_{BE}].$$

In this case V_{CQ} decreases directly with increasing β_0. The circuit would be very unstable with respect to changes in β_0. The base-bias circuit, with $R_E = 0$, is a good example of this. On the other hand, if $(\beta_0 + 1) R_E$ is much greater than R_{th}, Eq. (3.23) becomes

$$V_{CQ} = V_{CC} - \frac{\beta_0 R_C}{(\beta_0 + 1) R_E} [E_{th} - V_{BE}] = V_{CC} - \frac{\alpha_0 R_C}{R_E} [E_{th} - V_{BE}]. \tag{3.24}$$

Since α_0 varies only 1 or 2 percent for very large variations in β_0, the collector voltage is quite stable in this case. Stability requirements can be satisfied for the emitter-bias stage if R_{th} is limited to a fraction of $(\beta_0 + 1) R_E$. Some circuits requiring less parameter stability allow R_{th} to become larger, but, in general, the upper limit on R_{th} is $0.2(\beta_0 + 1) R_E$. For germanium transistors we have seen that an upper limit of $R_{th} = 0.1(\beta_0 + 1) R_E$ is required for temperature stability; thus parameter stability automatically results in this case.

We should consider why the emitter-bias circuit is so much more stable than the base-bias stage. In the base-bias stage, the base current is constant at a given temperature at

$$I_{BQ} = \frac{V_{CC} - V_{BE}}{R_B}.$$

If β_0 is doubled, the collector current will also double. In the emitter-bias circuit the base current is

$$I_{BQ} = \frac{E_{th} - V_{BE} - I_{EQ} R_E}{R_{th}}.$$

If β_0 increases, I_{EQ} increases; but this decreases the value of I_{BQ} by raising the emitter voltage. In this case I_{BQ} is lowered as β_0 increases, and this offsets the increase in I_{CQ} due to increased β_0.

Example 3.5. In the emitter-bias circuit of Fig. 3.5(b), $R_{th} = 10\ k\Omega$, $R_E = 1\ k\Omega$, $R_C = 5\ k\Omega$, $V_{CC} = +12\ V$, $E_{th} = 1.6\ V$, and $V_{BE} = 0.5\ V$. Calculate the quiescent collector voltage when $\beta_0 = 80$. Compare this to the case when $\beta_0 = 160$.

Solution. From Eq. (3.23) for $\beta_0 = 80$,

$$V_{CQ} = 12 - \frac{80 \times 5 \times 10^3[1.6 - 0.5]}{10^4 + 8.1 \times 10^4} = 7.16\ V.$$

If β_0 is now changed to 160, the quiescent voltage becomes

$$V_{CQ} = 12 - \frac{160 \times 5 \times 10^3[1.6 - 0.5]}{10^4 + 16.1 \times 10^4} = 6.85\ V.$$

The change in V_{CQ} due to the doubling of β_0 is only 0.31 V. Recall that in the base-bias circuit, using a 20 V supply, the doubling of β_0 changed V_{CQ} from 15 V to 10 V.

Example 3.6. In the collector-base-bias circuit of Fig. 3.6, $R_F = 500\ k\Omega$, $R_C = 6\ k\Omega$, $V_{CC} = 12\ V$, $R_E = 1\ k\Omega$, and $V_{BE} = 0.5\ V$. Calculate the quiescent collector voltage when $\beta_0 = 100$. Compare this to the case when $\beta_0 = 200$.

Solution. Equation (3.20) is used to find the quiescent collector voltage for $\beta_0 = 100$:

$$V_{CQ} = \frac{12 + \frac{101}{500}(12 + 0.5 \times 6)}{1 + 101(\frac{1}{500}) + 101(\frac{6}{500})} = 6.24\ V.$$

If $\beta_0 = 200$, the voltage becomes

$$V_{CQ} = \frac{12 + \frac{201}{500}(12 + 0.5 \times 6)}{1 + 201(\frac{1}{500}) + 201(\frac{6}{500})} = 4.73\ V.$$

In this case the quiescent-point shift is -1.5 V, which is fairly large. However, if R_F is made smaller and R_E larger, the stability can be improved a great deal.

An important result of this section is that the same precautions that must be applied for good temperature stability generally result in good stability with respect to β_0 changes. The exception to this is the base-bias circuit using a silicon transistor, which can have reasonable temperature stability for large values of R_B. In the very important emitter-bias circuit we have found that a value of $R_{th} \leq 0.1(\beta_0 + 1)R_E$ is reasonable for germanium transistors, and that $R_{th} \leq 0.2(\beta_0 + 1)R_E$ gives reasonable values for silicon transistors.

Sometimes it is advantageous to calculate the maximum quiescent-point change that can occur, assuming that all variables reach their extreme values at the same time. If the combination of variables so chosen to give the maximum possible shift allow a satisfactory design, then all other values of variables will result in a satisfactory design. This is referred to as "worst-case" design. As an example of this, consider the equation for collector voltage in the emitter-bias case,

$$V_{CQ} = V_{CC} - \frac{\beta_0 R_C [E_{th} - V_{BE}]}{R_{th} + (\beta_0 + 1)R_E}. \tag{3.23}$$

Let us assume that $V_{CC} = 20$ V, $R_{th} = 10$ kΩ, $R_E = 1$ kΩ, $R_C = 5$ kΩ, and $E_{th} = 3.0$ V. If β_0 can vary from 100 to 200 (due either to temperature or interchanging of transistors) and $V_{BE} = 0.5$ V at 25°C, let us consider the extremes of V_{CQ} as the temperature varies from 25°C to 75°C. As T increases, V_{CQ} will decrease, due to changes in V_{BE}. The quiescent collector voltage V_{CQ} also decreases when β_0 increases. Thus the lower extreme of V_{CQ} will occur when β_0 is at its maximum value and $T = 75$°C. Since V_{BE} decreases by 2 mV/°C, $V_{BE} = 0.5 - (50)(0.002) = 0.4$ V at $T = 75$°C. The lower extreme of V_{CQ} is then

$$V_{CQ(min)} = 20 - \frac{200 \times 5 \times 10^3 [3.0 - 0.4]}{10^4 + 201 \times 10^3} = 7.7 \text{ V}.$$

The maximum value of V_{CQ} occurs when β_0 is at its minimum and V_{BE} is maximum ($T = 25$°C). Therefore

$$V_{CQ(max)} = 20 - \frac{100 \times 5 \times 10^3 [3.0 - 0.5]}{10^4 + 101 \times 10^3} = 8.7 \text{ V}.$$

The worst possible shift that can occur is a 1-V shift. All other combinations of β_0 and V_{BE} will result in a quiescent collector voltage falling between 7.7 and 8.7 V. In germanium transistors β_0 does not vary greatly with temperature; however, the same procedure should be followed for worst-case design, since β_0 varies greatly from one transistor to another of the same type. In addition, changes in I_{co} must be considered for germanium stages.

The concept of worst-case design is quite valuable in engineering practice. Circuit manufacturers can specify tolerance limits of important quantities based on worst-case calculations. Recognition of the importance of this type of consideration has resulted in this feature being included in several of the digital computer programs for automatic circuit analysis. As the complexity of a circuit grows, worst-case design becomes more difficult to perform manually; thus, the inclusion of this feature in programs such as ECAP, SCEPTRE, or others is of great practical significance.

An interesting point can be discussed here in relation to the results of this section. If β_0 can double and only cause a small bias-point shift (0.5–1.0 V), how accurately must β_0 be known to design an emitter-bias circuit? Quite obviously, in Example 3.5, if β_0 were estimated to be any number between 80 and 160, the value of V_{CQ} would be within 0.31 V of the desired value. The emitter-bias circuit causes the results of the design to depend more on element values than on transistor parameters. This is a very useful point and demonstrates that β_0 can be estimated very grossly and still be used to give accurate results with the emitter-bias circuit. This leads one to believe that the type of transistor used in an emitter-bias circuit is relatively unimportant. This conclusion is partially correct. For low-frequency, noncritical applications, several hundred types of transistors can be used to satisfy a given design. We shall see later that the power requirements of the stage will impose additional constraints on the circuit. Furthermore, high frequency and switching performance of a circuit is usually limited by the transistor; hence special transistors are required. Nevertheless, it is interesting to know that a great deal of design can be done with only a vague knowledge of the transistor characteristics.

There are other types of circuits that tend to stabilize transistor stages. The following section will consider methods of compensation for temperature variations. Chapter 6, which deals with feedback amplifiers, will cover some of the practical aspects of feedback stabilization.

3.8 TEMPERATURE COMPENSATION TECHNIQUES

There are various methods of compensation which will allow better temperature stability than that obtained by conventional circuits. One of the simpler methods is called diode compensation.

A. Diode Compensation

In a silicon transistor, the change in V_{BE} with temperature causes a change in I_{BQ}. The circuit of Fig. 3.7 shows a configuration that overcomes this problem. If a Thévenin equivalent circuit of the bias network is assumed to be as shown in Fig. 3.5(b), the equation for base current is

$$I_{BQ} = \frac{E_{th} - V_{BE} + V_0}{R_{th} + (\beta_0 + 1)R_E}.$$

For silicon, I_{co} is negligible and if the proper silicon diode is used so that $V_{BE} = V_0$,

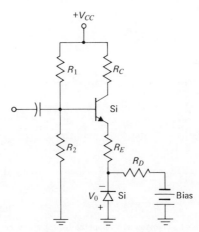

Fig. 3.7 Diode compensation for temperature effects in silicon transistors.

the equation becomes

$$I_{BQ} = \frac{E_{th}}{R_{th} + (\beta_0 + 1)R_E}.$$

Since the effect of temperature on the emitter-base junction and on the diode will be the same, the previous equation always holds. In practice it is difficult if not impossible to get perfect compensation, but a sizable improvement in stability over the uncompensated circuit usually results. Two properly biased diodes in series can compensate for increases in β_0 in addition to changes in V_{BE}.

Germanium transistors are more difficult to handle since the leakage current must also be considered. A reverse-biased diode properly placed in the circuit can compensate for changes in I_{co}. One method of I_{co} compensation is shown in Fig. 3.8. The input loop equation is written as

$$E_{th} = R_{th}(I_{BQ} + I_S) + (\beta_0 + 1)(I_{BQ} + I_{co})R_E + V_{BE},$$

where I_S is the diode leakage current. Solving for I_{BQ} and then using $I_{CQ} = \beta_0 I_{BQ} + (\beta_0 + 1)I_{co}$, we obtain

$$I_{CQ} = \frac{\beta_0(E_{th} - V_{BE})}{R_{th} + (\beta_0 + 1)R_E} + \frac{(\beta_0 + 1)(R_E + R_{th})I_{co} - \beta_0 R_{th} I_S}{R_{th} + (\beta_0 + 1)R_E}. \quad (3.25)$$

To eliminate the effect of I_{co} on collector current, I_S can be chosen to be

$$I_S = \frac{(\beta_0 + 1)(R_E + R_{th})}{\beta_0 R_{th}} I_{co}. \quad (3.26)$$

Practically, it would be almost impossible to select a diode with the exact leakage current required by Eq. (3.26). The usual method is to select a diode that has about

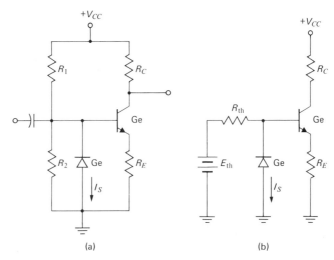

Fig. 3.8 Diode compensation for I_{co} changes with temperature: (a) actual circuit; (b) equivalent bias circuit.

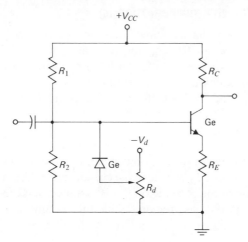

Fig. 3.9 Adjustable compensation circuit.

the same leakage current as the transistor, realizing that perfect compensation will not be obtained. To get better I_{co} compensation, the circuit of Fig. 3.9 could be used. In the figure, R_d is a potentiometer which determines the voltage appearing across the diode. Since I_S is a weak function of the diode voltage, it can be adjusted to equal I_{co}. It should be realized that changes in V_{BE} must also be considered for germanium, so that the overall circuit might appear as shown in Fig. 3.10.

In a multistage dc amplifier, the first stages are compensated carefully, since the drift in these stages is amplified by the remaining stages.

Chapter 5 will consider the use of a silicon resistor to compensate for temperature effects in differential stages.

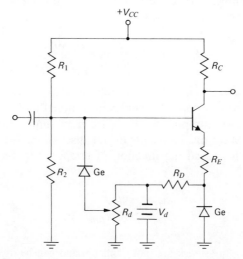

Fig. 3.10 Diode compensation for temperature effects in germanium transistors.

*3.9 VARIATION OF I_{co} WITH TEMPERATURE [5]

Throughout Chapter 3 we have assumed that the variation in I_{co} with temperature for germanium is $dI_{co}/dT = 0.09\ I_{co}$ and for silicon is $dI_{co}/dT = 0.15\ I_{co}$. These figures are quite approximate and vary with temperature. Furthermore, practical devices often tend to exhibit different I_{co} variations as a function of the construction process. Although surface leakage effects and impurities cause disagreement between practical and theoretical values of I_{co}, it is instructive to derive the variation of I_{co} with temperature.

For a reverse-biased diode with an appreciable applied voltage, the reverse current equals I_S, the reverse saturation current. For the transistor, the saturation or leakage current is called I_{co}. The leakage current can be related to the minority-carrier concentration in the base region if it is assumed that the collector is doped much more heavily in the base region. While this assumption is often invalid in practice, the results obtained apply equally well to all cases of doping.

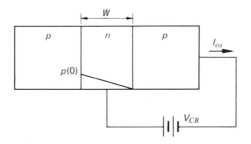

Fig. 3.11 Evaluation of reverse-biased leakage current.

Referring to Fig. 3.11, the reverse bias on the collector-base junction causes the concentration of holes to reach zero at the edge of the base nearest the collector. Applying the diffusion equation gives

$$I_{co} = -qD_p \frac{dp}{dx} = q \frac{D_p p(0)}{W}$$

assuming unity cross-sectional area. The relationship between $p(0)$ and p_{bo} has not been discussed. At this point it is sufficient to note that they differ by a multiplicative constant. For our purpose here, let us assume that $p_{bo} = p(0)$. This assumption has no effect on the results of our discussion. We can now write I_{co} in terms of n_i, which is a strong function of temperature. Since $p_{bo} = n_i^2/N_{DB}$, we can write

$$I_{co} = \frac{qD_p n_i^2}{WN_{DB}}. \tag{3.27}$$

---

* Sections preceded by an asterisk are more difficult, and can be used or avoided at the instructor's discretion.

Substituting the value for n_i given by Eq. (1.1) we find for leakage current

$$I_{co} = \frac{qD_p A^2 T^3}{WN_{DB}} e^{-E_g/kT} \qquad (3.28)$$

In this equation A is a constant not to be confused with cross-sectional area, which has been taken to be unity.

If the variation of D_p with temperature is taken as $D_p = k_1 T^{-3/2}$ the leakage current becomes

$$I_{co} = \frac{qk_1 A^2 T^{3/2}}{WN_{DB}} e^{-E_g/kT}. \qquad (3.29)$$

At this point we note that while the term $T^{3/2}$ varies with T, the dominant variation by far in I_{co} results from the exponential term. Thus, we can write for our purposes that

$$I_{co} = Ke^{-E_g/kT}.$$

Differentiating I_{co} with respect to temperature gives

$$\frac{dI_{co}}{dT} = \frac{E_g}{kT^2} Ke^{-E_g/kT} = \frac{E_g}{kT^2} I_{co}. \qquad (3.30)$$

For germanium near room temperature (300°K), $E_g = 0.68$ eV, so

$$\frac{dI_{co}}{dT} = 0.088 \, I_{co} \approx 0.09 \, I_{co} \qquad \text{(Ge)}. \qquad (3.31)$$

In silicon the value of E_g is taken to be 1.12 eV at 300°K giving

$$\frac{dI_{co}}{dT} \approx 0.15 \, I_{co} \qquad \text{(Si)}. \qquad (3.32)$$

Several points should be mentioned in relation to Eqs. (3.31) and (3.32). First of all, these are only theoretical values that do not always agree with actual device values. Second, as temperature increases, I_{co} increases and the values given for $T = 300°K$ no longer apply. The fractional changes at 300°K are

$$\frac{dI_{co}}{I_{co}} = 0.09 \, dT \qquad \text{(Ge)} \qquad (3.33)$$

and

$$\frac{dI_{co}}{I_{co}} = 0.15 \, dT \qquad \text{(Si)}. \qquad (3.34)$$

As a rule of thumb, I_{co} is said to double for a 10°C increase in temperature in germanium and for a 7°C increase in silicon. For more accurate values of I_{co} at elevated temperatures, Eq. (3.29) can be used.

*3.10 VARIATION OF V_{BE} WITH TEMPERATURE [5]

For a forward-biased diode (which corresponds to the emitter-base junction of a transistor) the applied voltage will vary with temperature. This can be seen by considering the diode equation

$$I = I_S(e^{qV/kT} - 1),$$

which reduces to

$$I = I_S e^{qV/kT} \tag{3.35}$$

for appreciable forward bias. In the case of the transistor, Eq. (3.35) can be written as†

$$I = I_{eo} e^{qV_{EB}/kT}, \tag{3.36}$$

where I_{eo} is the reverse-biased leakage current of the emitter-base junction. Solving for V_{EB} results in

$$V_{EB} = \frac{kT}{q} \ln \frac{I}{I_{eo}}, \tag{3.37}$$

or

$$V_{EB} = \frac{kT}{q} [\ln I - \ln I_{eo}].$$

The derivative of V_{EB} can be taken (assuming a constant I):

$$\frac{\partial V_{EB}}{\partial T} = \frac{k}{q} [\ln I - \ln I_{eo}] - \frac{d(\ln I_{eo})}{dT} \frac{kT}{q}$$

$$= \frac{k}{q} [\ln I - \ln I_{eo}] - \frac{kT}{q} \frac{1}{I_{eo}} \frac{dI_{eo}}{dT}. \tag{3.38}$$

Equation (3.38) can be written as

$$\frac{\partial V_{EB}}{\partial T} = \frac{V_{EB}}{T} - \frac{kT}{q} \frac{1}{I_{eo}} \frac{dI_{eo}}{dT} = \frac{V_{EB}}{T} - \frac{E_g}{qT}. \tag{3.39}$$

For typical values of V_{EB}, 0.2 V for germanium and 0.5 V for silicon, Eq. (3.39) gives the approximate values

$$\frac{\partial V_{EB}}{\partial T} = -1.6 \text{ mV/°C} \quad \text{(Ge at 300°K)} \tag{3.40}$$

and

$$\frac{\partial V_{EB}}{\partial T} = -2 \text{ mV/°C} \quad \text{(Si at 300°K)}. \tag{3.41}$$

While these quantities vary with temperature, the values given can be used to approximate the total change in V_{EB} over a fairly wide temperature range.

---

† This equation assumes that $I_{eo} = I_{ES}$, the emitter saturation current with collector shorted to base. This is not strictly true as shown later in Section 4.4A, but leads to no error in the derivation of temperature effects.

3.11 BIAS DESIGN OF FET AMPLIFIERS

Graphical design can often be quite effective in simple FET amplifiers. It was pointed out earlier that graphical transistor design is of little value in practice; however, the FET or MOSFET possesses certain characteristics that make graphical design more appropriate. The V–I characteristics at the drain of the FET are more uniform from one FET to another of the same type than are bipolar transistor characteristics. The key to the usefulness of graphical design procedures for the FET is the fact that the controlling quantity is gate voltage with negligible accompanying input current. The characteristic curves for the bipolar transistor are generated with input current as the parameter and require the use of a set of input characteristics to calculate voltage gain.

A. Graphical Calculations

Let us consider the simple FET amplifier along with the characteristic curves shown in Fig. 3.12.

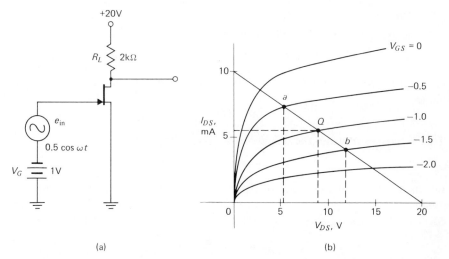

Fig. 3.12 FET amplifier: (a) simple amplifier; (b) FET characteristics.

The load-line is drawn with a slope equal to $-1/R_L$. The gate bias source sets the quiescent point at $V_{GSQ} = -1.0$ V, $V_{DSQ} = 9$ V, and $I_{DSQ} = 5.5$ mA. As e_{in} swings from zero to 0.5 V, the gate to source voltage moves from -1.0 V to -0.5 V, moving the operating point along the load-line to point a. This point swings back to point b as e_{in} reaches the negative peak of the cycle. The peak-to-peak output voltage is $12 - 5.5 = 6.5$ V, leading to a voltage gain of -6.5. Higher voltage gains generally result from larger values of load resistance especially if the drain supply is increased. Since the curves are unevenly spaced, larger signal swings result in a more distorted output voltage.

A self-bias resistance can be used to eliminate the necessity of a gate bias source. Figure 3.13 indicates a self-bias circuit with a bypassed source resistance.

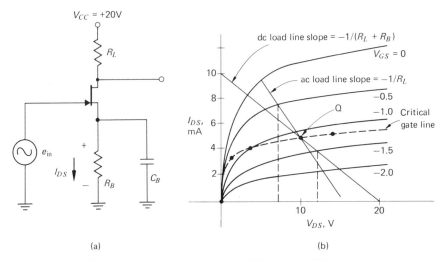

Fig. 3.13 Analysis of self-bias circuit: (a) self-bias circuit; (b) characteristics.

The expression for the dc load-line is

$$V_{DS} = V_{CC} - I_{DS}(R_L + R_B). \tag{3.42}$$

Quiescent drain-to-source current flows through R_B causing a voltage drop of value

$$V_B = I_{DS}R_B.$$

This voltage establishes the dc gate-to-source voltage at

$$V_{GS} = -V_B = -I_{DS}R_B. \tag{3.43}$$

Let us assume that the value of R_B used in the circuit is such to establish the quiescent level at point Q in Fig. 3.13(b). If the capacitor is large enough to effectively bypass R_B at all signal frequencies, the ac load-line equation is

$$\Delta V_{DS} = -\Delta I_{DS}R_L. \tag{3.44}$$

The ac load-line must intersect the dc load-line at the quiescent point, since this point must be approached as the amplitude of e_{in} is decreased to a negligibly small value. Voltage gain can be calculated by assuming a small variation of e_{in} and finding the resulting output voltage variation. For this circuit the voltage gain is approximately -6.

In designing a simple FET amplifier, R_L can be selected from voltage gain or output impedance requirements. An appropriate quiescent point can be selected and achieved by using R_B as calculated from Eq. (3.43). We will consider the selection of C_B later.

In analyzing a circuit with given values of V_{CC}, R_L, and R_B, both Eqs. (3.42) and (3.43) must be satisfied. There are three unknowns in these two expressions, namely V_{DS}, I_{DS}, and V_{GS}. However, the V–I characteristics represent the third relationship necessary to determine these three quantities. The intersection of the load-line and the critical gate line which is a plot of Eq. (3.43) determines the quiescent point of the circuit. Data for the critical gate line is obtained by assuming various values of V_{GS} and tabulating I_{DS}, as shown in Table 3.1 which assumes $R_B = 250\ \Omega$. As indicated by Table 3.1 it may be necessary to calculate I_{DS} for small changes in V_{DS} over the region where the critical gate line slope becomes constant.

Table 3.1. Data for critical gate line.

Assumed value V_{GS}, volts	Resultant I_{DS}, mA
−2.00	8
−1.50	6
−1.25	5
−1.00	4
−0.75	3
−0.5	2
0	0

If R_B is left unbypassed, the voltage gain drops considerably and it is not easy to evaluate this value graphically. When R_B is bypassed, the entire input signal drops across the gate and source terminals and the total variation in V_{GS} corresponds to the peak-to-peak value of e_{in}. For the unbypassed case, a portion of the input signal drops across R_B with the remaining portion dropping across the gate to source terminals. The variation in V_{GS} is not easily determined and thus, the graphical method is of little value. We will defer this calculation along with the calculation of the appropriate value of C_B to the chapter on equivalent circuits.

*B. **Selection of a Quiescent Point to Minimize Temperature Drift**

The temperature stability of the FET can be quite good if biased to the proper quiescent point. At this point the two major temperature effects tend to offset each other reducing temperature drift. Calculation of the proper bias point is based on the profile of impurity atoms, however, and the derivation to follow applies only to uniformly doped, abrupt junction units. More practical devices with different doping will exhibit a bias point that minimizes temperature drift, but it will differ from that of the ideal FET that will now be calculated.

For a specific gate-to-source bias voltage, the drain-to-source current will vary with temperature due primarily to (1) changes in carrier mobility, and (2) changes in the barrier voltage. The total derivative of I_{DS} with respect to temperature is

$$\frac{dI_{DS}}{dT} = \frac{\partial I_{DS}}{\partial \mu_n}\frac{d\mu_n}{dT} + \frac{\partial I_{DS}}{\partial \phi}\frac{d\phi}{dT}. \tag{3.45}$$

The partial derivatives can be evaluated from Eq. (2.51),

$$I_{DS} = I_{DSO}\left[1 - 3\left(\frac{V_{GS} - \phi}{V_0}\right) + 2\left(\frac{V_{GS} - \phi}{V_0}\right)^{3/2}\right]$$

which can also be expressed as

$$I_{DS} = -\frac{2baqN_D V_0 \mu_n}{3L}\left[1 - 3\left(\frac{V_{GS} - \phi}{V_0}\right) + 2\left(\frac{V_{GS} - \phi}{V_0}\right)^{3/2}\right]. \quad (3.46)$$

Equation (3.46) is obtained from Eq. (2.49) by substituting $1/q\mu_n N_D$ for the resistivity. From Eq. (3.46) we find that

$$\frac{\partial I_{DS}}{\partial \mu_n} = \frac{I_{DS}}{\mu_n}$$

and

$$\frac{\partial I_{DS}}{\partial \phi} = -g_m.$$

The total derivatives both depend on temperature and to some extent on doping. It can be shown that the barrier voltage varies with temperature in a manner similar to the forward-biased junction voltage. For an abrupt junction this variation is (see Section 3.10)

$$\frac{d\phi}{dT} = -\left(\frac{3k}{q} + \frac{E_g}{kT}\right) + \frac{\phi}{T}, \quad (3.47)$$

where k is Boltzmann's constant. The variation of $d\phi/dT$ with T is obvious from Eq. (3.47). The actual value of ϕ depends on the doping; thus, $d\phi/dT$ will not be constant for all devices and for all temperatures. We can, however, assume a reasonable average value for $d\phi/dT$ that allows us to compensate for temperature drift to a first-order approximation. For a silicon FET operating near room temperature we will assume that

$$\frac{d\phi}{dT} = -1.8 \times 10^{-3} \text{ V/°C}.$$

In silicon the electron mobility [2] is often taken as

$$\mu_n = \frac{K}{T^{5/2}}, \quad (3.48)$$

where K is generally assumed to be constant. From Eq. (3.48) we find that

$$\frac{d\mu_n}{dT} = -\frac{5}{2}KT^{-7/2} = -\frac{5}{2}\frac{\mu_n}{T}.$$

By experiment [2] it is found that a more accurate value for the mobility change with temperature is

$$\frac{d\mu_n}{dT} = -2.1\frac{\mu_n}{T}. \quad (3.49)$$

The total change in I_{DS} with temperature for an abrupt junction, n-channel FET is then approximately

$$\frac{dI_{DS}}{dT} = -\frac{2.1 I_{DS}}{T} + 1.8 \times 10^{-3} g_m. \quad (3.50)$$

Equating this value to zero near room temperature (300° K) results in the requirement that

$$\frac{I_{DS}}{g_m} = 0.257.$$

From Eqs. (2.51) and (2.52) this ratio can be expressed as

$$\frac{I_{DS}}{g_m} = -\frac{V_0}{3}\left[1 + \left(\frac{V_{GS}-\phi}{V_0}\right)^{1/2} - \left(\frac{V_{GS}-\phi}{V_0}\right)\right]. \quad (3.51)$$

It is found that I_{DS}/g_m will equal approximately 0.257 if

$$V_{GS} = V_P + 0.5 \text{ V}. \quad (3.52)$$

Note that for the n-channel FET both V_{GS} and V_P will have negative values. We see that the device should be biased somewhat near pinchoff in order to obtain low temperature drift in I_{DS}. Biasing the FET to such low values of I_{DS} may not always be appropriate since the output swing is somewhat limited. However, for most dc amplifiers, temperature drift of the input stage is more important than drift of succeeding stages. The first stage is generally not required to produce large output voltages; thus the results of this section are of practical significance.

Two points should be noted concerning temperature compensation of the FET. First of all, practical FETs are not uniformly doped, abrupt junction devices and Eq. (3.52) does not apply directly. It may be necessary to measure the temperature coefficient of I_{DS} at various bias points. The optimum bias point is generally found to be relatively close to the pinchoff voltage. Second, perfect compensation can occur at only one temperature, so only a first-order drift compensation can be expected.

3.12 INTEGRATED CIRCUIT BIASING

The popular emitter-bias scheme for bipolar transistors and the FET self-bias circuit both generally use a large bypass capacitor to achieve high ac gain. The unavailability of large capacitors in monolithic circuits and the lack of integrated inductors require that special techniques be used to establish bias currents for integrated amplifiers. Differential stages and complex feedback circuits are often used to obtain the correct bias in the integrated circuit amplifier. Since differential and feedback amplifiers are discussed in later chapters, only the simplest integrated circuit biasing schemes will be considered at this point.

The circuit of Fig. 3.14 shows one method of achieving proper bias on the transistor T_2 [4]. The transistor T_1 is used strictly for bias current generation and does not take part in signal amplification.

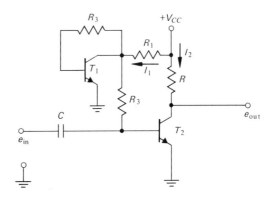

Fig. 3.14 Diode-biased stage.

While a coupling capacitor C is required for this biasing scheme, the value of this capacitor will be two orders of magnitude smaller than the bypass capacitor used in the emitter-bias circuit. The diode-biased current sink of Fig. 3.15 is closely related to the network of Fig. 3.14 and can be used to explain this type of bias technique.

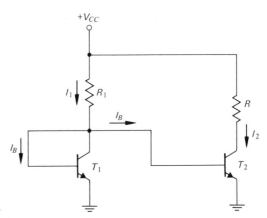

Fig. 3.15 Diode-biased current sink.

We note that this configuration has the base terminals in parallel and the emitters both return to ground; thus, $V_{BE1} = V_{BE2}$. Assuming that the transistors are integrated with the same geometry and are perfectly matched, the base currents, the emitter currents, and the collector currents of both transistors will be equal.

The collector current of T_2 can be expressed as

$$I_2 = I_1 - 2I_B. \tag{3.53}$$

The current I_1 is given by

$$I_1 = \frac{V_{CC} - V_{BE}}{R_1} \approx \frac{V_{CC}}{R_1}, \tag{3.54}$$

allowing us to write

$$I_2 = \frac{V_{CC}}{R_1} - 2I_B \approx \frac{V_{CC}}{R_1} = I_1. \tag{3.55}$$

Control of collector-bias current for T_2 is then accomplished by controlling the ratio of V_{CC} to R_1.

Although bias current for T_2 can easily be controlled with this configuration, it is difficult to apply the signal to the base of T_2 because this is a low impedance point. The transistor T_1 is near saturation and the low impedance path from collector to emitter of this transistor appears from the base of T_2 to ground. Insertion of the resistors R_3, as shown in Fig. 3.14, overcomes the impedance problem while maintaining a balanced dc circuit. Once the desired collector current of T_2 is selected, the values of R_1 and R_3 are found from Eq. (3.53) which is still valid and

$$I_1 = \frac{V_{CC} - V_{C1}}{R_1} \approx I_2. \tag{3.56}$$

The collector voltage of T_1 is

$$V_{C1} \approx \frac{V_{CC}R_3 + \beta_0 V_{BE} R_1}{\beta_0 R_1 + R_3}. \tag{3.57}$$

If βR_1 is much larger than R_3, then the bias current will be relatively independent of the value of β_0 and the collector current of T_2 will again be given by Eq. (3.55). The input impedance from the base of T_2 to ground will now consist of the resistance R_3 in parallel with the input impedance to the transistor.

Feedback methods and differential input stages are perhaps more practical than the diode-biased stage of Fig. 3.14. A discussion of these more complex methods will be deferred until the general theory of differential stages and feedback amplifiers can be considered.

Before leaving this topic we should note that it is possible to stabilize the temperature of the semiconductor chip by electronic means. This tends to minimize drift of the amplifier circuit by minimizing the chip temperature variation [1]. Adjacent to the amplifier circuit a forward-biased diode is placed to develop a control voltage that depends on temperature. This temperature-sensitive voltage controls a power transistor operating point resulting in less energy dissipation for higher temperatures and more dissipation for lower temperatures. As ambient temperature increases, tending to increase the chip temperature, the decreased

REFERENCES AND SUGGESTED READING

1. H. R. Camenzind, *Circuit Design for Integrated Electronics*. Reading, Mass.: Addison-Wesley, 1968, Chapter 6.
2. R. S. C. Cobbold and F. N. Trofimenkoff, "Theory and application of the field-effect transistor." *Proc. IEE* **111**, 12 (December, 1964); **112**, 4 (April, 1965).
3. M. S. Ghausi, *Electronic Circuits*. New York: Van Nostrand, 1971, Chapter 3.
4. A. B. Grebene, *Analog Integrated Circuit Design*. Cincinnati, Ohio: Van Nostrand-Reinhold, 1972, Chapter 4.
5. J. Millman and C. C. Halkias, *Integrated Electronics*. New York: McGraw-Hill, 1972, Chapter 19.

PROBLEMS

Sections 3.1–3.2

* **3.1** A 16 V source is used with a base-current bias stage to establish a base current of 20 μA in a silicon transistor. Calculate the value of bias resistance required when V_{BE} is neglected. Using this calculated value of bias resistance, find the base current that flows when V_{BE} is taken to be 0.5 V. Compare this value to the desired value of 20 μA.

3.2 Repeat Problem 3.1 for a germanium transistor with $V_{BE} = 0.2$ V. Is it reasonable to neglect V_{BE}? Explain.

3.3 In the base-bias circuit of Fig. 3.3, $V_{CC} = +20$ V, $R_C = 5$ kΩ, $\beta_0 = 100$, and the transistor is silicon with $V_{BE} = 0.5$ V. Select R_B to set V_{CQ} at
  a) the midpoint of the active region,
  b) $+16$ V,
  c) $+5$ V.

* **3.4** Repeat Problem 3.3, given that β_0 is changed to 200.

3.5 Work Problem 3.3 using a germanium transistor with $\beta_0 = 100$, $V_{BE} = 0.2$ V, and $I_{co} = 2$ μA.

3.6 Work Problem 3.3 using a silicon transistor with $\beta_0 = 50$ and $V_{BE} = 0.6$ V.

Sections 3.3–3.4

3.7 In the base-bias circuit of Fig. 3.3, $V_{CC} = +20$ V, $R_C = 5$ kΩ, $\beta_0 = 80$, $R_B = 560$ kΩ, and $V_{BE} = 0.5$ V (Si).
  a) Find the quiescent collector voltage.
  b) Assuming that V_{BE} changes to 0.3 V due to a temperature increase, find V_{CQ}.
  c) Repeat part (a) with β_0 changed to 160.
  d) Is this amplifier a good production circuit? Explain.

3.8 Evaluate the stability factor for the circuit of Problem 3.7(a).

* **3.9** In the base-bias circuit of Fig. 3.3, $V_{CC} = +20$ V, $R_C = 5$ kΩ, $\beta_0 = 80$, $R_B = 560$ kΩ, $I_{co} = 2$ μA, and $V_{BE} = 0.2$ V at 25°C (Ge).

a) Find the quiescent collector voltage at 25°C.
b) Evaluate I_{co} at 75°C.
c) Assuming that T is raised to 75°C, find V_{CQ}.

3.10 Repeat Problem 3.9 if β_0 is changed to 50.

3.11 Evaluate the stability factor for the circuit of Problem 3.9 at $T = 25°C$ and at $T = 75°C$. Why are these quantities different?

*3.12 Neglecting the variation of I_{co} and β_0 with temperature, find the maximum value of R_C that can be used in a base-bias circuit which must operate with a shift in quiescent collector voltage of 0.1 V or less. The temperature varies from 25° to 55°C, $\beta_0 = 90$, and $V_{BE} = 0.5$ V at 25°C. A 1 MΩ bias resistance is used.

3.13 Repeat Problem 3.12 if T varies from 25°C to 75°C.

Section 3.5

*3.14 Select R_1 for a silicon emitter-bias circuit with $R_C = 8.2$ kΩ, $R_E = 1$ kΩ, $R_2 = 20$ kΩ (see Fig. 3.5), and $V_{CC} = 12$ V. Assume that $\beta_0 = 100$ and $V_{BE} = 0.5$ V. Select R_1 to set V_{CQ} at the midpoint of the active region. Find the maximum symmetrical peak-to-peak output voltage that can be obtained before saturation or cutoff occurs.

3.15 In the circuit of Problem 3.14 select R_1 for a quiescent collector voltage of $V_{CQ} = +4$ V. Find the maximum symmetrical peak-to-peak output voltage that can be obtained before saturation or cutoff occurs.

3.16 In the circuit of Problem 3.14 select R_1 for a quiescent collector voltage of $V_{CQ} = +10$ V. Find the maximum symmetrical peak-to-peak output voltage that can be obtained before saturation or cutoff occurs.

3.17 Repeat Problem 3.14 if V_{CC} is changed to 20 V.

*3.18 The emitter-bias circuit of Fig. 3.5 has $R_C = 8.2$ kΩ, $R_E = 1$ kΩ, $V_{CC} = 20$ V, $\beta_0 = 80$, and $V_{BE} = 0.5$ V. Choose reasonable values of R_1 and R_2 to set V_{CQ} at the midpoint of the active region. If β_0 now changes to 160, to what value must R_1 be changed to maintain the same value of V_{CQ}?

3.19 Repeat Problem 3.18 for $V_{CC} = 12$ V.

*3.20 In the emitter-bias circuit of Fig. 3.5, $R_C = 8.2$ kΩ, $R_E = 1$ kΩ, $R_2 = 10$ kΩ, and $V_{CC} = 20$ V. Assume that $\beta_0 = 100$ and $V_{BE} = 0.5$ V.
a) Select R_1 to set $V_{CQ} = 14$ V.
b) If β_0 is lowered to 50 by changing transistors, what is the new value of V_{CQ}?
c) Find the maximum symmetrical peak-to-peak output voltage that can occur for each value of β_0 before cutoff or saturation occurs.

3.21 Repeat Problem 3.20 if R_2 is changed to 100 kΩ. Is this circuit less stable with respect to β_0 changes than the circuit of Problem 3.20? Explain.

*3.22 An emitter-bias circuit having $R_C = 6.8$ kΩ, $V_{CC} = +12$ V, and $R_E = 750$ Ω is to be capable of delivering a sinusoidal output signal of 3-V amplitude. Given that a silicon transistor is used with $\beta_0 = 60$ and $V_{BE} = 0.5$, find the upper and lower limits on V_{CQ} to satisfy the design. Choose reasonable values of R_1 and R_2 that would result in the lower of the two values of V_{CQ}.

3.23 Assuming that the output signal is to have an amplitude of 4 V, repeat Problem 3.22.

*3.24 Design the bias network for an emitter-bias stage that has $R_C = 5$ kΩ and $R_E = 0.5$ kΩ. The drift in V_{CQ} for an increase in T of 30°C should not exceed 0.8 V. Assume that

β_0 has a constant value of 80, that $I_{co} = 1 \ \mu A$ at 25°C and $V_{BE} = 0.2$ V at 25°C, and that the upper temperature limit is 55°C. Use a 12 V power supply.

3.25 Repeat Problem 3.24 given that β_0 can vary from 80 to 160.

Section 3.6

3.26 Given that $R_C = 5 \ k\Omega$, $R_E = 1 \ k\Omega$, and $V_{CC} = 20$ V in the collector-base-bias stage of Fig. 3.6, select R_F to set V_{CQ} at
a) the midpoint of the active region,
b) $+14$ V,
c) $+6$ V.
Assuming that $\beta_0 = 50$ and $V_{BE} = 0.5$ V (Si),
d) calculate S_V for the three bias points.

3.27 Repeat Problem 3.26 for the case where $\beta_0 = 120$.

*__3.28__ Given that $\beta_0 = 50$, $V_{BE} = 0.2$ V, $I_{co} = 2 \ \mu A$, and $V_{CC} = -20$ V, find the value of R_F for a collector-base-bias stage that gives a value of $S_V = 12$ mV/°C. Assume that $R_C = 8.2 \ k\Omega$ and $R_E = 2 \ k\Omega$. What is the quiescent collector voltage for this stage?

3.29 For a germanium collector-base-bias stage, there are conflicting requirements on R_F for good temperature stability. Equation (3.22) indicates that the first term is minimized when $R_F \gg (\beta_0 + 1)(R_E + R_C)$, while the second term is minimized for $R_F \ll (\beta_0 + 1)(R_E + R_C)$. Which condition would you expect to be most important in minimizing S_V? Explain why.

Section 3.7

*__3.30__ Given that $\beta_{0(min)} = 50$ and $\beta_{0(max)} = 150$, select R_1 and R_2 so that the quiescent collector voltage of the circuit shown in the figure is known to within ± 0.5 V of the desired level of 7 V for all possible values of β_0. Assume that $V_{BE} = 0.5$ V (Si).

3.31 Repeat Problem 3.30 for the case where $\beta_{0(min)} = 80$ and $\beta_{0(max)} = 200$. Calculate the temperature stability of the circuit at both extremes of β_0.

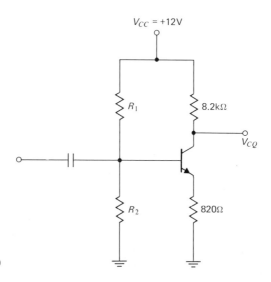

Problem 3.30

*3.32 Assume that $R_2 = 8.2$ kΩ and $R_1 = 82$ kΩ in Problem 3.30. What are the upper and lower limits of β_0 so that V_{CQ} is known to within ± 0.3 V of the quiescent value for $\beta_0 = 100$?

3.33 Repeat Problem 3.32 given that V_{CQ} must fall within ± 0.1 V of the quiescent value for $\beta_0 = 100$.

3.34 Assuming that the circuit of Problem 3.32 uses a germanium transistor with $V_{BE} = 0.2$ V and $I_{co} = 2$ μA at 25°C, and that β_0 can range from 80 to 200, find the greatest and least value assumed by V_{CQ} if the temperature can change from 25°C to 65°C. The change in β_0 does not take place due to temperature change, but it must be considered to make the transistors interchangeable.

*3.35 Repeat Problem 3.34 when β_0 ranges from 60 to 120.

Sections 3.9–3.10

3.36 For a germanium transistor the forbidden energy gap is $E_g = 0.68$ eV. Given that $I_{co} = 4$ μA at 300°K, use Eq. (3.29) to calculate I_{co} at 350°K. Compare this to the change in I_{co} calculated by using Eq. (3.31) and assuming that dI_{co}/dT is constant with temperature at 0.36 μA/°C.

*3.37 Use Eq. (3.37) to calculate the forward-biased base-emitter voltage drop of a silicon transistor at 350°K. Assume that $V_{BE} = 0.5$ V at 300°K. Compare this to the value calculated by using Eq. (3.41). Assume that I_{eo} is 0.1 nA at 300°K and 0.060 μA at 350°K.

Section 3.11

*3.38 If V_{CC} is changed to 18 V and $R_L = 3$ kΩ in Fig. 3.13, select R_B to give $V_{GS} = -1.0$ V.

3.39 If $V_{CC} = 18$ V, $R_L = 3$ kΩ, and $R_B = 500$ Ω in Fig. 3.13, find the quiescent levels of V_{GS} and V_{DS}.

Section 3.12

3.40 Derive Eq. (3.57).

*3.41 For silicon transistors with $V_{BE} = 0.6$ V, select values of R_1 and R_3 in the circuit of Fig. 3.14 to set the collector current of T_2 at 1 mA for $V_{CC} = 12$ V. Assume $\beta_0 = 75$.

EQUIVALENT CIRCUITS AND MODELING

The underlying physical principles governing the behavior of most electronic devices are fairly complex although the electrical behavior of the device may be quite straightforward. For example, while an understanding of solid-state physics is required to explain the behavior of a semiconductor diode, we can utilize this device effectively in an electrical network with only a knowledge of its volt-ampere characteristics. In circuit design an attempt is made to discard unnecessary detail and base the design on those fundamentals that are relative to the problem at hand. Thus, it is desirable to represent devices in the simplest possible terms.

In electrical networks certain elements that can be analyzed with little difficulty occur regularly. Voltage sources, passive elements such as resistors, capacitors, and inductors, and electrical switches are examples of well-known electrical components. When a semiconductor device is used in a network, the analysis can be carried out analytically by expressing the device characteristics in terms of mathematical equations. However, the resulting equations are in general too complex to be of practical value. In a large number of cases, the device can be replaced by an equivalent circuit prior to analysis. This equivalent circuit is made up of the elements mentioned previously, that is, resistors, capacitors, inductors, voltage or current sources, and switches or ideal diodes.

The equivalent circuit or model that replaces a device must exhibit the same properties that the device possesses. Often this means that the V–I characteristics of the device and those of the equivalent circuit must correspond at all terminals. Beyond this, there is generally little physical correspondence between the equivalent circuit and the device.

There are two important types of equivalent circuits that are utilized in network design. The first is a linear or small-signal equivalent circuit that represents a device only over a limited region of the characteristic curves. Fortunately, the large and important class of networks known as linear amplifiers can be treated as small-signal circuits. Large-signal or nonlinear networks are modeled by piecewise linear circuits that approximate the V–I characteristics with a series of linear or straight-line segments.

Many simplifying assumptions are made in developing an equivalent circuit in order to reduce the complexity of the resulting circuit. In most practical applications these approximations are quite acceptable and, for manual analysis, there is often little alternative to acceptance of these inaccuracies.

At first thought one might assume that computer-aided design programs eliminate all necessity of approximations. Certainly the computer should be able to carry out the additional calculations required by a precise model of the device. While there are several instances wherein this assumption is valid, there are many others to which this reasoning does not apply. If there are many devices in a network and the equivalent circuit for each requires several elements, storage capacity of the computer can be exceeded. Very accurate modeling can also lead to problems of excessive computer time under certain conditions. Some compromise between accuracy and complexity must be made for computer-aided design as well as for manual design, although the approximations may be more gross for the latter case. The first two sections of this chapter develop diode and transistor models for manual analysis, while the last section discusses computer-aided design models.

There are a number of networks that are designed to operate at frequencies far below the maximum operating frequencies of the devices used. In this instance a low-frequency model for the device is appropriate. The low-frequency equivalent circuit generally contains no frequency-dependent elements such as capacitors or inductors. At higher operating frequencies capacitors or inductors may be added to the model to represent frequency effects of the actual device. This type of model is called a high-frequency equivalent circuit.

4.1 THE DIODE EQUIVALENT CIRCUIT

There are many possible models for a diode depending on accuracy, frequency, or other requirements pertaining to the circuit at hand. This section begins with the most basic model and progresses to more complex models. The choice of the equivalent circuit to be used in a given problem rests with the designer.

A. Low-Frequency, Large-Signal Models

At low to moderate frequencies the equivalent circuit of the diode must reflect only the essential features of the diode characteristic curve as given in Fig. 2.7. The simplest model that can be proposed is the ideal diode or voltage-controlled switch. This switch is open if the applied voltage is negative (as defined by the actual diode polarity) and becomes a short circuit if the applied voltage tends to be positive. This model and the characteristics are shown in Fig. 4.1.

The ideal diode model is useful in many applications. Obviously, if the forward resistance of a diode is much less than the series resistance of the circuit and the back resistance is much greater than the series circuit resistance, the ideal diode model is appropriate. This assumes also that the small forward-bias voltage required to conduct appreciable diode current is negligible. If the forward offset voltage must be included in the analysis, the slightly more complex circuit of Fig. 4.2 can be used.

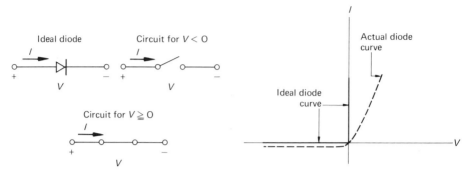

Fig. 4.1 The ideal diode and V–I characteristics.

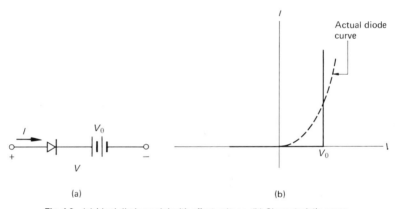

Fig. 4.2 (a) Ideal diode model with offset voltage. (b) Characteristic curve.

In addition to the practical utility of the ideal diode model, this equivalent circuit also serves as a component part of more complex models. For example, if an offset voltage and a finite slope for forward current is required, the diode model of Fig. 4.3 can be applied. The slope of the curve is determined by the choice of R_f.

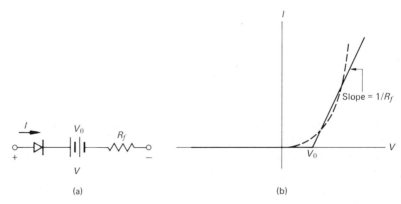

Fig. 4.3 (a) Ideal diode model with offset voltage and finite forward slope. (b) Characteristic curve.

More accuracy and additional complexity of the model can be achieved by combining parallel branches similar to that of Fig. 4.3. A two break point model is shown in Fig. 4.4. The slope for voltages above V_1 is equal to the reciprocal of the parallel equivalent resistance, $R_{f1} R_{f2}/(R_{f1} + R_{f2})$.

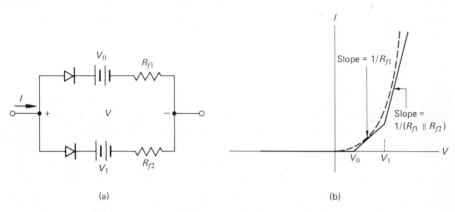

Fig. 4.4 (a) Two break point diode model. (b) Characteristic curve.

With the addition of new parallel branches, any degree of accuracy in representing the actual diode characteristics can be achieved. There are however, few applications that require this type of accuracy. It is worth noting that any general curve can be approximated in the manner outlined above. This fact is used to advantage in analog computers where a precise V–I curve is required. Twenty or more parallel paths may actually be constructed to generate the desired V–I curve. This approach is applied in the analog computer to the area of function generation.

The circuits previously discussed exhibit infinite back resistance. In some instances, a lower actual back resistance may lead to errors. To correct this situation, the model of Fig. 4.5 can be applied. The back resistance of the diode is equal to R_r in this model.

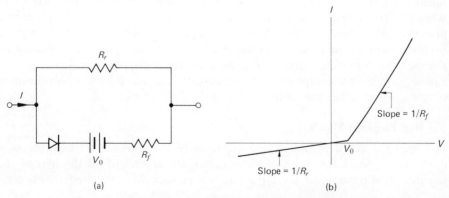

Fig. 4.5 (a) Diode model with finite back resistance. (b) Characteristic curve.

B. Low-Frequency, Small-Signal Models

The small-signal model applies only when the operating point of the diode is limited to a small region of the characteristic curve. In fact, the region must be small enough to allow the curve to be approximated by a straight-line segment over this region. In Section 2.1 we found that a dynamic resistance r_d can represent the diode in the forward direction. This resistance is called dynamic because it depends on the bias current flowing through the diode and is given by

$$r_d = \frac{kT}{qI} = \frac{0.026}{I} \quad \text{(at room temperature)}.$$

Once the bias current through a diode has been determined, the dc voltage sources can be considered to be short circuits for succeeding small-signal analysis. For example, the small-signal equivalent circuit of the diode network shown in Fig. 2.17 appears in Fig. 4.6. In Section 2.1 we found that the dc voltage source caused 1 mA of forward current to flow through the diode. This results in a resistance of 26 Ω for the model. The ac current that flows in the circuit is $0.077 \sin \omega t$ mA. While the dc source does not appear in the equivalent network, the input ac current depends on the value of this source since it determines r_d. If ohmic contact resistance is important it can be included in series with r_d. In the reverse direction the small-signal resistance of the diode is R_r. This value is fairly constant at some large value until breakdown is approached.

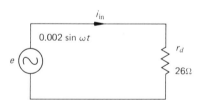

Fig. 4.6 Small-signal model of the diode.

It should be noted that the large-signal models which are called piecewise linear circuits can be used for small-signal analysis if desired. There are times when this can and should be done, although there are limitations to this procedure also. Piecewise linear models such as those shown in Figs. 4.3, 4.4, and 4.5 are more involved than the small-signal model of Fig. 4.6. The analysis is complicated by the use of such models. Furthermore, the accuracy of the piecewise linear circuits can be poor, since the slope of the curves may be determined by some compromise of two extreme resistance values.

C. High-Frequency Models

In Chapter 2 the depletion region capacitance and diffusion capacitance of a diode were considered. Both of these values are dependent on the bias of the junction. It is possible to develop a large-signal model for the diode, using analytical expressions for the capacitances, relating these values to bias. This model is too complex to apply in manual calculations, but is used in computer-aided design work. Further discussion of this large-signal equivalent circuit takes place in the last section of the chapter, when other computer models are treated. A small-signal, high-frequency equivalent circuit for the forward-biased region appears in Fig. 4.7.

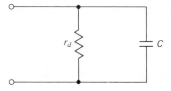

Fig. 4.7 High-frequency equivalent circuit for the forward-biased region.

The capacitance C is the sum of the diffusion capacitance, depletion region capacitance, and stray capacitance of contacts or header. The diffusion capacitance increases with forward current and is often much greater than the other two values. Thus, for most practical cases it is reasonable to replace C by the diffusion capacitance. A very important fact concerning the time constant of this diode circuit should now be discussed. The dynamic resistance of the diode varies inversely with bias current while the diffusion capacitance varies directly with this current (Eqs. 2.18 and 2.29). The product of these two element values is constant with current, although the individual values of both change with bias current. When a diode is driven by a current source or high resistance voltage source, the time constant is a large-signal value. The values for r_d and C_d may change considerably over the current swing, but the time constant remains constant. The following example demonstrates the usefulness of this result.

Example 4.1. Assume that the current of Fig. 4.8 is 1 mA until $t = 0$, at which time it switches to 3 mA. Find the voltage waveform across the diode. Assume that $C_d = 5000$ pF at $I = 1$ mA.

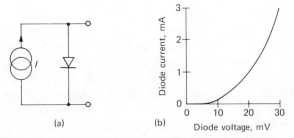

Fig. 4.8 Circuit used for Example 4.1: (a) diode circuit; (b) diode characteristics.

Solution. Since the circuit switches through a large current change, the small-signal analysis does not apply. We find that r_d varies from 26 to 8.7 Ω and C_d

changes from 5000 to 15,000 pF. The time constant of switching can be found as

$$\tau = r_d C_d = (26)(5000) \times 10^{-12} = 0.13 \ \mu s$$

and remains constant as I changes.

From the characteristic curve we see that the voltage across the diode changes from 20 mV to 30 mV, and has a time constant of 0.13 μs, as shown in Fig. 4.9. The fact that τ is independent of an operating point when the diode is forward-biased allows this parameter to be very useful in switching circuit design.

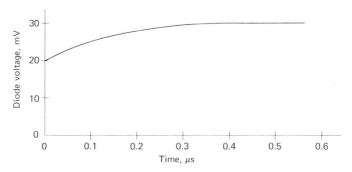

Fig. 4.9 Voltage waveform for Example 4.1.

4.2 BIPOLAR TRANSISTOR EQUIVALENT CIRCUITS

One of the largest users of transistors or integrated circuits for switching purposes is the computer industry. This and other digital fields make the study of switching circuits very important. The output signals of most digital circuits are quite large, often crossing into the cutoff or saturation region; hence large-signal analysis must be applied.

A. Low-Frequency, Large-Signal Transistor Models

We have discussed the piecewise linear approximation to the diode and have considered simple branches consisting of ideal diodes, resistors, and sources that can be utilized in practical diode models. To model the transistor we must develop circuits that approximate both the input and output V–I characteristics. The most widely used transistor configuration is the common-emitter stage. The input and output characteristics are shown in Fig. 4.10 along with piecewise linear approximations to these curves. Constant-parameter equivalent networks can now be proposed to synthesize the appropriate piecewise linear characteristics. These networks appear in Fig. 4.11. The input circuit of Fig. 4.11(a) is the equivalent circuit for a diode and has been discussed previously. We note in passing that the slope of the curve of the input branch increases at the break point. For the output circuit, the slope must decrease at a break point. The parallel combination of ideal diode, current source, and resistance leads to the desired type of characteristic.

4.2 BIPOLAR TRANSISTOR EQUIVALENT CIRCUITS

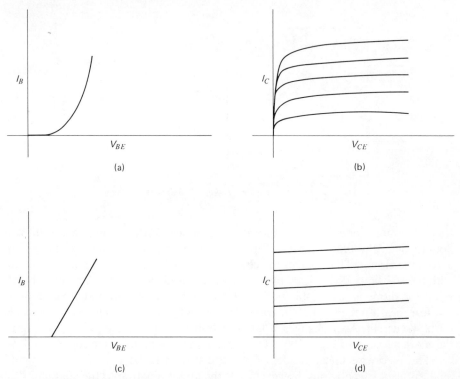

Fig. 4.10 Characteristics for a common-emitter stage: (a) actual input characteristics; (b) actual output characteristics; (c) piecewise-linear approximation to input; (d) piecewise-linear approximation to output.

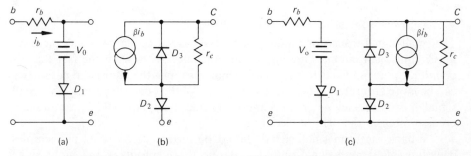

Fig. 4.11 (a) Input model of n-p-n transistor. (b) Output circuit. (c) Combined circuit.

The network of Fig. 4.12 demonstrates this feature. When the input current I is less than the source current I_0, a forward current equal to $I_0 - I$ must flow through the diode to satisfy Kirchhoff's current law at the input node. Since forward current is flowing, the diode is forward-biased with zero voltage drop across it. Thus, there is no voltage across the resistor and consequently no current flow through this element. As I increases above I_0, a current equal to $I - I_0$ must

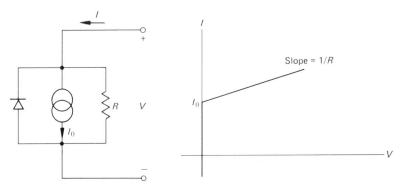

Fig. 4.12 Circuit having a decreased slope at the break point.

leave the input node through either R or the diode. Since no reverse current can flow through the diode, there will be a resistive current causing a voltage drop across the circuit. The slope of the characteristic below the break point is infinite with the resistance determined by the forward-biased ideal diode. Above the break point the diode is reverse-biased and the slope is determined by the value of R.

For the output circuit of the transistor, the value of the current source depends on base current. As i_b changes, a family of curves is generated corresponding to the characteristics of Fig. 4.10(d). The diode D_2 allows no negative current to flow.

The piecewise linear model of the transistor can be used for small-signal analysis also. It is obvious, however, that the approximation to the characteristic curves departs from the actual curves in some regions to a significant extent. This is to be expected when nonlinear functions are approximated by straight-line segments having equal slopes for each curve of a family.

B. Low-Frequency, Small-Signal Transistor Models

The small-signal transistor model is a special case of the large-signal, piecewise linear equivalent circuit. One important difference arises from the fact that the operating point is fixed in the small-signal case and the transistor parameters can be more accurately specified in the vicinity of the operating point. The small-signal or linear model is extremely important since it applies to the linear amplifier used so often in the field of electronics.

Perhaps the term small-signal should be further considered. The simplest definition of a small-signal amplifier is one that reproduces the input signal with negligible distortion. The output signal is directly proportional at all times to the applied input. There are no voltage limits that can be prescribed to guarantee small-signal operation. Each application or amplifier configuration may be different. One network may produce a 50 V output with no distortion while another may produce only a few millivolts before distortion limits the output swing. One application may allow much more distortion than another. Of course, when cutoff or saturation is reached in any stage of an amplifier, nonlinear distortion will take place. Excursions into these regions are easily identified. It is not always so obvious, however, when distortion takes place due to the small nonlinearities of the output characteristics. In amplifiers requiring very low distortion, measurements can be taken by a distortion analyzer to ensure proper design. With the exception of the fact that operation by any stage in the saturation or cutoff regions is unacceptable, each amplifier must be judged on the basis of distortion specifications.

We have seen previously that in developing a device model, the proposed equivalent circuit must (1) have the same number of terminals as the device, and (2) have V–I characteristics that duplicate the device characteristics at the terminals of interest. For small-signal operation we are concerned with dynamic characteristics at the input and output terminals. Since there are three possible configurations for a transistor stage we will develop a different model for each one.

Common-base configuration; first-order equivalent circuit

This configuration is not used as often as the common-emitter configuration, but once the common-base equivalent circuit is developed, it is simple to extend the principles to the common-emitter circuit. A common-base amplifier is shown in Fig. 4.13.

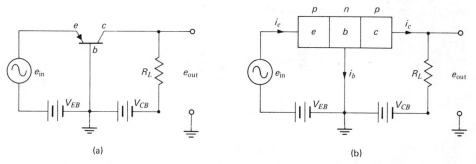

Fig. 4.13 Common-base amplifier: (a) circuit diagram; (b) alternate circuit diagram.

We first must develop the small-signal V–I relations at the input and output terminals. The ac source is applied to the emitter-base terminals, which is an operation similar to applying a source to a forward-biased diode. However, there are two differences that modify the behavior. First, there is an ohmic resistance associated with the contact to the base region. Since the base is very narrow, as mentioned before, an appreciable resistance results from the electrical connection to the region. Typical values of $r_{bb'}$, the ohmic base resistance, for present-day transistors range from 50 to 300 Ω. Certain special-purpose transistors can have values on either side of this range. The second departure from diode behavior is that the current injected from the p-region (emitter) does not leave through the n-region terminal (base). Instead, almost all of the injected current diffuses to the collector region, leaving only a very small fraction to exit through the base terminal. Again, α is used to represent the fraction of emitter current that is collected by the collector. In summary, the emitter current obeys the diode equation, since the emitter and base regions form a diode. But instead of the diode current exiting through the base terminal, most of it leaves through the collector. The key to transistor action is the narrow base region which allows the collector to absorb the holes after only a very short diffusion. A wider base region causes more recombination to take place and less holes reach the collector. This is why two back-to-back series diodes, which form a pnp-device, will not simulate transistor operation. Two ac equations can be written to describe the above behavior. One equation is the small-signal V–I relation of the emitter-base loop; the other is the characteristic of the collector terminal:

1. $e_{in} = i_e r_d + r_{bb'} i_b = i_e(r_d + (1 - \alpha)r_{bb'})$, (4.1)
2. $i_c = \alpha i_e$. (4.2)

The dynamic resistance of the emitter-base junction is represented by r_d. Note that the collector current depends only on the injected emitter current and not on collector voltage. Actually, there is a slight dependence on collector voltage, which will be mentioned later. The problem now is to devise an equivalent circuit that obeys the same equations as does the transistor. The circuit of Fig. 4.14 is one that has the proper characteristics.

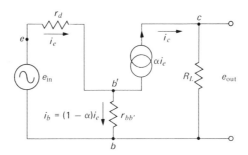

Fig. 4.14 Common-base equivalent circuit (first-order).

The dc sources are neglected, since they are apparent short circuits to ac signals. The equation for the input loop is

$$e_{in} = r_d i_e + (1 - \alpha) i_e r_{bb'},$$

which corresponds to Eq. (4.1). The output current equation is exactly the same as Eq. (4.2). So far as any V–I measurements on the input and output terminals are concerned, the given circuit could not be distinguished from the actual transistor. Figure 4.15 shows another circuit that is also equivalent.

The normal convention for transistor current flow is to assume that all terminal currents are positive when flowing into the device. This convention is not followed

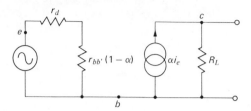

Fig. 4.15 Alternate first-order common-base equivalent circuit.

here in an effort to assign current directions corresponding more nearly to the actual model. It is always true that the base and collector currents sum together to give emitter current. In the active region, if base and collector currents enter the transistor, emitter current will leave. If base and collector currents are leaving the transistor, emitter current enters. These conditions prevail for all configurations of the stage. In Fig. 4.14, emitter current is chosen to be positive when entering the transistor; thus, base and collector currents are positive when leaving.

The circuit of Fig. 4.15 demonstrates two important points:

1. There can be more than one equivalent circuit for a given configuration.
2. The input loop is isolated from the output loop by changing the position and value of $r_{bb'}$. This amounts to a slight approximation, as will be seen later, but often leads to useful results.

Use of the equivalent circuit

Either of the two equivalent circuits can be used to evaluate the gain or input impedance to the common-base stage. This will be demonstrated by the following example.

Example 4.2. Assume that the circuit of Fig. 4.13 is biased such that $I_E = 1$ mA. Given that $R_L = 5$ kΩ, $r_{bb'} = 100$ Ω, and $\alpha = 0.99$, find the input impedance and the voltage gain of the amplifier.

Solution. The equivalent circuit of Fig. 4.15 is used (Fig. 4.16). The diode resistance is

$$r_d = \frac{0.026}{I_E} = 26 \text{ }\Omega.$$

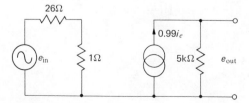

Fig. 4.16 Equivalent circuit used to evaluate gain and input impedance.

128 EQUIVALENT CIRCUITS AND MODELING

The input impedance is then

$$\frac{e_{in}}{i_{in}} = r_{in} = r_d + (1 - \alpha)r_{bb'} \qquad (4.3)$$
$$= 26 + 1 = 27 \, \Omega.$$

The voltage gain is

$$A_v = \frac{e_{out}}{e_{in}} = \frac{i_c R_L}{e_{in}} = \frac{\alpha i_e R_L}{e_{in}}.$$

Since $i_e = i_{in}$, the above expression can be written as

$$A_v = \frac{\alpha R_L}{r_d + (1 - \alpha)r_{bb'}}. \qquad (4.4)$$

In this case A_v is

$$A_v = \frac{0.99 \times 5 \times 10^3}{27} = 183.$$

The circuit used to evaluate the gain and input impedance is only a first-order circuit. A more exact equivalent circuit for the common-base amplifier will be discussed later on.

Common-emitter configuration; first-order equivalent circuit

A common-emitter amplifier is shown in Fig. 4.17. Basically, the only difference between the circuit shown in this figure and that of Fig. 4.13 is that in this figure the source has been moved around the emitter-base loop toward the base. The ac input voltage is still applied across the emitter-base junction, but base current rather than emitter current now flows through the source. The same equations apply to the common-emitter circuit as to the common base. That is,

$$e_{in} = i_e r_d + r_{bb'} i_b \qquad (4.1)$$

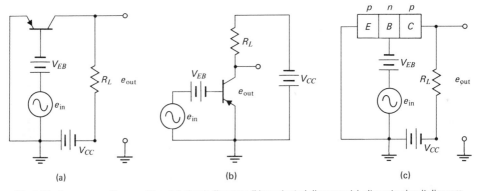

Fig. 4.17 Common-emitter amplifier: (a) circuit diagram; (b) reoriented diagram; (c) alternate circuit diagram.

and
$$i_c = \alpha i_e. \tag{4.2}$$
However, we are now interested in base current, since it is the input or source current; therefore, Eqs. (4.1) and (4.2) are put in terms of i_b. Since $i_e = i_b/(1 - \alpha)$,
$$e_{in} = \frac{r_d i_b}{1 - \alpha} + r_{bb'} i_b = r_d(\beta + 1) i_b + r_{bb'} i_b \tag{4.5}$$
and
$$i_c = \frac{\alpha}{1 - \alpha} i_b = \beta i_b. \tag{4.6}$$

A network must now be selected which satisfies the above two equations. The network of Fig. 4.18 can be shown to do so. Since base current is assumed to be positive when entering the transistor, collector current is also positive when entering and emitter current is positive when leaving. The direction of the current generator corresponds to this choice of directions. The equation for the input loop is
$$e_{in} = r_{bb'} i_b + r_d(\beta + 1) i_b.$$
The output current is
$$i_c = \beta i_b;$$
hence the circuit of Fig. 4.18 is a valid equivalent circuit. A more useful form is shown in Fig. 4.19. This circuit is called the hybrid-π circuit, when used for high-frequency analysis, because a capacitor links points b' and c to form a π-network. We will refer to both low- and high-frequency equivalents as the hybrid-π.

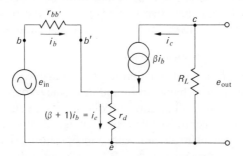

Fig. 4.18 Common-emitter equivalent circuit (first-order).

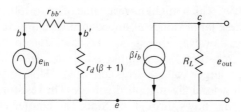

Fig. 4.19 Alternate common-emitter equivalent circuit (first-order).

Example 4.3. Assume that the circuit of Fig. 4.17 is biased such that $I_E = 1$ mA. Given that $R_L = 5$ kΩ, $r_{bb'} = 100$ Ω, and $\alpha = 0.99$, find the input impedance and the voltage gain of the amplifier.

Solution. From the equivalent circuit of Fig. 4.19, we see that the input impedance is

$$\frac{e_{in}}{i_{in}} = r_{in} = r_{bb'} + r_d(\beta + 1) = 100 + 26(100) = 2700 \ \Omega.$$

The voltage gain is

$$A_v = \frac{e_{out}}{e_{in}} = \frac{-\beta i_b R_L}{e_{in}} = \frac{-\beta R_L}{r_{bb'} + r_d(\beta + 1)}. \quad (4.7)$$

If both numerator and denominator of Eq. (4.7) are divided by $(\beta + 1)$, Eq. (4.7) becomes

$$A_v = \frac{-\alpha R_L}{r_d + r_{bb'}(1 - \alpha)},$$

which is exactly the same as the voltage gain in the common-base amplifier, except there is a phase inversion. Hence

$$A_v = \frac{-99 \times 5 \times 10^3}{100 + 26(100)} = -183.$$

It is to be expected that the voltage gain of the common-emitter and common-base amplifiers is the same, since the input voltage is applied across the same junction and the output is taken at the collector in both cases.

It will be shown in the next two sections that the first-order hybrid-π circuit can almost always be used in the analysis of practical circuits.

Accurate common-base equivalent circuit

It has previously been pointed out that as the voltage across the collector-base junction changes, there is a change in I_{co} and in I_E due to base-width modulation. Since the output ac voltage will modify the collector-base voltage, it will also cause I_{co} and I_E to change in accordance with this voltage change. These effects can be represented by an equivalent circuit with appropriate resistances, such as that shown in Fig. 4.20. The resistor r_v accounts for changes in I_{co} with output voltage. Its value is typically 1–10 MΩ. The emitter current changes with output voltage are accounted for by r_s, which might have a value of 50 to 150 kΩ. At first thought, one would think these large values of impedance would have little effect on the circuit performance; however, this is not the case, due to the feedback effects associated with the resistors. For example, additional current will flow through $r_{bb'}$ (via r_v), causing a larger voltage drop across $r_{bb'}$. This, in turn, will cause less emitter current to flow for a given applied voltage. The resistor r_s will increase the amount of current that flows through r_d. Since the emitter current is modified, the collector current is also modified. It is very difficult to work with the circuit of Fig. 4.20. Typical values of gain, input impedance, and output impedance

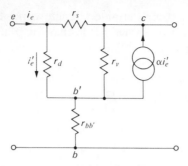

Fig. 4.20 Common-base equivalent circuit (second-order).

for this circuit are given in Table 4.1 and are compared to the corresponding values calculated from the approximate circuit. If we assume the same values as given in Example 4.2, along with the values of $r_v = 2$ MΩ and $r_s = 50$ kΩ, the values shown in Table 4.1 result. The errors in Table 4.1 are typical of those occurring when the approximate circuit is used.

Table 4.1 Comparison of accurate to approximate circuit for common-base amplifier.

Value	Calculated from		Error, percent
	Accurate circuit	Approximate circuit	
r_{in}	22 Ω	27 Ω	21
r_{out}	100 kΩ	∞	—
A_v	163	183	12
$R_{out} = r_{out} \| R_L$	4.76 kΩ	5 kΩ	5

Accurate common-emitter equivalent circuit

The same two effects of leakage-current and emitter-current change with collector voltage are also present in the common-emitter circuit. Again r_v and r_s are used to represent these effects as shown in Fig. 4.21. While the more accurate equivalent

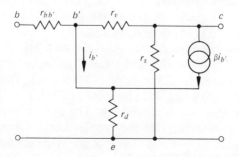

Fig. 4.21 Common-emitter equivalent circuit (second-order).

132 EQUIVALENT CIRCUITS AND MODELING

circuit is appropriate for computer-aided design work, it is quite difficult to use in manual analysis. It can be shown [3, 4] that for practical circuit design it is possible to neglect both r_v and r_s. The first-order equivalent circuit of Fig. 4.19 can then be used.

To compare the accuracy of the two common-emitter equivalent circuits, the single-stage amplifier of Fig. 4.22 can be used assuming typical parameter values of $\beta = 99$, $r_s = 50 \text{ k}\Omega$, $r_v = 2 \text{ M}\Omega$, $r_{bb'} = 100 \text{ }\Omega$, $r_d = 26 \text{ }\Omega$ ($I_E = 1 \text{ mA}$), $R_L = 5 \text{ k}\Omega$, and $R_g = 600 \text{ }\Omega$ (source resistance).

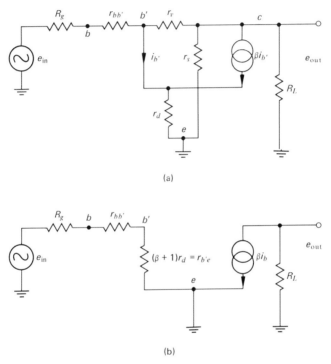

Fig. 4.22 (a) Accurate equivalent circuit. (b) First-order equivalent circuit.

Table 4.2 summarizes the results of using each equivalent circuit to analyze amplifier performance. In addition, the amplifier is considered when an external emitter resistance of 500 Ω is inserted.

In comparing the values calculated from the accurate circuit to those of the simple circuit, we find that the accurate value of voltage gain is -165 compared to -183 for the simple circuit. The error is 11 percent. The accurate value of input impedance is 2221 Ω compared to a value of 2700 Ω for the simple hybrid-π circuit. The error in this case is 22 percent. The output impedance is 32.3 kΩ compared to a value of infinity for the simple circuit. Generally we are more interested in the overall output impedance of the stage including the load resis-

Table 4.2 Comparison of accurate to approximate circuit for common-emitter amplifier.

Value	Calculated from		Error, percent
	Accurate circuit	Approximate circuit	
$R_E = 0$			
r_{in}	2221 Ω	2700 Ω	22
r_{out}	32.3 kΩ	∞	—
A	−165	−183	11
$R_{out} = r_{out} \| R_L$	4.33 kΩ	5 kΩ	15
$R_E = 500\ \Omega$			
r_{in}	38.3 kΩ	52.7 kΩ	38
r_{out}	389 kΩ	∞	—
A	−9.33	−9.39	0.7
$R_{out} = r_{out} \| R_L$	4.93 kΩ	5 kΩ	1.4

tance. This quantity is equal to the parallel combination of the transistor output impedance and the load resistance, that is,

$$R_{out} = r_{out} \| R_L.$$

The accurate circuit gives a value for R_{out} of 4.33 kΩ compared to 5 kΩ for the simple circuit: an error of 15 percent.

Most practical stages will include an emitter resistance for reasons that will be considered shortly. This resistance results in gain and output impedance expressions for the simple circuit that compare closely to the more accurate expressions.

If an external emitter resistance of 500 Ω is added to the circuit, the accurate circuit gives values of $A_v = -9.33$, $r_{in} = 38.3$ kΩ, and $r_{out} = 389$ kΩ. The overall output impedance R_{out} is 4.93 kΩ. Values calculated from the simple hybrid-π circuit are $A_v = -9.39$, $r_{in} = 52.7$ kΩ, $r_{out} = \infty$, and $R_{out} = 5$ kΩ. The gain figures disagree by only 0.7 percent and the overall output impedance by 1.4 percent. The input impedance values continue to show a rather large discrepancy.

The gain calculation from the simple circuit is quite accurate when R_E is present. The output impedance becomes much larger than normal values of R_L when R_E is used. Thus, the simple circuit can be applied in calculating these values. It is not so accurate for input impedance calculations. It must be pointed out, however, that the overall input impedance is determined by the bias resistors in parallel with r_{in}. The equivalent bias resistance is generally smaller than r_{in} when R_E is present. Thus, the overall input impedance calculation from the simple equivalent circuit is more reasonable than one might initially suspect.

The results of this section can be summarized as follows: If the simple equivalent circuit is used (Fig. 4.19), the expression resulting for voltage gain is a good approximation to the actual voltage gain. When an external emitter resistance is used, the accuracy becomes even better. The infinite output impedance resulting from the circuit is also a good approximation usually, and as R_E is added, the actual output impedance approaches a very large value. The greatest shortcoming of the simple equivalent circuit is that it predicts values of input impedance higher than that of the actual circuit. The error increases as R_L is increased, whether an external emitter resistance is used or not. However, when R_E is added, r_in becomes quite high. The overall input impedance to the circuit will be determined by r_in in parallel with the biasing resistances. In general, the bias network will determine the input impedance to the stage when R_E is present, because the bias network will offer a smaller resistance than r_in. In high-frequency designs, where R_E is often bypassed with a capacitor, the simple equivalent circuit will still give engineering accuracies, since R_L must be limited to a small value, due to frequency considerations. Because of these considerations and others to be mentioned in the following paragraphs, the equivalent circuits of Fig. 4.23 will be used throughout much of the remainder of the text.

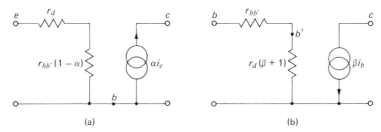

Fig. 4.23 Equivalent circuits to be used throughout text: (a) common-base equivalent; (b) common-emitter equivalent.

Justification for using the approximate circuits. The inaccuracies associated with the simple circuits of Fig. 4.23 are relatively small, but they become even less important due to the following reasons.

1. Standard component tolerances (resistances, capacitances) generally range from 5 to 20 percent. Since the values of the components used to construct the circuit are not known accurately, it is impractical to use extremely accurate equations for calculations.

2. In most practical applications an unbypassed external emitter resistance is used. It has been pointed out in Chapter 3 that emitter-bias is often used to achieve good dc stability. This resistance can be bypassed by a capacitor; however, chapter 5 will show that an unbypassed emitter resistance is required to obtain good ac gain stability with respect to changes in β.

3. For almost all engineering applications a design accuracy of 10 percent is considered quite good. For more accurate design, precision components are used in circuits that are made to depend on the component values rather than transistor parameters (feedback circuits).

4.2 BIPOLAR TRANSISTOR EQUIVALENT CIRCUITS

4. High-frequency and switching-circuit analysis is accurately and conveniently done with the hybrid-π model after adding certain reactive effects. Thus, it is reasonable to apply this circuit to the low-frequency area also.

5. Design work is limited quite severely by the complexity of the more accurate model.

6. The transistor parameters required to use the accurate circuit are very difficult to obtain. The manufacturer does not supply values for all elements. Furthermore, only typical values of transistor parameters can be supplied. Since these parameters vary widely from one transistor to another and also with operating point, the designer would have to measure values for each transistor used to utilize the accurate circuit. It is much simpler to obtain only the few values required for the simple circuit. When external emitter or base resistances are used, the required values can often be estimated to sufficient accuracy.

7. A physical feeling and understanding of the dominant mechanisms involved in transistor operation is more easily gained by studying the simpler circuit.

The arguments presented above should overwhelmingly demonstrate the appropriateness of the simple hybrid-π circuit for practical circuit design. While the accurate model may find application in computer-aided design, it serves little purpose in applications that require manual calculations.

Equivalent circuits with external resistance

In all practical cases the transistor amplifier will have resistors added to the equivalent circuit. Usually, these resistors can be combined with the resistances in the simple equivalent circuit for purposes of calculation. For example, if an emitter resistance is added to the circuit, it appears in series with r_d and therefore can be added to r_d in the gain and impedance equations, as was done in previous calculations. In the circuit of Fig. 4.24(a), the overall gain e_{out}/e_{in} can be found by using the equivalent circuit shown in part (b) of the figure. The equations for gain or

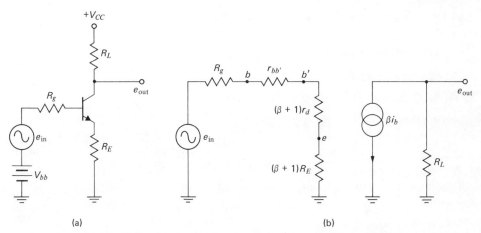

Fig. 4.24 Common-emitter amplifier: (a) actual circuit; (b) equivalent circuit.

impedance for the case where R_g and R_E are zero can be used if $r_{bb'}$ is replaced by $r_{bb'} + R_g$ and $(\beta + 1)r_d$ is replaced by $(\beta + 1)r_d + (\beta + 1)R_E$. Equation (4.7) becomes

$$A_v = \frac{-\beta R_L}{(R_g + r_{bb'}) + (\beta + 1)(r_d + R_E)}. \tag{4.8}$$

The impedance at point b (not including R_g) is approximately

$$r_{in} = r_{bb'} + (\beta + 1)r_d + (\beta + 1)R_E.$$

The common-base configuration can be handled equally well in this manner.

The emitter follower (common collector)

The emitter follower is very important in circuit design because it has a high input impedance and low output impedance. Even though its voltage gain is approximately unity, the impedance transformation properties make it a very useful circuit. Figure 4.25 shows the emitter-follower circuit. Figure 4.26 can be used to analyze this circuit.

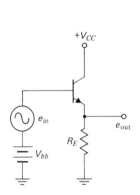

Fig. 4.25 Emitter follower.

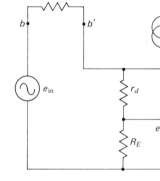

Fig. 4.26 Emitter-follower analysis.

The pertinent equations are:

1. $e_{out} = (\beta + 1)i_b R_E$,
2. $e_{in} = r_{bb'}i_b + r_d(\beta + 1)i_b + R_E(\beta + 1)i_b$.

Solving these equations for e_{out}/e_{in} gives

$$\frac{e_{out}}{e_{in}} = A_v = \frac{(\beta + 1)R_E}{r_{bb'} + (\beta + 1)r_d + (\beta + 1)R_E}. \tag{4.9}$$

The input impedance is found to be approximately

$$r_{in} = \frac{e_{in}}{i_b} = r_{bb'} + (\beta + 1)r_d + (\beta + 1)R_E. \tag{4.10}$$

The output impedance can be found by shorting the input source and assuming that there is a source at the output. The current through this output source is

$$i_{out} = i_{R_E} + i_e,$$

where i_{R_E} is simply e_{out}/R_E. The emitter current is found by using the following equations:

1. $i_e = i_b(\beta + 1)$,
2. $e_{out} = i_e r_d + i_b r_{bb'} = r_d(\beta + 1)i_b + i_b r_{bb'}$.

So

$$\frac{e_{out}}{i_e} = r_d + \frac{r_{bb'}}{\beta + 1}$$

and therefore

$$R_{out} = \frac{e_{out}}{i_{R_E} + i_e} = \left[R_E \,\middle\|\, \left(r_d + \frac{r_{bb'}}{(\beta + 1)} \right) \right]. \tag{4.11}$$

The output impedance of the transistor itself is just

$$r_{out} = r_d + \frac{r_{bb'}}{\beta + 1}.$$

A circuit which obeys the equations derived above and is easier to work with is shown in Fig. 4.27. Note that if $R_E \gg r_d$, then

$$r_{in} = (\beta + 1)R_E \quad \text{and} \quad A_v = 1.$$

These approximate equations are very useful in emitter-follower design. It should be remembered that the resistance r_v must often be included from point b' to ground in calculating the input impedance to the emitter follower. Since r_v is essentially in parallel with the input impedance of the circuit, it must be considered when r_{in} becomes large. Typically, r_v is 1 to 2 MΩ, and can be neglected when r_{in} is much smaller than this value.

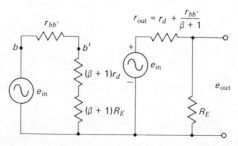

Fig. 4.27 Alternate emitter-follower equivalent circuit.

138 EQUIVALENT CIRCUITS AND MODELING 4.3

4.3 FET EQUIVALENT CIRCUITS

One of the first important applications of the FET was in amplifiers, especially for high input impedance stages. While the amplifier area continues to use the FET, this device now finds increasing usage in logic circuits.

A. Low-Frequency, Large-Signal FET Model

Although the FET characteristics show some similarity to bipolar transistor characteristics, there are three significant differences. The input or control parameter for the FET is voltage rather than current. Furthermore, when this input voltage is zero, an output current flow is present. The last significant difference between the two devices is the finite slope of the FET output characteristics to the left of the knee of the curves. The bipolar transistor characteristics can often be represented by a line of infinite slope below the knee of the curves. Fig. 4.28 shows the input and output characteristics for the FET.

Under normal operating conditions no signal current flows from gate to source in the FET. The gate-to-source resistance might vary from 1 MΩ to several

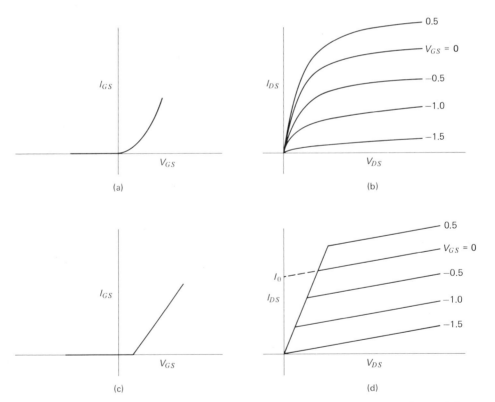

Fig. 4.28 Characteristics of an n-channel FET: (a) actual input characteristics; (b) actual output characteristics; (c) piecewise-linear approximation to input; (d) piecewise-linear approximation to output.

hundred megohms depending on the particular device used. The effect of V_{GS} on the current I_{DS} is indicated by the transconductance given by Eq. (2.52):

$$g_m = \frac{\partial I_{DS}}{\partial V_{GS}} = -\frac{3 I_{DS0}}{V_0}\left[1 - \left(\frac{V_{GS} - \phi}{V_0}\right)^{1/2}\right].$$

Variations in I_{DS} due to changes in V_{GS} can be represented by a voltage-dependent current source in the drain-to-source circuit.

Although we did not derive the expression for I_{DS} in the pinchoff region in Chapter 2, the characteristic curves offer additional information concerning the effect of V_{DS} on I_{DS}. The curves have a slight positive slope indicating that as V_{DS} increases so also does I_{DS}. This effect can be represented by a resistance from drain to source. This element typically ranges from 10 kΩ to 100 kΩ for the junction FET and the MOSFET.

The equivalent circuit of Fig. 4.29 will synthesize the characteristics of Fig. 4.28(c) and (d).

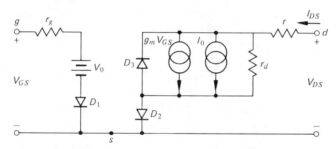

Fig. 4.29 Piecewise-linear model of the FET.

In general, the voltage from gate to source will be negative although for forward voltages less than V_0, little gate current will flow. The diode D_2 prevents negative current from flowing in the output circuit even if V_{GS} takes on a large negative value. This guarantees that the pinchoff behavior of the FET is satisfied by the model.

The characteristics of the output circuit can be determined by breaking the network into sections and considering each individual section. We will first demonstrate that the circuit of Fig. 4.30 satisfies the characteristics shown.

So long as input current is less than I_0, the diode is forward-biased, allowing no voltage drop across the current source. The entire applied voltage drops across the resistance r, which determines the slope of the curve. When the terminal voltage equals or exceeds $I_0 r$, the current through r equals I_0 and the diode becomes reverse-biased. Thus, the curve breaks downward to zero slope when the current reaches a value of I_0 as shown in Fig. 4.30(b). If a resistor of value r_d is placed in parallel with the current source, the value of slope above the break point is given by the reciprocal of R_{eq} where

$$R_{eq} = r + r_d.$$

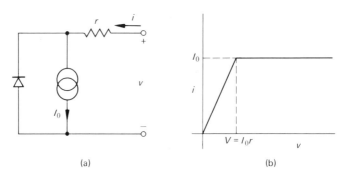

Fig. 4.30 (a) Simple circuit. (b) V–I characteristics.

Instead of zero slope occurring when the current reaches I_0, the slope will now be given by $1/(r + r_d)$.

Adding a current source $g_m V_{GS}$ in parallel to the constant source allows the entire family of curves to be generated. The model of Fig. 4.29 is seen to generate the characteristics of the FET and hence is an appropriate equivalent circuit.

Quite often, the input network can be approximated by an open circuit since the reverse-biased diode will exhibit a high impedance. This is always true for the MOSFET model and applies to the junction FET when the gate to source junction remains reverse-biased during circuit operation.

The piecewise linear model of the FET can be used in switching and large signal applications.

Example 4.4. Find the output waveform of the circuit in Fig. 4.31. Assume that $I_0 = 6$ mA, $g_m = 2$ m℧, $V_0 = 0.6$ V, $r = 500$ Ω, and $r_d = 40$ kΩ.

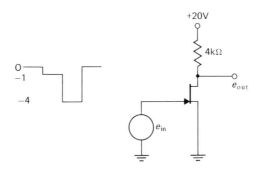

Fig. 4.31 FET stage for Example 4.4.

Solution. Using the model of Fig. 4.29 results in the equivalent circuit shown in Fig. 4.32. When e_{in} is at zero volts, the current through diode D_3 is 6 mA less the drain current. Since the load resistance limits the maximum possible drain current to something less than 6 mA, the diode must be forward-biased and can be

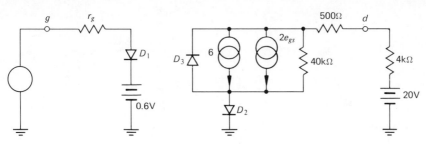

Fig. 4.32 Equivalent circuit for Example 4.4.

represented by a short circuit. Figure 4.33(a) indicates the resulting simplified equivalent circuit. The FET is saturated for this input voltage and the output voltage for $V_{GS} = 0$ V is found to be $V_{DS} = 2.22$ V.

Fig. 4.33(b) shows the circuit for $V_{GS} = -1$ V. In this case the net current generated by the two sources is $I_0 + g_m V_{GS} = 6 - 2 = 4$ mA. If D_3 were still a short circuit, the drain current would be given by $20/(4 + 0.5) = 4.44$ mA. Since this figure exceeds the 4 mA of the sources, the diode cannot be forward-biased. Taking a Thévenin equivalent of the 4 mA source and the 40 kΩ resistance leads to the circuit shown. The output voltage is now found to be 3.8 V.

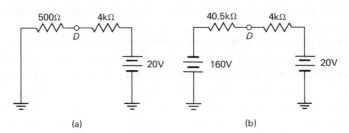

Fig. 4.33 Output circuit: (a) for $V_{GS} = 0$ V; (b) for $V_{GS} = -1$ V.

When e_{in} switches to -4 V, the net current generated by the sources would be $6 - 4g_m = -2$ mA. Diode D_2 now becomes reverse-biased, allowing no current to flow in the drain circuit. The output signal assumes a value of 20 V since the FET is cut off. When e_{in} swings back to 0 V, the output returns to 2.22 V.

B. Low-Frequency, Small-Signal FET Model

Figure 4.34 shows a simple FET amplifier with a bias supply along with the small-signal equivalent circuit. In order to use this model, the quantities g_m and r_{ds} must be given or evaluated from the characteristic curves. If the curves are used, these quantities should be evaluated near the quiescent operating point and are found from

$$g_m = \left. \frac{\Delta I_{DS}}{\Delta V_{GS}} \right|_{V_{DS}=\text{constant}} \tag{4.12}$$

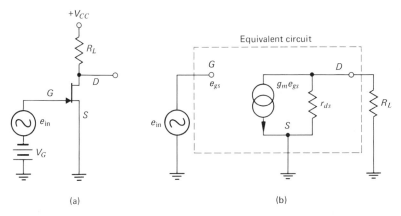

Fig. 4.34 FET amplifier: (a) actual circuit; (b) equivalent circuit.

and

$$r_{ds} = \frac{\Delta V_{DS}}{\Delta I_{DS}}\bigg|_{V_{GS}=\text{constant}} \quad (4.13)$$

It is emphasized that the circuit shown is an incremental or small-signal model valid only for incremental signals. It does not reflect dc levels, and all dc voltage sources such as V_G and V_{CC} are replaced with short circuits. DC calculations are made either graphically or from the piecewise-linear model discussed previously.

To demonstrate the interaction of dc and ac design, let us assume that the characteristics of Fig. 3.12 apply to the FET used and that $V_G = -1\,\text{V}, R_L = 2\,\text{k}\Omega$, and $V_{CC} = 20\,\text{V}$. The dc conditions can be established by drawing the load-line and locating the quiescent point. In this case $V_{DSQ} = 9\,\text{V}$ for $V_{GSQ} = -1\,\text{V}$. The transconductance at this point is estimated to be $g_m = 3.5\,\text{m}\mho$ and $r_{ds} = 20\,\text{k}\Omega$. From the equivalent circuit of Fig. 4.34(b) we can write

$$e_{in} = e_{gs}$$

$$e_{out} = -(g_m e_{gs})r_{ds}\|R_L = -g_m e_{in}\frac{r_{ds}R_L}{r_{ds}+R_L}.$$

The voltage gain is then given by

$$A_v = \frac{e_{out}}{e_{in}} = -g_m\frac{r_{ds}R_L}{r_{ds}+R_L}. \quad (4.14)$$

For the values specified the voltage gain is found to be

$$A_v = -3.5 \times 1.82 = -6.4.$$

If $e_{in} = 0.5\cos\omega t$, the ac output signal is $-6.4 \times 0.5\cos\omega t = -3.2\cos\omega t$. Combining the dc and ac signals gives

$$V_{GS} = V_G + e_{in} = -1 + 0.5\cos\omega t$$

and
$$V_{DS} = V_{DSQ} + e_{out} = 9 - 3.2 \cos \omega t.$$

The results developed here by the use of an equivalent circuit agree reasonably well with the graphical analysis. One might question the object of using the equivalent circuit when graphical methods can be applied so easily to this stage. There are several advantages of the analytical method using the equivalent circuit. Design using this method can be accomplished without the use of the V–I characteristics if dc levels are not critical. Typical values of g_m, r_{ds}, and V_P allow the equivalent circuit to be applied with sufficient accuracy for most design purposes. Multistage design is also accomplished more readily by analytical methods. A third advantage becomes apparent when the self-bias resistance is unbypassed. It is very difficult to use graphical means to calculate the voltage gain for this circuit. We will now develop an analytical expression for voltage gain when the self-bias resistance is unbypassed. Figure 4.35 shows this case. A Thévenin equivalent circuit of the current generator and r_{ds} is shown in Fig. 4.35(c). The output voltage is

$$e_{out} = \frac{-R_L}{R_L + r_{ds} + R_B} \mu e_{gs},$$

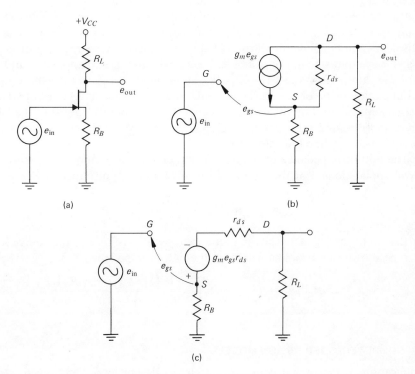

Fig. 4.35 FET amplifier with unbypassed R_B: (a) actual circuit; (b) equivalent circuit; (c) alternate equivalent circuit.

where $\mu = g_m r_{ds}$. In this circuit the gate-to-source voltage does not equal the input signal. We can write

$$e_{gs} = e_{in} - \frac{R_B}{R_L + r_{ds} + R_B} \mu e_{gs}.$$

Solving for e_{gs} gives

$$e_{gs} = e_{in} \frac{R_L + r_{ds} + R_B}{R_L + r_{ds} + (\mu + 1)R_B}.$$

Substituting this value of e_{gs} into the expression for e_{out} gives

$$A_v = \frac{e_{out}}{e_{in}} = \frac{-\mu R_L}{R_L + r_{ds} + (\mu + 1)R_B}. \tag{4.15}$$

As R_B approaches zero, Eq. (4.15) approaches the value given by Eq. (4.14) for the amplifier with no self-bias resistance.

The quantity μ is called the amplification factor and can be evaluated by

$$\mu = -\frac{\Delta V_{DS}}{\Delta V_{GS}}\bigg|_{I_{DS}=\text{constant}}. \tag{4.16}$$

More accuracy in evaluating μ can be obtained by using $\mu = g_m r_{ds}$.

It must be pointed out that the calculations for voltage gain assume that the circuit operates over a linear portion of the characteristic curves. The small-signal equivalent circuit includes no means of determining if the active region is exceeded for a given output signal. For example, a 2 V peak input signal applied to an FET of voltage gain -12 predicts a 24 V peak output signal. If $V_{CC} = 20$ V this output, of course, cannot be obtained. Thus, some idea of the active region boundaries must be kept in mind by the circuit designer. For computer analysis of the FET, a more appropriate circuit can be developed that behaves in the cutoff and saturation regions much like the actual FET.

The output impedance of the amplifier represents an important quantity in several applications. For the circuit of Fig. 4.34, the output impedance is

$$R_{out} = R_L \| r_{ds} = \frac{R_L r_{ds}}{R_L + r_{ds}}. \tag{4.17}$$

For the circuit of Fig. 4.35 showing an unbypassed resistance in the source lead, it can be shown (see Problem 4.46) that the output impedance is

$$R_{out} = R_L \| [r_{ds} + (\mu + 1)R_B] = \frac{R_L[r_{ds} + (\mu + 1)R_B]}{R_L + r_{ds} + (\mu + 1)R_B}. \tag{4.18}$$

*4.4 COMPUTER-AIDED DESIGN MODELS

Application of the digital computer to circuit design has, in certain instances, modified the approach taken to device modeling. For manual analysis the standard approach consists of applying small-signal equivalent circuits to linear problems and piecewise-linear circuits or graphical methods to nonlinear networks. With the tremendous capability of rapid calculation by the computer, it is possible to utilize analytical expressions for the device characteristics. The computer then applies some nonlinear solution method to calculate the network response. One very popular nonlinear transistor model, included in several computer-aided design programs, is the Ebers-Moll equivalent circuit discussed in the following section.

Nonlinear methods are not always appropriate for large-signal circuits even with the digital computer. Nonlinear problems often require excessive computer time and cost. It is appropriate in many instances to use piecewise-linear device models when possible. A reduction in computer cost is generally affected by this approach.

A. The Ebers-Moll Model [5]

The schematic representation of the minority-carrier distributions in a pnp transistor is given in Fig. 4.36.

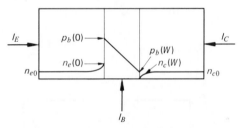

Fig. 4.36 Minority concentrations in a transistor.

We will assume an ideal transistor of uniform cross-sectional area and neglect recombination of carriers in the base region. These assumptions are unnecessary to derive the Ebers-Moll equations [3], but simplify the derivation considerably. All currents are assumed positive when flowing into the transistor to lead to conventional results.

The emitter current can be written in terms of the gradient of holes in the base region and the gradient of electrons at the right edge of the emitter. Thus,

$$I_E = \frac{AqD_n n_{e0}}{L_n}(e^{V_{EB}/\delta} - 1) + \frac{AqD_p}{W}[p_{b0}(e^{V_{EB}/\delta} - 1) - p_{b0}(e^{V_{CB}/\delta} - 1)],$$

where $\delta = kT/q$. The collector current can be expressed in like manner and is

$$I_C = \frac{AqD_n n_{c0}}{L_n}(e^{V_{CB}/\delta} - 1) + \frac{AqD_p}{W}[p_{b0}(e^{V_{CB}/\delta} - 1) - p_{b0}(e^{V_{EB}/\delta} - 1)].$$

Rearranging terms gives

$$I_E = Aq\left[\frac{D_p p_{b0}}{W} + \frac{D_n n_{eo}}{L_n}\right](e^{V_{EB}/\delta} - 1) - \frac{AqD_p p_{b0}}{W}(e^{V_{CB}/\delta} - 1) \quad (4.19)$$

and

$$I_C = \frac{-AqD_pp_{b0}}{W}(e^{V_{EB}/\delta} - 1) + Aq\left[\frac{D_pp_{b0}}{W} + \frac{D_nn_{c0}}{L_n}\right](e^{V_{CB}/\delta} - 1). \quad (4.20)$$

We can identify the term

$$Aq\left[\frac{D_pp_{b0}}{W} + \frac{D_nn_{e0}}{L_n}\right]$$

as the magnitude of emitter current that flows when the emitter-base junction is reverse-biased and the collector-base junction is shorted. This condition is depicted in Fig. 4.37. The emitter current that flows in this case is called I_{ES} and is

$$I_{ES} = -Aq\left[\frac{D_pp_{b0}}{W} + \frac{D_nn_{e0}}{L_n}\right]. \quad (4.21)$$

The collector current flowing when the collector-base junction is reverse-biased and the emitter is shorted to the base is

$$I_{CS} = -Aq\left[\frac{D_pp_{b0}}{W} + \frac{D_nn_{c0}}{L_n}\right]. \quad (4.22)$$

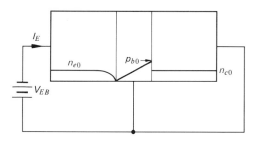

Fig. 4.37 Calculation of I_{ES}.

When the transistor operates in the normal active region, the current crossing the emitter-base junction is made up of two components. One results from holes injected from emitter to base which diffuse across the base region to the collector. The other component of current results from the electrons injected from base to emitter. Obviously, this component is not collected by the collector; thus, the ratio of collector to emitter current is lowered from the desired value of unity by this component. This explains why it is desirable to dope the emitter much more heavily than the base region.

The ratio of emitter current carried by holes (in a pnp transistor) to the total emitter current is called the emitter efficiency.

Assuming no recombination in the base region results in the emitter efficiency equaling normal short-circuit current gain from emitter to collector α_N. This value is

$$\alpha_N = \frac{\dfrac{D_p p_{b0}}{W}}{\dfrac{D_p p_{b0}}{W} + \dfrac{D_n n_{e0}}{L_n}}. \tag{4.23}$$

The short-circuit current gain from collector to emitter is called the inverted α and is

$$\alpha_I = \frac{\dfrac{D_p p_{b0}}{W}}{\dfrac{D_p p_{b0}}{W} + \dfrac{D_n n_{c0}}{L_n}}. \tag{4.24}$$

It is possible to construct transistors with equal values of α_N and α_I, but most modern transistors have lower values of α_I, typically near 0.5. Examining Eqs. (4.23) and (4.24) shows that

$$\alpha_N = \frac{\dfrac{AqD_p p_{b0}}{W}}{|I_{ES}|}$$

and

$$\alpha_I = \frac{\dfrac{AqD_p p_{b0}}{W}}{|I_{CS}|}$$

It follows that

$$\alpha_N |I_{ES}| = \alpha_I |I_{CS}|. \tag{4.25}$$

We can now rewrite Eqs. (4.19) and (4.20) in terms of I_{ES}, I_{CS}, α_N, and α_I. The results are

$$I_E = |I_{ES}|(e^{V_{EB}/\delta} - 1) - \alpha_I |I_{CS}|(e^{V_{CB}/\delta} - 1) \tag{4.26}$$

and

$$I_C = -\alpha_N |I_{ES}|(e^{V_{EB}/\delta} - 1) + |I_{CS}|(e^{V_{CB}/\delta} - 1). \tag{4.27}$$

Eqs. (4.25), (4.26), and (4.27) along with Kirchhoff's current law for the transistor

$$I_E + I_B + I_C = 0 \tag{4.28}$$

make up the well-known Ebers-Moll equations which can be applied to any possible conditions of the transistor. Before proceeding to a model for these equations let us briefly consider their utility in certain calculations.

It is easy to calculate I_{eo} and I_{co} from Eqs. (4.26) and (4.27). The quantity I_{eo} is the reverse-biased emitter current measured with the collector terminal open while I_{co} is the reverse-biased collector current with emitter open. For I_{eo} we assume that V_{EB} is large and negative, giving

$$I_{eo} = -|I_{ES}| - \alpha_I |I_{CS}|(e^{V_{CB}/\delta} - 1).$$

Since the collector terminal floats we do not immediately know the value assumed by V_{CB}. However, from Eq. (4.27) with I_C equated to zero we can solve for the quantity $(e^{V_{CB}/\delta} - 1)$. This results in

$$(e^{V_{CB}/\delta} - 1) = -\alpha_N \left|\frac{I_{ES}}{I_{CS}}\right|.$$

Substituting this value into the expression for emitter leakage current gives

$$I_{eo} = -|I_{ES}|(1 - \alpha_I \alpha_N). \tag{4.29}$$

It can likewise be shown that

$$I_{co} = -|I_{CS}|(1 - \alpha_I \alpha_N). \tag{4.30}$$

The floating potential of the emitter when evaluating I_{co} can be found from Eq. (4.27) by equating I_C to $-I_{co}$ and solving for V_{EB}. The result is

$$V_{EB} = \left(\frac{kT}{q}\right) \ln(1 - \alpha_N). \tag{4.31}$$

The model representing the Ebers-Moll equations is quite simple and is shown in Fig. 4.38. This model can be used in several well-known computer-aided design programs. The information derived from this model can be used in two ways by the computer. The program may generate response equations involving nonlinear terms and solve by some iterative method. Alternatively, the program itself may make a piecewise-linear approximation to the equations to simplify the response calculations. Programs such as SCEPTRE [3], ECAP II [2], and others will accept a description of the model in the form of a table. For example, the equations for I_F and I_R can be described to the computer as functions of V_{EB} and V_{CB}. Straight-line segments are then assumed to exist between adjacent data points.

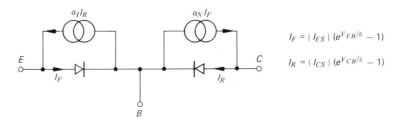

Fig. 4.38 Ebers-Moll model for pnp transistor.

Some programs store the Ebers-Moll model (NET−I) as the prime model to be used for transistors. Only certain parameters must then be specified for analysis of transistor circuit problems.

It must be further emphasized that a nonlinear model such as the Ebers-Moll model is not always the best choice for computer-aided design. While it is a simple matter for the designer to enter such a model, the computer may then make time-consuming calculations that are unnecessary. In certain large-signal problems, far simpler models may cut computer costs with sufficient accuracy for practical application. Examples of such models will be given in the following section.

B. Piecewise-Linear Circuit Models

In some computer-aided design programs, such as ECAP, models must be entered for each device used in a circuit. Both ECAP II and SCEPTRE offer the capability of storing several well-used models, thus eliminating the redundancy involved in ECAP. If a great deal of circuit design is performed in the area of switching circuits, perhaps a simple model involving an equivalent circuit that is either saturated or cut off can be stored in the computer and used for each transistor. While the models considered here will work with ECAP, many are appropriate for storage with SCEPTRE, ECAP II, or other programs offering this feature. For details of programming, a computer-aided design textbook can be consulted [1, 3].

Any static V–I characteristic of practical devices can be synthesized using constant-parameter elements. We have discussed two types of circuit that cause a break point in the V–I curve. In the case of parallel branches each consisting of a diode, a resistor, and a voltage source, the slope of the curve increases at each succeeding break point. Series branches consisting of a current source shunted by a diode and resistance cause the slope of the curve to decrease at the break point. These types of branch can be combined to synthesize more sophisticated V–I curves. Both types of branch are used in Fig. 4.39 to create the characteristics shown.

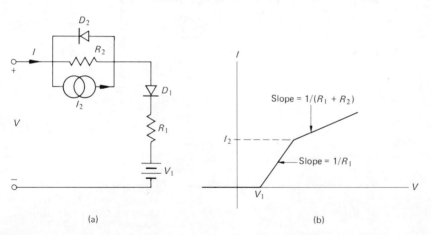

Fig. 4.39 (a) Circuit with increased slope in V–I curve after first break point and decreased slope after second break point. (b) V–I curve.

Diode D_1 will not allow current to flow until the input voltage exceeds V_1. Above this point until the current reaches a value of I_2, the slope is determined by R_1 since D_1 and D_2 now appear as a short circuit. At $I = I_2$, diode D_2 becomes

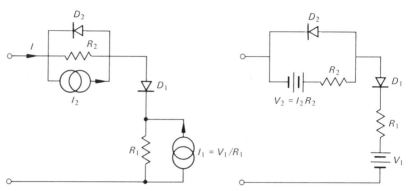

Fig. 4.40 Alternate circuits with V–I characteristics given by Fig. 4.39.

reverse-biased and the resistance of the shunt combination equals R_2. The slope of the curve above this point is determined by the sum of R_1 and R_2. The same characteristics can be realized by using current sources only or voltage sources only rather than by mixing these sources. These possibilities are shown in Fig. 4.40.

The basic branches can serve as building blocks for networks with very complex piecewise-linear characteristics. For example, the network of Fig. 4.41 possesses the rather unusual characteristic shown.

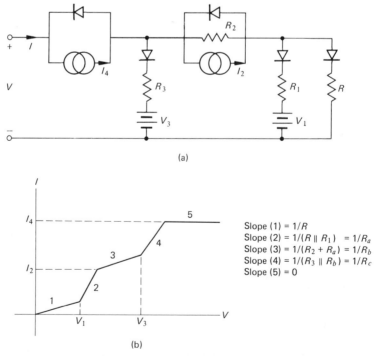

Slope (1) = $1/R$
Slope (2) = $1/(R \| R_1)$ = $1/R_a$
Slope (3) = $1/(R_2 + R_a)$ = $1/R_b$
Slope (4) = $1/(R_3 \| R_b)$ = $1/R_c$
Slope (5) = 0

Fig. 4.41 (a) Piecewise-linear network. (b) V–I characteristics.

It is possible to create negative slopes by specifying negative values of resistance. Most computer-aided design programs accept negative resistances as valid circuit elements. The tunnel diode characteristics of Fig. 4.42 are synthesized by the four-branch network shown.

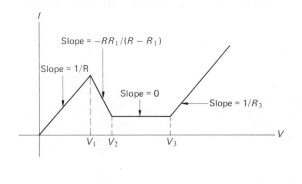

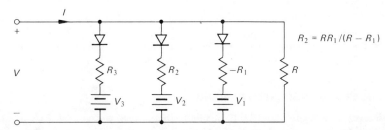

Fig. 4.42 (a) Piecewise-linear approximation of tunnel diode characteristics. (b) Tunnel diode model.

The entire V–I curve can be shifted in any direction by inserting a series voltage source or a shunt current source or both. Families of curves can be generated for devices such as transistors or vacuum tubes. Consider the situation shown in Fig. 4.43. An increase in E shifts the original curve horizontally to the

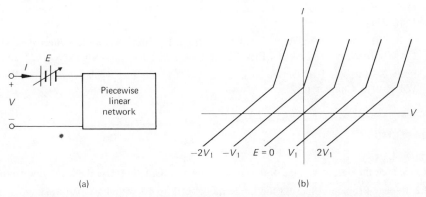

Fig. 4.43 (a) Piecewise-linear network with series voltage source. (b) V–I characteristics.

right while a decrease shifts the curve to the left. A family of curves can be generated with E as the parameter. The curve can be shifted vertically using the scheme shown in Fig. 4.44. Positive values of J shift the curve upward while negative values result in a downward shift. A combination series current source and voltage source can be used to shift the characteristics to any location of the V–I plane.

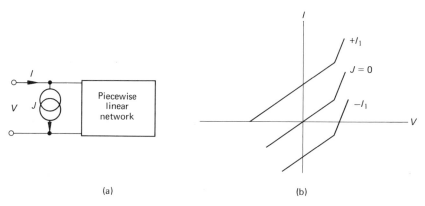

Fig. 4.44 (a) Piecewise-linear network with shunt current source. (b) V–I characteristics.

REFERENCES AND SUGGESTED READING

1. C. Belove, H. Schachter, and D. L. Schilling, *Digital and Analog Systems, Circuits, and Devices: An Introduction.* New York: McGraw-Hill, 1973, Chapters 4 and 5.
2. F. H. Branin, G. R. Hogsett, R. L. Lunde, L. E. Kugel, "ECAP II: an electronic circuit analysis program." *IEEE Spectrum* **8**, 6:14–25 (June 1971).
3. D. J. Comer, *Computer Analysis of Circuits.* Scranton, Penn.: International Textbook, 1971, Chapters 6 and 7.
4. D. J. Comer, *Introduction to Semiconductor Circuit Design.* Reading, Mass.: Addison-Wesley, 1968, Chapter 4.
5. D. T. Comer, *Large Signal Transistor Circuits.* Englewood Cliffs, N.J.: Prentice Hall, 1967, Chapter 5.
6. P. E. Gray, D. DeWitt, A. R. Boothroyd, and J. F. Gibbons, *Physical Electronics and Circuit Models of Transistors*, SEEC Vol. 2. New York: Wiley, 1964, Chapters 6, 7, and 8.
7. S. D. Senturia and B. D. Wedlock, *Electronic Circuits and Applications.* New York: Wiley, 1975, Chapter 11.

PROBLEMS

Section 4.1A

4.1 Sketch the output voltage of the circuit shown in the figure. Use the ideal diode model.

4.2 Repeat Problem 4.1 using the diode model of Fig. 4.2 with $V_0 = 0.4$ V.

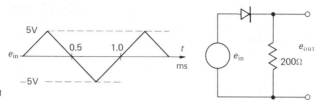

Problem 4.1

4.3 Repeat Problem 4.1 using the diode model of Fig. 4.3 with $V_0 = 0.4$ V and $R_f = 20\ \Omega$.

4.4 Repeat Problem 4.1 using the diode model of Fig. 4.4 with $V_0 = 0.2$ V, $V_1 = 0.6$ V, $R_{f1} = 80\ \Omega$, $R_{f2} = 20\ \Omega$.

* **4.5** In the circuit of Problem 4.1 the 200 Ω resistor is replaced by a 10 kΩ resistance. The input voltage is increased from a 5 V amplitude to a 50 V amplitude. How does the accuracy of the simple ideal diode model compare to the more complex models for this problem? Explain.

4.6 Using the diode model of Fig. 4.5(a) with $V_0 = 0.2$ V, $R_r = 2$ MΩ, and $R_f = 30\ \Omega$, sketch the output waveform for the circuit. (See the figure for Problem 4.6.) Now use the ideal diode model of Fig. 4.1 to find the output waveform. Compare results.

* **4.7** Repeat Problem 4.6 when the positions of the diode and the 10 kΩ resistor are interchanged using the ideal diode model only.

4.8 Repeat Problem 4.7 given that a 5-V dc source is inserted between points a and b with terminal a positive with respect to point b.

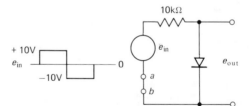

Problem 4.6

Section 4.1B

4.9 Evaluate the dynamic resistance of the diode at room temperature when the dc bias current is (a) 1 mA, (b) 2 mA, (c) 10 mA. If the series ohmic resistance is 2 Ω, what is the total small-signal resistance of the diode in each case?

* **4.10** The total small-signal diode resistance is measured to be 47 Ω at room temperature. If the ohmic resistance is 3 Ω, calculate the dc bias current.

4.11 Repeat Problem 4.10 if the measured resistance is 25 Ω.

Section 4.1C

4.12 If the diffusion capacitance of a diode is 500 pF at a dc current of 1 mA, find C_d for currents of (a) 2 mA, (b) 10 mA.

* **4.13** Evaluate the time constant ($\tau = r_d C_d$) for the diode of Problem 4.12 at 1 mA and 10 mA.

4.14 For the germanium diode shown in the figure $C_d = 5000$ pF. Assuming that the coupling capacitor appears as a short circuit to the input pulse, sketch the incremental output voltage as a function of time.

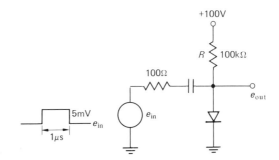

Problem 4.14

4.15 Repeat Problem 4.14 for $R = 50$ kΩ. Explain why the amplitude of the output voltage changes when R is changed.

Section 4.2A

*4.16 The transistor of the circuit can be modeled by the circuit of Fig. 4.11(c) with $r_b = 500$ Ω, $V_0 = 0.5$ V, $\beta = 100$, and $r_c = 40$ kΩ. Neglecting capacitive effects sketch the output waveform of the stage if $V_1 = 1$ V and $V_2 = 6$ V. (See the figure for Problem 4.16.)

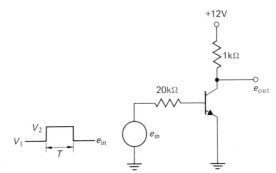

Problem 4.16

4.17 Repeat Problem 4.16 for $V_1 = -1$ V and $V_2 = 2$ V.

4.18 Repeat Problem 4.16 for $V_1 = 2$ V and $V_2 = 2.4$ V.

*4.19 Sketch the V–I characteristic for the circuit shown in the figure.

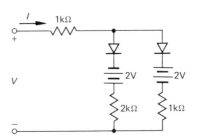

Problem 4.19

Section 4.2B

*4.20 Calculate the voltage gain of the circuit shown in the figure if $\alpha = 0.99$, $r_{bb'} = 200\ \Omega$, and $I_E = 2$ mA. Assume that C presents a short circuit for ac signals. Calculate the input impedance of the stage. What will be the maximum peak-to-peak output signal before cutoff is reached?

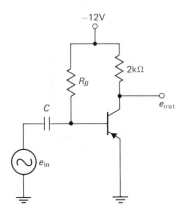

Problem 4.20

4.21 Repeat Problem 4.20 if I_E is changed to 1 mA.

4.22 If $r_{bb'} = 200\ \Omega$ and $\beta = \beta_0 = 100$ calculate the ac voltage gain and sketch the output voltage including the dc component. Assume that C presents a short circuit for ac signals. (See the figure for Problem 4.22.)

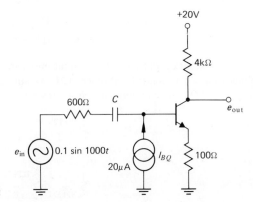

Problem 4.22

*4.23 If I_{BQ} is reduced to 5 µA in Problem 4.22 calculate the ac voltage gain. Compare to the gain when $I_{BQ} = 30$ µA.

4.24 Calculate the voltage gain of the circuit shown in the figure if $\alpha = 0.99$, $r_{bb'} = 200\ \Omega$, and $I_E = 2$ mA. Assume that C presents a short circuit for ac signals. Calculate the input and output impedances.

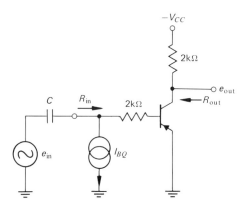

Problem 4.24

*4.25 Repeat Problem 4.24 if a 100 Ω resistance is inserted between emitter and ground.

4.26 Evaluate the voltage gain of the common-base stage shown in the figure. Assume that $\alpha = 0.98$, $r_{bb'} = 300\,\Omega$, and $r_d = 20\,\Omega$. Note that since $e_{in} = i_e(R_S + r_{in})$ and $e_{out} = i_c R_L$, the voltage gain can be expressed as

$$A_v = \frac{i_c R_L}{i_e(R_S + r_{in})} = \frac{\alpha R_L}{R_S + r_{in}} \approx \frac{R_L}{R_S}$$

for $R_S \gg r_{in}$. Compare this approximate value of gain to that calculated in the first part of the problem.

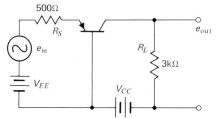

Problem 4.26

4.27 Calculate the quiescent output voltage, the input impedance from base to ground, and the ac voltage gain (see the figure for Problem 4.27). Assume that $V_{BE} = 0.5\,\text{V}$, $\beta_0 = \beta = 80$, and $r_{bb'} = 200\,\Omega$. The coupling capacitor presents a short circuit for frequencies of interest.

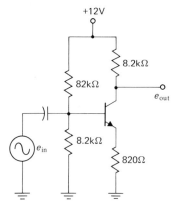

Problem 4.27

*4.28 Repeat Problem 4.27 when the 820-Ω emitter resistance is bypassed with a large capacitor.

4.29 Repeat Problem 4.27 when the 820-Ω emitter resistance is bypassed with a large capacitor and the 82-kΩ bias resistor is reduced to 60 kΩ.

4.30 Repeat Problem 4.27, given that a 10-kΩ resistor is placed in series with the input voltage source. Note that the bias resistors affect the voltage gain in this case.

*4.31 Select R_B in the figure to give the stage an ac voltage gain of -120. Assume $\beta_0 = \beta = 100$, $V_{BE} = 0.5$ V, and $r_{bb'} = 200\,\Omega$. What is the quiescent collector voltage?

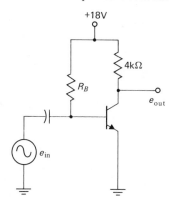

Problems 4.31 and 4.33

4.32 Repeat Problem 4.31, given that the load resistance is changed from 4 kΩ to 3 kΩ.

4.33 In the figure, using the transistor specifications of Problem 4.31, select R_B to give a quiescent output voltage of $+10$ V. What is the ac voltage gain of the circuit? What is the maximum sinusoidal output voltage that can appear without clipping occurring?

4.34 Repeat Problem 4.33 when the quiescent output voltage is required to be 4 V. Will the gain increase or decrease over the value found in Problem 4.33? Explain.

*4.35 Given that $\beta_0 = \beta = 80$, $V_{BE} = 0.5$ V, and $r_{bb'} = 100\,\Omega$, select R_E, R_1, and R_2 to give an ac voltage gain of -20 for the circuit in the figure. What is the stability factor of the stage? What is the maximum sinusoidal output voltage that can appear without clipping occurring?

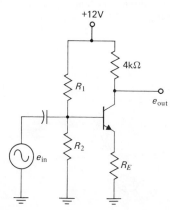

Problem 4.35

4.36 Repeat Problem 4.35 using a value of -40 for the required ac voltage gain. Evaluate the input and output impedances of the circuit.

158 EQUIVALENT CIRCUITS AND MODELING

4.37 Using a germanium transistor with $\beta_0 = \beta = 70$, $V_{BE} = 0.2$ V, $I_{co} = 2$ μA, and $r_{bb'} = 200$ Ω, design a common-emitter stage to have an ac voltage gain of -10 and a stability factor of $S_V \leq 10$ mV/°C. Assume that a 12-V power supply is to be used.

4.38 Using a value of -20 for the required ac voltage gain, repeat Problem 4.37.

*4.39 The silicon transistor in the circuit in the figure has $\beta_0 = \beta = 100$, $V_{BE} = -0.6$ V, and $r_{bb'} = 200$ Ω. Find the voltage stability S_V, the quiescent collector voltage, and the ac voltage gain of the stage. Assume that the capacitors appear as short circuits for all frequencies of interest, but are open circuits for dc voltages. Assuming that $r_{bb'} = 0$, calculate the ac voltage gain. Assuming $r_{bb'} + (\beta + 1)r_d = 0$, calculate the ac voltage gain. Compare the results to the value of gain calculated originally.

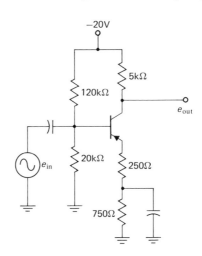

Problem 4.39

4.40 Repeat Problem 4.39, assuming that the silicon transistor is replaced by a germanium transistor with $\beta_0 = \beta = 90$, $V_{BE} = 0.2$ V, $I_{co} = 2$ μA, and $r_{bb'} = 150$ Ω.

*4.41 Calculate the quiescent output voltage of the circuit shown in the figure. Evaluate the ac voltage gain, the input impedance, and the output impedance. Use $\beta_0 = \beta = 100$, $r_{bb'} = 100$ Ω, and $V_{BE} = 0.6$ V. Assume that C is a short circuit to frequencies of interest.

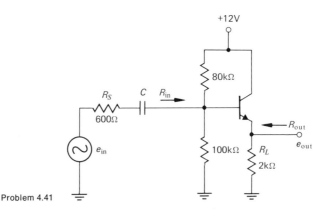

Problem 4.41

4.42 Repeat Problem 4.41 if R_L is changed to 4 kΩ.

Section 4.3A

4.43 In Example 4.4 calculate the gate voltage that brings the FET to the edge of the cutoff region.

4.44 Rework Example 4.4 if the load resistance is changed to 6 kΩ.

Section 4.3B

*__4.45__ Calculate the voltage gain and output impedance of the circuit of Fig. 4.34 if $R_L = 6$ kΩ, $g_m = 5$ m℧, and $r_{ds} = 50$ kΩ. Comment on the current gain of the stage.

4.46 Derive Eq. (4.18).

*__4.47__ Calculate the voltage gain and output impedance of the circuit of Fig. 4.35 if $R_L = 6$ kΩ, $g_m = 5$ m℧, $r_{ds} = 50$ kΩ, and $R_B = 750$ Ω.

Section 4.4A

4.48 Derive an expression for V_{EB} in terms of I_E, I_C, I_{eo}, α_I, and δ.

4.49 Derive an expression for V_{CB} in terms of I_E, I_C, I_{eo}, α_N, and δ.

4.50 Derive an expression for $V_{EC(\text{sat})}$ for a saturated pnp transistor. Calculate this value at room temperature if $\alpha_I = 0.2$, $\alpha_N = 0.99$, $I_E = 2$ mA, $I_C = -1.5$ mA, $I_{eo} = I_{co} = -0.1$ μA.

Section 4.4B

*__4.51__ Sketch the V–I curve generated by the network in the figure.

4.52 Sketch the V–I curve generated by the network in the figure.

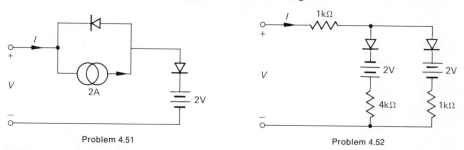

Problem 4.51   Problem 4.52

*__4.53__ Synthesize a circuit to realize the characteristics shown in the figure.

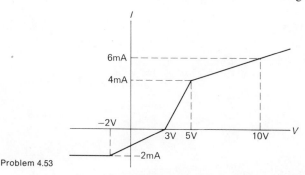

Problem 4.53

4.54 Synthesize a circuit to realize the characteristics shown in the figure. This curve represents an approximation to the tunnel diode characteristics.

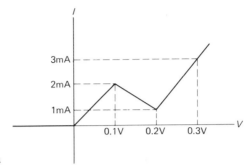

Problem 4.54

4.55 Synthesize a circuit to realize the characteristics shown in the figure.

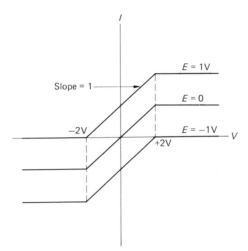

Problem 4.55

LOW-FREQUENCY AMPLIFIERS

The term "low frequency" can be misleading when applied to semiconductor amplifiers. There is no specific frequency range to which this term refers in all cases. Rather, the term refers to the frequency range that can be amplified with negligible reactive effects being introduced by the transistors of the amplifier. Since the current gain-bandwidth of many modern transistors can exceed 1000 MHz, low-frequency design could apply to a 20 MHz amplifier. At the other extreme are power stages which might allow low-frequency analysis to apply only up to a frequency of 20 kHz. This chapter will consider amplifiers that operate well within the range of frequencies over which reactive effects of the amplifying devices can be neglected.

Design of discrete amplifiers will first be considered. Although several principles of discrete amplifier design apply to integrated amplifier design, there are additional unique considerations that must be made in fabricating integrated amplifiers. An introduction to the integrated amplifier will be given in this chapter, while Chapter 6 will cover more advanced aspects of a popular integrated circuit: the operational amplifier.

5.1 VOLTAGE AMPLIFIERS

There are three basic types of amplifier: voltage, current, and power. While a given amplifier may exhibit gain of all three quantities, amplification of one particular quantity is often the desired result. The emitter follower is an example of a single-stage current amplifier. While this stage exhibits no voltage amplification (approximately unity voltage gain), there is an associated power gain. Current amplifiers are often used to couple high impedance sources to low or medium impedance loads. Coupling a photomultiplier tube to a 1 kΩ load is a typical application of a current amplifier.

Power amplifiers are used when an appreciable amount of power gain is required. Public address systems, hi-fi systems, and power transmitters are examples of systems requiring a power amplifier to boost the power of the signal to an acceptable output level. Special stages are often used in such applications and these circuits will be discussed in Chapter 12. The load impedance of the power stage is generally quite small.

Voltage amplifiers are the most universally used of the three types. Generally even power amplifier stages will be preceded by voltage amplification stages, sometimes called preamps. Common-base or common-emitter stages often are included in voltage amplifiers. Most preamps will exhibit current and power gain also, but the primary quantity of interest is voltage or voltage gain. The principles of voltage amplification will be emphasized in the following sections.

A. Impedance Considerations

In a voltage amplifier one is concerned with obtaining maximum voltage transfer from source to load. To achieve this, an ideal voltage amplifier would have an infinite input impedance (to transfer maximum voltage from source to amplifier input) and zero output impedance (to transfer maximum voltage from amplifier output to load). Usually, a sacrifice in amplifier gain is necessary to approach these ideal impedance conditions. Therefore a compromise between gain and impedance levels will often be required in order to deliver maximum voltage to the load.

Consider the ideal amplifier (infinite input and zero output impedances) shown in Fig. 5.1. The output voltage is given by $e_{out} = Ae_{in}$. The total applied voltage e_{in} reaches the input of the amplifier, since no drop is present across R_S. The output voltage e_{out} is applied directly across the load. In the practical case, the amplifier's input impedance will be finite and the output impedance will be nonzero. The effect of these nonideal impedances can be accounted for by adding an input resistance and an output resistance to the ideal stage, as shown in Fig. 5.2.

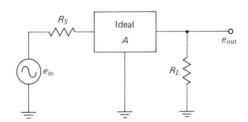

Fig. 5.1 Ideal amplifier.

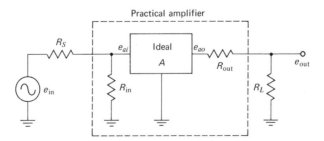

Fig. 5.2 Practical amplifier.

The practical amplifier is then made up of R_{in}, R_{out}, and an ideal amplifier of gain A. The total applied input voltage will not reach the amplifier input, due to the drop across R_S. The output voltage of the amplifier is not applied directly to R_L; rather, it is applied to R_{out} in series with R_L. The overall voltage gain of the circuit is defined as

$$A_o = e_{out}/e_{in}.$$

From Fig. 5.2, it is obvious that

1. $e_{ai} = \dfrac{R_{in}}{R_S + R_{in}} e_{in}$,

2. $\dfrac{e_{ao}}{e_{ai}} = A$,

3. $e_{out} = \dfrac{R_L}{R_L + R_{out}} e_{ao}$.

Combining these equations to find A_o results in

$$A_o = A \left(\frac{R_{in}}{R_{in} + R_S} \right) \left(\frac{R_L}{R_L + R_{out}} \right). \tag{5.1}$$

The first term in parentheses accounts for the loading of the input voltage caused by R_{in}. The second term accounts for the portion of output voltage that drops across R_{out}. Both terms are always less than or equal to unity. We note that the gain of the ideal amplifier A could be measured by applying a voltage e_{ai} to the input and measuring the output voltage by removing the load R_L. With no load there will be no drop across the output impedance of the amplifier and thus the output voltage will equal e_{ao} in this instance. This provides a useful method of evaluating the gain A of the stage. In fact, this value is often called the unloaded gain to emphasize that it corresponds to the gain measured after short-circuiting R_S and open-circuiting the load.

One might think that the problem of designing an amplifier is only slightly complicated by the presence of R_{in} and R_{out}. After all, Eq. (5.1) reflects these effects in a very simple equation. Unfortunately, the situation is more involved than Eq. (5.1) indicates for transistor stages because:

1. R_{in} depends on the biasing network in addition to transistor parameters;
2. the gain depends on both R_{in} and R_{out};
3. provisions for temperature stability can affect both R_{in} and R_{out}; and
4. in ac design, capacitors will be present, adding frequency effects to the circuit.

Example 5.1 will demonstrate some of the above points.

Example 5.1. Reduce the circuit of Fig. 5.3 to the form shown in Fig. 5.2, showing values of R_{in}, R_{out}, and A. Evaluate the overall gain, assuming that $\beta = 100$, $r_{bb'} = 200\ \Omega$, and $r_d = 30\ \Omega$. Assume that the capacitors are short circuits for ac signals.

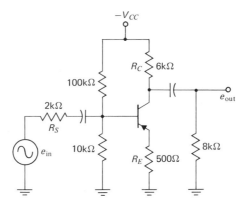

Fig. 5.3 Common-emitter stage for Example 5.1.

Solution. The small-signal equivalent circuit for this stage is shown in Fig. 5.4. Note that both biasing resistors connect to ac ground, since the dc supply is a short circuit for ac signals.

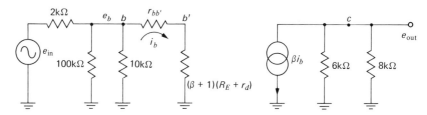

Fig. 5.4 Equivalent circuit of stage shown in Fig. 5.3.

The input impedance to the amplifier will consist of the combination of the three resistances connected from point b to ground:

$$R_{in} = (100 \text{ k}\Omega) \| (10 \text{ k}\Omega) \| r_{in},$$

where r_{in} is equal to $r_{bb'} + (\beta + 1)(r_d + R_E) = 53.7 \text{ k}\Omega$. The value of R_{in} is found to be $7.77 \text{ k}\Omega \simeq 7.8 \text{ k}\Omega$.

The ideal gain A can be found by assuming R_S is a short circuit and R_L is removed, giving

$$A = \frac{-\beta R_C}{r_{bb'} + (\beta + 1)(R_E + r_d)}.$$

We want to find the overall gain of the stage e_{out}/e_{in}; thus, we relate e_b to e_{in} and e_{out} to e_{ao}, giving

$$\frac{e_b}{e_{in}} = \frac{e_{ai}}{e_{in}} = \frac{R_{in}}{R_{in} + R_S}$$

and
$$\frac{e_{\text{out}}}{e_{ao}} = \frac{R_L}{R_C + R_L}.$$

We have used the fact here that $R_C = R_{\text{out}}$. The overall gain is

$$A_o = \frac{e_{ai}}{e_{\text{in}}} \frac{e_{ao}}{e_{ai}} \frac{e_{\text{out}}}{e_{ao}}$$

$$= -\frac{\beta R_C}{r_{bb'} + (\beta + 1)(R_E + r_d)} \frac{R_L}{R_C + R_L} \frac{R_{\text{in}}}{R_{\text{in}} + R_S}.$$

Using the values given, we determine a gain of

$$A_o = -\frac{100 \times 6000}{200 + 101 \times 530} \frac{8000}{6000 + 8000} \frac{7800}{7800 + 2000} = -5.1.$$

There is a second method of calculating overall gain that is particularly useful for transistor amplifiers. We can write the overall gain as

$$A_o = \frac{e_b}{e_{\text{in}}} \frac{e_{\text{out}}}{e_b}.$$

We have evaluated e_b/e_{in} previously. The quantity e_{out}/e_b is identified as the base to collector voltage gain with R_L present. The effective collector load consists of the parallel combination of R_C and R_L. The gain from base to collector is

$$\frac{e_{\text{out}}}{e_b} = -\frac{\beta R_{\text{eff}}}{r_{bb'} + (\beta + 1)(R_E + r_d)},$$

where

$$R_{\text{eff}} = \frac{R_C R_L}{R_C + R_L}.$$

The overall gain is

$$A_o = -\frac{\beta R_{\text{eff}}}{r_{bb'} + (\beta + 1)(R_E + r_d)} \frac{R_{\text{in}}}{R_{\text{in}} + R_S}. \tag{5.2}$$

Note that this equation can be put in the same form as Eq. (5.1) by writing R_{eff} in terms of R_C and R_L to give

$$A_o = -\frac{\beta R_C}{r_{bb'} + (\beta + 1)(R_E + r_d)} \frac{R_L}{R_C + R_L} \frac{R_{\text{in}}}{R_{\text{in}} + R_S}.$$

Since $R_{\text{out}} = R_C$ for this stage, this form corresponds to that of Eq. (5.1).

Equation (5.2) can be physically interpreted as follows.

1. If $R_S = 0$ (short circuit) and $R_L = \infty$ (open circuit), the gain is given by

$$A = -\frac{\beta R_C}{r_{bb'} + (\beta + 1)(R_E + r_d)} = -11.2.$$

2. The presence of R_L reduces the effective collector load or causes a drop across the output resistance, thereby reducing the gain. This effect is accounted for by the term $R_L/(R_C + R_L)$.
3. The presence of R_S further reduces the gain, since the input voltage is attenuated before it reaches the base. This effect is accounted for by the term $R_{in}/(R_{in} + R_S)$.

Before leaving this example, let us examine some of the possible interactions of element values with Eq. (5.1) or (5.2). If we want to increase the ideal gain A, we can decrease the emitter resistance R_E. However, this will not only increase A, but it will also decrease R_{in}. If R_C is lowered, the term $R_L/(R_L + R_C)$ increases, but the gain A decreases. The interactions extend to the dc design also. For example, a change in dc voltage stability can be effected by changing the bias resistors which, in turn, changes the value of R_{in}. We will see later that the low-frequency cutoff point is also a function of R_{in}. It is this interaction between impedance, stability, frequency response, voltage gain, and bias point that makes transistor design difficult.

B. Multistage Theory

The previous theory can be extended to multistage amplifiers such as that shown in Fig. 5.5.

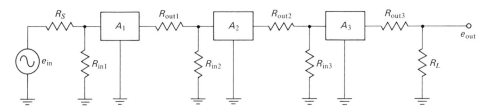

Fig. 5.5 Three-stage amplifier.

The expression for overall gain is

$$A_o = A_1 A_2 A_3 \frac{R_{in\,1}}{R_{in\,1} + R_S} \frac{R_{in\,2}}{R_{in\,2} + R_{out\,1}} \frac{R_{in\,3}}{R_{in\,3} + R_{out\,2}} \frac{R_L}{R_L + R_{out\,3}}, \quad (5.3)$$

where A_1, A_2, and A_3 are the unloaded gains of the individual stages, that is, the gains from input to output of the amplifier when no loads are applied to the output. With transistor stages, the output impedance of one stage is easily combined with the input impedance of the next stage to give an alternate form, as shown in Fig. 5.6.

The values for effective resistances at each collector are

$$R_{1\,eff} = R_{out\,1} \| R_{in\,2},$$
$$R_{2\,eff} = R_{out\,2} \| R_{in\,3},$$
$$R_{3\,eff} = R_{out\,3} \| R_L.$$

5.1 VOLTAGE AMPLIFIERS

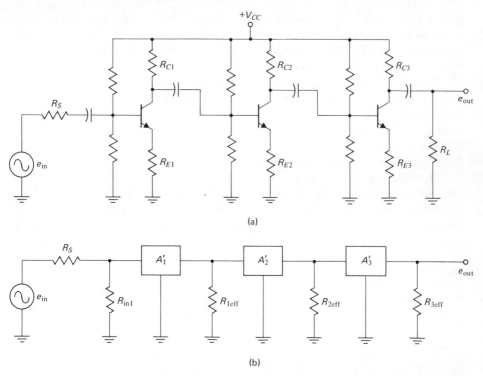

Fig. 5.6 Three-stage transistor amplifier: (a) actual three-stage amplifier; (b) reduced stages for calculations.

The gains A'_1, A'_2, and A'_3 now represent base-to-collector gains when the above values of collector loads are present. We see then that the input impedance to a stage appears as part of the collector load of the preceding stage. Of course, this lowers the gain of the preceding stage. Equation (5.4) expresses the overall gain of the amplifier in a more convenient form than that of Eq. (5.3). The gain is

$$A_o = A'_1 A'_2 A'_3 \frac{R_{\text{in 1}}}{R_{\text{in 1}} + R_S}. \tag{5.4}$$

For the circuit of Fig. 5.6, the gains A'_1, A'_2, and A'_3 are

$$A'_1 = -\frac{\beta_1 R_{1\text{ eff}}}{r_{bb'_1} + (\beta_1 + 1)(r_{d_1} + R_{E_1})},$$

$$A'_2 = -\frac{\beta_2 R_{2\text{ eff}}}{r_{bb'_2} + (\beta_2 + 1)(r_{d_2} + R_{E_2})},$$

$$A'_3 = -\frac{\beta_3 R_{3\text{ eff}}}{r_{bb'_3} + (\beta_3 + 1)(r_{d_3} + R_{E_3})}.$$

It should be obvious that Eq. (5.4) can be put into the form of Eq. (5.3) with very little manipulation. The difference between the two equations is simply that the primed values of gain have absorbed the loading effects of succeeding stages, while the unprimed values have not. Equation (5.3) accounts for these loading effects with the additional terms. Some examples given later in the chapter will demonstrate the application of these equations.

The only stages thus far discussed for voltage amplifiers are common-emitter stages. Other configurations can be used and, in fact, the common-collector or emitter-follower stage is often included in the practical amplifier. While this stage exhibits no voltage gain, a larger overall amplifier gain can result when the emitter follower is added. The loading of one stage on another is minimized to increase the overall gain. This fact will be demonstrated in a later section.

The FET has one important design advantage over the bipolar transistor in that input impedance is very large for the FET. Multistage FET amplifiers require very little attention to loading effects as a result of the input impedance level. When ac coupling is used between stages, a gate resistance is used to supply a ground reference for the gate, as shown in the amplifier of Fig. 5.7.

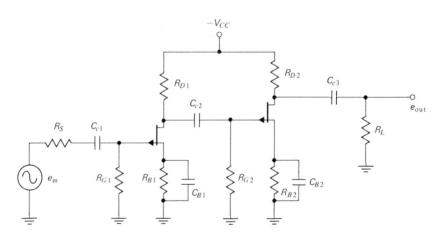

Fig. 5.7 A two-stage FET amplifier.

Applying Eq. (5.3) allows us to write the overall gain as

$$A_o = g_{m1} \frac{r_{ds1} R_{D1}}{r_{ds1} + R_{D1}} g_{m2} \frac{r_{ds2} R_{D2}}{r_{ds2} + R_{D2}} \frac{R_L}{R_L + R_{out\,2}},$$

where

$$R_{out\,2} = \frac{r_{ds2} R_{D2}}{r_{ds2} + R_{D2}}.$$

Normally the attenuation factors

$$\frac{R_{G1}}{R_{G1} + R_S}$$

and
$$\frac{R_{G2}}{R_{G2} + R_{\text{out 1}}}$$
would be included in the gain expression; however, the gate resistors for the FET are generally selected to be in the megohm range. This results in unity attenuation factors and leads to the simpler overall gain expression.

There are occasions when hybrid FET-bipolar transistor circuits are advantageous. Fig. 5.8 shows a high input impedance amplifier with higher overall gain than could be achieved using only FET stages.

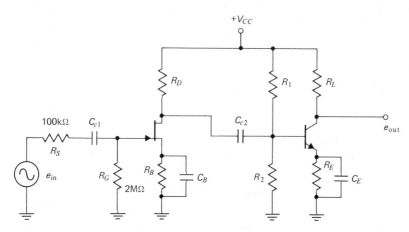

Fig. 5.8 A hybrid amplifier with both high overall gain and high input impedance.

In this amplifier, input loading and loading between stages must be considered, giving an overall gain expression of
$$A_o = g_m \frac{r_{ds}R_D}{r_{ds} + R_D} \frac{\beta R_L}{r_{bb'} + r_{b'e}} \frac{R_G}{R_G + R_S} \frac{R_{\text{in 2}}}{R_{\text{in 2}} + R_{\text{out 1}}},$$
where
$$R_{\text{in 2}} = R_1 \| R_2 \| (r_{bb'} + r_{b'e})$$
and
$$R_{\text{out 1}} = r_{ds} \| R_D.$$

5.2 LOW FREQUENCY FALLOFF DUE TO COUPLING CAPACITOR

In the preceding discussion we assumed that the coupling capacitors appeared as short circuits to all frequencies of interest. At low frequencies this assumption can no longer be justified, since the impedance of a capacitor becomes infinite at dc or zero frequency. As the impedance becomes larger at low frequencies, a greater

170 LOW-FREQUENCY AMPLIFIERS

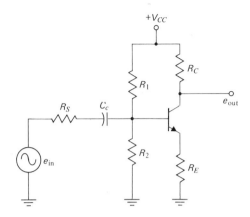

Fig. 5.9 Amplifier with coupling capacitor.

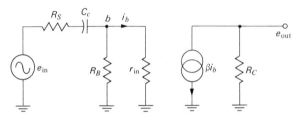

Fig. 5.10 Equivalent circuit of Fig. 5.9.

voltage drop occurs across the capacitor. This causes the voltage that reaches the transistor to become less, reducing the overall gain. Consider the circuit of Fig. 5.9. The equivalent circuit is shown in Fig. 5.10.

The overall voltage gain is again found by considering the base-to-collector gain and the relation of base voltage to e_{in}, that is,

$$e_b = \frac{R_{in}}{R_{in} + R_S + 1/j\omega C_c} e_{in}$$

and

$$A_{bc} = -\frac{\beta R_C}{r_{bb'} + (\beta + 1)(R_E + r_d)} = -\frac{\beta R_C}{r_{in}}.$$

These expressions are combined to give

$$A_o = \frac{R_{in}}{R_{in} + R_S + 1/j\omega C_c} A_{bc}. \tag{5.5}$$

As ω becomes smaller, the denominator of the above equation becomes larger. Physically, this is due to the fact that the impedance of the capacitor increases, dropping more of the applied voltage across C_c. The overall gain decreases at lower values of ω, approaching a value of zero when ω approaches zero. As ω increases, the gain increases to a point. At higher values of frequency, the term $1/j\omega C_c$ becomes very small, compared to $R_{in} + R_S$, and the expression for gain becomes

$$A_o = \frac{R_{in}}{R_{in} + R_S} A_{bc}$$

for all higher values of ω. In the earlier parts of this chapter, we always assumed that the input frequency was high enough to use the above expression for gain.

The magnitude of the overall gain is the quantity that must be considered in order to predict the magnitude of output voltage for a given input signal. The magnitude of A_o is

$$|A_o| = R_{in} \frac{|A_{bc}|}{\sqrt{(R_{in} + R_S)^2 + 1/\omega^2 C_c^2}}$$

$$= \frac{|A_{bc}| R_{in}}{(R_{in} + R_S)} \frac{1}{\sqrt{1 + 1/\omega^2 C_c^2 (R_{in} + R_S)^2}}. \quad (5.6)$$

The magnitude of the denominator was found by taking the square root of the sum of the squares of the real and imaginary parts. Since ω varies as we vary the input frequency, the magnitude of A_o will also change. A plot of the magnitude of the gain as a function of frequency is shown in Fig. 5.11.

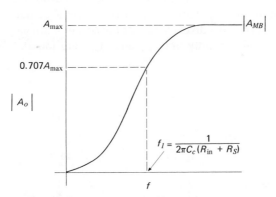

Fig. 5.11 Low-frequency falloff of gain due to C_c.

The frequency at which $|A_o|$ has dropped to 0.707 of the maximum value of $|A_o|$ is a key point. This frequency is called the 0.707-frequency, the 3-db frequency, or the corner frequency, and it will be denoted by f_1. At frequencies slightly above f_1, the gain reaches its maximum value, while below f_1, the magnitude of the gain steadily falls toward zero. We can determine f_1 by considering Eq. (5.6). At high frequencies, the gain is maximum and can be written

$$|A_o| = \frac{|A_{bc}| R_{in}}{R_{in} + R_S} = A_{max}.$$

As ω is lowered, the square-root term becomes larger, and at ω_1 the following is true:

$$1 + \frac{1}{\omega_1^2 C_c^2 (R_{in} + R_S)^2} = 2.$$

When this is true, $|A_o| = 0.707\,|A_{bc}|\,R_{in}/(R_{in} + R_S)$. The above equation can be solved for ω_1 to give

$$\omega_1 = \frac{1}{C_c(R_{in} + R_S)}, \tag{5.7}$$

and

$$f_1 = \frac{\omega_1}{2\pi} = \frac{1}{2\pi C_c(R_{in} + R_S)}. \tag{5.8}$$

Usually, the magnitude of the gain is plotted in decibels, so that

$$A_{db} = 20 \log_{10} |A_o|. \tag{5.9}$$

A db-plot of $|A_o|$ is shown in Fig. 5.12. In the decibel plot the falloff approaches a constant rate of 6 db per octave or 20 db per decade. The db-plot can be approximated by a straight-line segment of zero slope for frequencies above f_1 and a sloped line for frequencies below f_1. Only at frequencies approaching f_1 does the curve depart from the straight-line approximations. The dashed-line segments in Fig. 5.12 make up the asymptotic plot. This plot approximates the actual plot quite closely, with the maximum error of 3 db occurring at the corner frequency. The asymptotic plot can be made quite rapidly once f_1 and A_{max} are found. Above the corner

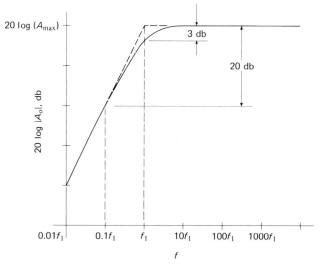

Fig. 5.12 Plot of A_{db} as a function of frequency.

frequency, the function has zero slope and a value of 20 log A_{max}. Below the corner frequency, the slope is 6 db per octave; that is, $|A_o|$ is halved each time the frequency is halved. The coupling capacitor, along with R_S and R_{in}, determine the corner frequency; therefore, the designer controls this value by the selection of C_c.

5.3 FREQUENCY RESPONSE

Since the frequency response and db-graph are so important in amplifier design, some of the major points in constructing these graphs will be developed in this section. Three forms of gain equations are of most interest for amplifier design. We shall analyze three passive networks which obey the equations of interest, and eventually relate the equations to the amplifier. Figure 5.13 shows the networks of interest.

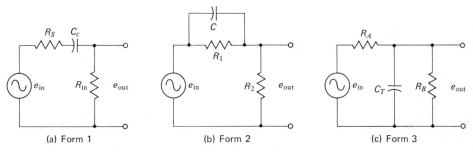

Fig. 5.13 Three basic forms.

FORM 1. This corresponds to the case of the coupling capacitor in the input circuit of an amplifier, as was previously discussed. The equation for e_{out}/e_{in} is found by noting that

$$e_{out} = \frac{R_{in}}{R_{in} + R_S + 1/j\omega C_c} e_{in}.$$

The transmission factor is then

$$G = \frac{R_{in}}{R_{in} + R_S + 1/j\omega C}.$$

The magnitude of G is

$$|G| = \frac{R_{in}}{R_{in} + R_S} \frac{1}{\sqrt{1 + 1/\omega^2 C_c^2 (R_{in} + R_S)^2}} = \frac{R_{in}}{R_{in} + R_S} \frac{1}{\sqrt{1 + f_1^2/f^2}}. \qquad (5.10)$$

To make a decibel plot we shall express $|G|$ in db,

$$G_{db} = 20 \log |G| = 20 \log \frac{R_{in}}{R_{in} + R_S} - 20 \log \sqrt{1 + f_1^2/f^2}.$$

The two log factors can be plotted separately and then added to arrive at a plot of G_{db}. The first term is a constant and is shown in Fig. 5.14(a). The second term is also constant at zero db for values of f much greater than f_1. The term reduces to

$$-20 \log \sqrt{1} = 0.$$

At very low frequencies, where f is much smaller than f_1, the term becomes

$$-20 \log \sqrt{1 + f_1^2/f^2} = -20 \log (f_1/f) = -20 \log f_1 + 20 \log f.$$

This term rises at 6 db per octave due to the $20 \log f$ term. When $f = f_1$, the term becomes

$$-20 \log \sqrt{2} = -3 \text{ db}.$$

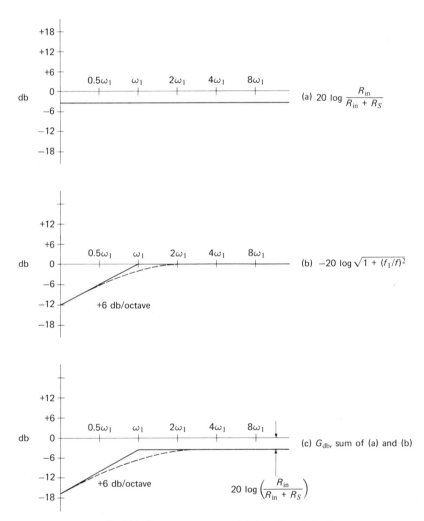

Fig. 5.14 Frequency-response plots for a Form-1 circuit.

The term $-20 \log \sqrt{1 + f_1^2/f^2}$ is plotted in Fig. 5.14(b). The sum of the two terms is shown in part (c). The key point on the curve occurs at $f = f_1$. This is the point at which the asymptotic plot no longer rises and has reached its maximum value. To make the asymptotic plot for a Form-1 circuit, we need only evaluate G_{max} and f_1 and note that below f_1, the function has a +6 db per octave slope.

Example 5.2. Evaluate the key quantities of the frequency response plot for the circuit of Fig. 5.13(a). Assume that $R_{in} = 10$ kΩ, $R_S = 5$ kΩ, and $C_c = 0.01$ μF.

Solution. The corner frequency is found from Eq. (5.8) to be

$$f = \frac{1}{2\pi \times 10^{-8} \times 1.5 \times 10^4} = 1.06 \text{ kHz.}$$

The maximum value of G_{db} is

$$G_{db} = 20 \log \frac{R_{in}}{R_{in} + R_S} = 20 \log \tfrac{2}{3} = -3.52 \text{ db.}$$

We can consider the Form-1 circuit on a more physical basis to verify the correctness of the equations relating to the frequency response. At high frequencies, the capacitor will appear as a short circuit and, by inspection of the circuit, we see that

$$e_{out} = \frac{R_{in}}{R_{in} + R_S} e_{in} \quad \text{or} \quad G_{max} = \frac{R_{in}}{R_{in} + R_S}.$$

The current through the circuit is also maximum for these frequencies. As the frequency is lowered, a point will be reached when the impedance of the capacitor equals the total resistance of the loop $R_S + R_{in}$. The magnitude of current is then

$$i = \left| \frac{e_{in}}{(R_{in} + R_S) - j(R_{in} + R_S)} \right| = \frac{0.707 e_{in}}{(R_{in} + R_S)}.$$

The voltage across R_{in} will now equal 0.707 of the maximum value. We conclude then that when

$$X_C = R_S + R_{in},$$

the frequency must equal the corner frequency. This condition amounts to

$$\frac{1}{\omega_1 C_c} = R_S + R_{in} \quad \text{or} \quad f_1 = \frac{1}{2\pi C_c(R_S + R_{in})}.$$

FORM 2. The expression for G of the Form-2 circuit is

$$G = \frac{R_2}{R_2 + R_1 || 1/j\omega C} = \frac{R_2(1 + j\omega C R_1)}{R_2 + R_1 + j\omega C R_1 R_2}.$$

The above equation can be written in a more appropriate form,

$$G = \frac{R_2}{R_1 + R_2} \frac{(1 + j\omega C R_1)}{(1 + j\omega C R_{eq})}, \tag{5.11}$$

where $R_{eq} = R_1 R_2/(R_1 + R_2)$. Again, when G is expressed in db, the factors are separated:

$$G_{db} = 20 \log \left(\frac{R_2}{R_1 + R_2}\right) + 20 \log |(1 + j\omega C R_1)| - 20 \log |(1 + j\omega C R_{eq})|.$$

The plots of these terms are shown in Fig. 5.15. Equation (5.11) can be rapidly sketched by noting that:

1. At frequencies approaching dc, the gain becomes $R_2/(R_1 + R_2)$; this can also be seen from inspection of the circuit, since C will approach an open circuit.

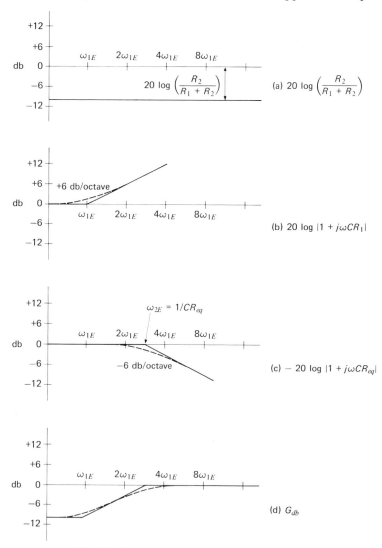

Fig. 5.15 Frequency-response plots for a Form-2 circuit.

2. At high frequencies the gain will approach unity, as C becomes a short circuit.
3. The lower corner frequency is given by

$$f_{1E} = \frac{1}{2\pi C R_1}. \tag{5.12}$$

4. The upper corner frequency is given by

$$f_{2E} = \frac{1}{2\pi C R_{eq}}, \tag{5.13}$$

where $R_{eq} = R_1 R_2/(R_1 + R_2)$.

If $R_1 \gg R_2$ so that $f_{2E} \gg f_{1E}$, then the asymptotic plot corresponds fairly closely to the actual plot. Otherwise, the asymptotic plot is somewhat inaccurate in the region of falloff.

FORM 3. The gain expression for the circuit of Fig. 5.13(c) is

$$G = \frac{R_B\|1/j\omega C_T}{R_A + (R_B\|1/j\omega C_T)} = \frac{R_B}{R_A + R_B + j\omega C_T R_A R_B}$$

$$= \frac{R_B}{R_A + R_B} \frac{1}{1 + j\omega C_T [R_A R_B/(R_A + R_B)]}. \tag{5.14}$$

A db-plot of Eq. (5.14) is very easily made, since

$$G_{db} = 20 \log \left(\frac{R_B}{R_A + R_B}\right) - 20 \log \left|1 + j\omega C_T \frac{R_A R_B}{R_A + R_B}\right|.$$

The plots of these terms are shown in Fig. 5.16. By inspection the low-frequency gain is seen to approach $R_B/(R_A + R_B)$. Taking a Thévenin equivalent circuit of the two resistors, we obtain the circuit shown in Fig. 5.17. The corner frequency in radians per second occurs when $R_{th} = X_C$ or

$$\omega_2 = \frac{1}{C_T R_{th}}.$$

This gives a frequency of

$$f_2 = \frac{1}{2\pi C_T R_{th}}, \tag{5.15}$$

where $R_{th} = R_A R_B/(R_A + R_B)$. Here, f_2 is called the upper corner frequency of the circuit.

Since the three forms of circuits discussed are passive with no resonant effects, the voltage gain is always less than unity. In the actual case of an amplifier, a gain of greater than unity can result. For example, Eq. (5.5) is a Form-1 equation, but is multiplied by a constant gain. The effect of this multiplication is to shift the db-plot upward by an amount $20 \log |A_{bc}|$. The shape of the curve remains the same; hence

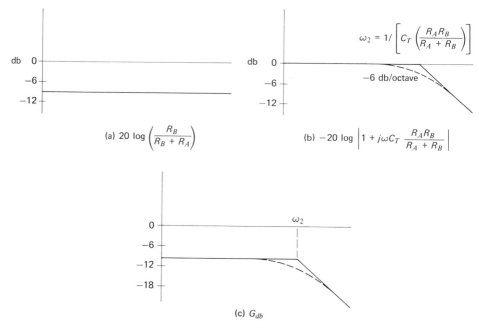

Fig. 5.16 Frequency-response plots for a Form-3 circuit.

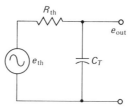

Fig. 5.17 Equivalent circuit of Form-3 circuit.

all principles developed for the Form-1 circuit apply to this amplifier. The occurrence of the Form-2 circuit will be demonstrated next. The Form-3 circuit is used in conjunction with high-frequency analysis and is covered in Chapter 7.

5.4 LOW FREQUENCY FALLOFF DUE TO EMITTER-BYPASS CAPACITOR

Consider the circuit of Fig. 5.18. The voltage gain of the stage as a function of frequency can be predicted by considering the extremes of a very high or a very low frequency input signal. At very high frequencies the capacitor will appear as a short circuit, causing the emitter impedance to approach zero. The base-to-collector gain will be quite high and is given by

$$A_{max} = \left| -\frac{\beta R_C}{r_{bb'} + (\beta + 1)r_d} \right|.$$

5.4 LOW FREQUENCY FALLOFF DUE TO EMITTER-BYPASS CAPACITOR

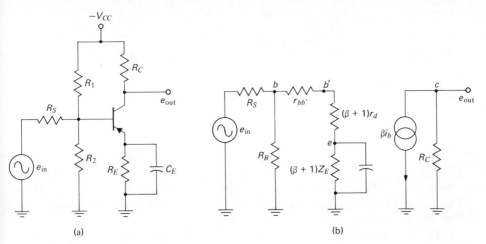

Fig. 5.18 Common-emitter amplifier with emitter-bypass capacitor: (a) actual circuit; (b) equivalent circuit.

At frequencies approaching dc, the capacitor appears as an open circuit and the base-to-collector gain is lowered to a value of

$$A_{\min} = \left| -\frac{\beta R_C}{r_{bb'} + (\beta + 1)r_d + (\beta + 1)R_E} \right|.$$

There will be a range of frequencies over which the base-to-collector gain increases from its minimum value up to the maximum value. We shall now proceed to determine this range of frequencies by considering the expression for overall gain as a function of frequency.

The circuit of Fig. 5.19 allows a calculation of the overall gain of the stage. The gain is (after considerable manipulation)

$$A_o = -\frac{\beta R_C \dfrac{R_B}{R_B + R_S}(1 + j\omega C_E R_E)}{[R_{th} + r_{bb'} + (\beta + 1)r_d + (\beta + 1)R_E]} \times \left[1 + j\omega C_E \frac{R_E[R_{th} + r_{bb'} + (\beta + 1)r_d]}{(\beta + 1)R_E + R_{th} + r_{bb'} + (\beta + 1)r_d}\right]. \quad (5.16)$$

We note that this equation corresponds to the Form-2 circuit and therefore will have the same general shape of frequency response as that shown in Fig. 5.15. Equation (5.16) is a very formidable equation but can be plotted to give the desired results. However, the same information can be obtained by using the inspection method, as applied to the Form-2 circuit. The high-frequency gain is found by assuming that the capacitor is a short circuit, completely bypassing R_E. This results in an overall gain

$$A_{\max} = \left| -\frac{\beta R_C R_B / (R_B + R_S)}{R_{th} + r_{bb'} + (\beta + 1)r_d} \right|. \quad (5.17)$$

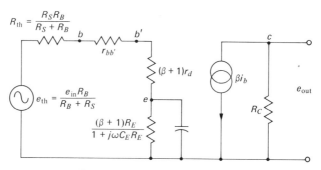

Fig. 5.19 Simplified equivalent circuit.

This value is found by using the equivalent circuit with $R_E = 0$ or by finding the limit of Eq. (5.16) as ω becomes very large.

The low-frequency gain is found by assuming that the capacitor is an open circuit. The gain expression then becomes

$$A_{min} = \left| -\frac{\beta R_C R_B/(R_B + R_S)}{R_{th} + r_{bb'} + (\beta + 1)r_d + (\beta + 1)R_E} \right|. \tag{5.18}$$

Comparing Eq. (5.16) to the Form-2 case given by Eq. (5.11), we can evaluate the two corner frequencies of the expression. Considering Eq. (5.12), we see that the lower of the two corner frequencies is

$$f_{1E} = \frac{1}{2\pi C_E R_E}, \tag{5.19}$$

Considering Eq. (5.13), we see that the other corner frequency is

$$f_{2E} = \frac{1}{2\pi C_E R_{eq}}, \tag{5.20}$$

where

$$R_{eq} = R_E \left\| \left[r_d + \frac{r_{bb'} + R_{th}}{\beta + 1} \right] \right..$$

It should be easy to remember the value for R_{eq}, since it is equal to that of the resistor R_E in parallel with the output resistance as seen looking back into the emitter. In other words, f_{2E} is determined by the equivalent resistance seen by the capacitor, as shown in Fig. 5.20. The asymptotic plot of frequency response is

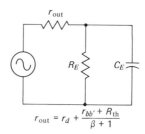

Fig. 5.20 Equivalent circuit presented to C_E.

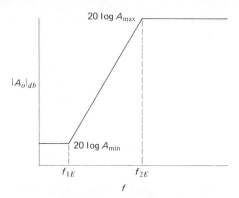

Fig. 5.21 Frequency response of the circuit in Fig. 5.18.

shown in Fig. 5.21. Usually, f_{2E} will be much larger than f_{1E} and the asymptotic plot is fairly accurate.

5.5 THE COMBINATION OF A COUPLING CAPACITOR AND AN EMITTER-BYPASS CAPACITOR

Often an amplifier such as that whose circuit is shown in Fig. 5.22 will be utilized. As the frequency of the input signal is decreased, the gain can start falling off due to either C_E or C_c, or to both of these capacitors. An expression can be written for the overall gain as a function of frequency, but this leads to a very complex equation. Generally, we are interested only in the corner frequency defined by the point at which the overall gain equals 0.707 of its maximum value. It will next be shown that this corner frequency is usually equal to f_{2E} for a practical amplifier.

We might first consider why C_E and C_c are used in the amplifier circuit. If they were not used, no low frequency reactive effects would be present, and the resulting design would be much simpler. However, we must use C_c to keep the ac source

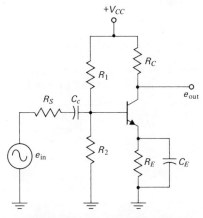

Fig. 5.22 The use of both C_c and C_E.

isolated from the bias circuit. Occasionally this is unnecessary, but for most cases C_c must be used. The value of R_E is often determined by stability requirements. Unfortunately, the presence of R_E reduces the ac gain of the stage. This gain can be restored by using an emitter-bypass capacitance to lower the ac emitter impedance. In designing the amplifier, we must know the frequency range over which a given amplification is to occur. For example, a public address amplifier might be required to extend over the audio range from 100 Hz to 10 kHz. It does not detract from the amplifier if signal frequencies below 100 Hz are not fully amplified; these frequencies are not present in the speech signal anyway. On the other hand, frequencies above 100 Hz must be amplified uniformly so that no distortion of the speech waveform takes place. Therefore C_c and C_E must be so chosen that the corner frequency falls below 100 Hz. This means both f_1 and f_{2E} must be less than 100 Hz. Now let us assume that we choose equal values for C_E and C_c. Then from Eq. (5.8), it follows that

$$f_1 = \frac{1}{2\pi C_c(R_{in} + R_S)},$$

and from Eq. (5.20),

$$f_{2E} = \frac{1}{2\pi C_E R_{eq}},$$

we see that f_1 will be much smaller than f_{2E}. This is due to the fact that the input impedance to the stage R_{in} will be much greater than R_{eq}, which is determined mainly by the output impedance of the stage seen at the emitter. Typically R_{in} is over 100 times greater than R_{eq}. In fact, C_c can be quite a lot smaller than C_E and still f_{2E} will be larger than f_1. We conclude that for cases where C_c is of the same order of magnitude as C_E, f_{2E} is greater than f_1. Thus the frequency at which the overall gain equals 0.707 of its maximum value is equal to f_{2E} in this case.

Over the frequency range where the capacitive effects can be neglected, the amplifier gain is referred to as the midband gain A_{MB}. This value corresponds to A_{max} in the previous discussion.

5.6 LOW FREQUENCY REACTIVE EFFECTS IN THE FET

A typical FET stage might appear as shown in Fig. 5.23.

The overall gain can be expressed as the product of two transfer functions; that is,

$$\frac{e_{out}}{e_{in}} = \frac{e_g}{e_{in}} \frac{e_{out}}{e_g}.$$

The transfer function e_g/e_{in} is

$$\frac{e_g}{e_{in}} = \frac{R_G}{R_G + R_S - j1/\omega C_c}. \qquad (5.21)$$

5.6 LOW FREQUENCY REACTIVE EFFECTS IN THE FET

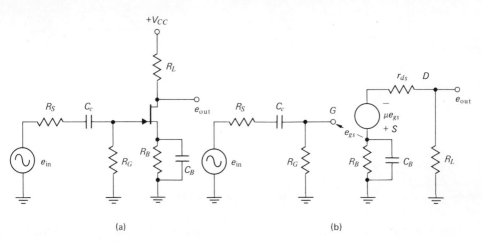

Fig. 5.23 FET amplifier: (a) actual circuit; (b) equivalent circuit.

From gate to output the voltage gain is

$$\frac{e_{out}}{e_g} = \frac{-\mu R_L}{R_L + r_{ds} + (\mu + 1)Z_B} = \frac{-\mu R_L}{R_L + r_{ds} + \dfrac{(\mu + 1)R_B}{1 + j\omega C_B R_B}}$$

$$= \frac{-\mu R_L}{R_L + r_{ds} + (\mu + 1)R_B} \frac{(1 + j\omega C_B R_B)}{\left(1 + \dfrac{j\omega C_B R_B (R_L + r_{ds})}{R_L + r_{ds} + (\mu + 1)R_B}\right)}. \quad (5.22)$$

This gain can also be expressed as

$$\frac{e_{out}}{e_g} = \frac{-\mu R_L}{R_L + r_{ds} + (\mu + 1)R_B} \frac{(1 + j\omega C_B R_B)}{(1 + j\omega C_B R_{EQ})}, \quad (5.23)$$

where

$$R_{EQ} = R_B \left\| \frac{R_L + r_{ds}}{\mu + 1} \right. = \frac{R_B (R_L + r_{ds})}{R_L + r_{ds} + (\mu + 1)R_B}. \quad (5.24)$$

Combining Eqs. (5.21) and (5.23) gives an overall voltage gain of (for $R_G \gg R_S$)

$$A_v = \frac{e_{out}}{e_{in}} = \frac{-\mu R_L}{R_L + r_{ds} + (\mu + 1)R_B} \frac{(1 + j\omega C_B R_B)}{(1 + j\omega C_B R_{EQ})} \frac{1}{1 - j1/\omega C_c R_G}. \quad (5.25)$$

An earlier section demonstrated methods of constructing a db plot for factors such as those of Eq. (5.25); therefore, we will not discuss this expression in detail. Figure 5.24 shows the asymptotic frequency response of the voltage gain. The corner frequency f_{2S} is given by

$$f_{2S} = \frac{1}{2\pi C_B R_{EQ}}. \quad (5.26)$$

184 LOW-FREQUENCY AMPLIFIERS

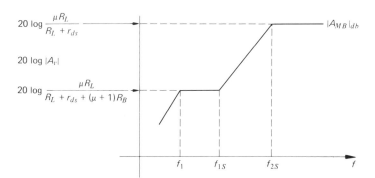

Fig. 5.24 Voltage gain as a function of frequency.

This frequency, as assumed in Fig. 5.24, is generally much higher than f_1 which is

$$f_1 = \frac{1}{2\pi C_c(R_G + R_S)}. \tag{5.27}$$

Often R_S is negligible compared to R_G in this equation. Since R_G can be very large, f_1 can be made almost arbitrarily small. The third critical frequency of the voltage gain is

$$f_{1S} = \frac{1}{2\pi C_B R_B} \tag{5.28}$$

which is always smaller than f_{2S} and usually exceeds f_1 by several orders of magnitude.

In practice the major point of interest is the frequency at which the voltage gain has dropped 3 db from the midband value. Thus, f_{2S} is an important design parameter for the normal case in which f_2 is much smaller than f_{2S}. The midband gain is given by

$$A_{MB} = \frac{-\mu R_L}{R_L + r_{ds}} = -g_m \frac{R_L r_{ds}}{R_L + r_{ds}}. \tag{5.29}$$

5.7 GAIN STABILITY

Chapter 3 covered stability of the quiescent operating point with respect to temperature and parameter variation. We can now consider the effect of these variations on the voltage gain of the transistor stage.

The midband voltage gain of the stage indicated in Fig. 5.25 is expressed as

$$A_{MB} = \frac{-\beta R_C}{r_{bb'} + (\beta + 1)(r_d + R_E)}.$$

If $(\beta + 1)R_E$ is much larger than $r_{bb'} + (\beta + 1)r_d$, the gain is

$$A_{MB} = -\frac{\beta R_C}{(\beta + 1)R_E} = -\frac{\alpha R_C}{R_E} \approx -\frac{R_C}{R_E}.$$

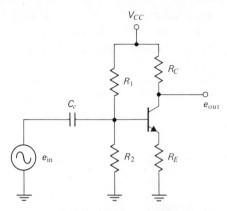

Fig. 5.25 A simple amplifier.

Since α will not change more than a few percent from one transistor to another, or when the temperature is varied, the voltage gain is very stable. The voltage gain in this case is approximately given by the ratio of the effective collector resistance to the emitter resistance. This is an important point that can be applied in estimating the gain of common-emitter stages. This result can be checked by physical reasoning. When an input voltage is applied to the base, this incremental value also appears at the emitter. The ac-emitter current due to this voltage is

$$i_e = \frac{e_{in}}{R_E}.$$

The incremental collector current is $i_c = \alpha i_e$. The output voltage developed across R_C by the collector current is

$$e_{out} = -i_c R_C = -\frac{\alpha e_{in} R_C}{R_E},$$

which gives the same voltage gain previously developed.

If R_E is not present or is bypassed by a capacitor, the voltage gain is

$$A_{MB} = -\frac{\beta R_C}{r_{bb'} + (\beta + 1)r_d}.$$

Usually, $(\beta + 1)r_d$ is much larger than $r_{bb'}$, so

$$A_{MB} \approx -\frac{\beta R_C}{(\beta + 1)r_d} = -\frac{\alpha R_C}{r_d} \approx -\frac{R_C}{r_d}.$$

If I_E is stable with temperature or β changes, one might think that the gain is also stable. This can be seen to be incorrect by considering the equation for r_d, which is

$$r_d = \frac{kT}{qI_E}.$$

The resistance r_d increases directly with temperature. If I_E is 1 mA, for example, r_d changes from 26 Ω at 27°C (300°K) to 30 Ω at 77°C (350°K). The gain changes about 16 percent over this temperature variation. If gain stability is important, an unbypassed emitter resistor must be used to cause the voltage gain to be less dependent on r_d. In Chapter 6 we shall see that feedback amplifiers can be used to obtain good voltage stability. While we have not considered it as such, an unbypassed emitter resistance amounts to negative feedback. This will be considered in more detail in Chapter 6. Practical amplifiers typically include circuits with gains ranging from 5 to 10 per stage. While the stages are capable of much higher gain, good production practice generally limits the gain per stage in order to achieve stable amplifiers.

5.8 COMMON-BASE AMPLIFIER

The common-base amplifier is not used as often as the common-emitter stage because the current gain is slightly less than unity. Since the voltage is not inverted through this stage, it is occasionally used when the output voltage is required to be in phase with the input. The circuit of Fig. 5.26 is a typical common-base stage. The ac gain can be easily derived as

$$A_{MB} = \frac{\alpha R_L}{R_{E1} + r_d + (1 - \alpha)r_{bb'}}. \tag{5.30}$$

If $R_{E1} \gg r_d + (1 - \alpha)r_{bb'}$, then $A_{MB} = \alpha R_L/R_{E1}$. The bias and the voltage stability considerations are the same as the common-emitter case with $R_E = R_{E1} + R_{E2}$. The biggest difference is that the input impedance is much lower for the common-base stage. This impedance is

$$R_{in} = R_{E2} \| [R_{E1} + r_d + (1 - \alpha)r_{bb'}]. \tag{5.31}$$

As R_{E1} approaches zero, r_{in} approaches r_d, which can be quite small. This means that if a source resistance is present, the input impedance of the stage can load the source quite heavily, attenuating the input voltage. As R_{E1} is increased to raise r_{in}, the voltage gain from Eq. (5.30) is seen to decrease. Normally, the common-base stage is used only when R_S is small.

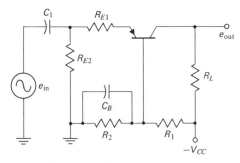

Fig. 5.26 Common-base amplifier.

5.9 EMITTER FOLLOWER

The emitter follower is used when a high-input impedance or a low-output impedance is required. This circuit is used more often than the common-base stage. The current gain of this stage can be quite high while the voltage gain is near, but never exceeds, unity. Although a low voltage gain is exhibited, the emitter follower can increase the voltage and power transferred from the source to the load. In Fig. 5.27 a source with 10 kΩ-output impedance is coupled to a 100 Ω-load. In part (a) the output voltage is

$$e_{out} = \frac{1}{101} e_S.$$

In part (b), where an impedance buffer is used, the output voltage is

$$e_{out} = \frac{10 \text{ k}\Omega}{20 \text{ k}\Omega} \frac{100 \text{ }\Omega}{200 \text{ }\Omega} e_S = \frac{1}{4} e_S,$$

which is twenty-five times greater than the first case. With no buffer the load current flows through the source, dropping much of the applied voltage across this resistance. The buffer stage draws little current from the source minimizing the voltage drop across the source resistance. This stage then supplies the necessary current to the load.

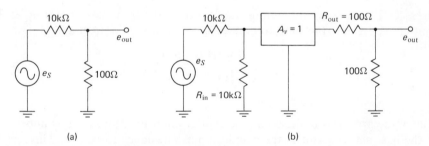

Fig. 5.27 Voltage transferred to load: (a) no impedance buffer; (b) emitter follower.

The circuit of Fig. 5.28 shows an emitter-follower stage. The quiescent emitter voltage is set somewhere near the center of the active region or $V_{CC}/2$. This usually results in R_1 and R_2 being close to the same size. The input impedance to the stage consists of the biasing resistors in parallel with the input impedance of the transistor. If r_v, the transistor resistance from base to collector, is large, the transistor input impedance is

$$r_{in} = r_{bb'} + (\beta + 1)r_d + (\beta + 1)(R_E \| R_L). \tag{5.32}$$

Since $R_E \| R_L$ is usually much larger than $r_d + (1 - \alpha)r_{bb'}$, then

$$r_{in} = (\beta + 1)(R_E \| R_L). \tag{5.33}$$

188 LOW-FREQUENCY AMPLIFIERS

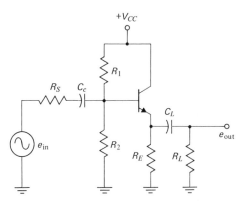

Fig. 5.28 Emitter follower.

The effective load determines the input impedance of the transistor. The overall input impedance R_{in} is then

$$R_{in} = R_B || [(\beta + 1)(R_E || R_L)]. \qquad (5.34)$$

The output impedance is (Section 4.2B)

$$R_{out} = R_E \left\| \left[r_d + \frac{r_{bb'} + R_S || R_B}{(\beta + 1)} \right], \right. \qquad (5.35)$$

the second term being the output impedance of the transistor appearing at the emitter. The gain from base to emitter is found from Eq. (4.9) and is

$$A_{MB} = \frac{(\beta + 1)(R_E || R_L)}{r_{bb'} + (\beta + 1)r_d + (\beta + 1)(R_E || R_L)} \approx 1. \qquad (5.36)$$

A. Problems of the Emitter Follower

One of the problems in designing the emitter follower is that the low impedance of the bias network causes the overall input impedance to be low. This can be overcome when necessary by using the "bootstrap" circuit of Fig. 5.29 (not to be confused with the bootstrap sweep circuit). Here R is chosen small enough to affect the bias design very little. The overall input impedance is r_{in} in parallel with the impedance offered by the path that includes R. Due to the feedback in the circuit, this path offers a very high effective impedance. The impedance of this path is e_{in}/i_R, where i_R is the current through R. The current through R is merely

$$\frac{e_{in} - e_{out}}{R} = \frac{e_{in} - A_{MB}e_{in}}{R}.$$

But since A_{MB} is almost unity, this current is very small. The impedance of this path is then

$$\frac{e_{in}/e_{in}(1 - A_{MB})}{R} = \frac{R}{1 - A_{MB}}.$$

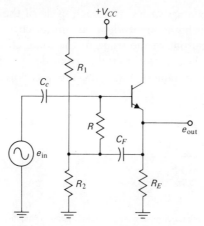

Fig. 5.29 Bootstrap circuit for high-input impedance.

If the gain is 0.99 and R is 20 kΩ, this value becomes

$$\frac{20 \text{ k}\Omega}{0.01} = 2 \text{ M}\Omega.$$

The overall input impedance is now

$$R_{in} = \frac{R}{1 - A_{MB}} \bigg\| r_{in}.$$

As was pointed out in Chapter 4, the upper limit of r_{in} is fixed by r_v. A practical bootstrap circuit is shown in Fig. 5.30.

Another problem of the emitter follower relates to larger signal swings in a stage with capacitor coupling to the load. When large emitter current variations

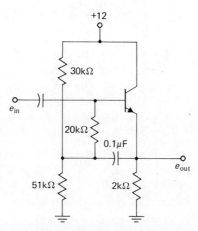

Fig. 5.30 Practical bootstrap circuit.

occur, the output impedance varies over each cycle of the signal. This is due to the change in r_d which is highly dependent on emitter current. If a large capacitor is used to couple the output of the stage to the load, clipping can often result. The clipping will take place as the emitter current decreases. If a large sinusoidal signal is applied to the circuit of Fig. 5.28, the waveforms shown in Fig. 5.31 could result.

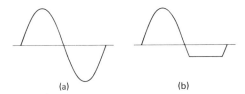

Fig. 5.31 Waveforms for large applied voltage: (a) applied signal; (b) output signal.

As the output impedance becomes less on the positive swing, the capacitor takes on a small amount of positive charge. When the signal swings negative, charge is removed more slowly, since r_d becomes greater. The resulting time lag in voltage at the emitter can cause the emitter-base junction to become reverse-biased, and clipping results. If large signals are to be amplified symmetrically, the complementary emitter follower of Fig. 5.32 can be used. In this figure, each stage is biased slightly on; T_1 passes the positive half cycle of input signal to the load, while T_2 passes the negative half cycle. The impedance presented to the capacitor is the same on both positive and negative swings.

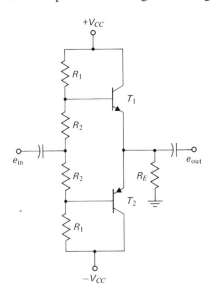

Fig. 5.32 Complementary emitter follower.

5.10 DESIGN EXAMPLES

A circuit designer is often called on to use the methods of analysis discussed in the preceding sections. Although an acceptable design is the goal of the designer's effort, analysis serves as the means to achieve this end. Because of the infinite combinations of specifications for an amplifier there can be no simple design procedure developed. Nor is there always a single solution that is optimum. There may be ten different configurations that could be used to satisfy a given problem. A solution is proposed, and the analysis is then carried out. If specifications are met, the design is acceptable; if not, revisions are made followed by another analysis. This procedure is continued until an acceptable solution is achieved.

One point that deserves emphasis relates to the fact that a given parameter change may affect several different specifications. This problem complicates the design process. One might lower the value of an unbypassed emitter resistance to increase the amplifier gain, but the input impedance will decrease while the lower corner frequency may increase. The best we can hope for is to develop a somewhat systematic design procedure to avoid the necessity of solving a great number of interacting equations. Such a procedure will be outlined after we consider the parameters that are generally specified in an amplifier design problem.

If we are to design an amplifier we will usually know the following details:

1. the required midband voltage gain,
2. the desired frequency range of the amplifier,
3. the source impedance,
4. the load impedance, and
5. the power supply voltage.

Some or all of these quantities may be specified indirectly or can be easily evaluated. For example, if we are to design a public address system, we can specify a bandwidth that corresponds to the audio range of frequencies, say, 100 Hz to 10 kHz. If high fidelity is required for the system, perhaps for the purpose of music reproduction, the bandwidth may be extended to cover the range 50 Hz to 20 kHz. The source impedance will correspond to the output impedance of the microphone to be used and the output impedance will be determined by the loudspeaker system. A knowledge of the magnitude of output voltage for the microphone and the required voltage to drive the loudspeakers will allow one to specify the midband voltage gain. The power-supply voltage may be chosen to be compatible with an existing unit or to allow application of a dry cell if the unit is to be portable. With the basic set of specifications listed above the amplifier design can then be undertaken.

Before we proceed further we should recall the important approximation for the gain of a transistor stage with an unbypassed emitter resistance. When R_E is much larger than $r_d + (1 - \alpha)r_{bb'}$ the midband voltage gain from base to collector is

$$A_{MB} \approx -\frac{R_C}{R_E}. \qquad (5.37)$$

Even when R_E is decreased and this expression becomes less accurate, it is useful in that an estimated value of gain can provide valuable information to the design process.

A very useful, practical circuit that should be noted before discussing the design procedure is indicated in Fig. 5.33.

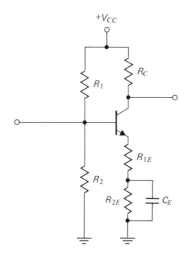

Fig. 5.33 A useful amplifier stage.

We have found earlier that an unbypassed emitter resistance tends to stabilize the voltage gain of the stage with variations in temperature or current gain β.

Bias stability requirements might dictate a reasonably large value of emitter resistance, however, and the resulting voltage gain might be quite low. The circuit of Fig. 5.33 is a compromise circuit offering good bias stability, reasonable gain stability, and reasonable voltage gain.

It is also possible to construct a simple volume control stage using this configuration. A potentiometer replaces the two emitter resistors with the outer terminals connected to the emitter and to ground. A capacitor is connected between ground and the wiper arm of the potentiometer. A fixed dc resistance appears between emitter and ground, but the unbypassed resistance can be varied between zero and the maximum value of potentiometer resistance. This varies the ac voltage gain of the stage since both the unloaded gain and the input impedance depends on the value of unbypassed emitter resistance.

Generally temperature stability of ac amplifier stages is not critical. As drift occurs the quiescent point may change enough to affect distortion or gain if no emitter resistance is used, however.

The following suggested procedure can be used as a design procedure for bipolar transistor amplifiers.

1. Choose values of R_E and R_C to satisfy ac voltage gain requirements. The reduction of gain due to R_L and R_S should be estimated and allowed for at this point. If a knowledge of r_d is required to calculate the gain, the choice of quiescent point should be made prior to the gain calculation.
2. Choose the biasing network to satisfy temperature and parameter stability requirements. At this point some adjustment in R_E may be necessary to satisfy the stability specification. A second emitter resistance is sometimes added and bypassed with a capacitor. The ac gain is unaffected, but the dc stability is improved. If the unbypassed emitter resistance requires adjustment, R_C may have to be modified to satisfy the gain requirement.
3. The capacitors are chosen to satisfy the frequency requirements of the amplifier.

The following examples will demonstrate the application of this procedure. In these examples β_0 will be assumed to equal β. While this is often a fairly inaccurate assumption, we found in Chapter 3 that a very large variation of β_0 affects the dc design only slightly in a properly designed stage.

Example 5.3. Design an amplifier that has a midband gain of -10 ± 10 percent. A 12-V dc power supply is available and a silicon transistor with $\beta = 80$ is used. The lower corner frequency should be 20 Hz. The stage is to couple a source of zero impedance to a 10 kΩ load.

Solution. The configuration is chosen to be that shown in Fig. 5.34. The load can appear in the collector circuit since ac coupling was not specified. The gain of a

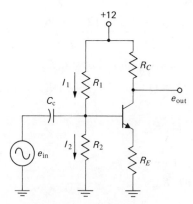

Fig. 5.34 Single-stage amplifier.

stage of this type is given by

$$A_{MB} = -\frac{R_C}{R_E}, \qquad (5.37)$$

where R_E is easily found to be

$$R_E = -\frac{R_C}{A_{MB}} = \frac{10}{10} = 1.0 \text{ k}\Omega.$$

In selecting the bias network we have a great deal of freedom since S_V was not specified. For Si a reasonable choice of R_2 (as explained in Chapter 3) is $R_2 = 20R_E$. Thus

$$R_2 = 20 \text{ k}\Omega.$$

To select R_1 the quiescent collector voltage must be found. Since the voltage at the collector can swing from $+12$ V down to $R_E/(R_E + R_C) \times 12 = 1.07$ V, the midpoint voltage is 6.5 V. In many applications this value might not be the optimum quiescent point, but since nothing was specified, we will pick this value. The quiescent collector current should then be

$$I_{CQ} = \frac{12 - 6.5}{10 \text{ k}\Omega} = 0.55 \text{ mA}.$$

The required base current is

$$I_{BQ} = \frac{I_{CQ}}{\beta} = \frac{0.55 \text{ mA}}{80} = 7 \text{ }\mu\text{A}.$$

Since $I_1 = I_2 + I_B = V_B/R_2 + I_B$, we must find V_B:

$$V_B = R_E I_E + V_{BE} = 1 \times 0.557 + 0.5 = 1.06 \text{ V};$$

$$I_1 = \frac{1.06}{20 \text{ k}\Omega} + 0.007 = 0.060 \text{ mA}.$$

We can now calculate R_1 as

$$R_1 = \frac{V_{CC} - V_B}{I_1} = \frac{12 - 1.06}{60 \text{ }\mu\text{A}} = 182 \text{ k}\Omega.$$

The design is now complete except for choosing C_c:

$$R_{in} = 182 \text{ k}\Omega \, || \, 20 \text{ k}\Omega \, || \, (\beta + 1)R_E = 14.9 \text{ k}\Omega,$$

$$f_1 = \frac{1}{2\pi C_c R_{in}};$$

so

$$C_c = \frac{1}{2\pi f_1 R_{in}}$$

$$= \frac{1}{2\pi \times 20 \times 14.9 \times 10^3} = 0.53 \text{ }\mu\text{F}.$$

The degree of accuracy carried in this problem is not necessary, since standard resistor values will be used. The completed design would have the following standard values: $R_C = 10\ \text{k}\Omega$, $R_E = 1\ \text{k}\Omega$, $R_2 = 20\ \text{k}\Omega$, $R_1 = 180\ \text{k}\Omega$, and $C_c = 0.50\ \mu\text{F}$.

Example 5.4. Using a silicon transistor, design a single-stage amplifier that has a midband gain of -10 ± 10 percent. The lower corner frequency should be 20 Hz. The available dc voltage is $+20$ V. The load resistor is 8.2 kΩ. The source resistance is 4 kΩ and the transistor has a $\beta = 100$. The voltage stability should be less than 25 mV/°C.

Solution. The configuration to be used is shown in Fig. 5.35. This problem is similar to the previous example except that

1. a voltage stability is specified, and
2. R_S is present.

The value of R_E cannot be chosen as simply in this example as in the previous example due to the presence of R_S. This requires the use of Eq. (5.2), which is

$$A_{MB} = A \frac{R_{\text{in}}}{R_{\text{in}} + R_S}.$$

The gain A is the base-to-collector gain, which is

$$-\frac{\alpha R_C}{r_{bb'}(1-\alpha) + r_d + R_E} \simeq -\frac{\alpha R_C}{R_E}.$$

The current gain α could be taken to be unity with little loss of accuracy. Now $R_{\text{in}} = R_2 \| R_1 \| r_{\text{in}}$, but since none of these values are known, we will choose reasonable values for them. If we let $R_B = R_2 \| R_1 = 20 R_E$ and assume that

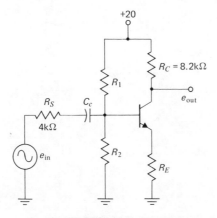

Fig. 5.35 Configuration for Example 5.4.

$r_{in} = (\beta + 1)R_E = 101R_E$, then R_{in} becomes $16.7R_E$. The gain is

$$A_{MB} = -\frac{\alpha R_C}{R_E} \frac{16.7R_E}{16.7R_E + R_S} = -\frac{(0.99)(8.2)(16.7)}{16.7R_E + 4}$$

We can now find R_E to be

$$R_E = \frac{\dfrac{(0.99)(8.2)(16.7)}{10} - 4}{16.7} = 570 \, \Omega.$$

The voltage stability can be checked by Eq. (3.15):

$$S_V = \frac{-(\beta R_C)(0.002)}{R_B + (\beta + 1)R_E} = -24 \text{ mV/}^\circ\text{C}.$$

The capacitor C_c is chosen as

$$C_c = \frac{1}{2\pi f_1(R_S + R_{in})} = \frac{1}{2\pi \times 20 \times 13.5 \times 10^3} = 0.59 \, \mu\text{F}.$$

The bias resistance must be chosen so that the proper I_{BQ} results and

$$\frac{R_1 R_2}{R_1 + R_2} = 20R_E = 11.4 \text{ k}\Omega.$$

If the quiescent collector current is chosen as 1 mA, then V_{CQ} will be $+12$ V. The base current should be

$$I_{BQ} = \frac{I_{CQ}}{\beta} = \frac{1 \text{ mA}}{100} = 10 \, \mu\text{A}.$$

The bias network can be drawn as shown in Fig. 5.36. The required E_{th} is found from

$$E_{th} = R_B I_B + V_{BE} + R_E I_E = 0.114 + 0.5 + 0.576 = 1.2 \text{ V}.$$

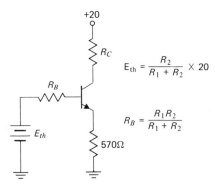

Fig. 5.36 Bias network.

We can now calculate R_2 and R_1 from the two equations given in Fig. 5.36:

$$R_1 = 190 \text{ k}\Omega, \qquad R_2 = 12.2 \text{ k}\Omega.$$

The completed circuit might have standard values as follows: $R_C = 8.2$ kΩ, $R_E = 560\ \Omega$, $R_2 = 12$ kΩ, $R_1 = 180$ kΩ, and $C_c = 0.62\ \mu$F. We could choose R_E slightly smaller than the calculated value, since r_d was neglected in the gain expression.

Example 5.5. Design a single-stage amplifier to have a midband voltage gain of -40 ± 10 percent. The lower corner frequency should be 30 Hz and the voltage stability should be less than 25 mV/°C. The amplifier is to couple a low impedance source to a 10 kΩ load. The load is to be ac coupled to the amplifier. A 20-V power supply is available, along with a germanium transistor having a β of 100 and $I_{co} = 1\ \mu$A.

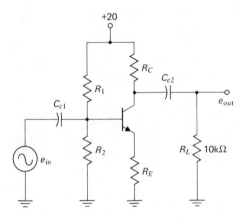

Fig. 5.37 Configuration for Example 5.5.

Solution. The configuration chosen is shown in Fig. 5.37. Here we choose R_C to be 10 kΩ, noting that if R_C is too large, poor stability may result; and if it is too small, R_{eff} will be low, and a low gain results. Noting that R_E must be fairly small to get such a high gain, we need to know the value of r_d in order to choose R_E. If we select an emitter current of 1mA, then $r_d = 26\ \Omega$. Then from Eq. (4.8),

$$A_{MB} = -\frac{\beta R_{\text{eff}}}{r_{bb'} + (\beta + 1)(r_d + R_E)} \simeq -\frac{\alpha R_{\text{eff}}}{r_d + R_E},$$

and R_E is calculated to be 98 Ω. For germanium, $R_2 = 0.1(\beta + 1)R_E = 10R_E$ is a reasonable choice for good voltage stability, so

$$R_2 = 980\ \Omega.$$

We could now choose R_1, but a rough check of stability is done first to see if a value of $R_B \leq 980\ \Omega$ is reasonable. From Eq. (3.16),

$$S_V = -\frac{0.0016\beta R_C}{R_B + (\beta + 1)R_E} - \frac{0.09 I_{co} R_C (R_E + R_B)(\beta + 1)}{R_B + (\beta + 1)R_E}$$
$$= -147\ \text{mV/}^\circ\text{C} - 9\ \text{mV/}^\circ\text{C} = -156\ \text{mV/}^\circ\text{C}.$$

This does not satisfy the specification. The fact that $R_B + (\beta + 1)R_E$ is so small causes the stability to be poor; however, R_E has to be small to satisfy the gain specification. The circuit of Fig. 5.38 shows one method to overcome these conflicting requirements. This circuit allows the dc value of R_E to be large, since

$$R_{E_{dc}} = R_{1E} + R_{2E},$$

while the ac value is $R_{E_{ac}} = R_{1E}$.

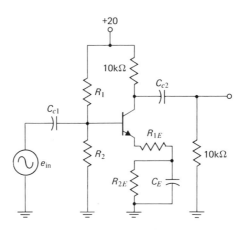

Fig. 5.38 Configuration for Example 5.5.

If R_{1E} is chosen as $98\ \Omega$, R_{2E} as $1\ \text{k}\Omega$, and $R_2 = 10\ \text{k}\Omega$, the stability is approximately (assuming $R_B \approx R_2$)

$$S_V = -23\ \text{mV/}^\circ\text{C}.$$

This satisfies the stability requirement, so R_1 can now be picked. Since $I_{CQ} = 1\ \text{mA}$, Eq. (2.36b) gives

$$I_{BQ} = \frac{I_{CQ}}{\beta} - \frac{(\beta_0 + 1)}{\beta} I_{co} = 9\ \mu\text{A}.$$

Now, $V_B = V_E + 0.2 = I_E R_E + 0.2 = 1.3\ \text{V}$, so $I_2 = V_B/R_2 = 130\ \mu\text{A}$. Hence

$$R_1 = \frac{18.7}{0.139} = 134\ \text{k}\Omega.$$

We must now choose C_E to give the proper corner frequency. Referring to Fig. 5.39, we see that the equivalent resistance presented to C_E is

$$R_{eq} = R_{2E}||(R_{1E} + r_{out})$$
$$= (1 \text{ k}\Omega)||[98 + r_d + (1 - \alpha)r_{bb'}]$$
$$= 1 \text{ k}\Omega||124 \text{ }\Omega = 110 \text{ }\Omega.$$

Equation (5.20) is used to give

$$C_E = \frac{1}{2\pi f_{2E} R_{eq}} = 48 \text{ }\mu\text{F}.$$

We shall choose C_{c1} so that f_1 is much smaller than 30 Hz. A value of 10 μF should satisfy this requirement. From previous discussion it should be obvious that the corner frequency introduced by C_{c2} is

$$f_1 = \frac{1}{2\pi C_{c2}(R_C + R_L)}.$$

A value of 10 μF will cause this frequency to be much less than 30 Hz also. The final element values are: $R_L = 10$ kΩ, $R_C = 10$ kΩ, $R_{1E} = 100$ Ω, $R_{2E} = 1$ kΩ, $R_2 = 10$ kΩ, $R_1 = 130$ kΩ, $C_E = 56$ μF, $C_{c1} = 10$ μF, and $C_{c2} = 10$ μF.

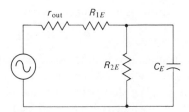

Fig. 5.39 Resistance presented to C_E.

We can note from these examples that a knowledge of $r_{bb'}$ is not often required. For normal applications, $r_{bb'}$ has a negligible effect on amplifier performance. This quantity becomes more important in high-frequency amplifiers and is often given by the manufacturer for high-frequency transistors.

Example 5.6. Design an amplifier to give an overall amplification of 50, coupling a source of 500 Ω impedance to a 1-kΩ load. The lower corner frequency should be 40 Hz or less. Assume that a 20-V power supply is to be used and that npn-silicon transistors are available with $\beta = 100$.

Solution. Since the source impedance is 500 Ω, the input impedance of the first stage should be considerably greater than this value. If $R_{in} \approx 0.1(\beta + 1)R_E$ is estimated to be 5 kΩ, then R_E must be 500 Ω. Using trial and error methods, it appears that two stages will not give sufficient voltage gain. Three stages will work, however, and all these stages could be common-emitter stages. On the other

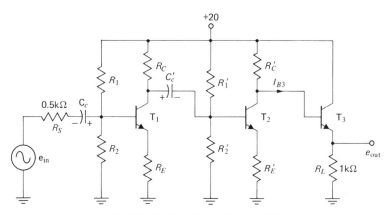

Fig. 5.40 Configuration for Example 5.6.

hand, if the third stage is an emitter follower, the effective collector load of stage two can be increased to a value giving the required gain. This configuration is shown in Fig. 5.40.

Using an emitter follower avoids the need for a coupling capacitor and a bias network for the third stage. The next step in the design is to select the individual collector and emitter resistances. This is often most easily done by working from the load back toward the source. The input impedance to the emitter follower is

$$r_{\text{in 3}} = (\beta + 1)R_E = 101 \text{ k}\Omega.$$

If the output impedance of stage two is much smaller than this value, the voltage transfer from stage two to the load will be quite efficient. A value of 10 kΩ for R'_C will satisfy this condition. We shall select a value of $R'_E = 1$ kΩ to give this stage an unloaded gain of approximately 10; R'_2 is chosen as $0.1(\beta + 1)R'_E \approx 10$ kΩ; and R'_1 can be selected to set the proper quiescent voltage. If V_{C2} is chosen as 11 V, the current required through R'_C is

$$I_{R'_C} = \frac{20 - 11}{10 \text{ k}\Omega} = 0.9 \text{ mA}.$$

Of this 0.9 mA, I_{B3} will be

$$I_{B3} = \frac{I_{E3}}{(\beta + 1)} = \frac{V_{C2} - V_{BE3}}{(\beta + 1)R_E} = 0.104 \text{ mA}.$$

Therefore, I_{C2} should be $0.9 - 0.1 = 0.8$ mA.

We choose R'_1 in the normal way and calculate it to be 139 kΩ. The input impedance of stage two is

$$R_{\text{in 2}} = 139 \text{ k}\Omega \| 10 \text{ k}\Omega \| 101 \text{ k}\Omega = 8.5 \text{ k}\Omega.$$

If R_C is selected to be 10 kΩ, then R_E, R_2, and R_1 can be calculated on the basis of the overall gain requirements. If we assume that $R_1 \| R_2 = R_B$ is equal to

$10R_E$, the input impedance to stage one is

$$R_{\text{in }1} = 10R_E \| 101\,R_E = 9.1R_E.$$

Using Eq. (5.4), we get

$$A_o = A'_1 A'_2 A'_3 \frac{R_{\text{in }1}}{R_{\text{in }1} + R_S}.$$

The gain of the emitter follower is $A'_3 = 1$. The base-to-collector gain of stage two, A'_2, is

$$A'_2 = -\frac{\alpha R_{2\text{ eff}}}{R'_E} = -\frac{0.99 \times (10\text{ k}\Omega \| 101\text{ k}\Omega)}{1\text{ k}\Omega} = -9.0.$$

The base-to-collector gain of stage one is

$$A'_1 = -\frac{\alpha R_{1\text{ eff}}}{R_E} = -\frac{0.99 \times (10\text{ k}\Omega \| 8.5\text{ k}\Omega)}{R_E} = -\frac{4.6}{R_E}.$$

The gain equation is

$$A_o = \left(\frac{4.6}{R_E}\right)(9.0)(1.0)\left(\frac{9.1R_E}{9.1R_E + 0.5}\right) = \frac{376}{9.1R_E + 0.5}.$$

For A_o to be 50, R_E must be 765 Ω. We select R_1 and R_2 to give a proper bias point and so that

$$R_B = \frac{R_1 R_2}{R_1 + R_2} = 10R_E.$$

Values of $R_1 = 114$ kΩ and $R_2 = 8.3$ kΩ give a bias of $I_{CQ1} = 1$ mA.

The capacitors are selected to give corner frequencies below 40 Hz from the following equations:

$$f_1 = \frac{1}{2\pi C_c (R_S + R_{\text{in }1})} \quad \text{and} \quad f'_1 = \frac{1}{2\pi C'_c (R_{\text{out }1} + R_{\text{in }2})}.$$

Standard element values used in this design would be: $R_L = 1$ kΩ, $R'_C = 10$ kΩ, $R'_E = 1$ kΩ, $R'_2 = 10$ kΩ, $R'_1 = 150$ kΩ, $R_C = 10$ kΩ, $R_E = 750$ Ω, $R_2 = 8.2$ kΩ, $R_1 = 100$ kΩ, $C'_c = 5$ μF, and $C_c = 5$ μF.

Before leaving this example, it should be noted that an FET stage could be used to advantage in this amplifier. The loading of the second stage on the first stage is rather severe in the circuit of Fig. 5.40. If an FET is used for the second stage with the load appearing in the drain, a third stage is unnecessary. This solution to the design problem is shown in Fig. 5.41.

If the first stage remains unchanged and the gate bias resistor is 1 MΩ, the gain from input to FET gate is

$$\frac{R_{\text{in }1}}{R_{\text{in }1} + R_S}\left(\frac{-\alpha R_C}{R_E}\right) = \frac{-6.83}{7.33}\left(\frac{0.99 \times 10}{0.75}\right) = -12.7$$

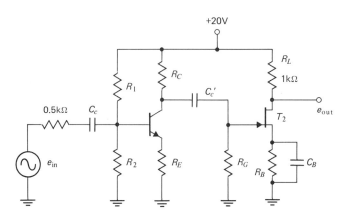

Fig. 5.41 A two-stage hybrid amplifier.

The FET stage need only supply a gain of $50/12.7 \approx 3.9$ to satisfy the gain specification. If r_{ds} is 20 kΩ then g_m must exceed a value of only 420 $\mu\mho$ to reach the required overall gain. The lack of significant loading between the stages allows this rather modest value of transconductance to satisfy the design. If a larger value of transconductance is used, the overall voltage gain can be decreased to the design value by lowering R_C. The reduction from three stages to two stages in this amplifier may be important in some applications.

There are several other designs that can satisfy the given specifications. It was pointed out earlier that nonunique solutions to design problems are to be expected. As the designer gains more experience, the selection of the configuration becomes easier and only rapid trial-and-error techniques are involved.

A specific point worth noting is that the emitter follower can often be direct-coupled to the preceding stage. This eliminates the need of the bias network which results in a higher input impedance to the emitter follower without bootstrapping since the parallel bias resistors are eliminated.

5.11 DC AMPLIFICATION

There are applications of amplifiers wherein a dc signal must be amplified along with an ac signal. These circuits are called dc amplifiers and cannot use coupling capacitors between stages. Monolithic broadband amplifiers also avoid using coupling capacitors since the large capacitances required are difficult to integrate. Two major problems arise in designing amplifiers without coupling capacitors:

1. The biasing of each stage is dependent on adjacent stages.
2. The temperature stability is often poor since the drifts of the first stages are amplified by succeeding stages.

Let us consider the two common-emitter stages shown in Fig. 5.42.

In an ac amplifier the quiescent operating points could be set independently and would not be changed if the stages are coupled together with a capacitor.

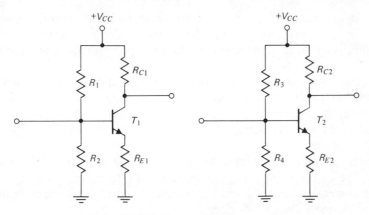

Fig. 5.42 Two amplifier stages.

Typical dc voltage levels in the two circuits might be

$$V_{B1} = 1\text{ V}, \qquad V_{C1} = 5\text{ V}, \qquad V_{B2} = 1\text{ V}, \qquad V_{C2} = 7\text{ V}, \qquad V_{CC} = 12\text{ V}.$$

If an attempt is made to couple the collector of stage one to the base of stage two directly, the quiescent levels of the second stage will shift considerably. This results from the rather high value of V_{C1} which will force V_{B2} to a higher level and may saturate T_2. There are several methods that can be used to overcome this difficulty. Figure 5.43 indicates two approaches to the problem, each requiring additional circuitry or expense.

In the circuit of Fig. 5.43(a) the collector voltage of stage one can be set near 7 V to cause a reasonable voltage to appear across the base-emitter junction of T_2. The 6 V supply placed in series with R_{E2} reduces the forward bias on T_2 to a level that will result in a reasonable value of collector current flow for this stage. A

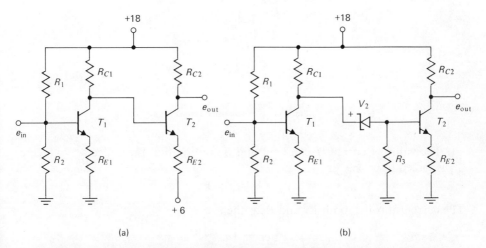

Fig. 5.43 Two dc amplifiers: (a) dc amplifier using two supplies; (b) Zener-coupled dc amplifier.

serious disadvantage of this method for an amplifier of several stages is that several power supply voltages must be available. The collector voltage of each succeeding stage is, in general, higher than the previous stage requiring higher source voltages for later stages.

The circuit of Fig. 5.43(b) overcomes this disadvantage and can also be used in monolithic circuits. So long as enough current flows through the Zener to keep the diode in the voltage regulating region, the voltage at the base of T_2 will be $(V_{C1} - V_Z)$. Properly choosing V_Z allows the quiescent voltage of T_2 to be set at the desired point.

Figure 5.44 shows a dc level-shift stage that can also be used for integrated circuits. The output voltage V_{out} can be related to the input signal V_{in} by noting that the current through R_1 will approximately equal the current through R_2. Summing currents through R_1 gives

$$I = I_{E1} + I_{C2}.$$

Since base current of T_2 equals collector current of T_1 we can write

$$I_{B2} \approx I_{E1}.$$

Thus,

$$I = I_{B2} + I_{C2},$$

but this also is the expression for emitter current of T_2 which flows through R_2.

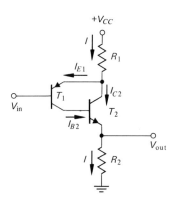

Fig. 5.44 A dc level-shift stages.

The value of current I can be found by noting that the collector voltage of T_2 is forced to be

$$V_{C2} = V_{in} + V_{BE}.$$

The current through both R_1 and R_2 is then

$$I = \frac{V_{CC} - V_{in} - V_{BE}}{R_1}.$$

Solving for V_{out} gives

$$V_{out} = IR_2 = \frac{R_2}{R_1}(V_{CC} - V_{in} - V_{BE}). \tag{5.38}$$

The small-signal voltage gain can be found by differentiation to be

$$A = \frac{dV_{out}}{dV_{in}} = -\frac{R_2}{R_1}. \tag{5.39}$$

Although the voltage gain of the stage may be quite small, a dc level shift from a rather high voltage to a reasonable value can be effected by this circuit. For example, if the quiescent level of a stage is 9 V and the power supply V_{CC} is 12 V, this stage could shift the level to 2.5 V volts if R_2 and R_1 are equal. The signal voltage in this case would be transferred with unity gain. The output of this level-shift circuit could then reasonably be dc coupled to another transistor stage.

The problem of temperature drift in dc amplifiers is much more serious than in ac amplifiers. Since the change in component temperature is a rather slow process, the drift in voltage approaches a low-frequency or dc signal variation. In ac amplifiers this very low-frequency signal is blocked by the coupling capacitors. In dc amplifiers the drift in voltage or current of the first stages is amplified by the gain of succeeding stages resulting in a large drift signal at the output. The absolute value of drift in output voltage is not especially meaningful since it tells little about the size of the input signal which can be differentiated from the drift voltage. For this reason the specification of drift is referred to the input by dividing the output drift by the overall gain of the amplifier. If this drift figure is 10 μV/°C, and the amplifier is to operate over a temperature range of 15°C, the total drift would be 150 μV. When an output voltage change occurs it is impossible to determine whether it is a result of a temperature change or a slowly varying input signal of less than 150 μV magnitude. If an input signal of 150 μV magnitude is important to a given application, great care must be taken to reduce the drift of this amplifier.

Reduction of drift can be accomplished by using the compensation methods of temperature stabilization discussed in Chapter 3. Obviously, it is more important to stabilize the early stages in the amplifier. Very critical designs may justify the expense of a temperature oven in which the circuit can be placed. A control system maintains the oven at a constant temperature in spite of ambient temperature variations. It was also mentioned in Chapter 3 that an integrated circuit amplifier might apply these automatic control techniques to stabilize the substrate temperature and thereby minimize drift.

Another important circuit used in dc amplifier design is the differential input stage. The following section will cover this very useful stage.

5.12 THE DIFFERENTIAL AMPLIFIER

The basic differential stage consists of two balanced amplifiers as shown in Fig. 5.45. The two amplifiers can be connected in several different configurations. If the same signal is applied to both inputs, the circuit is said to operate in common

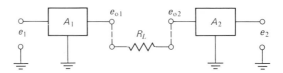

Fig. 5.45 Configuration of a differential stage.

mode. If the amplifiers have exactly equal gains, the signals e_{o1} and e_{o2} will be equal. The output signal normally used is the difference signal

$$e_{\text{out}} = e_{o2} - e_{o1}$$

and will be zero in the ideal differential stage operating in common mode. In practice the two amplifier gains will not be identical; thus, the common mode output signal will have a small value. The common mode gain A_{CM} is defined as

$$A_{CM} = \frac{e_{o2} - e_{o1}}{e} = A_2 - A_1, \qquad (5.40)$$

where e is the voltage applied to both inputs. In the general case, the signals e_1 and e_2 will not be equal and the gain in this case is

$$A_{DA} = \frac{e_{o2} - e_{o1}}{e_2 - e_1} = \frac{e_{\text{out}}}{e_2 - e_1}. \qquad (5.41)$$

Other combinations of input signals can be used. The output can be taken between either output and ground and is then called a single-ended output. The normal case finds the double-ended output with a single-ended input.

The great advantage of the differential stage is in the cancellation of drift at the output. Temperature drifts in each stage are common mode signals and if the stages are closely matched and temperatures of the components exhibit similar variations, very little output drift will be noted. The integrated differential amplifier can perform considerably better than its discrete counterpart since component

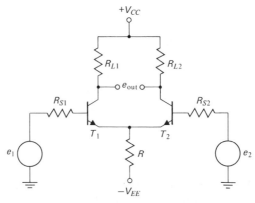

Fig. 5.46 Differential amplifier.

matching is more accurate and a relatively uniform temperature prevails throughout the chip.

The circuit of Fig. 5.46 shows a simple differential stage. For good temperature stability the resistor R should be quite large, as will be shown later.

A. Small-Signal Voltage Gain

The small-signal voltage gain of the stage can be found by considering the individual stage gain. Once e_{c1} and e_{c2} (the collector-to-ground voltages of T_1 and T_2) are found, the output voltage can be found to be $e_{out} = e_{c2} - e_{c1}$. Both input voltages e_1 and e_2 will affect the collector voltage e_{c1}, and both will affect e_{c2}. The effect of e_1 on e_{c1} can be found and then added to the effect of e_2 on e_{c1}. The voltage e_1 will be amplified by T_1, since this stage is a common-emitter stage. The impedance seen looking into the emitter of stage two appears as the external-emitter impedance of stage one. In addition, T_1 acts as an emitter follower, supplying a signal proportional to e_1 to the emitter of stage two. Then T_2 amplifies this signal as a common-base amplifier. The four circuits of Fig. 5.47 summarize this operation. Note that T_2 acts as a common-emitter amplifier for the input signal e_2, and also passes this signal to the emitter of T_1. Here, T_1 is a common-base

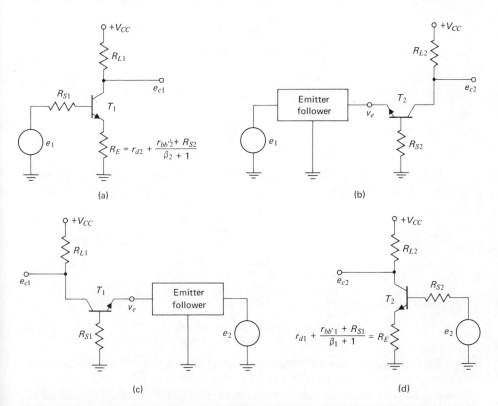

Fig. 5.47 Circuits used to calculate the collector voltages due to e_1 and e_2: (a) circuit used to find e_{c1}/e_1; (b) circuit used to find e_{c2}/e_1; (c) circuit used to find e_{c1}/e_2; (d) circuit used to find e_{c2}/e_2.

amplifier for this signal. The resistor R is so large it can be neglected in the gain calculations.

If it is assumed that the transistors are perfectly matched, it is not necessary to carry the subscripts on the transistor parameters. Using Eq. (4.8), which applies to a common-emitter stage, the desired relations for Figs. 5.47(a) and (d) are found to be

$$\frac{e_{c1}}{e_1} = \frac{-\beta R_L}{(r_{b'e} + r_{bb'} + R_S) + (\beta + 1)R_E} = \frac{-\beta R_L}{2(r_{b'e} + r_{bb'} + R_S)} \quad (5.42)$$

and

$$\frac{e_{c2}}{e_2} = \frac{-\beta R_L}{2(r_{b'e} + r_{bb'} + R_S)}. \quad (5.43)$$

The gain of both emitter followers is (from Eq. 4.9)

$$\frac{v_e}{e_1} = \frac{v_e}{e_2} = \frac{(\beta + 1)R_E}{r_{bb'} + R_S + r_{b'e} + (\beta + 1)R_E} = \frac{1}{2}.$$

The gain of both common-base stages in Figs. 5.47(b) and (c) from emitter to collector is

$$\frac{e_{c1}}{v_e} = \frac{e_{c2}}{v_e} = \frac{\beta R_L}{r_{b'e} + r_{bb'} + R_S}.$$

The values e_{c1}/e_2 and e_{c2}/e_1 can now be found:

$$\frac{e_{c1}}{e_2} = \frac{\beta R_L}{2(r_{b'e} + r_{bb'} + R_S)} \quad (5.44)$$

and

$$\frac{e_{c2}}{e_1} = \frac{\beta R_L}{2(r_{b'e} + r_{bb'} + R_S)}. \quad (5.45)$$

The collector voltages can be calculated from the above equations when both input signals are present as

$$e_{c1} = \frac{\beta R_L}{2(r_{b'e} + r_{bb'} + R_S)}(e_2 - e_1) \quad (5.46)$$

and

$$e_{c2} = \frac{\beta R_L}{2(r_{b'e} + r_{bb'} + R_S)}(e_1 - e_2). \quad (5.47)$$

The overall gain is then

$$A_{DA} = \frac{e_{c2} - e_{c1}}{e_2 - e_1} = -\frac{\beta R_L}{(r_{b'e} + r_{bb'} + R_S)}. \quad (5.48)$$

Since Eq. (5.48) shows that the output voltage is proportional to the difference of the input signals, it is obvious that the circuit can be used to subtract one signal from another. If one of the signals is inverted, the stage can also function as an adder. That is, if e_1 is inverted before applying it to the input of T_1, the output voltage is proportional to $e_1 + e_2$. In many cases emitter resistances will be added from each emitter to form a junction with the resistance R. This modifies the gain equations by changing $r_{b'e}$ to $r_{b'e} + (\beta + 1)R_E$ in the denominators of Eqs. (5.42) through (5.48).

There are other possible modes of operation of the differential amplifier. For example, e_1 can be zero, reducing the circuit to a simple amplifier with gain

$$A_{DA} = \frac{e_{c2} - e_{c1}}{e_2} = \frac{-\beta R_L}{r_{b'e} + r_{bb'} + R_S}.$$

DC stability of differential stages

In Chapter 3 it was shown that changes in I_{co}, β, and $V_{BE(on)}$ were responsible for the drift in output voltage with temperature change. All three causes of drift can be represented by a change in base voltage; that is, a transistor stage can be represented as shown in Fig. 5.48. The transistor parameters can now be assumed to remain constant with temperature, but $V(T)$ varies with T to account for the drift of the stage. In a differential amplifier, the circuit including temperature effects is as shown in Fig. 5.49.

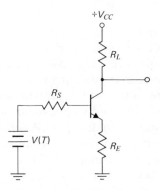

Fig. 5.48 Equivalent circuit for calculating temperature drift.

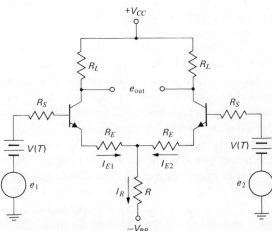

Fig. 5.49 Differential amplifier including temperature effects.

If the transistors are perfectly matched, the output voltage will not change with temperature, even though I_{C1} and I_{C2} will increase, due to an increased $V(T)$. Both collector voltages will decrease, but $e_{c1} - e_{c2}$ will not change. Unfortunately, it is impossible to match the transistor parameters exactly even for integrated circuits, so an output drift will result. Therefore, it becomes necessary to minimize variations of I_C with temperature. Furthermore, there are applications where a single-ended output is required; that is, the output voltage is referenced to ground. In this case e_{c1} or e_{c2} becomes the output voltage, and again the collector current variation with temperature must be minimized. Since the collector current is related to the emitter current by $I_C = \alpha I_E$ for silicon, a stable collector current will result if I_E is stabilized. If both transistors in Fig. 5.49 are assumed to have equal emitter currents flowing, the emitter current is

$$I_E = \frac{I_R}{2}.$$

When $V(T)$ changes with temperature, the emitter current change will be

$$\Delta I_E = \frac{\Delta I_R}{2},$$

resulting from an increased voltage appearing across R. The current change through R can be minimized by making R as large as possible. To do this, however, would result in very low collector currents. In noncritical applications, we would use a reasonably large emitter resistor and an emitter voltage source whose polarity is opposite to that of the collector supply. For critical applications, the emitter resistance is replaced with a constant current source offering a very high incremental resistance to the circuit. A perfect current source would completely stabilize variations in I_E. In practice, small variations still occur, but the overall drift in output voltage is minimized. The circuit of Fig. 5.50 shows a typical stage with a current source in the emitter circuit.

The constant current stage must be temperature compensated in order to ensure low drift of I_R. Since T_3 must be compensated, one might wonder what has been gained by adding this stage, when temperature compensation could have been applied to T_1 and T_2. Often this is done; that is, T_1 and T_2 are designed to have low drift. However, the application of most compensation techniques reduces the voltage gain of the circuit. For example, if R_E is increased, the voltage gain decreases. Since the voltage gain of T_3, the constant current source, is unimportant, very sophisticated compensation techniques can be applied to stabilize I_R. Thus the overall voltage gain of the differential stage is unaffected, while the stability is increased. One practical method of stabilization of I_R for discrete circuits is to use a silicon resistor or other temperature-dependent resistance for the emitter resistance of T_3. The silicon resistor has a positive temperature coefficient of 0.7% per °C. The base-emitter circuit of T_3 reduces to that shown in Fig. 5.51. As V_{BE} decreases with temperature, causing I_C to increase, the resistance of R increases to offset the increase in I_C. We can calculate the value of R that

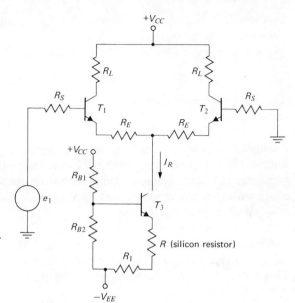

Fig. 5.50 Low-drift differential amplifier.

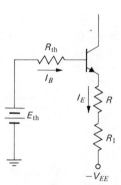

Fig. 5.51 Equivalent circuit for T_3.

exactly offsets the increase in I_C. Assuming that T_3 is a silicon transistor and that R_{th} is much smaller than $(\beta + 1)(R + R_1)$, the collector current is

$$I_C = \frac{\beta[E_{th} - V_{BE} + V_{EE}]}{(\beta + 1)(R + R_1)} = \frac{\alpha[E_{th} - V_{BE} + V_{EE}]}{R + R_1}.$$

The change in I_C depends only on changes in V_{BE} and is independent of changes in β. If a silicon resistor is used for R, having a temperature coefficient of 0.7% per °C, the change in I_C with temperature is found by differentiation to be

$$\frac{dI_C}{dT} = \frac{0.002\alpha(R + R_1) - 0.007R\alpha[E_{th} - V_{BE} + V_{EE}]}{(R + R_1)^2}.$$

To cause $dI_C/dT = 0$, the numerator of this equation can be set equal to zero if the following condition is satisfied,

$$R = \frac{R_1}{\frac{7}{2}[E_{th} - V_{BE} + V_{EE}] - 1}. \tag{5.49}$$

The collector current should be very stable if this value of R is chosen for the silicon resistor.

Example 5.7. Design a differential amplifier having a single-ended, noninverted output. The voltage gain should be $+20$, the load resistance is 2 kΩ, and the source resistance is 1 kΩ. Assume that npn silicon transistors with $\beta = 50$ are available along with two power supplies of 12 V each. A maximum output voltage swing of 4 V peak to peak is required.

Solution. The circuit configuration of Fig. 5.52 is used. To allow the required output voltage swing, the collector currents of T_1 and T_2 are chosen to be 2 mA. This fixes r_d at 13 Ω. Equation (5.45) is applied to calculate R_E:

$$A = \frac{e_{c2}}{e_1} = \frac{\beta R_L}{2(R_S + r_{bb'} + (\beta + 1)r_d + (\beta + 1)R_E)} = 20.$$

Approximating $r_{bb'}$ as 200 Ω, R_E is found to be 12.5 $\Omega \approx 12 \Omega$. With this value of R_E, the gain will be $+20$.

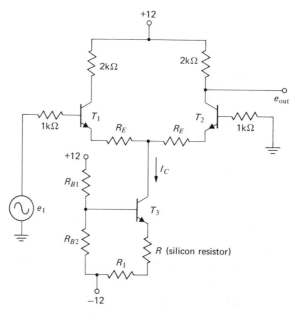

Fig. 5.52 Differential amplifier for Example 5.7.

Since both emitter currents must flow into the collector of T_3, this transistor must be designed for a collector current of 4.08 mA. In selecting R_{B1}, R_{B2}, R, and R_1 there are two equations to satisfy: the bias equation and Eq. (5.49), giving the proper value of R. First consider the bias. If $R_{B1} = 6$ kΩ and $R_{B2} = 4$ kΩ, the Thévenin equivalent of the bias network reduces to $E_{th} = -2.4$ V and $R_{th} = 2.4$ kΩ. Assuming that $(\beta + 1)(R + R_1)$ is much greater than this value, I_C becomes

$$I_C = \frac{\alpha[E_{th} - V_{BE} + V_{EE}]}{R + R_1} = \frac{0.98[-2.4 - 0.5 + 12]}{R + R_1},$$

and $R + R_1$ is found to be 2.2 kΩ. Using Eq. (5.49), we obtain

$$R = R_1/30.9.$$

The values $R_1 = 2130$ Ω and $R = 70$ Ω satisfy both equations.

Monolithic differential stages generally use more active devices than discrete stages since fabrication of these devices is so easily accomplished. The circuit of Fig. 5.53 shows one popular configuration of a differential amplifier with a single-ended output [3].

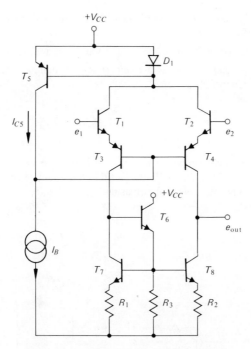

Fig. 5.53 A popular monolithic differential stage.

Transistors T_1 and T_2 are emitter followers working into common-base stages. The loads presented to the common-base stages T_3 and T_4 are the impedances looking into the collectors of T_7 and T_8. The use of active loads allows the operating point to be set at a level resulting in a large output voltage swing. These loads also present high ac impedances to the common-base stages giving good voltage gain for the amplifier. The output signal represents a voltage proportional to the difference between e_2 and e_1 even though it is a single-ended stage. The emitter follower T_6 is a unity gain stage that converts the collector signal at T_6 into a base-drive signal for T_8. Transistor T_8 inverts this signal at the output. Voltage gains in excess of 60 db can be obtained with this circuit.

The operating points of T_3 and T_4 are stabilized by the diode D_1 and transistor T_5. As currents through T_1 and T_2 increase, the diode voltage also increases, driving T_5 harder. The increased collector current of T_5 lowers the bias current to T_3 and T_4 since the bias drive equals $I_B - I_{C5}$. This reduction in bias current tends to offset the increase in current due to a temperature change.

Transistors T_1, T_2, T_3, T_4, and T_5 can be replaced with just two FET input stages for exceptionally high input impedance. The poor matching characteristics of the FET leads to other problems with this configuration.

REFERENCES AND SUGGESTED READING

1. P. M. Chirlian, *Electronic Circuits: Physical Principles, Analysis, and Design.* New York: McGraw-Hill, 1971, Chapter 9.
2. M. S. Ghausi, *Electronic Circuits.* New York: Van Nostrand, 1971, Chapters 3 and 4.
3. A. B. Grebene, *Analog Integrated Circuit Design.* Cincinnati, Ohio: Van Nostrand-Reinhold, 1972, Chapter 5.

PROBLEMS

Section 5.1A

5.1 Consider the amplifier shown in the figure where $R_B = R_1 \| R_2 = 5 \text{ k}\Omega$, $R_C = 6 \text{ k}\Omega$, $R_E = 750 \, \Omega$, $R_S = 5 \text{ k}\Omega$, $R_L = 10 \text{ k}\Omega$, and $\beta = 99$. Evaluate the overall midband voltage gain of the stage. Assume proper biasing.

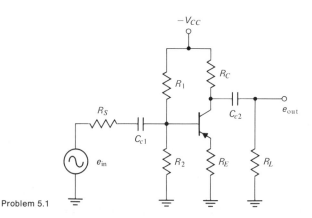

Problem 5.1

* **5.2** Repeat Problem 5.1 for the case where $R_L = \infty$ and $R_S = 5$ kΩ.
5.3 Repeat Problem 5.1 for the case where $R_L = 10$ kΩ and $R_S = 0$.
5.4 Repeat Problem 5.1 for the case where $R_L = \infty$ and $R_S = 20$ kΩ.
5.5 Given that $R_L = 8$ kΩ and $R_S = 2$ kΩ in the circuit of Problem 5.1, choose values of R_C and R_E to give an overall midband voltage gain of -10. Assume that $I_E = 1$ mA, if necessary, and that $\beta = 50$ and $R_B = R_1 \| R_2 = 15 R_E$. Calculate the voltage stability of the stage, assuming that a silicon transistor is used.
* **5.6** If R_L is changed to 20 kΩ in Problem 5.5, what values of R_C and R_E will give an overall midband voltage gain of -20?
5.7 Design a common-emitter amplifier using a silicon transistor with $\beta = 99$, which has a gain of -20 when $R_S = 0$ and $R_L = \infty$. When $R_S = 0$ and $R_L = 8$ kΩ, the gain drops to -15. Specify all resistor values, including the bias resistors. Assume $V_{CC} = 12$ V.
5.8 Design a common-emitter amplifier using a silicon transistor with $\beta = 99$, which has a gain of -20 when $R_S = 0$ and $R_L = \infty$. When $R_S = 10$ kΩ and $R_L = \infty$, the gain drops to -8. Specify the value of R_C and R_E assuming that the parallel combination of bias resistors is ten times the value of emitter resistance.
* **5.9** A common-emitter stage has a base-to-collector voltage gain of -16 when $R_L = \infty$ and $R_S = 0$. Given that $R_{in} = 15$ kΩ and $R_{out} = 8$ kΩ, use Eq. (5.1) to evaluate the overall voltage gain when $R_S = 5$ kΩ and $R_L = 10$ kΩ. In order to compare this answer with that given by Eq. (5.2), find the effective collector load and base-to-collector gain when R_L is present.

Section 5.1B
5.10 Given that both of two stages are identical to that of Problem 5.9, find the overall midband gain if these stages are cascaded (capacitor coupling) to couple a 5-kΩ source and a 10-kΩ load.
*__5.11__ Repeat Problem 5.10 for three identical cascaded stages coupling a 5-kΩ source to a 10-kΩ load.

Section 5.2
5.12 For the circuit of Problem 5.1 assume that $R_S = 1$ kΩ, $R_{in} = 10$ kΩ, $R_C = 10$ kΩ, and $R_L = 10$ kΩ. Select C_{c1} to cause

$$f_1 = \frac{1}{2\pi C_{c1}(R_S + R_{in})} = 50 \text{ Hz}.$$

Select C_{c2} to cause

$$f'_1 = \frac{1}{2\pi C_{c2}(R_C + R_L)} = 20 \text{ Hz}.$$

Given that the midband overall voltage gain is -10, write an analytic expression for the overall gain as a function of frequency. Sketch an asymptotic frequency response for this amplifier. What is the overall lower corner frequency?
*__5.13__ Repeat Problem 5.12, assuming that $R_L = 40$ kΩ and $A_{MB} = -16$.
5.14 Repeat Problem 5.12 for the case where $f'_1 = f_1 = 50$ Hz. How many db below the midband value is A_0 at 50 Hz?

216 LOW FREQUENCY AMPLIFERS

5.15 Given that an n-stage amplifier contains n-coupling capacitors, all of which are chosen to give an individual lower corner frequency of f_1, prove that the overall lower corner frequency f_1^o is

$$f_1^o = \frac{f_1}{\sqrt{2^{1/n} - 1}}$$

How many db will the gain at f_1 be down compared to the midband value of gain?

Sections 5.3–5.4

5.16 Consider the circuit shown in the figure using a silicon transistor with $\beta = 50$, $r_{bb'} = 200\ \Omega$, and $r_d = 30\ \Omega$. If $R_{1E} = 0$, $R_{2E} = 500\ \Omega$, $R_C = 8\ k\Omega$, $R_B = R_1 R_2/(R_1 + R_2) = 10\ k\Omega$, and $R_S = 2\ k\Omega$, what is the midband voltage gain? Select C_E to give a lower corner frequency of 80 Hz. Evaluate the dc gain. Find f_{1E}. Sketch the asymptotic frequency response.

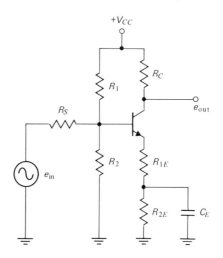

Problem 5.16

5.17 Repeat Problem 5.16, given that R_{2E} is changed to 750 Ω. Assume that r_d remains equal to 30 Ω.

*5.18 Calculate f_{2E} in Problem 5.16, assuming that the dc emitter current is doubled.

5.19 Given that the stage of Problem 5.16 uses a coupling capacitor in series with R_S, and a germanium transistor with $\beta = 100$ and $r_{bb'} = 100\ \Omega$, select values of R_1, R_2, R_{1E}, R_{2E}, and C_E to give the amplifier a midband voltage gain of -30, using an 18-V supply and $R_C = 6\ k\Omega$. The lower corner frequency (the point where $A_0 = -0.707 \times 30$) should be 40 Hz or less. R_S remains at 2 $k\Omega$.

Section 5.6

*5.20 In the circuit of Fig. 5.23, $R_S = 10\ k\Omega$, $C_c = 1\ \mu F$, $R_G = 1\ M\Omega$, $R_B = 300\ \Omega$, $C_B = 20\ \mu F$, $R_L = 2\ k\Omega$, $r_{ds} = 30\ k\Omega$, and $g_m = 3\ m\mho$. Calculate f_1, f_{1S} and f_{2S}. Which of these frequencies determines the lower 3-db frequency of the amplifier? Calculate the midband gain.

5.21 In Problem 5.20 select the capacitors C_c and C_B to cause the lower corner frequency of the amplifier to be 80 Hz.

Sections 5.7–5.10

*5.22 The amplifier of Fig. 5.38 uses a silicon transistor with $\beta = 80$. Assume that $R_{1E} = 100\ \Omega$, $R_{2E} = 1\ \text{k}\Omega$, $R_2 = 10\ \text{k}\Omega$, $R_1 = 130\ \text{k}\Omega$, $C_E = 68\ \mu\text{F}$, $C_{c1} = 10\ \mu\text{F}$, and $C_{c2} = 10\ \mu\text{F}$. Evaluate the midband voltage gain and the lower corner frequency.

5.23 Repeat Problem 5.22 for $R_{1E} = 500\ \Omega$ and $R_{2E} = 600\ \Omega$. Can you devise an ac gain control using a potentiometer for R_E? Sketch such a circuit.

5.24 The transistor used for the common-base amplifier of Fig. 5.26 is a silicon transistor with $\beta = 60$. Given that the power supply voltage is -20 V, select reasonable values of R_L, R_{E1}, R_{E2}, R_1, and R_2 to give the stage a midband voltage gain of $+10$. What is the current gain?

5.25 Repeat Problem 5.24 assuming that a midband voltage gain of $+8$ is required along with a voltage stability of $S_V \leq 5\ \text{mV}/^\circ\text{C}$.

5.26 A circuit is required to couple a 1 kΩ-impedance source to a 4 kΩ load. The load can be dc-coupled to the amplifier output. What is the maximum value of overall voltage gain possible if a common base stage is used? Show that a common-emitter stage can yield a higher voltage gain than the common-base stage in this case. Explain why.

*5.27 If an amplifier is to couple a 1 kΩ-impedance source to a 2 kΩ load with an overall voltage gain of ± 1000, and common-emitter stages with reasonable stability are to be used, how many stages are required?

5.28 Repeat Problem 5.27 given that $R_S = 1\ \text{k}\Omega$ and $R_L = 10\ \text{k}\Omega$.

5.29 Repeat Problem 5.27 assuming that $R_S = 1\ \text{k}\Omega$, $R_L = 10\ \text{k}\Omega$ and $A_{MB} = \pm 400$.

5.30 Repeat Problem 5.27 given that $A_{MB} = \pm 10{,}000$.

5.31 Show the configuration necessary to satisfy the requirements of Problem 5.27. Specify values of R_C and R_E, assuming that the bias network of each stage has an impedance of $R_B = 10\ R_E$. Assume that R_C for the last stage is the 2 kΩ-load resistor.

5.32 Repeat Problem 5.31 given that $R_S = 1\ \text{k}\Omega$ and $R_L = 10\ \text{k}\Omega$.

5.33 Repeat Problem 5.31 assuming that $R_S = 1\ \text{k}\Omega$, $R_L = 10\ \text{k}\Omega$, and $A_{MB} = \pm 400$.

5.34 Repeat Problem 5.31 given that $A_{MB} = \pm 10{,}000$.

5.35 An amplifier with reasonable stability is required to couple a 1 kΩ-impedance source to a 100 Ω load with an overall midband gain of -20. If the available silicon transistors have a $\beta = 80$, design the amplifier, showing all chosen values. Use the minimum number of stages.

5.36 If the gain in Problem 5.35 must be ± 1000, how many stages are needed?

*5.37 Evaluate the overall midband voltage gain of the circuit shown in Fig. 5.40. Assume that

$$R_S = 2\ \text{k}\Omega, \quad R_E = 500\ \Omega, \quad \frac{R_1 R_2}{R_1 + R_2} = 8\ \text{k}\Omega, \quad R_C = 6\ \text{k}\Omega, \quad R'_E = 500\ \Omega,$$

$$\frac{R'_1 R'_2}{R'_1 + R'_2} = 10\ \text{k}\Omega, \quad R'_C = 8\ \text{k}\Omega, \quad R_L = 500\ \Omega, \quad \text{and} \quad \beta = 100$$

for all transistors. Assume that the collector currents of T_1 and T_2 are 1 mA and that the emitter current of T_3 is 4 mA.

5.38 The emitter resistor of T_1 in Problem 5.37 is bypassed by a capacitor. Evaluate the overall midband gain.

5.39 Repeat Problem 5.38 given that the emitter resistors of both T_1 and T_2 are bypassed.

5.40 If $C_{c1} = 1\ \mu\text{F}$ and $C_{c2} = 1\ \mu\text{F}$ in Problem 5.37, what is the lower corner frequency?

*5.41 Given that a $1\ \mu\text{F}$-capacitor bypasses the emitter resistance of T_1 in Problem 5.37, evaluate the lower corner frequency.

5.42 In the circuit shown in the figure, the source V_{BB} is adjustable to provide a dc emitter current of 2 mA for the emitter-follower. Plot the ac power delivered to the load as a function of transistor β as β ranges from 20 to 200. Assume that $e_{in} = 2$ V, rms. Also plot the power delivered to the transistor base.

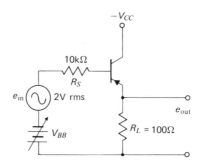

Problem 5.42

5.43 Using the same circuit as Problem 5.42, plot the ac output voltage as a function of β as β ranges from 20 to 200.

5.44 In the circuit of the figure assume that $\beta = 100$ and that proper bias is established. Plot the ac power delivered to R_L as R_L varies from 10 Ω to 1 kΩ. Plot the ac output voltage over this same range of R_L. Neglect r_d.

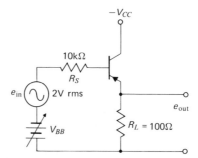

Problem 5.44

5.45 Design an emitter-follower to have a minimum input impedance of 50 kΩ, including the bias network. The load resistor is 2 kΩ. Specify β_{min}, V_{CC}, and all element values.

5.46 Design an emitter-follower which uses a silicon transistor with a β of 80 to have a minimum input impedance of 40 kΩ. The load resistance is 1.5 kΩ, and a 12-V power supply is used.

5.47 Rework Example 5.6 using only common-emitter stages. The gain can be ± 50.

5.48 Design an amplifier to give an overall amplification of 100 coupling a source of 200-Ω impedance to a 10-kΩ load. The lower corner frequency should be 100 Hz or less. Assume that a 20-V power supply is used and that pnp silicon transistors are available with $r_{bb'} = 150\ \Omega$ and $\beta = 100$.

Section 5.11

***5.49** Given that a dc input source with $R_S = 10$ kΩ is coupled directly to the input of the amplifier of Fig. 5.43(a), choose values of R_1, R_2, R_{E1}, R_{C1}, R_{E2}, and R_{C2} to give an overall voltage gain of $+40$ and a proper quiescent output voltage. Assume that silicon transistors with $\beta = 50$ are used. Note that dc current will flow through R_S.

5.50 Design a dc level shift stage to shift from a voltage of $+8$ V to $+2$ V. The power supply is 12 V and $V_{BE} = 0.6$ V.

Section 5.12

5.51 Design a differential amplifier having a gain of 40. The load resistors are to be 4 kΩ; the two power supplies are 12-V sources. Use a common resistor R to obtain the proper bias. The source impedance is 1 kΩ. Assume that $r_{bb'} = 100\ \Omega$ and $\beta = 70$ for the silicon transistors used.

5.52 Repeat Problem 5.51 assuming that 6-kΩ load resistors are used.

5.53 Repeat Example 5.7 for the case where $V_{BE(on)} = 0.6$ V, $\beta = 100$, $r_{b'e} = 2$ kΩ, and $r_{bb'} = 100\ \Omega$.

*****5.54** Redesign the constant current stage of Example 5.7 given that 6 mA of collector current must be supplied by this stage.

FEEDBACK AMPLIFIERS

Active devices are generally characterized by inaccurately defined device parameters and temperature variable parameters. For example, the transistor current gain from base to collector is highly variable from one transistor to another of the same type, and it also depends on temperature. Likewise, the transconductance of the FET depends strongly on device and temperature. It follows then that most amplifiers using these devices will not exhibit a constant, accurately specified gain unless the design minimizes the influence of device parameters on amplifier gain. On the other hand, passive networks containing resistors, capacitors, and inductors can usually be constructed with arbitrary accuracy for most engineering applications. The passive circuit, however, has no capability of power gain.

It is possible to combine the accuracy of passive circuits with the power gain of active circuits to produce stable amplifiers with precise gain figures. The overall accuracy may not reach the level of the passive component accuracy and the overall gain may not attain the maximum possible value, but a compromise in accuracy and gain can be effected by the use of feedback. Many production circuits require accuracy and reproducibility; thus, feedback is almost universally applied in commercial circuits.

There are other advantages that can be obtained in feedback amplifiers that are sometimes more important than the consideration of accuracy. For example, negative feedback can increase bandwidth, control input and output impedances, and lower distortion in the output signal due to nonlinearities or noise introduced by elements of the amplifier. Major disadvantages of feedback amplifiers are additional complexity and an increased tendency for the amplifier to oscillate and thereby generate unwanted output signals.

6.1 THE IDEAL FEEDBACK AMPLIFIER

To introduce the feedback concept we will assume that we are dealing with an ideal amplifier of gain A, infinite input impedance, and zero output impedance. This amplifier is considered to have no delay or phase shift between input and output signals. Figure 6.1 shows one possible configuration for a feedback amplifier.

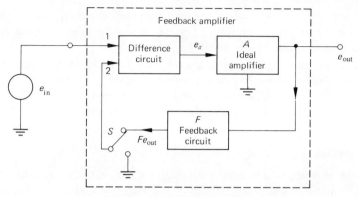

Fig. 6.1 Negative feedback amplifier.

We will consider voltage relationships in this amplifier, but later we will see that currents in feedback amplifiers may also be of interest.

The difference circuit is constructed such that the output signal is the algebraic difference between the signals at input 1 and input 2. If input 2 is grounded via the switch, the signal reaching the amplifier input is

$$e_a = e_{in} - 0 = e_{in}.$$

In this instance, the overall open loop gain G_o is equal to the ideal amplifier gain A; that is,

$$G_o = \frac{e_{out}}{e_{in}} = \frac{e_{out}}{e_a} = A.$$

When the switch connects the difference circuit to the output of the feedback circuit, a portion of the output signal is subtracted from the input signal. The gain of the feedback circuit is F and is normally less than unity. The signal reaching the ideal amplifier is reduced to

$$e_a = e_{in} - Fe_{out},$$

and the output voltage will consequently be smaller than the open loop value. The output voltage is

$$e_{out} = A(e_{in} - Fe_{out}) = Ge_{in}.$$

Solving for the overall or closed loop gain G gives

$$G = \frac{e_{out}}{e_{in}} = \frac{A}{1 + AF}. \tag{6.1}$$

The closed loop gain G is always less than the open loop gain G_o for the negative feedback circuit.

We can visualize the reason for improved gain stability by considering the input signal presented to the ideal amplifier. In the open loop case the signal e_a is equal to e_{in} and remains constant regardless of the output signal. The output voltage varies directly with A; therefore, temperature variations or device replacement will then affect the output voltage. When the loop is closed, the signal presented to the input of the ideal amplifier is $e_a = e_{in} - Fe_{out}$. The output voltage is again given by Ae_a. If the gain were to increase, the output voltage would also increase. However, as e_{out} rises the signal e_a is reduced; thus, the increase in A is partially offset by the decrease in e_a to reduce variations in e_{out}. The following section will consider the gain stabilization of feedback amplifiers in a quantitative manner.

A. Gain Stability

The gain stability can be discussed most readily if related to the sensitivity function which compares the fractional change of a dependent variable to the fractional change of an independent variable. The sensitivity of e_{out} with respect to ideal amplifier gain A is

$$S_{e_{out}, A} = \frac{\partial e_{out}/e_{out}}{\partial A/A} = \frac{A}{e_{out}} \frac{\partial e_{out}}{\partial A}. \tag{6.2}$$

If the dependent variable exhibits a direct variation with the independent variable a sensitivity of unity will result. This means that a 20% variation of the independent variable leads to a 20% variation of the dependent variable.

In an open loop amplifier the sensitivity of output voltage with respect to gain is

$$\text{(Open loop)} \quad S_{e_{out}, A} = \frac{A}{e_{out}} \frac{\partial e_{out}}{\partial A} = \frac{A}{Ae_{in}} e_{in} = 1.$$

This figure can be compared to that of the closed loop amplifier. Since

$$e_{out} = \frac{A}{1 + AF} e_{in},$$

then

$$\frac{\partial e_{out}}{\partial A} = \frac{e_{in}}{(1 + AF)^2}$$

and

$$\text{(Closed loop)} \quad S_{e_{out}, A} = \frac{A(1 + AF)}{Ae_{in}} \frac{e_{in}}{(1 + AF)^2} = \frac{1}{1 + AF}.$$

The fractional change in e_{out} is now less than the fractional change in A by a factor of $1/(1 + AF)$. For example, if $1 + AF = 100$, a 10% change in A results in an approximate change in e_{out} of 0.1%.

The sensitivity of G with respect to A is equal to the closed loop value of $S_{e_{out}, A}$; that is,

$$S_{G, A} = \frac{1}{1 + AF}. \tag{6.3}$$

The overall gain becomes very stable as $(1 + AF)$ approaches large values.

We might consider an interesting limiting case in the calculation of gain stability. If AF is much greater than unity, the gain equation is

$$G = \frac{A}{1 + AF} \approx \frac{A}{AF} = \frac{1}{F}. \tag{6.4}$$

Under this condition, the overall amplifier gain depends only on the feedback factor and is independent of the ideal amplifier gain A. The sensitivity for this case approaches zero and the amplifier is insensitive to changes in A. Overall gain now depends on F, but this feedback factor is generally determined by passive components with almost arbitrary preciseness. Off-the-shelf resistors and capacitors are accurate enough for many applications while precise, low temperature coefficient elements can be secured for highly critical amplifiers.

We note that sensitivity of G with respect to A decreases as $(1 + AF)$ takes on larger values. The overall gain G simultaneously decreases with increases in $(1 + AF)$. Thus, gain stability is achieved at the expense of overall gain.

We can also calculate the gain sensitivity with respect to changes in the feedback factor F. This value is found to be

$$S_{G,F} = \frac{-AF}{1 + AF}. \tag{6.5}$$

Numerically, the sensitivity of G with respect to F is greater than the sensitivity of G with respect to A. However, as temperature changes occur, G may be affected equally by changes in A and changes in F. The total derivative of G with respect to temperature T is

$$\frac{dG}{dT} = \frac{\partial G}{\partial A}\frac{dA}{dT} + \frac{\partial G}{\partial F}\frac{dF}{dT}.$$

This derivative can also be expressed as

$$\frac{dG}{dT} = \frac{G}{A} S_{G,A} \frac{dA}{dT} + \frac{G}{F} S_{G,F} \frac{dF}{dT}.$$

The sensitivity of G with respect to temperature is a meaningful quantity and can be written as

$$S_{G,T} = \frac{T}{G}\frac{dG}{dT} = S_{G,A} S_{A,T} + S_{G,F} S_{F,T}. \tag{6.6}$$

Although $S_{G,F}$ is greater than $S_{G,A}$, the change in A with temperature is generally much greater than the corresponding change in F. Thus, gain changes may be as significant as feedback factor changes.

From a practical standpoint it is often convenient to relate variations in G to the temperature coefficients of A and F. Temperature coefficients are generally specified by the manufacturer and can be used directly in calculating temperature effects. The temperature coefficient of G relates the fractional change in G to the change in T (rather than the fractional change in T); thus,

$$TC_G = \frac{1}{G}\frac{dG}{dT} = \frac{1}{T}S_{G,T}. \qquad (6.7)$$

Using Eqs. (6.6) and (6.7) and the definition of temperature coefficients allows us to write

$$TC_G = S_{G,A}TC_A + S_{G,F}TC_F. \qquad (6.8)$$

Since TC_F usually depends on passive components while TC_A is a function of active devices, TC_F is much smaller than TC_A. An example will demonstrate the relative importance of these terms.

Example 6.1. A feedback amplifier is constructed such that $F = 1/100$ and $A = 10{,}000$. The temperature coefficient of the feedback network is ± 25 ppm/°C (parts per million per degree) and TC_A is 3000 ppm/°C. All quantities are room temperature values. Calculate the closed loop gain and the maximum temperature coefficient of this gain.

Solution. The closed loop gain is calculated from Eq. (6.1) to be $G = 99$. We note that the gain can be approximated with 1% accuracy by $G = 1/F$. The sensitivities are found from Eqs. (6.3) and (6.5). These values are $S_{G,A} = 9.9 \times 10^{-3}$ and $S_{G,F} = -9.9 \times 10^{-1}$. From Eq. (6.8) the temperature coefficient of closed loop gain is

$$TC_G = (9.9 \times 10^{-3})(3 \times 10^3) + (9.9 \times 10^{-1})(25) = 54.5 \text{ ppm/°C}.$$

Changes in the feedback factor and the ideal gain A with temperature contribute almost equally to TC_G in this example. Note that the negative value of TC_F was used to find the maximum value of TC_G.

B. Signal-to-Noise Ratio

A signal containing distortion is often characterized by the signal-to-noise ratio. This ratio can be expressed in terms of rms voltages or powers. Let us assume that we have an input voltage that can be expressed as

$$e_{in} = e_s + e_n,$$

where e_s is the pure signal of interest and e_n is a noise component resulting from any of the possible electronic noise sources. The noise component could be introduced by resistors in the circuit, by pickup of extraneous signals, or by the active devices used to generate the signal e_s. The noise component e_n may contain white noise, harmonic or nonlinear distortion, or an extraneous signal of almost any shape. Obviously, it may often be difficult to measure e_n in view of the possible complex makeup of this component. Assuming that e_n can be measured allows us to measure the signal-to-noise ratio of e_{in} which is given by

$$\text{SNR} = \frac{e_s}{e_n}. \qquad (6.9)$$

If e_{in} is now presented to a feedback amplifier that is assumed to add no noise to the amplified signal, the output voltage also consists of a noise component and the pure signal component. Both signal and noise components of e_{in} are amplified equally leading to the same SNR for both e_{in} and e_{out}. We conclude here that the SNR of an input signal is not improved by amplification.

Most amplifiers of significant gain will introduce a noise component into the signal that is being amplified. This noise may be white noise, $1/f$ noise, nonlinear distortion, or extraneous noise picked up from unwanted sources (60 Hz hum and *rf* signals are examples of common extraneous pickup). In general, the SNR at the amplifier output is decreased over that measured at the input. In fact, because most of the noise components introduced increase with signal levels, the noise generated by the amplifier often is the dominant noise term.

If we designate the noise term introduced within the amplifier by e_{an}, the output signal of an open loop amplifier becomes

$$e_{out} = Ae_{in} + e_{an} = Ae_s + Ae_n + e_{an}.$$

To examine the effect of feedback on amplifier noise we will neglect the noise contained in the input signal. In this case the SNR at the amplifier output is

$$\text{SNR} = \frac{Ae_s}{e_{an}}.$$

If feedback is now applied, the output signal decreases from the value without feedback. Since noise introduced by the amplifier depends on the magnitude of the output voltage, a meaningful comparison of the open loop and closed loop amplifiers can only be made for equal output voltages. Hence, after feedback is applied we assume that the input signal is increased in magnitude, without affecting the input SNR, until the output signal equals the open loop value. The noise introduced by the amplifier will again equal e_{an}, but a portion of the output noise will be fed back and subtracted from the input, resulting in a lower value of output noise e_{gn}. The noise fed back to the input is $-Fe_{gn}$. We can solve for e_{gn} since

$$e_{gn} = e_{an} - AFe_{gn},$$

giving

$$e_{gn} = \frac{e_{an}}{1 + AF}. \tag{6.10}$$

The output noise introduced by the amplifier is now less than the open loop value by the same factor as the reduction in gain. Again neglecting noise in the input signal and recalling that this signal has been increased enough to yield an output voltage of Ae_s, we find that

$$\text{SNR} = \frac{Ae_s}{e_{an}}(1 + AF). \tag{6.11}$$

The signal-to-noise ratio of the feedback amplifier is better than that of the open loop amplifier by the factor $(1 + AF)$.

At this point we must examine the validity of the assumption that the input signal can be increased without increasing the SNR of the input signal. Basically, the gain of the closed loop amplifier is less than the open loop value. To obtain equal output signals in the two cases requires that a preamplifier be inserted between the input signal and the feedback amplifier. This preamplifier will in general introduce noise that will degrade the SNR of the signal. Thus, the improvement in SNR indicated by Eq. (6.11) is overly optimistic and represents the maximum possible improvement. If the noise at the output is primarily nonlinear distortion or a component that results from large output signal swings, the preamplifier noise may be negligible. Equation (6.11) may then accurately describe the SNR of the amplifier. On the other hand, if the noise is introduced primarily within the preamplifier, the SNR may be improved little over the open loop amplifier. Typical of this case is the preamplifier that requires a very high input impedance leading to greater values of resistor noise and extraneous pickup. One must be rather careful when applying feedback to improve SNR, giving prime consideration to the points at which noise is introduced within the amplifier stages.

C. Bandwidth Improvement

Certain configurations of feedback amplifier lead to considerable improvements in bandwidth over the open loop amplifier. For the negative feedback amplifier of Fig. 6.1, let us assume that the ideal amplifier has a radian bandwidth of ω_2 and approaches a rolloff of 6 db/octave above this value. The open loop gain can then be written as

$$A = \frac{A_{MB}}{1 + j\omega/\omega_2},$$

where A_{MB} is the midband gain of the amplifier. The gain-bandwidth product is defined as $GBW = A_{MB}\omega_2$.

When the loop is closed we can apply Eq. (6.1) to calculate the gain G as a function of frequency, giving

$$G = \frac{A}{1 + AF} = \frac{\dfrac{A_{MB}}{1 + j(\omega/\omega_2)}}{1 + \dfrac{A_{MB}F}{1 + j(\omega/\omega_2)}}.$$

After manipulating we can write this gain as

$$G = \frac{A_{MB}}{1 + A_{MB}F} \frac{1}{1 + j[\omega/\omega_2(1 + A_{MB}F)]}. \tag{6.12}$$

The bandwidth with feedback is $\omega_2(1 + A_{MB}F)$ increasing over the open loop case by the same factor as the gain reduction. The gain-bandwidth product is then constant as the loop is closed, allowing gain and bandwidth to be interchanged

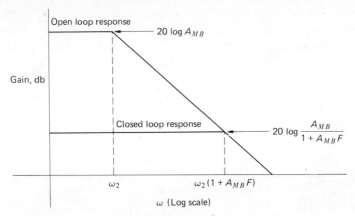

Fig. 6.2 Frequency responses of open loop and closed loop amplifiers.

directly by feedback. Figure 6.2 compares the frequency responses of the open loop and closed loop amplifiers.

The extension of bandwidth indicated by Eq. (6.12) applies only to the ideal feedback stage of Fig. 6.1. In practical cases, F may also be frequency dependent and loading may not be negligible. Generally, the bandwidth given by Eq. (6.12) should be considered as an upper limit approached in few cases. An outstanding example of an amplifier that very nearly approximates the ideal case is the operational amplifier considered later in this chapter.

Not all feedback configurations will extend the voltage bandwidth of the amplifier. An example of this occurs in a pentode amplifier with an unbypassed cathode resistance. The cathode resistance introduces negative feedback which reduces the gain of the stage, but bandwidth is not improved. In this case, however, the feedback voltage developed at the cathode does not exhibit the same frequency dependence as does the output signal. The feedback voltage is constant with frequency while the capacitance that shunts the pentode load resistance leads to a frequency dependent output voltage. A rather intuitive criterion for increasing the bandwidth with negative feedback is that both the signal fed back to the input and the output voltage must exhibit similar frequency dependence. A more quantitative discussion of this point will be undertaken after we consider the difference between voltage and current feedback.

D. Types of Feedback

The feedback signal can be derived from one of two output quantities: either output voltage or output current. Furthermore, the feedback signal can be subtracted from the input in two ways to form a difference signal at the input of the ideal amplifier. It is possible to subtract a feedback voltage from the input voltage by including the feedback signal in series with the input loop. This results in series feedback. It is also possible to use shunt feedback wherein the amplifier input is shunted by a network that decreases the input current by an amount

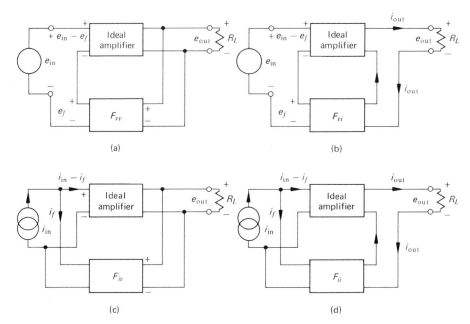

Fig. 6.3 Four possible feedback types: (a) voltage-series or voltage-ratio feedback; (b) current-series or transimpedance feedback; (c) voltage-shunt or transadmittance feedback; (d) current-shunt or current-ratio feedback.

proportional to the feedback signal to form a difference current. Thus there are four basic types of feedback: voltage-series, voltage-shunt, current-series, and current-shunt. Figure 6.3 indicates these combinations.

In general, the feedback network will be different for each amplifier configuration. The transfer function F of this network has different units for each case. For the voltage-series feedback F generates a feedback voltage e_f proportional to the output voltage. In this case F is a voltage-ratio transfer function. For current-series feedback F must generate a feedback voltage in proportion to output current. The transfer function F is now a transimpedance. Voltage-shunt feedback requires F to be a transadmittance generating a feedback current i_f that is proportional to output voltage. Current-shunt feedback requires F to be a current-ratio transfer function developing a feedback current i_f in proportion to the output current i_{out}.

It is instructive to consider the closed loop gain expressions for the four types of amplifier. The voltage-series type corresponds closely to the idealized amplifier of Fig. 6.1 for which the gain expression has been developed. The low frequency gain is

$$G = \frac{A}{1 + AF_{vv}}, \qquad (6.13)$$

where A is the open loop or ideal amplifier voltage gain. The feedback factor is $F_{vv} = e_f/e_{\text{out}}$.

For the current-series case we must relate the output current to the input voltage. The gain is then a transadmittance which we shall designate G_{iv}. If F_{vi} is defined by

$$F_{vi} = \frac{e_f}{i_{out}},$$

then the transadmittance of the amplifier is

$$G_{iv} = \frac{i_{out}}{e_{in}} = \frac{A_{iv}}{1 + A_{iv}F_{vi}}, \qquad (6.14)$$

where A_{iv} is the transadmittance of the ideal amplifier.

Equation (6.14) has the same form as Eqs. (6.1) and (6.13); however, the output quantity associated with the former equation is current rather than voltage. Noting this difference, we might extend some previous conclusions applying to the voltage amplifier to the transadmittance amplifier. The arguments for improved stability, improved bandwidth, and reduction of noise with feedback are valid for the transadmittance amplifier, but current is the output quantity to which the improvements apply rather than voltage. Equations developed earlier for frequency response, sensitivity, and noise performance are easily extended to the transadmittance amplifier with this change in output variable and appropriate units defining A and F.

It is often useful to know the voltage gain of the transadmittance amplifier. This quantity can be found by expressing the output voltage as

$$e_{out} = Ae_a = A(e_{in} - e_f),$$

where A is the voltage gain of the ideal amplifier. The feedback signal is

$$e_f = F_{vi}i_{out} = F_{vi}A_{iv}e_a.$$

Combining the two preceding equations results in

$$G = \frac{A}{1 + A_{iv}F_{vi}}. \qquad (6.15)$$

The voltage gain is reduced by the same factor as the transadmittance when feedback is applied.

Equation (6.15) does not correspond to Eq. (6.1) since the gain of the numerator is a voltage gain while the denominator contains a transadmittance. We cannot then apply the stability, bandwidth, and noise results to the output voltage even though these results apply to the output current as previously mentioned. Generally, the output voltage will be less sensitive to parameter changes when feedback is applied, although it is easy to visualize examples wherein the sensitivity of output voltage is only slightly affected by feedback while the output current becomes very stable. Suppose an amplifier drives a load resistance that is highly temperature-dependent. The output current may be stabilized with current feedback, yet the output voltage amplitude would vary with temperature change.

A similar situation occurs in relation to bandwidth improvement. The output current bandwidth of a transadmittance amplifier is $\omega_2(1 + A_{iv}F_{vi})$, where ω_2 is the current bandwidth of the ideal amplifier. The output voltage bandwidth may not exhibit the same bandwidth unless the output voltage is proportional to output current for all frequencies up to the upper corner frequency. In general $e_{out} = Z_L i_{out}$, where Z_L is the load impedance. Often the reactive component of Z_L (due perhaps to shunt capacitance) will significantly lower the magnitude of load impedance at frequencies below $\omega_2(1 + A_{iv}F_{vi})$. Hence voltage bandwidth can be less than current bandwidth.

The overall transimpedance of the voltage-shunt feedback amplifier is found to be

$$G_{vi} = \frac{e_{out}}{i_{in}} = \frac{A_{vi}}{1 + A_{vi}F_{iv}}, \tag{6.16}$$

where A_{vi} is the transimpedance of the ideal amplifier. The transimpedance amplifier has a feedback factor F_{iv} which represents a transadmittance.

The current-shunt or current amplifier has an overall current gain given by

$$G_i = \frac{i_{out}}{i_{in}} = \frac{A_i}{1 + A_i F_{ii}}, \tag{6.17}$$

where A_i is the current gain of the ideal amplifier.

In summary, the four types of amplifier each obey the closed loop gain expression of Eq. (6.1). This gain may correspond to a voltage gain, a current gain, a transadmittance, or a transimpedance depending on the configuration. All equations relating to bandwidth, stability, and distortion apply to the output and input quantities defined by the gain equation for the amplifier in question.

It is often more meaningful from a practical standpoint to discuss overall voltage gain of these amplifiers. Equation (6.15) indicates such an expression for the transadmittance amplifier. As pointed out earlier in relation to this equation, the bandwidth, stability, and noise equations based on Eq. (6.1) do not necessarily hold for voltage gain of the transadmittance amplifier. We will discuss later the voltage gain of transimpedance and current amplifiers.

One point that is rather significant in discussing the ideal feedback amplifiers of Fig. 6.3 is that we have implied certain impedance levels in our derivations that should be considered more fully. The feedback network of the voltage amplifier shown in Fig. 6.3(a) must have negligible output impedance compared to the ideal amplifier input impedance and must also have a high input impedance compared to R_L. If such were not the case, loading effects would cause the gain equation to be inaccurate. In Fig. 6.3(b) the input impedance to the feedback network must be small compared to R_L and the output impedance must be small compared to the input impedance of the ideal amplifier. In parts (c) and (d) the output impedances of the feedback network should be very large. The input impedance of the network in Fig. 6.3(c) should be large compared to R_L while that of part (d) should exhibit a low input impedance to the feedback network. Departures from these ideal conditions can be treated as shown in the following section, but they lead to somewhat more complex equations.

E. Effect of Feedback on Impedance Levels

The input and output impedances of an amplifier can be changed greatly when feedback is applied. Occasionally feedback is used to obtain some desired impedance condition, although in many other instances the effect of feedback on impedance level may be of secondary interest.

Input impedance. In general we can say that the input impedance of an amplifier will be increased with series feedback and decreased with shunt feedback. We can demonstrate the basis for the conclusions by referring to the amplifiers of Fig. 6.4.

The basic amplifier is now assumed to consist of an ideal amplifier with finite input and output resistances, R_{ia} and R_{oa} respectively. Before feedback is applied the input resistance is then R_{ia} for the basic amplifier. When series feedback is

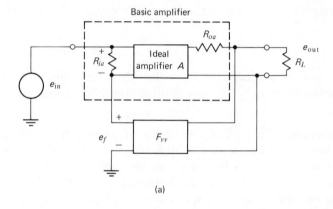

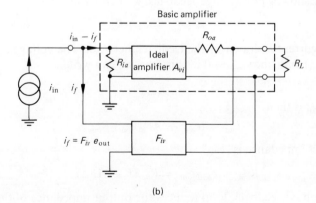

Fig. 6.4 Feedback amplifiers: (a) series; (b) shunt.

applied, the input current i_{in} is given by

$$i_{in} = (e_{in} - e_f)/R_{ia} = (e_{in} - F_{vv}e_{out})/R_{ia}$$
$$= (e_{in} - GF_{vv}e_{in})/R_{ia} = \frac{e_{in}}{R_{ia}(1 + AF_{vv})}.$$

The input impedance to the series feedback amplifier is

$$R_{in} = e_{in}/i_{in} = R_{ia}(1 + AF_{vv}). \tag{6.18}$$

Normally the factor $(1 + AF_{vv})$ is much greater than unity, indicating that the input impedance has greatly increased. The same result applies to the transadmittance amplifier wherein current-series feedback is applied.

For the shunt feedback case we can write

$$e_{in} = (i_{in} - i_f)R_{ia} = (i_{in} - F_{iv}e_{out})R_{ia}$$
$$= (i_{in} - G_{vi}F_{iv}i_{in})R_{ia} = \frac{i_{in}R_{ia}}{(1 + A_{vi}F_{iv})}.$$

The input impedance to the shunt feedback amplifier is

$$R_{in} = e_{in}/i_{in} = \frac{R_{ia}}{(1 + A_{vi}F_{iv})}. \tag{6.19}$$

This impedance is normally much smaller than the impedance of the basic amplifier without feedback. The same result applies to the current-shunt or current amplifier.

Output impedance. The output impedance of an amplifier will decrease for voltage feedback and increase for current feedback. Consider the stages of Fig. 6.5. The output impedance can be calculated by finding the Thévenin equivalent resistance at the output terminals. To do this we assume that the input source is shorted and that the voltage e_{out} appears at the output terminals, resulting in a current i_{out} as shown. For voltage feedback we can write

$$i_{out} = (e_{out} - e_{a2})/R_{oa},$$

assuming no current is drawn by the feedback network. The voltage e_{a2} is simply $A(e_{in} - F_{vv}e_{out})$, but since $e_{in} = 0$ in this case, we can write

$$e_{a2} = -AF_{vv}e_{out}.$$

The current i_{out} is then

$$i_{out} = (e_{out} + AF_{vv}e_{out})/R_{oa}$$

and the output impedance is

$$R_{out} = e_{out}/i_{out} = R_{oa}/(1 + AF_{vv}). \tag{6.20}$$

Closing the voltage feedback loop reduces the output impedance of the amplifier by the factor $(1 + AF_{vv})$. The same result applies to the voltage-shunt or transimpedance amplifier.

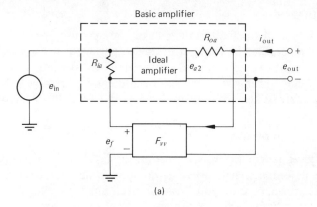

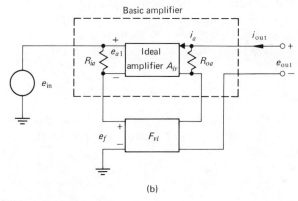

Fig. 6.5 Feedback amplifiers: (a) voltage feedback; (b) current feedback.

For the current feedback stage we can write

$$i_{\text{out}} = e_{\text{out}}/R_{oa} + i_a,$$

assuming negligible voltage develops across the input to the feedback network. The current i_a is

$$i_a = -A_{iv}e_{a1} = -A_{iv}F_{vi}i_{\text{out}}.$$

The output current is now expressed as

$$i_{\text{out}} = (e_{\text{out}}/R_{oa}) - A_{iv}F_{vi}i_{\text{out}}$$

and

$$R_{\text{out}} = e_{\text{out}}/i_{\text{out}} = R_{oa}(1 + A_{iv}F_{vi}). \tag{6.21}$$

Closing the current feedback loop increases the output impedance by the factor $(1 + A_{iv}F_{vi})$. The same result applies to the current-shunt or current amplifier.

If high output impedance is desirable, current feedback may be used, while voltage feedback is appropriate when low output impedance is a major requirement. Higher input impedance levels result from series feedback, while shunt feedback reduces the input impedance.

F. AC and DC Feedback

DC feedback is used to stabilize dc gain or the operating point with respect to changes in temperature or changes in device parameters. AC feedback is used to stabilize ac gain, extend the bandwidth, decrease distortion, or control impedance levels. Often an amplifier will use both dc and ac feedback, but either can be used separately. When both are used, the ac and dc feedback factors may differ.

6.2 PRACTICAL FEEDBACK AMPLIFIERS

The theory developed in the previous section applies to the ideal feedback amplifier with an ideal feedback network and an ideal difference circuit. While finite impedances can occur in the ideal case, a given impedance must be either small enough or large enough to be negligible, depending on the location of this impedance in the circuit. Some practical amplifiers approximate the ideal case closely enough to be analyzed with the ideal theory. Others can be analyzed with only slight extensions of the theory. Still other amplifiers require special considerations to analyze completely. We will consider examples of each case in the following sections.

A. Voltage Amplifier

The amplifier of Fig. 6.6 indicates a three-stage network with voltage-series feedback. We will assume that R_B, V_Z, and R_Z are selected to produce suitable dc biasing.

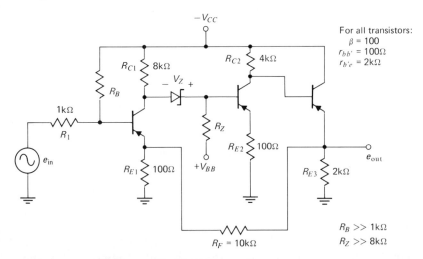

Fig. 6.6 Voltage amplifier.

The calculations can be simplified if the circuit is redrawn as shown in Fig. 6.7. We note that the feedback resistance R_F does not return to ground as shown in Fig. 6.7, but actually returns to the emitter of stage 1. It is easy to show,

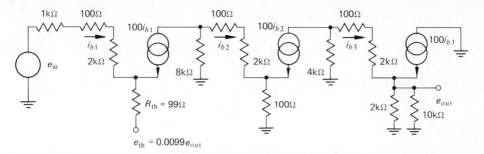

Fig. 6.7 Equivalent circuit for amplifier of Fig. 6.6.

however, that since the output voltage applied to one end of R_F is much greater than the feedback voltage appearing at the other end of R_F, we can assume that this resistance is returned to ground. A Thévenin equivalent of R_F, R_{E1}, and e_{out} is taken to evaluate the feedback voltage. The amplifier of Fig. 6.7 can be drawn to correspond to the ideal stages discussed earlier. Figure 6.8 shows the appropriate identifications.

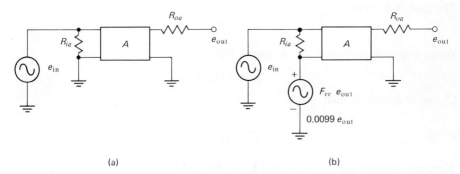

Fig. 6.8 Circuits for feedback calculations: (a) open loop amplifier; (b) closed loop amplifier.

The open loop amplifier of Fig. 6.8(a) differs from the closed loop circuit only by the fact that the feedback voltage is shorted for the open loop case. Thus, we need only evaluate the quantities A, F_{vv}, R_{ia}, and R_{oa} to calculate the overall voltage gain G and the input and output impedances. It must be emphasized that the open loop circuit used for calculations may not correspond to the actual open loop amplifier. For example, to open the feedback loop physically we could simply remove the 10 kΩ feedback resistance. The open loop gain calculated in this case might differ from the required value A found from the equivalent circuit of Fig. 6.7 which includes any loading effects of R_F.

We can now calculate the desired quantities. The input impedance to the open loop amplifier R_{ia} is

$$R_{ia} = R_1 + r_{bb'} + r_{b'e} + (\beta + 1)R_{th} = 13.1 \text{ k}\Omega.$$

The impedance R_{oa} is given by

$$R_{oa} = R_{E3} \| R_F \| \frac{r_{b'e} + r_{bb'} + R_{C2}}{\beta + 1} = 58 \, \Omega.$$

The open loop voltage gain is

$$A = A_{v1} A_{v2} A_{v3},$$

where

$$A_{v1} = -\frac{\beta R_{C1} \| [r_{bb'} + r_{b'e} + (\beta + 1) R_{E2}]}{R_1 + r_{bb'} + r_{b'e} + (\beta + 1) R_{th}} = -37,$$

$$A_{v2} = -\frac{\beta R_{C2} \| [r_{bb'} + r_{b'e} + (\beta + 1)(R_{E3} \| R_F)]}{r_{bb'} + r_{b'e} + (\beta + 1) R_{E2}} = -32,$$

and

$$A_{v3} = \frac{(\beta + 1) R_{E3} \| R_F}{r_{bb'} + r_{b'e} + (\beta + 1) R_{E3} \| R_F} = 0.99.$$

The gain A is then

$$A = 1170.$$

When feedback is applied, the feedback factor is

$$F_{vv} = 0.0099.$$

Closed loop gain is then

$$G = \frac{A}{1 + AF_{vv}} = 93.$$

The input impedance is found from Eq. (6.18) to be

$$R_{in} = R_{ia}(1 + AF_{vv}) = 165 \, k\Omega.$$

Output impedance is calculated from Eq. (6.20) and is

$$R_{out} = R_{oa}/(1 + AF_{vv}) = 4.7 \, \Omega.$$

It is often useful to approximate the closed loop gain without making detailed calculations. We can do this by noting that $G \approx 1/F_{vv}$ and

$$F_{vv} = \frac{R_{E1}}{R_{E1} + R_F} = 0.0099,$$

giving

$$G \approx 1/0.0099 = 101.$$

An error of eight percent results from applying this approximation.

B. Transadmittance Amplifier

A very simple transadmittance amplifier is shown in Fig. 6.9. We can easily find the transadmittance of this stage by conventional circuit analysis to be

$$G_{iv} = \frac{i_{out}}{e_{in}} = \frac{-\beta}{R_1 + r_{bb'} + r_{b'e} + (\beta + 1)R_E} \frac{R_C}{R_C + R_L}.$$

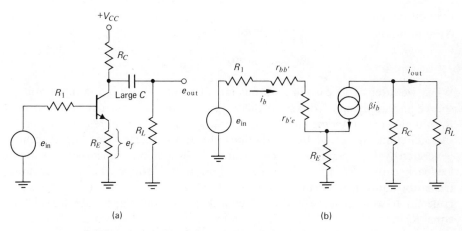

Fig. 6.9 Transadmittance amplifier: (a) actual circuit; (b) equivalent circuit.

The feedback method is more complex than conventional analysis in this instance, but is quite useful for multistage feedback amplifiers. We will analyze this simple stage merely to demonstrate the method. The feedback signal e_f appears across the emitter resistance; thus, if R_E is removed no feedback is present. The open loop gain is

$$A_{iv} = \frac{-\beta}{R_1 + r_{bb'} + r_{b'e}} \frac{R_C}{R_C + R_L}.$$

The feedback factor after inserting R_E is

$$F_{vi} = \frac{e_f}{i_{out}} = \frac{(\beta + 1)i_b R_E}{i_{out}} = \frac{-(R_C + R_L)R_E}{\alpha R_C}.$$

Applying Eq. (6.14) results in

$$G_{iv} = \frac{i_{out}}{e_{in}} = \frac{\dfrac{-\beta}{R_1 + r_{bb'} + r_{b'e}} \dfrac{R_C}{R_C + R_L}}{1 + \dfrac{\beta}{R_1 + r_{bb'} + r_{b'e}} \dfrac{R_C}{R_C + R_L} \dfrac{R_E(R_C + R_L)}{\alpha R_C}}$$

$$= \frac{-\beta}{R_1 + r_{bb'} + r_{b'e} + (\beta + 1)R_E} \frac{R_C}{R_C + R_L}.$$

The input impedance is

$$R_{in} = R_{ia}(1 + A_{iv}F_{vi}) = (R_1 + r_{bb'} + r_{b'e}) \frac{R_1 + r_{bb'} + r_{b'e} + (\beta + 1)R_E}{R_1 + r_{bb'} + r_{b'e}}$$

$$= R_1 + r_{bb'} + r_{b'e} + (\beta + 1)R_E.$$

We must be careful in calculating the output impedance of this stage. The current that flows through R_E to develop the feedback voltage is emitter current rather than output current. Under operating conditions the emitter current bears a fixed relationship to i_{out}. In fact we can write

$$i_{out} = -\alpha i_e \frac{R_C}{R_C + R_L}.$$

When we attempt to calculate output impedance by shorting the input voltage source and applying a voltage to the output terminals, a current flows through R_C with no corresponding emitter current flow. Thus, i_{out} is not related to i_e under these circumstances and Eq. (6.21) does not apply. In fact, since no current is generated by the dependent collector source, the output impedance with or without feedback equals R_C. In practice this is not strictly true as the more accurate transistor equivalent circuit would include resistors from collector to base and collector to emitter also. Without feedback the output impedance would then equal R_C in parallel with some effective impedance of the transistor. With feedback the transistor impedance would increase sharply and the overall output impedance would again approach R_C.

We are often more interested in voltage gain of an amplifier even though the configuration leads to classification of the circuit as a transadmittance stage. For the circuit of Fig. 6.9 we can easily calculate the open and closed loop voltage gains by noting that $e_{out} = R_L i_{out}$ leading to

$$A = A_{iv} R_L$$

and

$$G = G_{iv} R_L.$$

The closed loop voltage gain becomes

$$G = \frac{-\beta}{R_1 + r_{bb'} + r_{b'e} + (\beta + 1)R_E} \frac{R_C R_L}{R_C + R_L}.$$

It can easily be shown that the voltage feedback factor in this case is

$$F = \frac{e_f}{e_{out}} = \frac{F_{vi}}{R_L}.$$

C. Transimpedance Amplifier

It is possible to avoid using the difference circuit indicated in Fig. 6.1 if the basic amplifier is an inverting amplifier. A difference signal can then be generated by using a resistive adder which combines the input and inverted output signals.

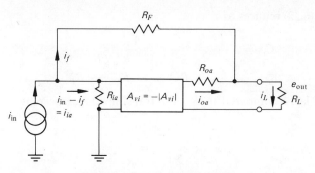

Fig. 6.10 Transimpedance amplifier.

Figure 6.10 shows a transimpedance amplifier using an inverting element. The quantity A_{vi} refers to the unloaded value of transimpedance for the basic amplifier.

To find the overall transimpedance we begin by writing

$$e_{out} = -|A_{vi}|(i_{in} - i_f) - R_{oa}i_{oa}.$$

The current i_{oa} equals $e_{out}/R_L - i_f$, giving

$$e_{out} = \frac{-|A_{vi}|i_{in} + (|A_{vi}| + R_{oa})i_f}{\left(1 + \dfrac{R_{oa}}{R_L}\right)}.$$

The feedback current can be expressed as

$$i_f = \frac{e_{in} - e_{out}}{R_F} = \frac{R_{ia}(i_{in} - i_f) - e_{out}}{R_F}.$$

Isolating i_f in this equation results in

$$i_f = \frac{R_{ia}i_{in} - e_{out}}{R_F + R_{ia}}.$$

Combining the equations for e_{out} and i_f allows the closed loop transimpedance to be expressed as

$$G_{vi} = \frac{e_{out}}{i_{in}} = \frac{-|A_{vi}|\left(\dfrac{R_E - R_{oa}R_{ia}/|A_{vi}|}{R_F + R_{ia}}\right)}{1 + \dfrac{R_{oa}}{R_L} + \dfrac{|A_{vi}| + R_{oa}}{R_F + R_{ia}}}. \tag{6.22}$$

Equation (6.22) is rather imposing and not of great value in the practical case. The transimpedance amplifier is of value primarily in the general area of operational amplifiers and in this application, the closed loop transimpedance can often be simplified. The feedback resistance is normally much larger than

R_{oa} and Eq. (6.22) reduces to

$$G_{vi} = \frac{-|A_{vi}|}{1 + \dfrac{R_{ia} + |A_{vi}|}{R_F} + \dfrac{R_{oa}}{R_L}\left(\dfrac{R_F + R_{ia}}{R_F}\right)}. \tag{6.23}$$

When R_L is large compared to R_{oa} the transimpedance becomes

$$G_{vi} = \frac{-|A_{vi}|}{1 + \dfrac{R_{ia} + |A_{vi}|}{R_F}}. \tag{6.24}$$

The figure of most interest in operational amplifier design is overall voltage gain rather than transimpedance. Generally a resistance will appear in series with the input as shown in Fig. 6.11.

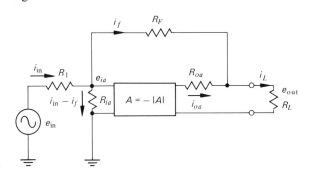

Fig. 6.11 Operational amplifier.

The gain A is the unloaded voltage gain of the basic amplifier. The output voltage can be written as

$$e_{out} = -|A|\, e_{ia} - R_{oa} i_{oa}.$$

The current $i_{oa} = i_L - i_f$, which can also be written as

$$i_{oa} = \frac{e_{out}}{R_L} - \frac{e_{ia} - e_{out}}{R_F}.$$

With this substitution, the output voltage becomes

$$e_{out} = \frac{e_{ia}\left(-|A| + \dfrac{R_{oa}}{R_F}\right)}{1 + \dfrac{R_{oa}}{R_L} + \dfrac{R_{oa}}{R_F}}. \tag{6.25}$$

The input voltage can be expressed as

$$e_{in} = e_{ia} + R_1 i_{in} = e_{ia} + R_1(i_F + e_{ia}/R_{ia})$$

$$= e_{ia}\left[1 + \frac{R_1}{R_F} + \frac{R_1}{R_{ia}}\right] - \frac{R_1}{R_F} e_{out}. \tag{6.26}$$

Through the use of Eq. (6.25) to eliminate e_{ia}, the closed loop gain expression is found to be

$$G = \frac{-|A| + \dfrac{R_{oa}}{R_F}}{1 + \dfrac{R_1}{R_{ia}} + \dfrac{R_{oa}}{R_L} + \dfrac{R_{oa}}{R_F} + \dfrac{R_1 R_{oa}}{R_F R_L} + \dfrac{R_1 R_{oa}}{R_{ia} R_L} + \dfrac{R_1 R_{oa}}{R_{ia} R_F} + \dfrac{(|A|+1)R_1}{R_F}}. \tag{6.27}$$

Often the impedance R_{oa} will be much smaller than R_F and the circuit is used to drive a high impedance load. For these practical conditions, the closed loop voltage gain becomes

$$G = \frac{-|A|}{1 + \dfrac{R_1}{R_{ia}} + \dfrac{(1+|A|)R_1}{R_F}}. \tag{6.28}$$

A further simplification occurs when R_1 is much smaller than R_{ia}, giving

$$G = \frac{-|A|}{1 + \dfrac{(1+|A|)R_1}{R_F}}. \tag{6.29}$$

Equation (6.29) can also be found by considering the feedback factor and the general relationship between open and closed loop gains. If R_{ia} is large enough to be negligible and R_{oa} is much smaller than R_F, the input voltage presented to the basic amplifier can be found from Fig. 6.12. The voltage e_{ia} is given by

$$e_{ia} = e_{in} - (e_{in} - e_{out}) \frac{R_1}{R_1 + R_F} = e_{in} \frac{R_F}{R_1 + R_F} + e_{out} \frac{R_1}{R_1 + R_F}. \tag{6.30}$$

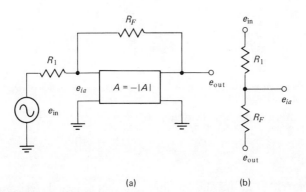

Fig. 6.12 Feedback amplifier using resistive adder: (a) amplifier; (b) resistive adder.

In the ideal feedback amplifier of Fig. 6.1, the voltage presented to the basic amplifier input was

$$e_a = e_{in} - F e_{out}.$$

When the nonideal adder and inverting stage are used, the basic amplifier input voltage given by Eq. (6.30) is

$$e_{ia} = \frac{R_F}{R_1 + R_F} e_{in} + F e_{out},$$

where $F = R_1/(R_1 + R_F)$. Neglecting the phase inversion of e_{out} in this case, the only difference in the two voltages is the attenuation factor $R_F/(R_1 + R_F)$. The output voltage for the amplifier of Fig. 6.12 is

$$e_{out} = -|A| e_{ia} = \frac{-|A| e_{in} R_F}{R_1 + R_F} - \frac{|A| e_{out} R_1}{R_1 + R_F}.$$

Solving for e_{out}/e_{in} gives

$$G = \frac{-|A|}{1 + |A| R_1/(R_1 + R_F)} \frac{R_F}{R_1 + R_F} = \frac{-|A|}{1 + |A| F} \frac{R_F}{R_1 + R_F}. \qquad (6.31)$$

The adder attenuates the input voltage by the factor $R_F/(R_1 + R_F)$ and reduces the overall gain by this same factor compared to the ideal stage of Fig. 6.1. A small amount of algebraic manipulation will demonstrate the equality of Eqs. (6.29) and (6.31).

A method called the reflected impedance method can also be used to calculate voltage gain or input impedance for this case. In Fig. 6.13(a) the amplifier is shown with a slightly different orientation.

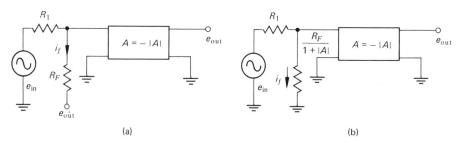

Fig. 6.13 Operational amplifier: (a) alternate schematic; (b) feedback resistance reflected to input.

The current through the feedback resistance is

$$i_f = \frac{e_{ia} - e_{out}}{R_F} = \frac{e_{ia}(1 + |A|)}{R_F}.$$

This same current will be drawn through the reflected resistance of value $R_F/(1 + |A|)$ shown in Fig. 6.13(b).

The effect of the feedback resistance is reflected to the input side of the amplifier. For input impedance and forward gain calculations this method is quite useful. The input impedance is now

$$R_{in} = R_1 + R_F/(1 + |A|) \qquad (6.32)$$

and the voltage gain can easily be calculated as

$$G = -|A|\frac{R_F/(1+|A|)}{R_1 + R_F/(1+|A|)}, \qquad (6.33)$$

which can be manipulated to yield Eq. (6.29). If R_{ia} is not large enough to ignore, it can be included in parallel with the reflected resistance in these calculations.

Example 6.2. Calculate the midband voltage gain and input impedance of the circuit of Fig. 6.14. Assume that the stages are properly biased and $\beta = 80$, $r_{bb'} = 100\ \Omega$, and $r_{b'e} = 1.5\ \text{k}\Omega$ for both stages. Neglect effects of biasing resistors.

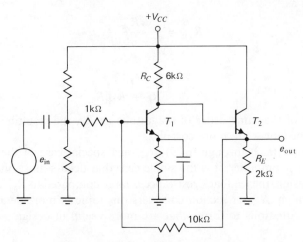

Fig. 6.14 Feedback amplifier for Example 6.2.

Solution. The amplifier circuit corresponds to the diagram of Fig. 6.13(a) with $R_1 = 1\ \text{k}\Omega$ and $R_F = 10\ \text{k}\Omega$. The output impedance of the emitter follower is small in comparison to R_F and can be neglected. The input impedance to the basic amplifier is $R_{ia} = r_{bb'} + r_{b'e} = 1.6\ \text{k}\Omega$. The voltage gain from the base of T_1 to the output is

$$A_v = \frac{-\beta[R_C \| (\beta+1)R_E]}{r_{bb'} + r_{b'e}} = -290.$$

This calculation assumes unity gain for the emitter follower.

Reflecting the effects of the feedback resistance to the input of the amplifier results in the circuit of Fig. 6.15.

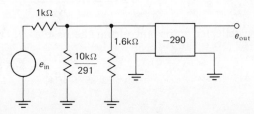

Fig. 6.15 Equivalent amplifier circuit.

The parallel combination of R_{ia} and $R_F/(1 + |A_v|)$ is 33.6 Ω. The input impedance is

$$R_{in} = 1000 + 33.6 = 1034 \, \Omega,$$

while the voltage gain is found from Eq. (6.33) to be

$$G = -290 \frac{33.6}{1034} = -9.4.$$

Checking this result by Eq. (6.29) or (6.31) yields the same value of voltage gain.

It can easily be shown for the preceding case that the output impedance can be expressed as

$$R_{out} = \frac{R_{oa}}{1 + |A_v| F_1},$$

where

$$F_1 = \frac{R_1 \| R_{ia}}{R_F + R_1 \| R_{ia}}.$$

If R_{oa} is of the same order of magnitude as R_F, the resulting equations for gain and impedances are more complex.

While Eqs. (6.23) through (6.29) are for a special case of the more general transimpedance amplifier, it is the single case that occurs most often in practice. In op amp design, the feedback resistance is large compared to R_{oa} and the voltage gain is very high. A later section considers this topic in more detail and extends the analysis equations to forms that are more useful in design work.

D. Current Amplifier

Figure 6.16 shows a two-stage amplifier with current feedback utilizing the shunt configuration. While the current gain could be calculated for this amplifier, we will again calculate voltage gain as the more important parameter. The reflected impedance method is conveniently applied to this problem. We note that the feedback resistance is connected to the emitter of T_2 rather than to the output. To reflect the effect of R_F to the input side requires that we calculate the voltage gain from the base of T_1 to the emitter of T_2 since R_F appears between these points. Assuming $\beta = 100$ for both transistors we calculate this gain to be

$$A_1 = -\frac{\alpha R_{1\,eff}}{R_{E1}} \approx -\frac{R_{1\,eff}}{R_{E1}}$$

$$= -\frac{R_{C1} \| (\beta + 1) R_{E2}}{R_{E1}} = -16.7.$$

The reflected impedance from the base of T_1 to ground is

$$\frac{R_F}{1 + |A_1|} = 565 \, \Omega.$$

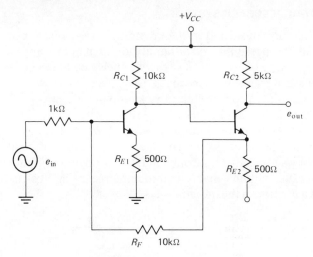

Fig. 6.16 Current amplifier.

The voltage gain of the second stage is

$$A_2 = -\frac{\alpha R_{C2}}{R_{E2}} \approx -\frac{R_{C2}}{R_{E2}} = -10.$$

The circuit of Fig. 6.17 applies to this case. The input impedance R_{ia} is approximately 50 kΩ and can be considered an open circuit since $R_F/(1 + |A_1|)$ is much smaller. The overall voltage gain is then

$$G = \frac{565}{1565} \times 167 = 60.3.$$

Overall input impedance including R_1 is

$$R_{in} = 1000 + 565 = 1565 \ \Omega.$$

There are obvious similarities between this configuration and that of the transimpedance amplifier and in both cases the reflected impedance method can be useful.

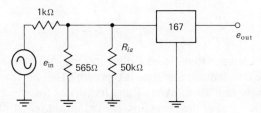

Fig. 6.17 Circuit showing R_F reflected to input.

6.3 STABILITY OF FEEDBACK SYSTEMS

The feedback stages considered in the preceding sections have been idealized to the extent that the possibility of instability or oscillations appears to be nonexistent. In neglecting high-frequency reactive effects we have chosen to ignore an important source of problems in feedback design. We must now consider methods of treating the stability problem in feedback amplifier design.

A feedback amplifier applies negative feedback while an oscillator uses positive feedback. Before considering the means by which unwanted positive feedback is introduced in an amplifier, let us consider the meaning of positive feedback. Figure 6.18 shows a positive feedback circuit that might exemplify an oscillator. In this instance, the output signal is used to reinforce the input signal by summing the input and a fraction of the output.

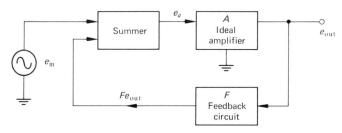

Fig. 6.18 Positive feedback circuit.

The expression for overall gain is found to be

$$G = \frac{A}{1 - AF}. \qquad (6.34)$$

If the return ratio AF approaches unity the overall gain approaches infinity, implying that an output signal can exist with no corresponding input. For sinusoidal oscillators the term AF is designed to approach unity at one particular frequency and at the same time the phase shift of this term is zero or multiples of 360°. The circuit then produces a sinusoid at this frequency with no external ac input. We will consider the specific case of oscillator design in Chapter 11.

Returning to the feedback amplifier of Fig. 6.1 using a difference circuit, we should examine the gain expression in more detail. Overall gain is

$$G = \frac{A}{1 + AF}, \qquad (6.1)$$

but this expression assumes that A and F are constant with frequency. If the return ratio happens to be frequency-dependent it is possible that positive feedback might occur. Let us assume that there exists some frequency for which the phase shift of the term AF is 180° and at the same time the magnitude equals or exceeds unity. Positive feedback is then present and oscillations at that frequency can result.

While it is possible that unwanted oscillations might occur with only two active stages, it is highly probable that three or more stages will result in oscillations unless special precautions are taken. The reactive elements that are often negligible in multistage design can become significant when a feedback loop is added. Depletion region capacitances, stray capacitance, stray inductance, and other reactive elements are potential sources of instability problems.

A. Prediction of Stability

Stability can be predicted in several ways given that the overall gain or the quantity AF can be expressed as a function of frequency. Perhaps the most straightforward method available to indicate stability is to plot frequency responses of both magnitude and phase of AF. The Nyquist criterion is also useful, but adds no additional information beyond that contained in the frequency responses. A typical plot of magnitude and phase of AF as a function of frequency is shown in Fig. 6.19.

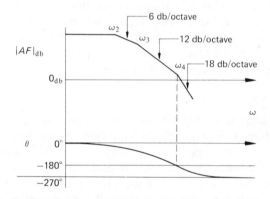

Fig. 6.19 Magnitude and phase responses of return ratio AF.

The critical point occurs when the phase of AF has reached $-180°$ leading to positive feedback at this frequency rather than the desired negative feedback. If the magnitude of the return ratio is less than unity, the amplifier will not oscillate even though positive feedback exists. If $|AF|$ equals or exceeds unity the circuit will oscillate. A sinusoidal output will result when $|AF|$ is only slightly larger than unity while the waveform will contain more distortion for larger values of $|AF|$.

For many practical feedback amplifiers, stability can be predicted simply by finding the frequency at which the phase of AF equals $\pm 180°$ and then calculating the magnitude of AF at this frequency. There is a situation referred to as conditional stability that can occur when the frequency variation of AF becomes more complex. In this instance the phase shift may reach $180°$ before $|AF|$ has dropped to unity. At higher frequency, however, the phase shift decreases and when $|AF|$ equals unity, the phase shift may equal only $170°$. At still higher frequencies as

$|AF|$ decreases below unity the phase again exceeds 180°. This amplifier is said to be conditionally stable and in general is not a useful amplifier.

We will defer a discussion of compensation methods until we have discussed the operational amplifier.

6.4 OPERATIONAL AMPLIFIERS

The operational amplifier or op amp is an amplifier that exhibits very high ac and dc gain. Generally these amplifiers have a high input impedance and a low output impedance. This circuit derives its name from the traditional application of performing mathematical operations such as derivation, integration, summing, and differencing. For several years the analog computer has required high quality, high precision, and high cost op amps. Recent advances in microcircuit and integrated circuit techniques have reduced the size and cost of op amps to the point that a greater variety of practical applications are now feasible.

Op amps are now used in the fields of servomechanisms and automatic control, active network synthesis, instrumentation, and others. Signal conditioning, regulation, and waveshaping are only a few of the many functions now performed quite effectively by the op amp. Because of the growing importance of this component in electronics and related areas, it is appropriate to consider the theory and applications of the op amp in some detail.

A. The Ideal Op Amp

A truly ideal op amp would exhibit infinite input impedance, zero output impedance, infinite voltage gain, and infinite bandwidth. Since these specifications are unattainable we will define the ideal op amp as a circuit having negligibly high input impedance (no loading of preceding stages), negligibly low output impedance (compared to the load), very high voltage gain, and a bandwidth exceeding that required by the particular application. We will further restrict our discussion to op amps having both inverting and noninverting inputs, since this type of op amp is presently very popular. Figure 6.20 shows the schematic representation of an ideal differential op amp.

In practice it is easy to obtain op amps with gains ranging from 10^4 to 10^7, open loop bandwidths ranging from a few hertz to several kilohertz, input imped-

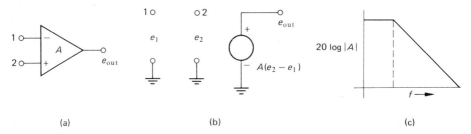

Fig. 6.20 An ideal op amp: (a) schematic representation; (b) equivalent circuit; (c) frequency response.

ances in the megohm range, and output impedances in the ohm range. The asymptotic falloff of gain with frequency is often 6 db/octave or 12 db/octave. While the orders of magnitude of output impedance and bandwidth may seem to invalidate the concept of the ideal op amp, such is not the case. The bulk of applications requires the use of negative feedback and/or frequency compensation to extend the available bandwidth and decrease the output impedance of the amplifier. We will now consider some applications of the op amp.

Figure 6.21 shows the op amp being used as an inverting amplifier. If the open loop gain is given by

$$A = \frac{A_{MB}}{1 + j\omega/\omega_2}, \tag{6.35}$$

then the gain with feedback can be found from Eq. (6.31) as

$$G = \frac{-A}{1 + AF} \frac{R_F}{R_1 + R_F}$$

$$= \frac{-A_{MB}}{1 + A_{MB}F} \frac{R_F}{R_1 + R_F} \frac{1}{1 + j[\omega/\omega_2(1 + A_{MB}F)]}. \tag{6.36}$$

In this case the feedback factor is

$$F = \frac{R_1}{R_1 + R_F}.$$

If we assume that the return ratio $A_{MB}F$ is very large compared to unity, the closed loop gain becomes

$$G = \frac{-R_F}{R_1} \frac{1}{1 + j[\omega/\omega_2(1 + A_{MB}F)]}. \tag{6.37}$$

The closed loop midband gain is determined by the ratio of the feedback resistance to the resistor R_1. The amplifier bandwidth is

$$\omega_{2F} = \omega_2(1 + A_{MB}F), \tag{6.38}$$

a considerable increase over the open loop bandwidth ω_2. For values of closed loop midband gain considerably greater than unity, ensuring that $R_F \gg R_1$, midband gain and bandwidth are directly exchanged. The closed loop midband

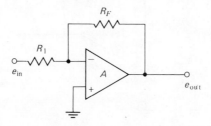

Fig. 6.21 An inverting amplifier.

gain will be related to the open loop gain by the factor $1/(1 + A_{MB}F)$ while the closed loop bandwidth will be $(1 + A_{MB}F)$ times the open loop value. In effect the overall gain-bandwidth product of the amplifier is constant so long as R_F is much larger than R_1. We should further note that we have restricted our considerations thus far to amplifiers exhibiting a 6 db/octave rolloff with frequency.

As an example, let us consider the case in which $R_F/R_1 = 1000$, $A_{MB} = 10^5$, and $\omega_2 = 2\pi \times 10^3$ rad/sec. When feedback is applied the closed loop gain is expressed quite accurately by

$$G = \frac{-10^3}{1 + j[\omega/2\pi \times 10^5]}$$

The midband gain is equal to -1000 while the bandwidth is $2\pi \times 10^5$ rad/sec.

The impedance seen by a voltage source is very nearly equal to R_1 for the inverting amplifier. In the midband region the loading effect of R_F on the amplifier input can be represented by the reflected impedance $R_F/(1 + A_{MB})$. The input impedance to the circuit (as indicated in Fig. 6.13b) is $R_1 + R_F/(1 + A_{MB}) \approx R_1$ for practical values of A_{MB}. It is noteworthy that the effective impedance between the amplifier input and ground is approximately zero. The voltage at this point is consequently extremely small in most instances and in fact can be considered a virtual ground for purposes of calculation.

It now becomes a simple matter to calculate the current through R_1 and R_F. All current through R_1 must also flow through R_F since this path presents an approximate short circuit path and the amplifier itself has negligibly high input impedance. The current through both R_1 and R_F is

$$i_1 = e_{in}/R_1. \tag{6.39}$$

Loading on the source is determined by R_1 in the case of the inverting amplifier.

A noninverting stage is shown in Fig. 6.22. While this stage does not invert the input signal, negative feedback is still being applied since the output is fed back to the inverting input. The output voltage is proportional to the difference between e_{in} and e_m; thus,

$$e_{out} = A(e_{in} - e_m).$$

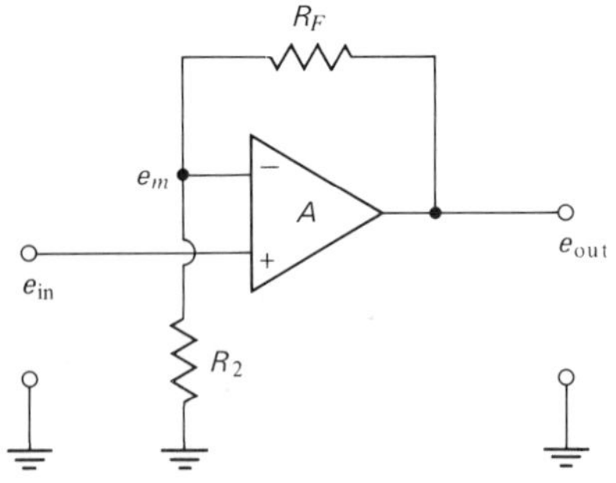

Fig. 6.22 A noninverting amplifier.

The voltage e_m is related to the output voltage by

$$e_m = Fe_{out},$$

where $F = R_2/(R_F + R_2)$. The closed loop gain is found to be

$$G = \frac{e_{out}}{e_{in}} = \frac{A}{1 + AF}.$$

If $A = A_{MB}/(1 + j\omega/\omega_2)$ the gain G can be expressed as

$$G = \frac{A_{MB}}{1 + A_{MB}F} \frac{1}{1 + j[\omega/\omega_2(1 + A_{MB}F)]}. \tag{6.40}$$

The bandwidth with feedback is equal to that for the inverting amplifier and represents a considerable improvement over the open loop bandwidth. If the return ratio $A_{MB}F$ is very high compared to unity, the midband closed loop gain is

$$G_{MB} = 1/F = (R_F + R_2)/R_2 = 1 + R_F/R_2. \tag{6.41}$$

For the numerous applications requiring a relatively high gain the ratio R_F/R_2 approximates the midband gain. This circuit allows gain and bandwidth to be exchanged directly for all values of gain including very low values. The gain-bandwidth product is constant with variation of G_{MB}.

The input impedance is a function of the input impedance of the op amp. This point will be discussed in more detail later.

B. Nonideal Effects in the Op Amp

Some of the practical effects that account for departures from ideal operation include finite input impedance, nonzero output impedance, dc drift or offset, noise, instability, and greater than 6 db/octave asymptotic falloff with frequency.

Impedance effects. A suitable equivalent circuit of the op amp with sufficient accuracy for almost any application is shown in Fig. 6.23. The resistors r_1 and r_2 are generally in the range of 10 MΩ to 1000 MΩ for transistor or integrated circuit differential stages. The resistance r, appearing between the input terminals, is typically between 10 kΩ and 1 MΩ while r_{out} may range from a few ohms to a few kilohms. For most op amps r_1 and r_2 will be approximately equal and are specified in terms of the common mode impedance r_{cm}. This impedance corresponds to the

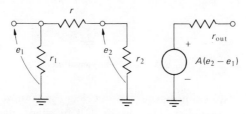

Fig. 6.23 Equivalent circuit for the op amp.

input impedance of the amplifier when the input terminals are connected together. Thus, in terms of r_1 and r_2,

$$r_{cm} = \frac{r_1 r_2}{r_1 + r_2}. \tag{6.42}$$

The values r_1 and r_2 can be estimated to be twice the common mode impedance.

Some manufacturers do not bother to list a value for r_{cm} indicating that perhaps r_1 and r_2 are insignificant. This is true to a large extent since r will normally have a greater influence on input impedance in most configurations. In the case in which feedback is used to increase input impedance considerably, r_1 or r_2 may become more significant. The equivalent circuit of Fig. 6.24 is then sufficient for all but the preceding case.

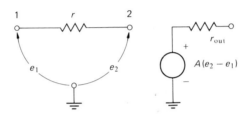

Fig. 6.24 A practical equivalent circuit for the op amp.

When the op amp is used as an inverting amplifier as shown in Fig. 6.21 the effect of r is negligible. The impedance r actually appears in parallel with the reflected impedance which has a value of $R_F/(1 + A_{MB}) \approx 0$. The negative terminal appears as a virtual ground; consequently the parallel path to ground presented by r is ignored. For the noninverting amplifier the input impedance is influenced by r. Consider the amplifier of Fig. 6.25 with the op amp replaced by its equivalent circuit.

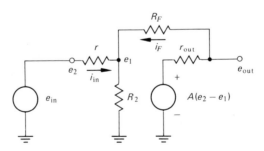

Fig. 6.25 Noninverting amplifier.

We will see that we cannot accurately neglect the common mode impedance for the noninverting amplifier. The circuit is a two-loop network which yields the equations

$$e_{in} = e_2 = i_{in}(r + R_2) + i_F R_2,$$
$$A(e_2 - e_1) = A[e_2 - R_2(i_{in} + i_F)] = i_{in}R_2 + i_F(R_F + r_{out} + R_2).$$

Solving for input current gives

$$i_{in} = \frac{e_2(R_F + r_{out} + R_2)}{rR_F + rr_{out} + rR_2(A + 1) + R_2 R_F + R_2 r_{out}}.$$

The input impedance is then

$$R_i = e_{in}/i_{in} = \frac{r[R_F + r_{out} + R_2(A + 1)] + R_2(R_F + r_{out})}{R_F + r_{out} + R_2}. \quad (6.43)$$

If the output impedance is much smaller than R_F, the input impedance becomes

$$R_i = \frac{r[R_F + R_2(A + 1)] + R_2 R_F}{R_F + R_2}.$$

Since A is so large, R_i can be approximated by

$$R_i \approx rA \frac{R_2}{R_F + R_2} = rAF. \quad (6.44)$$

We will generally be concerned with the midband case; thus, A_{MB} will be substituted for A. It is easy to see that R_i can take on a very large value; therefore, we must return in this case to the assumption that r_2 can be neglected. Since r_2 appears in parallel with the noninverting terminal, and since $rA_{MB}F$ may take on a very large value, a more accurate value of input impedance is given by

$$R_i = (2r_{cm}) || rA_{MB}F. \quad (6.45)$$

This resistance often approaches $2r_{cm}$ as a limiting value.

When voltage feedback is used as in the inverting and noninverting amplifiers discussed, the output impedance of the circuits is decreased considerably. As shown in Section 6.1, the overall output impedance is

$$R_o = r_{out}/(1 + A_{MB}F). \quad (6.46)$$

It is obvious from this discussion of impedance effects that only critical designs will require the use of the accurate equivalent circuit of Fig. 6.23. The usual approximations of zero output impedance for either inverting or noninverting amplifiers, zero impedance from the inverting terminal to ground for the inverting amplifier, and infinite impedance from the noninverting terminal to ground for the noninverting amplifier are generally quite reasonable.

Drift and offset. Drift and offset represent two problems that have great practical significance. Ideally zero input to the op amp would result in zero output. The actual open loop amplifier, however, will almost always show an output signal with no input present. Changes in voltage drops across junctions, reverse-bias leakage current changes, and changes in device current gains in early stages with temperature or aging result in this nonzero output voltage. Relatively small changes in these variables in the first few stages are greatly amplified before reaching the output. At a given temperature, application of a very small input signal will offset these device effects and zero the output voltage. This applied signal is defined as the offset signal. The offset signal can be applied as a current or voltage. The voltage (applied through zero source impedance) necessary to zero the output is called the offset voltage, while the offset current is that current (applied from a true current source with infinite impedance) necessary to zero the output.

Unfortunately, the offset signal is a function of temperature and aging since the effects leading to a nonzero output are functions of these variables. Thus, temperature changes cause the output voltage to vary from its zero value. The change in offset signal required to zero the output as temperature varies is normally called drift. Note that both offset and drift are referenced to the input of the amplifier rather than the output. This removes the necessity of considering gain and allows one to compare meaningfully the drift and offset of amplifiers with different values of gain.

In order to assess the effects of offset on performance of the op amp, the equivalent circuit of Fig. 6.26 has been proposed [1]. The sources e_{os} and i_{os} represent the effects within the amplifier that lead to offset voltage and current. Let us consider the influence of these sources on the behavior of the inverting amplifier, such as that shown in Fig. 6.27. Neglecting output and common mode

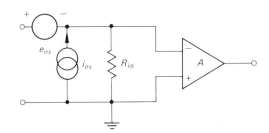

Fig. 6.26 Equivalent circuit reflecting offset effects.

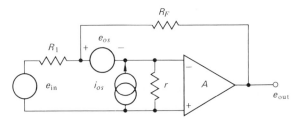

Fig. 6.27 Inverting amplifier with offset effects.

impedances the output voltage can be found to be

$$e_{out} = \frac{-ArR_F e_{in} - A(rR_F + rR_1)e_{os} - ArR_F R_1 i_{os}}{R_1 R_F + rR_F + (A + 1)rR_1}. \quad (6.47)$$

The closed loop gain from source to output is

$$G = \frac{-ArR_F}{R_1 R_F + rR_F + (A + 1)rR_1}. \quad (6.48)$$

We now have an expression relating e_{os} and i_{os} to output voltage, but it is more meaningful to refer offset effects to the input. Hence, if we divide the output quantities by closed loop gain G, all signals will be referred to the input or source voltage. The total applied voltage then appears to be

$$v_s = e_{in} + (1 + R_1/R_F)e_{os} + R_1 i_{os}. \quad (6.49)$$

The terms involving e_{os} and i_{os} are error terms. An input signal of e_{in} is applied, but the apparent voltage at the input consists of e_{in} plus terms directly proportional to e_{os} and i_{os}. Since R_F will generally be equal to R_1 (unity gain) or larger than R_1, the contribution to v_s from e_{os} will be a factor of one to two times the offset voltage. The contribution to v_s from i_{os} is proportional to R_1. This can lead to problems if a high input impedance to the amplifier is required. For this case R_1 must be large leading to a large error term involving i_{os}. In a specific application a bias voltage can be applied to cancel the errors due to e_{os} and i_{os}. However, as temperature changes these values drift, leading to finite errors.

As an example of the relative importance of the drift values, let us consider a stage with drift coefficients of 25 μV/°C and 1 nA/°C at room temperature. If R_1 is selected to be 1 kΩ, the drift in v_s will be

$$25 \ \mu V/°C + (1 \ nA/°C)(1 \ k\Omega) = 26 \ \mu V/°C.$$

The drift in v_s is determined mainly by drift in e_{os}. If R_1 is 100 kΩ, the drift in v_s is

$$25 \ \mu V/°C + (1 \ nA/°C)(100 \ k\Omega) = 125 \ \mu V/°C,$$

resulting mainly from drift in i_{os}. It is often true that some compromise between high input impedance and low drift must be made for the inverting amplifier.

The noninverting amplifier results in errors that are of the same order of magnitude as the inverting case. In particular, the error in v_s due to i_{os} is again given by $R_1 i_{os}$. However, the noninverting stage does not require that R_1 be large in order to achieve high input impedance. Low values of R_1 can be used in this case, leading to low drift without substantially decreasing the high input impedance.

We have noted in Chapter 5 that when differential stages are used, the resistances presented to both input terminals should be equal to cancel drift effects. The circuit of Fig. 6.28 shows one configuration that compensates somewhat for current drift effects if R_C is chosen to equal $R_1 R_F/(R_1 + R_F)$ [1]. The capacitor

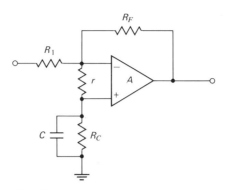

Fig. 6.28 Inverting amplifier with compensation.

is used to restore midband gain just as an emitter-bypass capacitor is used in a common-emitter stage.

When the op amp is used in a high gain configuration to amplify small signals, it is often necessary to adjust the offset to zero at some typical operating temperature. The configuration of Fig. 6.29 shows one method of zeroing the inverting stage. We note that since the apparent impedance from the inverting terminal to ground is very small, loading of R_B is negligible for typical values.

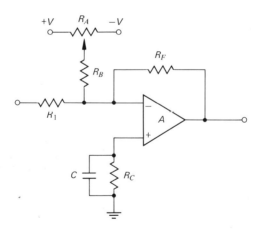

Fig. 6.29 Bias circuit for offset adjustment.

There are of course many applications wherein it is unnecessary to consider offset effects. Circuits with small overall gains generally require little attention to offset.

Common mode rejection. When we apply the op amp in a differential configuration, we assume that the open loop gain from the negative input to output is equal in magnitude to the gain from the other input to output. If this were the case, a very large signal could be applied simultaneously to both inputs with no output signal resulting. When a signal is applied simultaneously to the actual op amp, a small output signal results. A comparison of the resulting signal to the input signal is done in terms of the common mode rejection ratio

$$CM_{\text{rej}} = 20 \log \frac{e_{\text{in}}}{e_{\text{out}}}. \tag{6.50}$$

It is assumed that e_{in} is applied to both inputs in this equation. Typically this figure is 60 to 100 db. This figure is of interest in precise difference circuits and also gives a measure of common noise or temperature drift rejection.

Frequency response and compensation. One of the most obvious departures from ideal behavior exhibited by the practical op amp is the tendency to oscillate unless some sort of compensation is used. In fact, it is typical of an uncompensated amplifier to oscillate with no external feedback present. It is informative to investigate the reasons leading to this oscillatory tendency.

We have assumed in the previous discussions that the op amp has an open loop gain that is flat from dc to some corner frequency ω_2. Above ω_2 it was assumed that the magnitude of the gain fell at the rate of 6 db/octave. In general the actual frequency response of the op amp may appear as shown in Fig. 6.30. This frequency response results from an open loop gain expressed as

$$A = \frac{A_{MB}}{(1 + j\omega/\omega_2)(1 + j\omega/\omega_3)(1 + j\omega/\omega_4)}. \tag{6.51}$$

When negative feedback is applied, the gain for the inverting stage is

$$\frac{-A}{1 + AF} \frac{R_F}{R_1 + R_F}.$$

If the loop gain AF attains a phase shift of 180° before the magnitude drops below unity, the circuit will satisfy the conditions for oscillations. When it was assumed

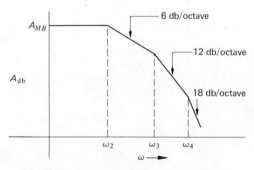

Fig. 6.30 Frequency response of a practical op amp.

that the open loop gain could be expressed as

$$A = \frac{A_{MB}}{1 + j\omega/\omega_2},$$

there was no stability problem so long as F was real. In this instance, 90° is the maximum possible phase shift. For the gain of Eq. 6.51, even though F may not be complex, the quantity AF can have 180° phase shift. Unless appropriate measures are taken, the practical op amp will oscillate when feedback is applied. In fact, the circuit will generally oscillate when no feedback is applied, since stray capacitances link input and output. Some of the popular compensation schemes, designed to stabilize the circuit against oscillations, will now be considered.

Many op amps are compensated by the manufacturer, but some require an additional compensating network. The basic idea is to add a circuit that forces the magnitude of the loop gain to fall below unity at some frequency before the rolloff rate reaches 12 db/octave. This guarantees that AF will never have a phase shift of 180° when the magnitude exceeds unity. At frequencies above the point where $|AF|$ has dropped to unity, the falloff can exceed 12 db/octave without affecting the stability of the circuit.

Compensating networks can be added to the appropriate points of an integrated op amp by means of terminals provided by the manufacturer. The compensating terminal is connected to some node within the amplifier as shown in Fig. 6.31. The squares represent amplifying stages within the op amp. The capacitor C is selected such that a new corner frequency is introduced far below the corner frequency of the uncompensated stage. Figure 6.32 shows the response of both cases. At low frequencies we note that the loop gain in db can be written as

$$20 \log AF = 20 \log A - 20 \log (1/F), \tag{6.52}$$

giving

$$20 \log A = 20 \log AF + 20 \log (1/F). \tag{6.53}$$

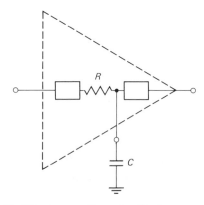

Fig. 6.31 Compensating node within the op amp.

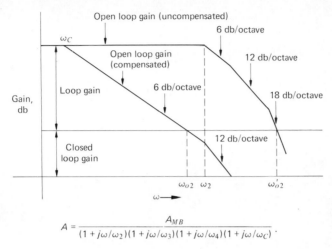

$$A = \frac{A_{MB}}{(1 + j\omega/\omega_2)(1 + j\omega/\omega_3)(1 + j\omega/\omega_4)(1 + j\omega/\omega_C)}.$$

Fig. 6.32 Frequency responses of uncompensated and capacitor compensated amplifiers.

Since closed loop gain is usually accurately approximated by

$$G = 1/F,$$

the db sum of $1/F$ and AF is equal to A as shown in Fig. 6.32. The point in frequency where $|AF| = 1$ is also the point where $|A| = 1/|F|$. This point is referred to as the crossover frequency. If we assume that $|F|$ is constant, Fig. 6.32 clearly locates the critical points of the uncompensated and compensated amplifiers. The closed loop gain has a magnitude of $1/|F|$ and the intersection of this line with the open loop gain (uncompensated) occurs at ω'_{o2}. The falloff of the open loop gain at this point is 18 db/octave and a phase shift of 180° is possible. In practice oscillations are almost certain for this case. With compensation added the new corner frequency of the open loop gain occurs at ω_{02} and is much lower than the corner frequency of the uncompensated amplifier. In this instance, the intersection of the open loop and closed loop gains occurs at ω_{02} and the falloff is only 6 db/octave at this point. At this point in frequency, even though $|AF| = 1$, oscillations cannot occur because the phase shift of AF will be near 90°.

The frequency ω_C is determined by

$$\omega_C = \frac{1}{CR_{eq}},$$

where R_{eq} is the equivalent resistance presented to the compensating capacitor. Manufacturers generally supply graphical information relating several values of C to corresponding corner frequencies.

If the closed loop gain is selected such that the intersection of the open loop and closed loop gains occurs between ω_C and ω_2, operation of the amplifier is stable. However, if the intersection is extended beyond ω_2, instability can take place.

260 FEEDBACK AMPLIFIERS

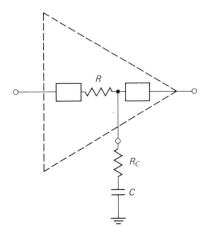

Fig. 6.33 Compensation using RC network.

It is possible to extend the region of stable operation by the lag compensator of Fig. 6.33. The compensated open loop gain becomes

$$A = \frac{A_{MB}}{(1 + j\omega/\omega_2)(1 + j\omega/\omega_3)(1 + j\omega/\omega_4)} \frac{1 + j\omega/\omega_{C2}}{1 + j\omega/\omega_{C1}}, \tag{6.54}$$

where

$$\omega_{C2} = 1/R_C C \tag{6.55}$$

and

$$\omega_{C1} = 1/(R + R_C)C. \tag{6.56}$$

The above equations assume that loading due to the stage following the compensating network is negligible. Selecting ω_{C2} such that

$$\omega_{C2} = \omega_2 \tag{6.57}$$

results in a gain of

$$A = \frac{A_{MB}}{(1 + j\omega/\omega_{C1})(1 + j\omega/\omega_3)(1 + j\omega/\omega_4)}. \tag{6.58}$$

The uncompensated and compensated cases are shown in Fig. 6.34. Note that since the asymptotic falloff reaches 12 db/octave at ω_3, stable operation can be expected for crossover frequencies nearing ω_3. Again most manufacturers will graphically relate values of C, R_C, and ω_{C1} to aid in the proper selection of the compensating network.

It should be noted that when the manufacturer builds the compensating network into the op amp, some flexibility is sacrificed. For many low frequency applications this flexibility is entirely unnecessary; however, as high frequency work is undertaken, access to a compensating terminal may prove to be very useful.

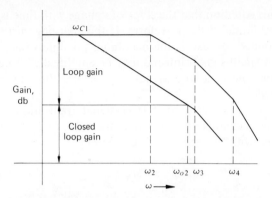

Fig. 6.34 Frequency responses of uncompensated and RC compensated amplifiers.

Rather than compensating the open loop gain of the amplifier, additional elements can be used in the feedback loop to achieve stable operation. A capacitance placed in parallel with the feedback resistance of the inverting stage will stabilize the amplifier if appropriately selected. In this case, the feedback transfer function F increases at high frequencies as A decreases in magnitude. If the feedback resistance is selected such that $1/R_F C = \omega_F$ is lower than the corner frequency of the uncompensated stage, more stable operation will result.

Operational amplifiers are becoming extremely important in the fields of circuit and system design. The op amp reduces considerably the number of discrete-element amplifiers that are now used. In addition op amps are used as summers, integrators, or digital to analog converters. They are used in active filter design and precision sweep circuitry. Many of these applications will be considered in Chapter 13.

Slew rate. We often specify an amplifier's frequency performance in terms of bandwidth. The 3 db bandwidth, however, applies primarily to the small-signal case and may be misleading in large-signal or transient applications. Generally, the output capacitance of a stage will have some effect on the time constant of output voltage. The rate of change of voltage on a capacitor is given by

$$\frac{dv}{dt} = \frac{i}{C}, \tag{6.59}$$

where i is the charging current and C is the value of capacitance.

For a sinusoidal signal given by

$$v = A \sin \omega t,$$

the maximum value of dv/dt occurs at $t = 0$ and is given by

$$\frac{dv}{dt}(\max) = \omega A \cos \omega t \big|_{t=0} = \omega A. \tag{6.60}$$

We note from this equation that the slope of voltage with time is directly proportional to the amplitude of the waveform. If the charging current of the output capacitance is limited, so also is the maximum possible value of dv/dt. While a low amplitude or small-signal sinusoid may be amplified with negligible attenuation due to output effects, a large signal may become distorted as a result of the limitation on dv/dt.

Since many op amps are used in large-signal applications, it is necessary to specify a slew rate to characterize the maximum value of dv/dt. This measurement is often performed with the circuit of Fig. 6.35 [1].

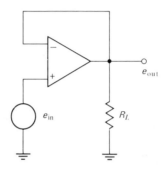

Fig. 6.35 Test circuit for measurement of slew rate.

The circuit is overdriven with a square wave and dv/dt is measured for both positive and negative transitions of the output level. Often the slew rate will have different absolute values for positive and negative transitions. The lower value is generally specified as the actual figure. Fast amplifiers may exceed slew rates of 1500 V/μsec while values as low as 0.1 V/μsec are reached by op amps designed for dc and low frequency applications [3].

From Eq. (6.60) we can find the maximum frequency of full output swing f_{om} by allowing A to take on the greatest value possible. Then f_{om} becomes

$$f_{om} = \frac{\text{slew rate}}{2\pi A_{max}}. \tag{6.61}$$

This frequency will always be less than the small-signal bandwidth of the same circuit.

An IC op amp. Figure 6.36 shows the schematic diagram of the well-known 741 op amp [2]. The input stages (T_1 to T_{10}) correspond to the circuit on Fig. 5.53. Transistors T_{16} and T_{17} are an emitter follower and a voltage gain stage respectively, with the multiple collector transistor T_{13} serving as the load. The output stage is a class B configuration which will be discussed in Chapter 12. The 741 is internally compensated.

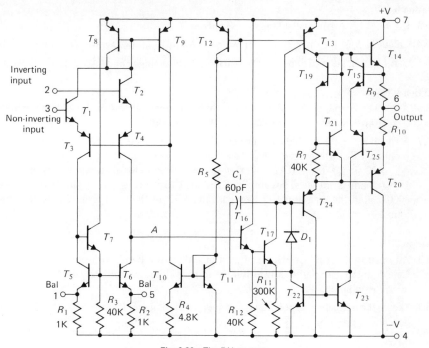

Fig. 6.36 The 741 op amp.

REFERENCES AND SUGGESTED READING

1. Burr-Brown Research Corp., *Operational Amplifiers: Design and Applications*. New York: McGraw-Hill, 1971.

2. A. B. Grebene, *Analog Integrated Circuit Design*. Cincinnati, Ohio: Van Nostrand-Reinhold, 1972, Chapter 5.

3. E. R. Hnatek, *A User's Handbook of Integrated Circuits*. New York: Wiley, 1973, Chapter 11.

4. S. D. Senturia and B. D. Wedlock, *Electronic Circuits and Applications*. New York: Wiley, 1975, Chapter 5.

PROBLEMS

Section 6.1

* **6.1** The open loop gain of the ideal amplifier of Fig. 6.1 is 1000. Calculate the closed loop gain for (a) $F = 0.1$, (b) $F = 0.01$, and (c) $F = 0.001$. Compare the closed loop gain in each case to the value of $1/F$. What condition on AF allows the closed loop gain to be approximated by $1/F$?

6.2 The open loop gain of the ideal amplifier of Fig. 6.1 is 1000. If the closed loop gain is 50 find the value of F.

264 FEEDBACK AMPLIFIERS

Section 6.1A

6.3 If $x = 101y + 10y^2$, calculate $S_{x,y}$ when $y = 2$.

* **6.4** Calculate $S_{G,T}$ for the amplifier of Example 6.1 at room temperature (300°K).

6.5 Can Eqs. (6.3) and (6.5) be used to calculate large changes in G? Explain.

Section 6.1B

6.6 A 1000 Hz signal contains second and third harmonic distortion. Calculate the SNR if $e = 10 \cos 2000\pi t + 0.2 \cos 4000\pi t + 0.1 \cos 6000\pi t$.

* **6.7** A signal $e_{in} = 0.1 \sin 2000\pi t$ is presented to an amplifier with open loop gain $A = 200$. Nonlinear distortion is introduced by the output stage, resulting in an output signal of $e_{out} = 20 \sin 2000\pi t + 0.3 \cos 4000\pi t$. If a loop is now closed around the amplifier with $F = 0.1$ and a noiseless preamplifier is added to cause $e_{out} = 20 \sin 2000\pi t + B \cos 4000\pi t$, find the magnitude B. Compare the SNR of the open loop case to the closed loop case.

Section 6.1C

* **6.8** An open loop amplifier exhibits a gain of 1000 and a bandwidth of $\omega_2 = 10^7$ rad/sec. Calculate the gain and bandwidth if the loop is closed with $F = 1/20$. Assume a 6 db/octave gain falloff.

6.9 Sketch the frequency response of the gain of the amplifier of Problem 6.8 if F is frequency-dependent given by

$$F = \frac{0.05}{1 + j(\omega/10^7)}.$$

Assume a 6 db/octave gain falloff.

Section 6.1D

6.10 Identify the types of feedback used in the stages shown in the figures shown here and at the top of the following page.

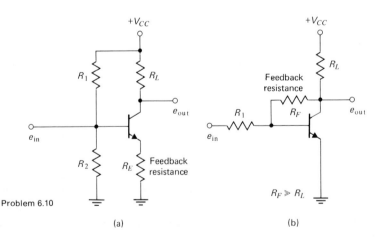

Problem 6.10

(a)   (b)

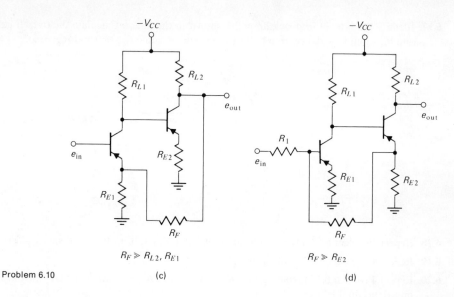

Problem 6.10

Section 6.1E

6.11 In Fig. 6.4(a), $R_{ia} = 1$ kΩ, $R_{oa} = 2$ kΩ, $F_{vv} = 0.05$, and $A = 100$. Calculate the input and ouput impedance of the overall amplifier.

*6.12 Repeat Problem 6.11 for the amplifier of Fig. 6.4(b). In this circuit $R_{ia} = 1$ kΩ, $R_{oa} = 2$ kΩ, $F_{iv} = 5 \times 10^{-5}$ $\mho$, and $A_{vi} = 1 \times 10^6$ Ω.

6.13 Repeat Problem 6.11 for the amplifier of Fig. 6.5(b). In this circuit $R_{ia} = 1$ kΩ, $R_{oa} = 2$ kΩ, $F_{vi} = 400$ Ω, and $A_{iv} = 3 \times 10^{-4}$ $\mho$.

Section 6.2

6.14 Identify the type of feedback amplifier shown in the figure. Calculate the closed loop gain, input impedance at point 1, and output impedance.

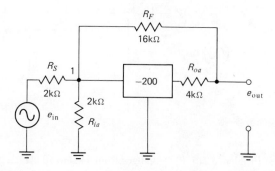

Problem 6.14

*6.15 Repeat Problem 6.14, given that $R_F = 10$ kΩ and $R_S = 5$ kΩ.

6.16 Repeat Problem 6.14, given that $R_F = 20$ kΩ.

266 FEEDBACK AMPLIFIERS

6.17 Identify the type of feedback amplifier shown in the figure. Calculate the overall gain and the input impedance from point 1 to ground. Assume $R_{oa} = 0$ and $R_{ia} = \infty$.

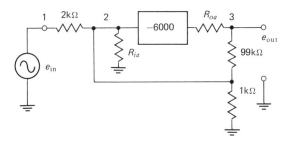

Problem 6.17

6.18 Repeat Problem 6.17 if the gain from point 2 to point 3 drops from -6000 to -600.

6.19 Repeat Problem 6.17 assuming that $R_{ia} = 6 \text{ k}\Omega$.

6.20 Repeat Problem 6.17 assuming that $R_{ia} = 6 \text{ k}\Omega$ and $R_{oa} = 6 \text{ k}\Omega$. Calculate the output impedance in this case.

*6.21 Identify the type of feedback amplifier shown in the figure. Evaluate the input impedance to the open loop amplifier between point 1 and ground ($R_F = \infty$). If $R_F = 4 \text{ k}\Omega$, what is the closed loop gain and input impedance at point 1? Assume that $I_E = 1 \text{ mA}$, $r_{bb'} = 200 \text{ }\Omega$, and $\beta = 80$ for both transistors.

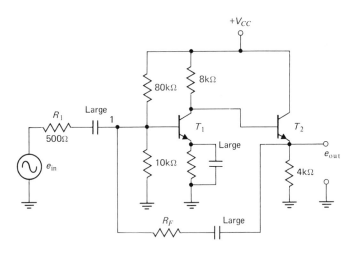

Problem 6.21

6.22 Repeat Problem 6.21 if R_F is changed from 4 kΩ to 10 kΩ.

6.23 Repeat Problem 6.21 assuming that R_1 is decreased to zero. Comment on your result.

*6.24 Given that the emitter resistor of T_2 is replaced by a 4-kΩ potentiometer with the wiper connected to R_F, evaluate the gain and input impedance of the closed loop amplifier when the wiper is at the midpoint. Assume that the element values of Problem 6.21 apply.

6.25 Given that $\beta = 100$, $r_{bb'} = 200$, and $r_{b'e} = 2$ kΩ, evaluate both the ac gain and the dc gain of the circuit shown in the figure.

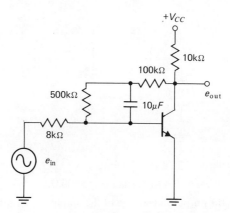

Problem 6.25

6.26 Assume that the amplifier has ac coupling between stages (not shown) and that all stages are properly biased. Identify the type of feedback amplifier shown in the figure. If $r_{b'e} = 1$ kΩ, $r_{bb'} = 200$ Ω, and $\beta = 80$ for all transistors, evaluate the voltage gain and input impedance between point 1 and ground.

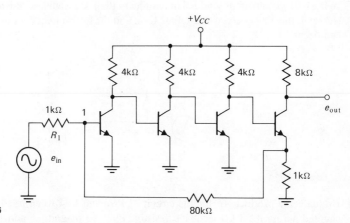

Problem 6.26

6.27 Repeat Problem 6.26 if R_1 is changed to 8 kΩ.

*6.28 Repeat Problem 6.26 if R_1 is changed to 200 Ω.

6.29 Calculate the output impedance of the amplifier of Problem 6.26.

6.30 Using the same transistors and same bias as in Problem 6.26, evaluate the gain and input impedance of the amplifier shown in the figure on the following page. What type of amplifier is this?

268 FEEDBACK AMPLIFIERS

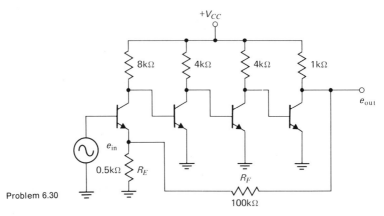

Problem 6.30

6.31 Repeat Problem 6.30 if R_E is changed to 2 kΩ.

6.32 Repeat Problem 6.30 if R_E is changed to 100 Ω.

6.33 Calculate the output impedance of the amplifier of Problem 6.30.

Section 6.3

*__6.34__ a) If the gain of the ideal amplifier shown is

$$A = \frac{1000}{(1 + j\omega/10^6)(1 + j\omega/10^6)},$$

can the circuit become unstable?

b) If a 0.01 μF capacitor is added in parallel to the 1-kΩ feedback resistor, can the circuit shown in the figure become unstable? If so, find the lowest frequency at which instability can occur.

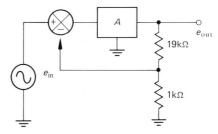

Problem 6.34

6.35 Evaluate the product AF for the circuit of Problem 6.34 when a 0.01 μF capacitor is shunted across the 1 kΩ feedback resistor. Plot the magnitude and phase of AF as a function of frequency. Plot both quantities on the same graph and identify the points of instability.

6.36 Repeat Problem 6.34(a), given that the gain A is

$$A = \frac{10,000}{(1 + j\omega/10^7)(1 + j\omega/2 \times 10^7)(1 + j\omega/4 \times 10^7)}.$$

What is the lowest frequency at which the circuit can become unstable?

Section 6.4

*6.37 An op amp has a midband gain of 10^5 and a bandwidth of 1000 Hz. Design an inverting amplifier with a gain of -500 and an input impedance of 1 kΩ. What is the bandwidth of the stage?

6.38 Select all element values in Fig. 6.29 when $V = 12$ V to result in a gain of -180 and an input impedance of 4 kΩ. The bias network should allow a maximum of ± 80 nA to flow into the inverting node.

*6.39 Assume that $A_{MB} = -1000$, $F = 1$, $\omega_2 = 10^4$, $\omega_3 = 4 \times 10^5$, and $\omega_4 = 4 \times 10^6$, in Eq. (6.51). Determine the frequency ω_C in Fig. 6.32 to lead to a phase shift of $-135°$ when $|AF| = 1$.

6.40 Repeat Problem 6.39 if A_{MB} is changed to -4000.

6.41 Using the amplifier of Problem 6.39, determine R_C and C in the compensation network of Fig. 6.33 to lead to a phase shift of $-135°$ when $|AF| = 1$. Assume $R = 1000$ Ω.

HIGH-FREQUENCY AMPLIFIERS

For the stages analyzed in the previous chapters we have not been concerned with reactive effects of the transistor. We are aware, however, of the existence of depletion layer capacitance and diffusion capacitance in semiconductor devices. We must now consider how to treat these reactive effects over the range of frequencies in which they are important. As a corollary topic we will also determine the range of frequencies for which these effects are significant enough to be considered.

The term high frequency can be very ambiguous. The normal breakdown of the frequency spectrum of electromagnetic waves gives one the impression that high frequencies mean those signals falling within the range extending from 3 MHz to 30 MHz. The classification of frequencies are [7]:

up to 30 kHz (VLF)	very low frequency
30 kHz to 300 kHz (LF)	low frequency
300 kHz to 3 MHz (MF)	medium frequency
3 MHz to 30 MHz (HF)	high frequency
30 MHz to 300 MHz (VHF)	very high frequency
300 MHz to 3 GHz (UHF)	ultrahigh frequency
3 GHz to 30 GHz (SHF)	superhigh frequency
30 GHz to 300 GHz (EHF)	extremely high frequency

These strict definitions of frequency ranges cannot be directly applied to transistor design. Some low-power transistors can amplify signals having frequencies from dc up to 1–2 GHz. Small-signal tunnel diode amplifiers can reach the 2–3 GHz frequency range. On the other hand, some high-power transistors will amplify only signals of frequency no higher than the 20–100 kHz range. The parameters of a given transistor determine the high-frequency range for that particular transistor. Thus, throughout this chapter we will use the term high frequency to denote the range of frequencies for which the transistor under consideration exhibits significant reactive effects. While the techniques developed here are general and will apply to any range of frequencies, most examples used are concerned with low-power devices with fairly high-frequency characteristics.

Two types of amplifier will be discussed in this chapter. The first is the wideband or video amplifier which is designed to amplify a large range of frequencies from some lower limit up to an upper cutoff point that will be several orders of magnitude greater than the lower cutoff point. The second type is the narrowband or tuned amplifier having a bandwidth that is only a small percentage of the resonant frequency. The term video amplifier arises from the use of wideband amplification in television receivers. A bandwidth extending from dc to approximately 4 MHz is required to carry the video information necessary to accurately reconstruct a black and white picture. Narrowband amplification is required in radio or television receivers to select and amplify a small bandwidth centered about a particular carrier frequency. Following the initial tuned amplifier, most receivers include several intermediate frequency amplifiers that amplify only a narrow range of frequencies while discriminating against frequencies outside this range.

7.1 THE TRANSISTOR AT HIGH FREQUENCIES

At high frequencies certain effects in the transistor that have previously been neglected must be considered. In the hybrid-π circuit these effects can be taken into account by additional capacitors, even though the effects are due to more complicated mechanisms. At lower frequencies the capacitors can be neglected, reducing the equivalent circuit to the low-frequency hybrid-π. The question of when these capacitive elements can be neglected must be answered in terms of the particular transistor of interest. For frequencies below the beta cutoff frequency f_β, the low-frequency circuit can normally be used with good accuracy. Near f_β and above, the reactive elements must be added to the equivalent circuit. The f_β of a transistor is not an absolute measure of the transistor's frequency performance, however; one circuit might allow the transistor to amplify at frequencies far above f_β while another circuit arrangement would allow amplification only of frequencies much lower than f_β.

A. Reactive Effects

The reactive effects associated with the transistor are the same as those discussed in relation to a diode in Chapter 2. The charge that must be removed or supplied to the depletion region, when the voltage across the region is changed, gives rise to an effective depletion-region capacitance. There will be a depletion-region capacitance associated with the collector-base junction called $C_{b'c}$ and a depletion-region capacitance of the emitter-base junction called $C_{b'e}$. From Eq. (2.23) this capacitance is seen to be a function of the voltage applied to the particular junction. Under small-signal conditions, the dc junction voltage will be much larger than the ac component of voltage. The capacitance then will be determined by the dc bias voltage and will be unaffected by the signal voltage. For large-signal (switching) circuits an average value of capacitance can be calculated for each junction.

Typical values of depletion-layer capacitance might range from 1 picofarad (1 pF = 10^{-12} F) for very-high-frequency transistors to 50 pF for low-frequency transistors.

In addition to the depletion-layer capacitances, there is a diffusion capacitance that must be included in the equivalent circuit also. This capacitance appears across the emitter-base junction and accounts for the change in charge that must accompany the change in minority-carrier gradient when the junction voltage is modified. The diffusion capacitance was derived for the diode in Chapter 2, but will be derived here for the transistor.

The hole distribution in the base region is shown in Fig. 7.1 for a transistor in the active region. As the emitter-base junction voltage increases or decreases, the hole concentration at the junction does likewise, as shown by the dashed lines. The total charge in the base region can be written as

$$Q_B = \tfrac{1}{2} q W p(0) A, \qquad (7.1)$$

where W is the width of the effective base region and A is the cross-section of the base region (assumed uniform). Since

$$p(0) = p_{b0} \exp\left(\frac{qV_{EB}}{kT}\right),$$

Eq. (7.1) can be written

$$Q_B = \tfrac{1}{2} q W A p_{b0} \exp\left(\frac{qV_{EB}}{kT}\right). \qquad (7.2)$$

The capacitance is then

$$C_D \equiv \frac{dQ_B}{dV_{EB}} = \tfrac{1}{2} q W A p_{b0} \frac{q}{kT} \exp\left(\frac{qV_{EB}}{kT}\right). \qquad (7.3)$$

In general the emitter-base bias voltage is unknown, but the emitter-bias current is known. It is more meaningful then to relate the diffusion capacitance to the dc

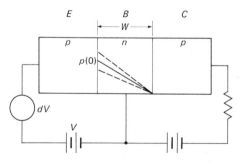

Fig. 7.1 Minority-carrier distribution in pnp-transistor base (holes).

emitter current rather than to the voltage. From Eq. (2.33) the emitter current is (assuming only hole conduction)

$$I_E = \frac{qD_p}{W} p_{b0} A \exp\left(\frac{qV_{EB}}{kT}\right). \tag{7.4}$$

Using this equation to eliminate the term involving $\exp(qV_{EB}/kT)$ from Eq. (7.3), we obtain

$$C_D = \frac{W^2}{2D_p} \frac{qI_E}{kT} = \frac{W^2}{2D_p} \frac{1}{r_d}. \tag{7.5}$$

The diffusion capacitance is seen to depend on the base width of the transistor quite strongly and also on the dc emitter current. A narrow base width is very important to high-frequency operation, since C_D must be made small. The impurity diffusion technique has permitted a great advancement in base-width control which has resulted in higher-frequency transistors.

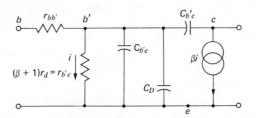

Fig. 7.2 High-frequency equivalent circuit.

The high-frequency equivalent circuit can be constructed as shown in Fig. 7.2. There are two immediate modifications that can be made to this circuit. Noting that C_D might range from 100 to several thousand picofarads for various transistors, $C_{b'e}$ is neglected, since it will be so much smaller. The second modification is done as a matter of convenience. Rather than leave the current generator in terms of the current through $r_{b'e}$, we write its value in terms of the voltage $v_{b'e}$. Since $i = v_{b'e}/(\beta + 1)r_d$, the current generator will have a value of $\beta i = (\alpha/r_d)v_{b'e}$. The modified circuit is shown in Fig. 7.3.

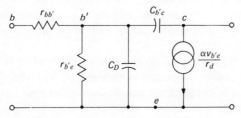

Fig. 7.3 Equivalent circuit to be used in this text.

As the frequency increases and the impedance of the capacitors becomes less, $v_{b'e}$ is decreased and the collector current becomes less. We might note at this point that no output resistance is shown in the equivalent circuit. Even though r_{out} will be present, it is almost always negligible in the high-frequency circuit because R_L will necessarily be small.

It should be realized that the high-frequency equivalent circuit of Fig. 7.3 also applies to the common-emitter circuit with an external emitter resistor that is bypassed by a capacitor. In terms of ac signals the emitter is also at ground potential in this case.

B. Simplification of the High-Frequency Circuit

The circuit of Fig. 7.3 is quite difficult to analyze when a source and load are applied. It then becomes a three-loop problem with $C_{b'c}$ linking the input and output loops. It is very difficult and time-consuming to analyze a three-loop problem manually. Thus, it seems reasonable to simplify the circuit, if such is possible with little loss of accuracy. It is appropriate to mention at this point that if the digital computer is used to analyze the network there is little need for further simplification. Analysis of a three-loop problem requires only slightly more time, and accuracy is guaranteed by the computer. For hand calculation, the time required to solve three simultaneous loop equations is considerably more than that required for the solution of two equations. Furthermore, the chance of error is much greater for the three-loop problem.

It is possible to simplify the circuit by reflecting the effect of $C_{b'c}$ to the input loop. That is, the current actually drawn by $C_{b'c}$ for a given applied input voltage can be found; then an additional element can be placed in the input loop to draw this same current. The troublesome element $C_{b'c}$ can then be removed with its effect now present due to the added element. The circuit becomes a two-loop problem that can more easily be analyzed. Before we conclude that this method of simplification is entirely valid we must note that $C_{b'c}$ has some direct loading effect on the output circuit. If we reflect $C_{b'c}$ to the input side, no loading of the output circuit appears. It has been shown [2, 7] that the loading on the output circuit is negligible for video stages since normal values of load resistance are small. In narrowband applications wherein the resonant collector impedance is large, it is more useful to reflect the effects of $C_{b'c}$ to the output loop of the circuit. In the following sections on wideband amplification we will reflect the effects of $C_{b'c}$ to the input circuit, noting that loading by $C_{b'c}$ on the output loop is negligible. Output circuit effects will be treated in the section on tuned amplifiers.

Effect of $C_{b'c}$ on the input circuit. When a voltage is applied to the input, a certain charge must be supplied by the source to charge $C_{b'c}$. This charge can be calculated and the proper value of capacitance can be connected between b' and e to replace $C_{b'c}$. This is very similar to the Miller effect in vacuum tubes; hence the equivalent capacitance will be denoted by C_M. The circuit of Fig. 7.4 is helpful in finding the value of C_M when R_L is present.

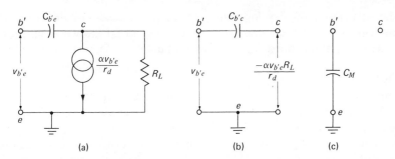

Fig. 7.4 Circuits used to find C_M: (a) equivalent circuit to the right of points b' and e; (b) voltage appearing across $C_{b'c}$; (c) equivalent Miller capacitance.

When the voltage between b' and ground equals $v_{b'e}$ the collector voltage will be

$$v_c = -\frac{\alpha v_{b'e} R_L}{r_d}.$$

The total voltage across the capacitor is $v_{b'c} = v_{b'} - v_c$ or

$$v_{b'c} = v_{b'e}\left(1 + \frac{\alpha R_L}{r_d}\right).$$

The voltage $v_{b'e}$ causes a voltage of $v_{b'e}(1 + \alpha R_L/r_d)$ to appear across the capacitor, because the collector voltage is dependent on the voltage $v_{b'e}$. The charge that must be supplied to $C_{b'c}$ is

$$Q_{b'c} = C_{b'c} v_{b'c} = C_{b'c} v_{b'e}\left(1 + \frac{\alpha R_L}{r_d}\right).$$

This charge is exactly the same as that delivered to a capacitor across the points b' and e of value

$$C_M = C_{b'c}\left(1 + \frac{\alpha R_L}{r_d}\right). \tag{7.6}$$

In terms of the input circuit, the equivalent circuit can then be drawn as shown in Fig. 7.5.

Since the input loop is isolated from the output loop, the circuit is called unilateral. It is sometimes referred to as the hybrid-y equivalent circuit.

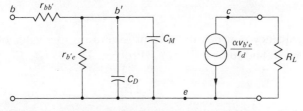

Fig. 7.5 Unilateral equivalent circuit.

C. Short-Circuit Current Gain

An examination of the behavior of the short-circuit current gain as frequency is varied is made by the manufacturer. From this examination, the beta cutoff frequency f_β or the current gain-bandwidth figure f_t can be specified. The measurements are taken with $R_L = 0$ and using a current source input, as shown in Fig. 7.6.

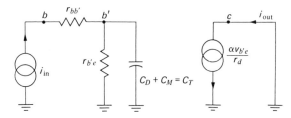

Fig. 7.6 Measurement of short-circuit current gain.

The behavior of i_{out}/i_{in} is easily predicted from the equivalent circuit. Since

$$v_{b'e} = i_{in} \frac{r_{b'e}}{1 + j\omega r_{b'e} C_T},$$

the output current is

$$\frac{\alpha}{r_d} v_{b'e} = \frac{\beta}{1 + j\omega r_{b'e} C_T} i_{in}.$$

Because $R_L = 0$, the Miller capacitance is $C_M = C_{b'c}$, which can usually be neglected when being compared to C_D. The short-circuit current gain is then

$$\frac{i_{out}}{i_{in}} = \frac{\beta}{1 + j\omega r_{b'e} C_D}. \tag{7.7}$$

At low frequencies the current gain is β, while at high frequencies the gain drops asymptotically at 6 db per octave. The graph of Fig. 7.7 shows this variation. The frequency where the gain is down 3 db is defined as the beta cutoff frequency:

$$f_\beta = \frac{1}{2\pi r_{b'e}(C_D + C_{b'c})} \approx \frac{1}{2\pi r_{b'e} C_D}. \tag{7.8}$$

The point in frequency at which the current gain drops to a value of unity (0 db) is f_t. This value is found to be β times f_β (Problem 7.9) or

$$f_t = \beta f_\beta \approx \frac{1}{2\pi r_d C_D}. \tag{7.9}$$

Usually, f_t is the quantity supplied by the manufacturer, rather than f_β.

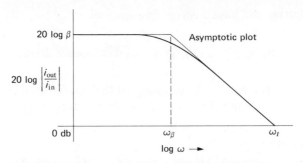

Fig. 7.7 Short-circuit current gain as a function of frequency.

D. The D-Factor

The circuit of Fig. 7.6 has no load resistor and, of course, could not be used as a voltage amplifier. When a load resistance is placed in the collector circuit, the Miller-effect capacitance is increased considerably. The total capacitance from b' to e becomes

$$C_T = C_D + C_M = C_D + \left(1 + \frac{\alpha R_L}{r_d}\right) C_{b'c}.$$

Here r_d is quite small compared to R_L, and the total capacitance is given fairly accurately by

$$C_T = C_D + \frac{\alpha R_L}{r_d} C_{b'c}.$$

This equation can be manipulated and put in a form that is easier to use:

$$C_T = C_D \left(1 + \frac{\alpha R_L C_{b'c}}{r_d C_D}\right) = C_D(1 + \alpha R_L C_{b'c} \omega_t). \tag{7.10}$$

The quantity in brackets is called the D-factor; it allows the total capacitance to be written as

$$C_T = DC_D, \tag{7.11}$$

where

$$D = (1 + \alpha R_L C_{b'c} \omega_t). \tag{7.12}$$

There are two reasons for using the D-factor. One is that the overall corner frequency of an amplifier is easily related to this factor. The second reason is that it is expressible in terms of specified quantities. The collector depletion-layer capacitance and ω_t will be given by the manufacturer; R_L will be chosen by the designer, and α is always within one or two percent of unity.

When R_L is added to the circuit of Fig. 7.6, the output voltage becomes

$$e_{out} = \frac{-i_{in} R_L \beta}{1 + j\omega r_{b'e} DC_D}.$$

The output voltage will have a corner frequency of

$$\omega_2 = \frac{1}{Dr_{b'e}C_D} = \frac{\omega_\beta}{D} \quad \text{(current source drive)}.$$

Therefore the presence of R_L has lowered the bandwidth due to the increase in Miller-effect capacitance. The D-factor accounts for this very easily.

E. The Broadband Factor

Even with a load resistor, the circuit of Fig. 7.6 is not a practical circuit. The input source is a current source which has infinite impedance. In practice, a voltage source with a finite source resistance will be used. This might be an actual signal generator or a previous stage. In any case, the circuit will appear as shown in Fig. 7.8. The voltage gain can be written as a function of frequency and the corner frequency can be found from this equation. Since

$$e_{out} = \frac{-\alpha v_{b'e}}{r_d} R_L$$

and

$$v_{b'e} = \frac{e_{in}r_{b'e}}{(r_{b'e} + R_S + r_{bb'})[1 + j\omega D C_D r_{b'e} || (R_S + r_{bb'})]},$$

the voltage gain is

$$A_o = \frac{-\beta R_L}{(r_{b'e} + r_{bb'} + R_S)[1 + j\omega D C_D r_{b'e} || (R_S + r_{bb'})]}. \tag{7.13}$$

Equation (7.13) can be arrived at more simply by evaluating the midband gain, and by then finding the corner frequency of $v_{b'e}$. Since the output current and voltage are proportional to $v_{b'e}$, the voltage gain will have the same corner frequency as $v_{b'e}$. The midband gain is

$$A_{MB} = \frac{-\beta R_L}{r_{b'e} + r_{bb'} + R_S}.$$

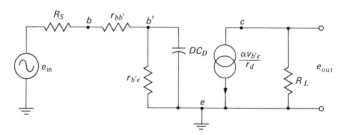

Fig. 7.8 Equivalent circuit for practical amplifier.

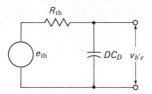

Fig. 7.9 Equivalent circuit of input loop.

The corner frequency of $v_{b'e}$ is found by taking a Thévenin equivalent of $R_S + r_{bb'}$ and $r_{b'e}$. The resulting circuit is shown in Fig. 7.9. The equivalent resistance is

$$R_{th} = \frac{r_{b'e}(R_S + r_{bb'})}{r_{b'e} + r_{bb'} + R_S}.$$

The corner frequency occurs when $X_c = R_{th}$ or when

$$\omega_2 = \frac{1}{\dfrac{DC_D r_{b'e}(R_S + r_{bb'})}{r_{b'e} + R_S + r_{bb'}}} = \frac{B\omega_\beta}{D} \qquad (7.14)$$

or

$$f_2 = \frac{Bf_\beta}{D},$$

where

$$B = \frac{r_{b'e} + R_S + r_{bb'}}{R_S + r_{bb'}}. \qquad (7.15)$$

Equation (7.13) can be written using the B-factor as

$$A_o = \frac{-\beta R_L}{(r_{b'e} + r_{bb'} + R_S)[1 + j(\omega D/B\omega_\beta)]}. \qquad (7.16)$$

The physical significance of the B-factor should be apparent. When the input circuit is driven by a current source of infinite resistance, the time constant of $v_{b'e}$ is simply

$$\tau = r_{b'e} DC_D = \frac{D}{\omega_\beta}.$$

As the source resistance is lowered to a finite value, the time constant is decreased to

$$\tau = \frac{r_{b'e}(R_S + r_{bb'})}{r_{b'e} + r_{bb'} + R_S} DC_D = \frac{r_{b'e} DC_D}{B} = \frac{D}{B\omega_\beta}.$$

The time constant of the circuit is the reciprocal of the radian bandwidth, so as τ decreases, the corner frequency increases.

F. Use of the High-Frequency Circuit

It is often mistakenly thought that a knowledge of the beta cutoff frequency is all that is required to predict the corner frequency. This is incorrect, as seen in the last two sections, due to:

1. the Miller effect, which is accounted for by the *D*-factor, and
2. the effect of the source resistance on the input-circuit bandwidth, which is accounted for by the *B*-factor.

The hybrid-π circuit is quite useful not only in small-signal design, but in switching-circuit design also. Even though both $r_{b'e}$ and C_D depend on the emitter current, the time constant and bandwidth are relatively independent of I_E. The product of $r_{b'e}$ and C_D is independent of I_E, since $r_{b'e}$ varies inversely, and C_D varies directly, with I_E. The actual value of C_D seldom must be found, since ω_β, *D*, and *B* are all that is needed to find ω_2.

The collector depletion-layer capacitance is given by the manufacturer and is measured at a specified voltage. In the amplifier circuit, the transistor might be used at a collector-base voltage different from that for which $C_{b'c}$ is given. For most high-frequency transistors, which are manufactured by a diffusion process, the capacitance is approximated by

$$C_{b'c} = \frac{k}{(V_{CB})^{1/3}}. \tag{7.17}$$

The constant *k* can be evaluated at the specified voltage and $C_{b'c}$ can then be calculated for the value of V_{CB} that is to be used in the circuit.

Example 7.1. A transistor has the following parameters: $\alpha = 0.99$, $r_{bb'} = 200\ \Omega$. $C_{b'c} = 5$ pF at the collector-base voltage used for the stage, and $f_t = 10$ MHz. The transistor is used in the common-emitter configuration with $R_L = 1\ \text{k}\Omega$ and is driven by a source with a resistance of 500 Ω. Given that I_E is 1 mA, find the high-frequency 3-db point (corner frequency) and the midband voltage gain.

Solution. The equivalent circuit is shown in Fig. 7.10. The midband gain is

$$A_{MB} = \frac{-\beta R_L}{r_{b'e} + r_{bb'} + R_S} = \frac{-99 \times 10^3}{2600 + 200 + 500} = -30.$$

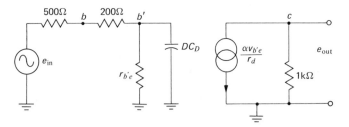

Fig. 7.10 Equivalent circuit for Example 7.1.

The B-factor and D-factor must be found to calculate f_2:

$$B = \frac{r_{b'e} + r_{bb'} + R_S}{r_{bb'} + R_S} = \frac{3300}{700} = 4.71,$$

$$D = 1 + \alpha R_L C_{b'c} \omega_t = 1 + 0.99 \times 10^3 \times 5 \times 10^{-12} \times 2\pi \times 10^7 = 1.31.$$

The corner frequency is then

$$f_2 = \frac{B}{D} f_\beta = \frac{B}{D} \frac{f_t}{\beta} = \frac{4.71 \times 10^7}{1.31 \times 99} = 360 \text{ kHz}.$$

G. High-Frequency FET Model

There are several reactive effects in the FET that limit the high-frequency performance of the device. There is a depletion-region capacitance associated with the gate-channel junction. The effects of this capacitance are represented by the gate-to-source capacitance C_{gs} and the gate-to-drain capacitance C_{gd}. These elements depend to some extent on the gate-to-source and gate-to-drain voltages. A drain-to-source capacitance C_{ds} also exists which is generally smaller than C_{gs} or C_{gd}. At very high frequencies the transit time of the carriers from source to drain becomes significant compared to the period of the input signal. This time lag between input and response can be represented by an effective capacitance, but is often a second-order effect and simply neglected.

While more accurate equivalent circuits can be developed, the model of Fig. 7.11 shows a practical circuit that offers a reasonable compromise between complexity and accuracy. The resistance r_g is ohmic resistance and can often be neglected.

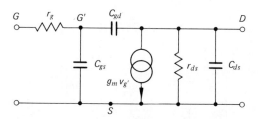

Fig. 7.11 Approximate high-frequency FET equivalent circuit.

The gate-to-source capacitance C_{gs} is much larger than the gate-to-drain capacitance C_{gd}. For most FET units the gate-to-source interface contains a larger area than the gate-to-drain interface. In addition, the width of the depletion region near the source is much less than the region near the drain. Both of the preceding factors tend to make C_{gs} larger than C_{gd}. The capacitance C_{ds} is of the same order of magnitude as C_{gs}. The areas between source and substrate and drain and substrate lead to additional drain-to-source capacitance. Typical values for these three capacitors are $C_{gd} = 0.5$ pF, $C_{gs} = 5$ pF, and $C_{ds} = 5$ pF. Although C_{gd}

is an order of magnitude smaller than the other two capacitors, it is by no means negligible. In amplifying applications where gain between gate and drain occurs, the Miller effect increases the apparent value of this element.

The high-frequency circuit can be unilateralized by reflecting the Miller capacitance to the input side of the amplifier. The procedure follows that of the bipolar transistor. Assuming a small value of load resistance, as will normally be encountered in high-frequency design, the unilateral circuit will appear as shown in Fig. 7.12. The unilateral model can be used for the calculation of forward gain and input impedance, but output impedance will not be accurately predicted by this model. For a single-stage amplifier driven by a source resistance the voltage gain becomes

$$A = \frac{-g_m R_{eq}}{(1 + j\omega R_S C_{in})(1 + j\omega R_{eq} C_{ds})}, \quad (7.18)$$

where

$$C_{in} = C_{gs} + C_{gd}(1 + g_m R_L) \quad (7.19)$$

and

$$R_{eq} = \frac{r_{ds} R_L}{r_{ds} + R_L}. \quad (7.20)$$

If R_S is large, which is often true for FET amplifiers, the upper corner frequency is determined by the input circuit and is

$$f_2 = \frac{1}{2\pi R_S C_{in}}. \quad (7.21)$$

If R_S is small, the corner frequency is determined by the output circuit and is

$$f_2 = \frac{1}{2\pi C_{ds} R_{eq}}. \quad (7.22)$$

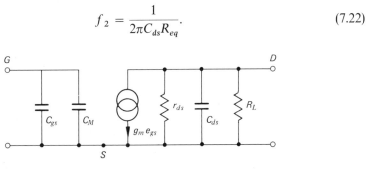

Fig. 7.12 Unilateral high-frequency FET equivalent circuit.

7.2 SINGLE-STAGE BROADBANDING TECHNIQUES

The single-stage amplifier discussed in the preceding section may be appropriate for a number of applications. There will be several other amplifiers requiring more bandwidth than is achievable with the simple common-emitter stage. The bandwidth of a single-stage amplifier can be extended by the proper selection and placement of additional elements, a procedure referred to as broadbanding. The extension of bandwidth is generally accompanied by a loss of gain; consequently, broadbanding is applied to circuits with a bandwidth smaller than that required, but with excess gain. Gain can then be exchanged for bandwidth. If both actual gain and bandwidth cannot be made to equal or exceed specified design values, then either a more effective device or additional stages must be used. The broadbanding techniques discussed here will be extended to multistage amplification in the next section.

With transistor amplifiers, a given broadbanding technique can be classified as one of three types. As the bandwidth is extended by a certain factor the midband gain is either

1. reduced by the same factor,
2. reduced by a value greater than this factor, or
3. reduced by a value less than this factor.

We can define a gain-bandwidth product that can determine the efficiency of a given broadbanding technique. If the gain-bandwidth product is

$$\text{GBW} = A_{MB}\omega_2,$$

then the three types of broadbanding listed above correspond to the mathematical expressions

1. $A_{MB}\omega_2 = A'_{MB}\omega'_2,$
2. $A_{MB}\omega_2 > A'_{MB}\omega'_2,$
3. $A_{MB}\omega_2 < A'_{MB}\omega'_2,$

where the unprimed values of midband gain and bandwidth refer to the amplifier before extension of the bandwidth, and the primed quantities refer to the amplifier after broadbanding. For very critical design, type 3 is most desirable with type 1 as the second choice. Type 2 broadbanding is least efficient, but can be used in less critical situations.

A. Common-Emitter Stage

Before considering some simple broadbanding methods we will calculate the GBW of the common-emitter amplifier of Section 7.1E.

From Eq. (7.16) the midband voltage gain is

$$A_{MB} = \frac{-\beta R_L}{r_{b'e} + r_{bb'} + R_S}.$$

The bandwidth is

$$\omega_2 = \frac{B\omega_\beta}{D} = \frac{\omega_\beta}{D}\left(\frac{r_{b'e} + r_{bb'} + R_S}{r_{bb'} + R_S}\right).$$

The product of these two is

$$\text{GBW} = \frac{\omega_\beta}{D} \frac{\beta R_L}{r_{bb'} + R_S} = \frac{\omega_t}{D} \frac{R_L}{r_{bb'} + R_S}. \tag{7.23}$$

There are several variables that can be controlled to extend the bandwidth of the common-emitter stage. In a given application, not all of these elements can be varied. For example, decreasing R_S increases the bandwidth and also increases midband gain resulting in a higher GBW after broadbanding. There is a lower limit on this element determined by the source resistance. Thus, a reasonable first step in any single-stage design procedure might consist of assuming the exact value of R_S that must be used. If there is some choice of values available, the lowest possible value should be selected.

If $C_{b'c}$ is decreased, the bandwidth will also increase with no corresponding loss of gain. This element can be lowered for a given transistor only by increasing the reverse bias of the collector-base junction. Here too, there is a limiting value imposed by other circuit requirements. Perhaps the power supply voltage is specified or the collector voltage may be limited by the maximum rating, thus setting $C_{b'c}$ at its minimum value. Again this element will be fixed and is not a convenient element to change for broadbanding purposes.

The gain-bandwidth product ω_t has been treated as a fixed quantity, but Problem 7.10 indicates that this figure drops off at lower values of emitter current. At very high values of emitter current ω_t also decreases. Figure 7.13 shows a typical variation of ω_t with emitter current. While there is a range of emitter current over which ω_t is relatively constant, this figure decreases rapidly at low emitter currents. It is obviously advantageous to bias the stage to the current that leads to maximum ω_t.

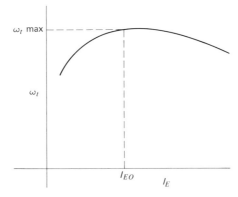

Fig. 7.13 Variation of ω_t with emitter current.

Returning to the equations for A_{MB} and ω_2 we see that still another parameter that can increase the bandwidth is $r_{b'e}$. If this element is increased the bandwidth increases while the gain decreases. The GBW remains constant if we assume that

ω_t is constant. An early proposal [1] for multistage design suggested that bias current be varied to control $r_{b'e}$, thus controlling the bandwidth. This method requires a low emitter current to achieve high bandwidths. Although Eq. (7.23) indicates that the GBW remains constant, we must remember that this equation assumes a constant value for ω_t. As indicated by Fig. 7.13, ω_t does not remain constant at low currents; hence the GBW falls off considerably as the bandwidth is increased. An effective broadbanding scheme should allow I_E to be selected such that ω_t is optimized and should not require the bias to change from this value.

The last remaining variable over which we have control in some applications is the load resistance R_L. In certain applications a load to be driven is specified, but this value can be decreased by a parallel resistance. In other applications, the value might be allowed to vary over quite a wide range. It is perhaps more realistic to use R_L to control bandwidth rather than any other element of the common-emitter stage. As the preceding paragraphs indicate, there are valid reasons for fixing these other element values.

As R_L is made smaller the D-factor decreases and bandwidth increases. The D-factor, given by

$$D = (1 + \alpha R_L \omega_t C_{b'c}),$$

does not vary directly with R_L and in fact becomes a weak function of R_L as this function approaches its minimum value of unity. Thus, as D approaches unity the bandwidth increases much more slowly than R_L decreases. On the other hand, the midband gain decreases directly with R_L and the GBW decreases as bandwidth is extended. This can be seen by expressing D in terms of R_L in Eq. (7.23) to give

$$\text{GBW} = \frac{\omega_t}{(1 + \alpha R_L \omega_t C_{b'c})} \frac{R_L}{r_{bb'} + R_S}$$

$$= \frac{\omega_t}{(1/R_L + \alpha C_{b'c} \omega_t)} \frac{1}{r_{bb'} + R_S}.$$

As R_L is lowered to extend the bandwidth, the denominator of the previous equation becomes larger resulting in a lower GBW.

Although large values of R_L result in very low bandwidths, perhaps we should consider the effects of increasing R_L on the GBW. As R_L becomes larger the D-factor will approach a value of $\alpha R_L \omega_t C_{b'c}$. The GBW will be given by

$$\text{GBW} = \frac{\omega_t}{\alpha R_L \omega_t C_{b'c}} \frac{R_L}{r_{bb'} + R_S}$$

$$\approx \frac{1}{C_{b'c}(r_{bb'} + R_S)}. \tag{7.24}$$

This equation gives the theoretical upper limit on GBW for the common-emitter stage, but two items must be noted that prevent the attainment of this limit in practice. The first item has been discussed; that is, large bandwidths cannot be obtained when R_L is large. The second item is a result of the fact that there is always a certain amount of stray capacitance that will shunt the load. For small values of R_L this capacitance is generally negligible, but as R_L takes on larger values, this capacitance can limit the actual bandwidth to a figure lower than that calculated by Eq. (7.14). Since bandwidth is now limited by these stray capacitance effects at large values of R_L, the GBW figure decreases as R_L increases, rather than remaining constant as indicated by Eq. (7.24).

We can plot the GBW figure as a function of bandwidth, noting that R_L is varied to change the bandwidth. This information is displayed in Fig. 7.14. It is obvious from this figure that the common-emitter stage is less efficient at higher bandwidths.

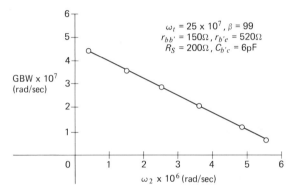

Fig. 7.14 GBW as a function of bandwidth.

B. Base Compensation

One method of broadbanding that allows the GBW to remain constant is that of base compensation, or the use of a parallel resistor-capacitor circuit in the base lead. This is referred to as a compensating circuit. In switching circuits, the capacitor C_1 is called a "speed-up" capacitor, due to the resulting improvement in switching speeds. This circuit is shown in Fig. 7.15. When R_S and $r_{bb'}$ are neglected, the input circuit reduces to that of an attenuator probe. Figure 7.16 shows the

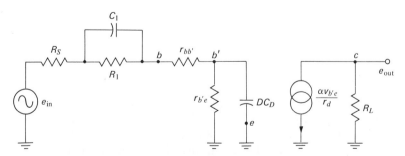

Fig. 7.15 Equivalent circuit for a common emitter with compensation.

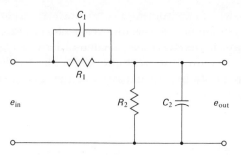

Fig. 7.16 Attenuator probe circuit.

schematic of such a probe. In the figure, R_2 and C_2 might represent the input resistance and capacitance to the vertical amplifier of an oscilloscope. Resistance R_1 is added to the circuit to attenuate the input signal, or to give the circuit a higher input impedance. When R_1 is present, the output signal will be frequency-dependent. If C_1 is properly chosen, the output voltage will not be frequency-dependent and will therefore have an infinite bandwidth. When $C_1 = 0$, the time constant of the circuit is

$$\tau = \frac{R_1 R_2}{R_1 + R_2} C_2.$$

The bandwidth equals the reciprocal of τ:

$$\omega_2 = \frac{R_1 + R_2}{R_1 R_2 C_2}.$$

When C_1 is included, the output voltage of Fig. 7.16 is

$$e_{out} = \frac{Z_2}{Z_1 + Z_2} e_{in} = \frac{R_2/(1 + j\omega R_2 C_2)}{R_2/(1 + j\omega R_2 C_2) + R_1/(1 + j\omega R_1 C_1)} e_{in}.$$

If C_1 is chosen such that $R_1 C_1 = R_2 C_2$, the above equation becomes

$$e_{out} = \frac{R_2}{R_1 + R_2} e_{in},$$

which is independent of frequency. The circuit now has an infinite bandwidth and zero rise time.

In the input loop of a transistor, $r_{b'e}$ and DC_D would correspond to R_2 and C_2 of the probe. The condition for compensation is

$$R_1 C_1 = r_{b'e} DC_D = \frac{D}{\omega_\beta}.$$

The capacitor C_1 is picked to be

$$C_1 = \frac{D}{R_1 \omega_\beta}. \tag{7.25}$$

The selection of R_1 will come from gain considerations. Unfortunately, the transistor will not have an infinite bandwidth because R_S and $r_{bb'}$ are not zero, as we assumed and we have neglected certain second-order effects that would become prominent as higher bandwidths are approached. The bandwidth can be found by analyzing the circuit of Fig. 7.15 and applying the condition for compensation. The output voltage is

$$e_{out} = \frac{-\alpha v_{b'e}}{r_d} R_L,$$

and $v_{b'e}$ is found to be

$$v_{b'e} = \frac{r_{b'e}/(1 + j\omega r_{b'e} D C_D)}{r_{b'e}/(1 + j\omega r_{b'e} D C_D) + R_1/(1 + j\omega C_1 R_1) + R_S + r_{bb'}} e_{in}.$$

Combining these two equations and remembering that $R_1 C_1 = r_{b'e} D C_D$ gives

$$A_o = \frac{e_{out}}{e_{in}} = -\beta R_L / [r_{b'e} + R_1 + R_S + r_{bb'}] \left[1 + \frac{j\omega r_{b'e} D C_D (R_S + r_{bb'})}{r_{b'e} + R_1 + R_S + r_{bb'}} \right].$$

The bandwidth is seen to be

$$\omega_2 = \frac{\omega_\beta}{D} \left(\frac{r_{b'e} + R_1 + R_S + r_{bb'}}{R_S + r_{bb'}} \right). \tag{7.26}$$

This can easily be compared to the case where R_1 and C_1 are not present. From Eq. (7.14), we have

$$\omega_2 = \frac{\omega_\beta}{D} B = \frac{\omega_\beta}{D} \left(\frac{r_{b'e} + R_S + r_{bb'}}{R_S + r_{bb'}} \right).$$

Since R_1 can be chosen to be several factors larger than $r_{b'e} + R_S + r_{bb'}$, the bandwidth from Eq. (7.26) can be much larger than that of the uncompensated case. The use of R_1 and C_1 to extend the bandwidth results in a corresponding decrease in gain. The midband gain is now

$$A_{MB} = \frac{-\beta R_L}{r_{b'e} + R_1 + R_S + r_{bb'}}. \tag{7.27}$$

The gain decreases by the same factor as the bandwidth increases in this case. The GBW is

$$\text{GBW} = \frac{\omega_\beta}{D} \frac{\beta R_L}{(R_S + r_{bb'})} = \frac{\omega_t}{D} \frac{R_L}{(R_S + r_{bb'})},$$

which is the same as the uncompensated case of Eq. (7.23). Note that since R_1 affects both gain and bandwidth, but does not appear in the GBW-equation, gain and bandwidth can be directly interchanged by this method. A high bandwidth can be achieved without decreasing the GBW-figure.

Example 7.2. If the transistor in Fig. 7.17 has the following parameters: $f_t = 100$ MHz, $\beta = 100$, $r_{bb'} = 100\ \Omega$, $r_d = 13\ \Omega$ ($I_E = 2$ mA), and $C_{b'c} = 5$ pF at the

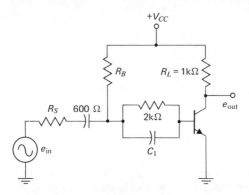

Fig. 7.17 Circuit for Example 7.2.

bias point, what are the midband gain and upper corner frequency for the cases where
a) $R_1 = 2$ kΩ, and C_1 has the correct value for compensation; and
b) $R_1 = C_1 = 0$ (assume the same I_E)?

Solution. (a) The midband gain is found from Eq. (7.27) to be

$$A_{MB} = \frac{-100 \times 10^3}{1313 + 2000 + 600 + 100} = -25.$$

From Eq. (7.26), we obtain

$$f_2 = \frac{f_\beta}{D}\left(\frac{r_{b'e} + R_S + R_1 + r_{bb'}}{r_{bb'} + R_S}\right).$$

Since

$$D = 1 + \alpha R_L \omega_t C_{b'c} = 1 + 0.99 \times 10^3 \times 2\pi \times 10^8 \times 5 \times 10^{-12} = 4.1$$

and

$$f_\beta = f_t/\beta = 1 \text{ MHz},$$

the bandwidth is

$$f_2 = \frac{1}{4.1}\left(\frac{4013}{700}\right) \text{ MHz} = 1.40 \text{ MHz}.$$

The correct value of C_1 is

$$C_1 = \frac{D}{R_1 \omega_\beta} = \frac{4.1}{2 \times 10^3 \times 2\pi \times 10^6} = 302 \text{ pF}.$$

(b) Without R_1 and C_1, the midband gain is

$$A_{MB} = \frac{-\beta R_L}{r_{b'e} + R_S + r_{bb'}} = -\frac{10^5}{2013} = -50.$$

The bandwidth is

$$f_2 = \frac{f_\beta}{D} B,$$

where

$$B = 2013/700 = 2.88,$$

so

$$f_2 = \frac{1}{4.1} \times 2.88 \text{ MHz} = 0.7 \text{ MHz}.$$

A comparison of the efficiency of the base-compensated stage to the uncompensated stage is made in Fig. 7.18. As higher bandwidths are required, the GBW figure, and therefore the gain, is higher than the uncompensated stage. It should also be noted that the bandwidth can be varied without regard to the bias current. Consequently, emitter current can be set to maximize ω_t and broadbanding can take place without affecting this value.

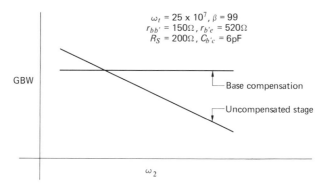

Fig. 7.18 GBW as a function of bandwidth for base compensation.

C. Emitter Compensation

There are two cases of emitter compensation. The first case uses only a resistor in series with the emitter lead. The second case, using a shunt resistor-capacitor combination in the emitter lead, gives somewhat better results than the first case.

CASE 1. *Extending the bandwidth with R_E.* This method extends the bandwidth of the stage because of the negative current feedback through R_E. The circuit of Fig. 7.19 can be used to explain this effect.

The input voltage is given by the sum of v_1, $v_{b'e}$, and v_3. The collector current is proportional to $v_{b'e}$ and hence decreases as $v_{b'e}$ decreases. When R_E is not present, $v_{b'e}$ decreases with frequency, since the impedance of $r_{b'e}/(1 + j\omega r_{b'e}DC_D)$ becomes less. The voltage across $r_{bb'}$ increases to make $v_1 + v_{b'e} = e_{in}$ at all frequencies. When R_E is added, $v_3 = (i_b + i_c)R_E$. At low frequencies, $v_3 = (\beta + 1)i_b R_E$. The voltage between the base and emitter terminals is now $v_1 + v_{b'e} =$

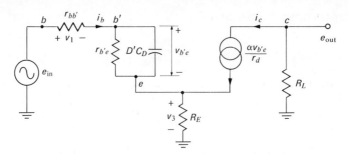

Fig. 7.19 Equivalent circuit with an external emitter resistor.

$e_{in} - v_3$. This voltage is smaller than that for the case where $R_E = 0$, and less output voltage is expected. However, it can also be seen that the base-to-emitter voltage decreases more slowly with frequency than it did in the previous case. Since v_3 will decrease at higher frequencies as i_c decreases, the voltage $e_{in} - v_3$ must become larger with frequency. Therefore $v_{b'e}$ will not decrease as rapidly when R_E is present. In other words, as i_e decreases, the voltage applied across the base-emitter terminals will increase if R_E is present. This amounts to current feedback and was discussed in more detail in Chapter 6.

We should note that the D-factor will not be the same in this case as that of the grounded emitter, since the Miller-effect capacitance no longer appears directly across C_D. For the present time we shall use D' as the D-factor in this case, although we have no means of calculating this value as yet.

For the circuit of Fig. 7.19, the output voltage can be written as

$$e_{out} = \frac{-\alpha v_{b'e} R_L}{r_d}.$$

The next step is to find $v_{b'e}$. The input loop equation is (assuming source resistance)

$$e_{in} = (R_S + r_{bb'})i_b + \frac{r_{b'e}}{1 + j\omega r_{b'e} D' C_D} i_b + R_E \left(1 + \frac{\beta}{1 + j\omega r_{b'e} D' C_D}\right) i_b.$$

The input current is

$$i_b = \frac{e_{in}(1 + j\omega r_{b'e} D' C_D)}{R_S + r_{bb'} + r_{b'e} + (\beta + 1)R_E + j\omega r_{b'e} D' C_D (R_S + r_{bb'} + R_E)},$$

and $v_{b'e}$ is found by multiplying i_b by $r_{b'e}/(1 + j\omega r_{b'e} D' C_D)$. The gain is then

$$A_o = -\beta R_L / [R_S + r_{bb'} + r_{b'e} + (\beta + 1)R_E]$$
$$\times \left\{ 1 + j\omega \left[\frac{r_{b'e} D' C_D (R_S + r_{bb'} + R_E)}{R_S + r_{bb'} + r_{b'e} + (\beta + 1)R_E} \right] \right\}. \quad (7.28)$$

The midband gain is

$$A_{MB} = \frac{-\beta R_L}{R_S + r_{bb'} + r_{b'e} + (\beta + 1)R_E}, \quad (7.29)$$

and the upper corner frequency is

$$\omega_2 = \frac{\omega_\beta}{D'}\left(\frac{R_S + r_{bb'} + r_{b'e} + (\beta + 1)R_E}{R_S + r_{bb'} + R_E}\right). \quad (7.30)$$

The GBW for this case is

$$\text{GBW} = \frac{\omega_\beta}{D'}\left(\frac{\beta R_L}{R_S + r_{bb'} + R_E}\right) = \frac{\omega_t}{D'}\left(\frac{R_L}{R_S + r_{bb'} + R_E}\right). \quad (7.31)$$

From Eq. (7.30), we see that the bandwidth will be increased as R_E is increased. This is seen to lower the GBW, assuming that D' is relatively constant with R_E; therefore the gain must decrease faster than ω_2 increases with R_E. This corresponds to the second type of broadbanding.

CASE 2. *Extending the bandwidth with R_E and C_E.* The addition of a shunt capacitor in the emitter circuit often extends the bandwidth further than it was in the preceding case. However, the value required for C_E in applications of very high frequency is sometimes too small to be practical. The equivalent circuit for this case is shown in Fig. 7.20.

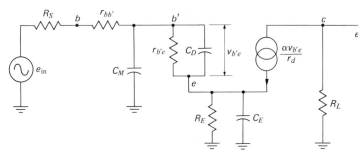

Fig. 7.20 Use of R_E and C_E for the extension of the bandwidth.

The value of the Miller capacitance in this case is

$$C_M = \left(1 + \frac{\alpha R_L}{R_E + r_d}\right) C_{b'c}.$$

Analyzing the stage of Fig. 7.20 is simplified considerably if C_E is picked such that

$$\omega_t = \frac{1}{C_E R_E}. \quad (7.32)$$

When this condition is fulfilled, the equivalent circuit to the right of point b' can be replaced by one that can easily absorb the Miller-effect capacitance.

The first step in finding the new equivalent circuit is to neglect C_M and evaluate the voltage gain. The two appropriate equations are

$$e_{\text{out}} = \frac{-\alpha v_{b'e} R_L}{r_d}$$

and

$$e_{in} = (R_S + r_{bb'})i_b + \frac{r_{b'e}}{1 + j\omega r_{b'e}C_D}i_b + \frac{R_E}{1 + j\omega R_E C_E}\left(1 + \frac{\beta}{1 + j\omega r_{b'e}C_D}\right)i_b.$$

Using Eq. (7.32), we find that this expression becomes

$$e_{in} = i_b\left[\frac{R_S + r_{bb'} + r_{b'e} + (\beta + 1)R_E + j\omega r_{b'e}C_D(R_S + r_{bb'})}{1 + j\omega r_{b'e}C_D}\right].$$

The gain is then

$$A'_o = -\beta R_L/[R_S + r_{bb'} + r_{b'e} + (\beta + 1)R_E]$$
$$\times \left[1 + j\omega \frac{r_{b'e}C_D(R_S + r_{bb'})}{R_S + r_{bb'} + r_{b'e} + (\beta + 1)R_E}\right].$$

The circuit of Fig. 7.21 can be made to yield the same expression for voltage gain if R_1 and C_1 are chosen properly. The voltage gain of this circuit is

$$A''_o = \frac{-\beta R_L}{(R_S + r_{bb'} + R_1)\left[1 + j\omega \frac{R_1 C_1 (R_S + r_{bb'})}{(R_S + r_{bb'} + R_1)}\right]}.$$

Voltage gain A''_o will equal A'_o if

$$R_1 = r_{b'e} + (\beta + 1)R_E$$

and

$$C_1 = \frac{r_d}{r_d + R_E} C_D.$$

The advantage of using the circuit of Fig. 7.21, rather than that of Fig. 7.20, is that C_M can now be combined with C_1, since it is in parallel with C_1. The total capacitance is then

$$C_T = C_1 + C_M = C_D \frac{r_d}{r_d + R_E} + \left(1 + \frac{\alpha R_L}{r_d + R_E}\right)C_{b'c}.$$

The D'-factor for this circuit is then

$$D' = \frac{C_1 + C_M}{C_1} = 1 + (r_d + R_E + R_L)\omega_t C_{b'c}. \tag{7.33}$$

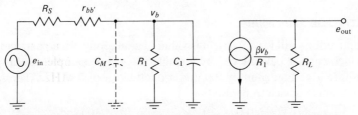

Fig. 7.21 A circuit which is equivalent to that of Fig. 7.20 when C_M is neglected.

The gain expression for the circuit of Fig. 7.20, including the Miller effect, is then

$$A_o = -\beta R_L/(R_S + r_{bb'} + r_{b'e} + (\beta + 1)R_E)$$
$$\times \left[1 + j\omega \frac{D'r_{b'e}C_D(R_S + r_{bb'})}{R_S + r_{bb'} + r_{b'e} + (\beta + 1)R_E}\right], \quad (7.34)$$

where D' is defined by Eq. (7.33). The midband gain is

$$A_{MB} = \frac{-\beta R_L}{R_S + r_{bb'} + r_{b'e} + (\beta + 1)R_E}, \quad (7.35)$$

and the bandwidth is

$$\omega_2 = \frac{\omega_\beta}{D'} \frac{(R_S + r_{bb'} + r_{b'e} + (\beta + 1)R_E)}{R_S + r_{bb'}}. \quad (7.36)$$

Comparing this to the case where $C_E = 0$ [Eqs. (7.29) and (7.30)] shows that the midband gain is the same, but the bandwidth is greater when the proper value of C_E is present. The GBW-figure is

$$\text{GBW} = \frac{\omega_t}{D'} \frac{R_L}{R_S + r_{bb'}}. \quad (7.37)$$

The GBW is slightly smaller for this stage than for the uncompensated stage if R_L has the same value for both circuits. The D'-factor will have a larger value than D. In order to achieve a high bandwidth with the uncompensated stage, we must decrease R_L to lower the D-factor. This, in turn, lowers the gain and the GBW. For the compensated stage, the value of R_L can remain constant, while R_E and C_E are adjusted to trade gain and bandwidth. Thus, when high bandwidths are required, the GBW of the compensated stage can exceed that of the uncompensated stage. Therefore the gain is greater for the compensated stage than for the uncompensated stage at high frequencies.

The value of C_E required for compensation is

$$C_E = \frac{1}{R_E \omega_t}, \quad (7.38)$$

and is often in the 1–10 pF range for high-frequency circuits.

This circuit uses feedback to obtain broadbanding, and so should rightfully be discussed in the chapter that covers feedback amplifiers. However, the circuit is easier to analyze by circuit analysis than by feedback methods; thus it is treated here.

7.3 MULTISTAGE AMPLIFIERS

Many applications call for a particular value of gain along with a particular bandwidth, which cannot be achieved by a single stage. For example, an application might require a voltage gain of 100 and a bandwidth of 10 MHz. The necessary GBW for a single device to be used is

$$\text{GBW} = 100 \times 10^7 \times 2\pi = 6.28 \times 10^9 \text{ rad/s}.$$

If the only available transistors have a GBW of 3.0×10^9 rad/s, then more than one stage must be used. The problem is quite complicated for transistor design because of two reasons. If we cascade stages, bandwidth shrinkage will take place, in addition to a certain amount of interaction between stages. It is not sufficient to cascade two stages, both with individual gains of 10 and bandwidths of 10 MHz. When the loading effect of stage two on stage one is considered, the overall gain may only be 30. The bandwidth may now have dropped to 5 MHz. These factors will now be considered.

A. Bandwidth Shrinkage in Iterative Stages

It is instructive to consider a cascade of identical, noninteracting stages such that if the individual gains are

$$A_1 = A_2 = A_3 = \cdots = A_n = \frac{A_{MB}}{1 + j(\omega/\omega_2)},$$

the overall gain will be

$$A_o = \frac{A_{MB}^n}{[1 + j(\omega/\omega_2)]^n} = \frac{A_{MBo}}{[1 + j(\omega/\omega_2)]^n}. \tag{7.39}$$

Pentode and field-effect transistor stages often satisfy the preceding equation due to the high input impedance of these devices. While the loading in bipolar transistor stages is such that they cannot be considered to be noninteracting, there are certain configurations having a gain equation corresponding to Eq. (7.39). The following theory can apply to these stages also.

The magnitude of the overall gain is a function of frequency and can be written in db as

$$20 \log |A_o| = n \, 20 \log \left| \frac{A_{MB}}{1 + j(\omega/\omega_2)} \right|.$$

It is obvious from the previous equation, since each stage is down 3 db at $\omega = \omega_2$, that the overall amplifier gain will be down $3n$ db at this frequency. The overall 3-db point will be lower than the single-stage 3-db point. The overall 3-db point is the frequency of interest in designing an amplifier and can be expressed in terms of n and ω_2. At low and midband frequencies Eq. (7.39) becomes

$$A_{MBo} = A_{MB}^n.$$

At higher frequencies the denominator increases dropping the overall gain A_o. We must find the point in frequency that causes the magnitude of the denominator to equal $\sqrt{2}$ resulting in a gain such that

$$|A_o| = \frac{A_{MB}^n}{\sqrt{2}} = \frac{A_{MBo}}{\sqrt{2}}.$$

This frequency is the overall 3-db frequency designated ω_{2o}. If we equate the magnitude of the denominator to $\sqrt{2}$ and solve for ω the resulting value will be

ω_{2o}. Thus,

$$\left\| \left[1 + j\frac{\omega_{2o}}{\omega_2} \right]^n \right\| = \sqrt{2} = \left\| \left[1 + j\frac{\omega_{2o}}{\omega_2} \right] \right\|^n.$$

Squaring both sides of the equation gives

$$2 = \left\| \left[1 + j\frac{\omega_{2o}}{\omega_2} \right] \right\|^{2n}.$$

The magnitude of the complex term is $\sqrt{1 + \omega_{2o}^2/\omega_2^2}$, so

$$2 = \left(1 + \frac{\omega_{2o}^2}{\omega_2^2} \right)^n$$

and

$$\omega_{2o} = \omega_2 \sqrt{2^{1/n} - 1}. \tag{7.40}$$

The overall 3-db frequency for a multistage amplifier is considerably smaller than that for a single-stage amplifier. For example, if $n = 2$, the overall 3-db point equals $0.64\omega_2$. If $n = 3$, the overall 3-db point is $0.51\omega_2$. Adding more stages to increase the overall gain reduces the bandwidth of the amplifier. Equation (7.40) is called the bandwidth shrinkage equation.

An interesting problem encountered in iterative stage design is the following: Given a specified overall gain and bandwidth, what number of stages should be used to achieve the required gain and bandwidth? If a high value of single-stage gain is used, only a few stages are needed to realize the gain. Assuming a constant GBW with changing gain, the single-stage bandwidth will be small to achieve high gain per stage. The overall bandwidth may then be smaller than required. If single-stage gain is decreased in an effort to increase the individual stage bandwidth, more stages will be required to achieve the overall gain. The term under the radical sign in Eq. (7.40) may now limit the overall bandwidth to an unacceptable value. One would expect that there exists some optimum value of gain per stage or some optimum number of stages that results in the required overall gain and maximizes the overall bandwidth.

It is possible to calculate this optimum value of gain for iterative stages after making the approximation

$$\omega_{2o} = \omega_2 \sqrt{2^{1/n} - 1} \approx \frac{\omega_2}{1.2\sqrt{n}} \quad \text{for} \quad n \geq 3. \tag{7.41}$$

The approximation of Eq. (7.41) is valid when the number of stages of the amplifier equals or exceeds three.

Let us now review the problem along with any constraints under which we must operate. We are given some specified value of required overall gain A_{MBo}. Using iterative stages we are to find the individual stage gain and hence the total number of stages that maximize the overall bandwidth. We will assume that the GBW product is a constant in this derivation.

The overall bandwidth can be written as

$$\omega_{2o} = \frac{\omega_2}{1.2n^{1/2}}.$$

Noting that GBW $= A_{MB}\omega_2$ allows us to rewrite the overall bandwidth as

$$\omega_{2o} = \frac{\text{GBW}}{1.2n^{1/2}A_{MB}} = \frac{\text{GBW}}{1.2n^{1/2}A_{MBo}^{1/n}}. \tag{7.42}$$

We can now differentiate ω_{2o} with respect to n and equate this result to zero to find the optimum value of n. The differentiation is simplified if we let $A_{MBo} = e^k$ where $k = \ln A_{MBo}$. We can now write

$$A_{MBo}^{1/n} = e^{k/n}$$

and

$$\omega_{2o} = \frac{\text{GBW}}{1.2n^{1/2}e^{k/n}}. \tag{7.43}$$

Differentiating gives

$$\frac{d\omega_{2o}}{dn} = \frac{\text{GBW}}{1.2n^{3/2}e^{k/n}}\left[\frac{k}{n} - \frac{1}{2}\right]. \tag{7.44}$$

Setting this derivative equal to zero determines the value of n to be

$$n = 2k = 2\ln A_{MBo}. \tag{7.45}$$

The midband gain becomes

$$A_{MB} = A_{MBo}^{1/n} = e^{k/n} = e^{1/2} = 1.65. \tag{7.46}$$

A quick check will prove that this value of gain results in a maximum rather than a minimum value of bandwidth. Thus, the bandwidth is maximized when the individual stage gain is 1.65. The total number of stages required is given by Eq. (7.45).

It should be noted that the same results are achieved if we assume that the overall bandwidth is fixed at some required value and the overall gain is to be maximized. Again the optimum single-stage gain is 1.65. Single-stage bandwidth is calculated by dividing the GBW by this value of gain. The number of stages to be used is then found from the bandwidth shrinkage equation.

Example 7.3. An amplifier requires a gain of 1000. If the transistors used have a GBW of 2×10^8 rad/sec, calculate the maximum overall bandwidth that can be achieved and the number of stages that must be used.

Solution. To achieve the maximum overall bandwidth a single-stage gain of 1.65 must be used. An overall gain of 1000 requires that

$$1.65^n = 1000$$

or

$$n = \frac{\ln 1000}{\ln 1.65} = 2 \ln 1000 = 13.8.$$

Since n must equal an integer, 14 stages will be used. The individual stage bandwidth is

$$\omega_2 = \text{GBW}/1.65$$

and the overall bandwidth is

$$\omega_{2o} = \frac{\omega_2}{1.2 \times \sqrt{14}} = \frac{\text{GBW}/1.65}{1.2 \times \sqrt{14}} = 27 \times 10^6 \text{ rad/sec}.$$

B. Noniterative Stages

There are several amplifiers that do not allow the use of iterative stages. For example, a three-stage amplifier might have some fixed load resistance, a fixed source resistance, and an interstage collector resistance different from these values. The three stages may exhibit three different values of GBW and are therefore noniterative. It has been shown [4] that the overall bandwidth of the amplifier will be maximized by selecting the individual stage bandwidths to be equal. This is an important result that again assumes that the GBW remains constant as gain and bandwidth are exchanged.

If we designate the GBW of the ith stage GBW_i, then it can be shown that for n stages to realize a specified overall gain A_o, the individual stage bandwidth is given by

$$\omega_2 = \left[\frac{\text{GBW}_1 \text{GBW}_2 \cdots \text{GBW}_n}{A_{MBo}}\right]^{1/n}. \tag{7.47}$$

We can demonstrate the usefulness of this result by assuming that we have three noninteracting stages with $\text{GBW}_1 = 10^8$, $\text{GBW}_2 = 4 \times 10^8$, and $\text{GBW}_3 = 8 \times 10^8$. An overall gain of 1000 is first realized by setting $A_{MB1} = A_{MB2} = A_{MB3} = 10$. For these values of gain, the bandwidths will be $\omega_{21} = 10^7$ rad/sec, $\omega_{22} = 4 \times 10^7$, and $\omega_{23} = 8 \times 10^7$. The overall amplifier gain can be written

$$A_o = \frac{A_{MBo}}{\left(1 + j\frac{\omega}{\omega_{21}}\right)\left(1 + j\frac{\omega}{\omega_{22}}\right)\left(1 + j\frac{\omega}{\omega_{23}}\right)}$$

$$= \frac{1000}{\left(1 + j\frac{\omega}{10^7}\right)\left(1 + j\frac{\omega}{4 \times 10^7}\right)\left(1 + j\frac{\omega}{8 \times 10^7}\right)}.$$

The 3-db point of the overall response occurs at approximately 0.96×10^7 rad/sec. Now let us assume that we adjust the gains and bandwidths of each stage in accordance with Eq. (7.47). Each bandwidth is then

$$\omega_2 = \left[\frac{(1 \times 10^8)(4 \times 10^8)(8 \times 10^8)}{1000}\right]^{1/3} = 3.18 \times 10^7 \text{ rad/sec}.$$

Applying the bandwidth shrinkage formula gives an overall bandwidth of

$$\omega_{2o} = 1.62 \times 10^7 \text{ rad/sec},$$

which is considerably greater than the overall bandwidth achieved with unequal single-stage bandwidths.

Alternatively, if overall bandwidth is specified along with a fixed number of stages to be used, overall gain is maximized if equal individual stage bandwidths are used. An example will demonstrate this point.

Example 7.4. Sketch the asymptotic gain as a function of frequency for the circuit of Fig. 7.22. Assume that the emitter resistors are completely bypassed at all frequencies of interest and that both transistors have the same parameters: $\beta = 100$,

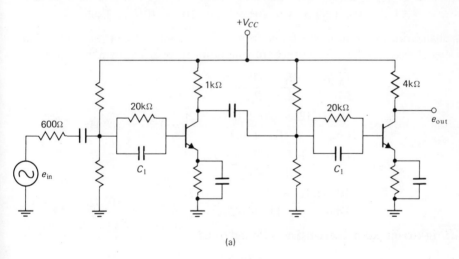

(a)

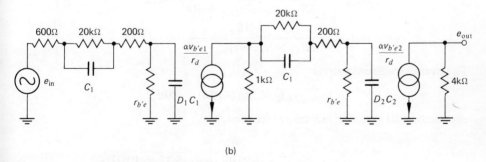

(b)

Fig. 7.22 Amplifier of Example 7.4: (a) actual circuit; (b) equivalent circuit.

$\omega_t = 2\pi \times 10^7$, $r_{bb'} = 200\ \Omega$, and $C_{b'c} = 10\ \text{pF}$ at the bias points used. Both stages are base-compensated. Neglect the loading effects of biasing and base-compensating resistors and assume that $I_{E1} = I_{E2} = 2\ \text{mA}$.

Solution. The first steps consist of evaluating $r_{b'e}$ and the D-factor for each stage. Since both emitter currents are set at 2 mA,

$$r_{b'e1} = r_{b'e2} = 101 \times 13 = 1313.$$

To find the D-factor we recognize that the resistive load of the first stage consists of the 1 kΩ resistance in parallel with the biasing resistances and the base compensation plus input resistance of the second stage. These resistances are generally much larger than the collector-load resistance and are neglected here. They can be included if necessary. The D-factor for stage 1 is

$$D_1 = 1 + 0.99 \times 10^3 \times 2\pi \times 10^7 \times 10^{-11} = 1.62.$$

For the second stage the D-factor is

$$D_2 = 1 + 0.99 \times 4 \times 10^3 \times 2\pi \times 10^7 \times 10^{11} = 3.49.$$

The bandwidth of each stage is found from Eq. (7.26) noting that the collector load of stage 1 assumes the role of source resistance for stage 2. The bandwidths are

$$f_{21} = \frac{f_\beta}{D_1} \frac{(r_{b'e} + R_1 + R_s + r_{bb'})}{R_s + R_{bb'}}$$

$$= \frac{10^5}{1.62} \frac{(1.313 + 20 + 0.6 + 0.2)}{0.6 + 0.2} = 1.70 \times 10^6\ \text{Hz}.$$

$$f_{22} = \frac{f_\beta}{D_2} \frac{(r_{b'e} + R_1 + R_{L1} + r_{bb'})}{R_{L1} + r_{bb'}}$$

$$= \frac{10^5}{3.49} \frac{(1.313 + 20 + 1.0 + 0.2)}{1.0 + 0.2} = 5.37 \times 10^5\ \text{Hz}.$$

The "in-circuit" gain of each stage is found to be

$$A_{MB1} = \frac{-\beta R_{L1}}{R_S + r_{bb'} + r_{b'e} + R_1} = -4.52,$$

$$A_{MB2} = \frac{-\beta R_{L2}}{R_{L1} + r_{bb'} + r_{b'e} + R_1} = -17.8.$$

The overall midband gain is

$$A_{MBo} = (-4.52)(-17.8) = 80.5,$$

and the overall gain as a function of frequency is

$$A_o = \frac{80.5}{[1 + jf/(1.7 \times 10^6)][1 + jf/(5.37 \times 10^5)]}.$$

The frequency response is shown in Fig. 7.23.

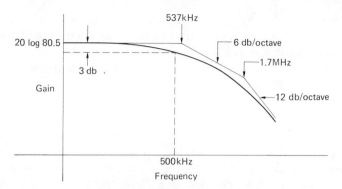

Fig. 7.23 Frequency response.

The overall corner frequency is found to be approximately

$$f_{2o} = 500 \text{ kHz}.$$

Let us now assume that we want to achieve this same overall bandwidth with equal individual bandwidths. The base-compensating networks can then be modified to obtain the desired bandwidths, giving new values of midband gain. The desired individual bandwidth is

$$f_2 = \frac{f_{2o}}{\sqrt{2^{1/2} - 1}} = \frac{500 \text{ kHz}}{0.644} = 776 \text{ kHz}.$$

The individual values of GBW are

$$\text{GBW}_1 = 4.52 \times 2\pi \times 1.7 \times 10^6 = 4.82 \times 10^7 \text{ rad/sec}$$
$$\text{GBW}_2 = 17.8 \times 2\pi \times 5.37 \times 10^5 = 6.00 \times 10^7 \text{ rad/sec}.$$

If gain and bandwidth are interchanged to give individual bandwidths of 776 kHz, the individual gains become

$$A_{MB1} = \frac{\text{GBW}_1}{\omega_{21}} = \frac{4.82 \times 10^7}{2\pi \times 7.76 \times 10^5} = 9.9$$

$$A_{MB2} = \frac{\text{GBW}_2}{\omega_{22}} = \frac{6.00 \times 10^7}{2\pi \times 7.76 \times 10^5} = 12.3.$$

The overall gain is now

$$A_{MBo} = 9.9 \times 12.3 = 122.$$

This is the maximum gain that can be obtained with these two stages using base compensation to realize the specified overall bandwidth.

Calculations of single-stage gain and bandwidth were easily made for the base-compensated amplifier. This follows from the fact that the loading of the base-compensating network on the preceding collector resistance is generally negligible at frequencies of interest. If simple common-emitter stages are used with the GBW controlled by R_L, the above results do not apply. That is, maximum gain may not result for equal individual bandwidths since the GBW of each stage changes as bandwidth is varied by R_L. It would be very difficult to calculate the desired bandwidths in this case of noniterative stages with the necessity of controlling the bandwidth by R_L.

C. Iterative Stages

Iterative amplifiers consist of long chains of identical stages. This is a rather idealistic model from a practical standpoint due to at least two reasons. First of all, the source impedance of the first stage and the load impedance of the last stage may, of necessity, have different values from the interstage load resistances. In this instance all stages are iterative except the end stages (first and last). There is also a practical limitation to obtaining iterative stages as a result of the variation of parameters from one device to another of the same kind.

In view of the practical departures from the ideal case, one might question the value of investigating iterative stages. Such a study is justified by noting that: (1) there are occasions when the end-stage resistances can be selected to have the same value as interstage load resistances; (2) while the parameters vary from one device to the next, the effects of this variation can often be overcome by slight adjustments in the resulting design; (3) it serves as the basis for a design procedure that can be extended to the noniterative end-stage case; and (4) it is impossible to develop a manual design procedure that accounts for all departures from ideal.

Common-emitter stages. Figure 7.24 shows an iterative amplifier comprised of common-emitter stages. Biasing networks and coupling capacitors are not included in order to simplify the ensuing discussion. We note that the stages will be iterative only if $R_S = R_L = R_{Li}$ and if $R = r_{bb'} + r_{b'e}$. If this is the case, all D-factors will be equal. The midband gain $e_1/e_{in} = e_2/e_1 = \cdots = e_n/e_{n-1}$ can be written

$$A_{MBi} = \frac{-\beta R_{Li}}{R_{Li} + r_{bb'} + r_{b'e}}.$$

We note that the voltages $e_1, e_2, \ldots, e_n$ cannot be measured physically since we do not have access to this point. It is possible and convenient, however, to use these voltages in our gain calculations. The overall gain can then be expressed as

$$A_{MBo} = (A_{MBi})^n \frac{R}{R_L + R}, \tag{7.48}$$

where the last term of the expression accounts for the loading effect of R.

The bandwidth of each stage is given by Eq. (7.14) as

$$\omega_{2i} = \frac{\omega_\beta B}{D} = \frac{\omega_\beta}{D} \frac{(R_{Li} + r_{bb'} + r_{b'e})}{(R_{Li} + r_{bb'})}.$$

The D-factor is given by

$$D = 1 + \alpha \omega_t C_{b'c} R_{eq},$$

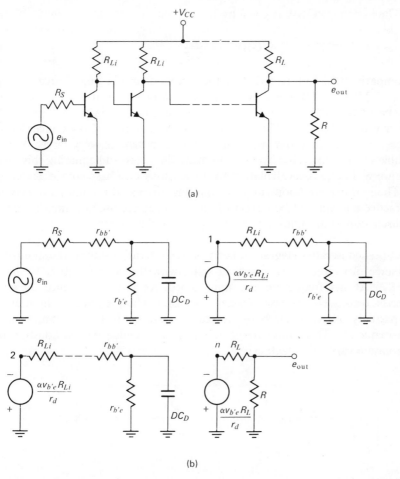

Fig. 7.24 Iterative amplifier: (a) actual amplifier neglecting biasing resistances and coupling capacitors; (b) equivalent circuit.

where

$$R_{eq} = R_{Li} \| (r_{bb'} + r_{b'e}).$$

The quantity R_{eq} is the effective collector load. At frequencies approaching and exceeding the individual stage bandwidth, the input impedance to the transistor becomes capacitive. In general, this range of frequencies is considerably higher than the overall bandwidth of the amplifier; thus, the consideration of changes in D-factor at high frequencies is unnecessary.

The overall bandwidth of the amplifier is

$$\omega_{2o} = \frac{\omega_{2i}}{1.2n^{1/2}} = \frac{\omega_\beta B}{1.2Dn^{1/2}}. \tag{7.49}$$

The interstage GBW is given by

$$\text{GBW}_i = \frac{\omega_t}{D} \frac{R_{Li}}{R_{Li} + r_{bb'}}. \tag{7.50}$$

In comparing this iterative GBW to the common-emitter single-stage value given by Eq. (7.23) we find that R_{Li} has replaced the source resistance R_S. For the single-stage case, the GBW was maximized as R_L became very large and approached a value given by Eq. (7.24). There is an optimum value of load resistance that maximizes the interstage GBW also, but it is not, in general, very large. While it is possible to find this value by differentiating Eq. (7.50) and equating this result to zero, we will defer this calculation to the discussion of base-compensated stages.

The iterative common-emitter stages just discussed are not extremely useful in practice and will not be further discussed. They are useful primarily as a basis for discussion of the analytical methods used in iterative stages.

Shunt-peaked iterative stages. A fairly popular iterative stage is the shunt-peaked interstage. An inductor is placed in series with the collector load as shown in Fig. 7.25. At high frequencies the input impedance and consequently the effective collector resistance decrease, lowering the gain. The presence of the inductance can partially offset this decrease in load since the inductive reactance increases with frequency. The corner frequency is then extended by the addition of the appropriate value of inductance.

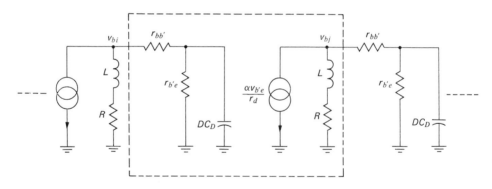

Fig. 7.25 Shunt-peaking.

A single-stage analysis results in an equation in which there is one factor involving ω in the numerator, and a quadratic equation in ω in the denominator. This problem can easily be handled by dividing the numerator factor out of both the numerator and the denominator, resulting in a very simple gain expression. We should note that this does not give optimum GBW, but it does increase this value over the uncompensated case. We also assume that L does not affect the D-factor. This is generally true at frequencies below the overall bandwidth. The output voltage at point bj is

$$v_{bj} = \frac{-\alpha v_{b'e}}{r_d}\left\{(\omega L + R) \,\middle\|\, \left(r_{bb'} + \frac{r_{b'e}}{1 + j\omega r_{b'e}DC_D}\right)\right\}.$$

The voltage $v_{b'e}$ will be

$$v_{b'e} = v_{bi} \frac{r_{b'e}/(1 + j\omega r_{b'e}DC_D)}{r_{bb'} + r_{b'e}/(1 + j\omega r_{b'e}DC_D)}.$$

Combining these two equations, we obtain

$$A_i = \frac{v_{bj}}{v_{bi}} = \frac{-\beta(j\omega L + R)}{-\omega^2 L r_{b'e}DC_D + j\omega(r_{b'e}DC_D(R + r_{bb'}) + L) + R + r_{bb'} + r_{b'e}}.$$

At this point we have an equation that can be put into the form

$$A_i = \frac{-k(j\omega + P)}{-\omega^2 + j\omega Q + T}.$$

where P, Q, and T are constants. Since we can determine the values of P, Q, and T by choosing particular values of R and L, we should make an attempt to obtain a larger GBW than exists in the case where $L = 0$. There has been much work done in determining these coefficients to get a particular type of response. One method that considerably simplifies the expression for gain (although it does not optimize the GBW) is to choose values of R and L such that the term $(j\omega + P)$ can be divided out of the numerator and denominator, leaving a simple term in the denominator. In other words, if the denominator were factored, we could write

$$A_i = \frac{-k(j\omega + P)}{(j\omega + E)(j\omega + F)}.$$

If we choose R and L to cause $E = P$, then the expression reduces to

$$A_i = \frac{-k}{(j\omega + F)}.$$

The bandwidth could now be written as

$$\omega_2 = F.$$

For maximum bandwidth, we should be sure that $E < F$. To cancel the numerator factor, we divide the numerator and denominator by $j\omega L + R$. The denominator becomes

$$\text{Den.} = \left(j\omega r_{b'e}DC_D + \frac{R + r_{bb'} + r_{b'e}}{R}\right) + \frac{j\omega r_{b'e}DC_D r_{bb'} - j(\omega L/R)(r_{bb'} + r_{b'e})}{j\omega L + R}.$$

The remainder will be zero if

$$r_{b'e}DC_D r_{bb'} = \frac{L}{R}(r_{bb'} + r_{b'e}).$$

This imposes the condition on L that

$$L = \frac{r_{b'e}r_{bb'}}{r_{b'e} + r_{bb'}} DC_D R = \frac{r_{bb'}R}{r_{bb'} + r_{b'e}} \frac{D}{\omega_\beta}. \tag{7.51}$$

The gain expression relating v_{bi} and v_{bj} is now

$$\frac{v_{bj}}{v_{bi}} = -\beta R \bigg/ \left[(R + r_{bb'} + r_{b'e})\left(1 + j\frac{\omega r_{b'e}DC_D R}{R + r_{bb'} + r_{b'e}}\right)\right].$$

The bandwidth is then

$$\omega_2 = \frac{\omega_\beta}{D} \frac{R + r_{bb'} + r_{b'e}}{R}. \tag{7.52}$$

The midband voltage gain per stage is easily evaluated as

$$A_{MBi} = \frac{-\beta R}{R + r_{bb'} + r_{b'e}}. \tag{7.53}$$

The GBW per stage is then

$$\text{GBW}_i = \frac{\omega_\beta}{D} \beta = \frac{\omega_t}{D}. \tag{7.54}$$

In comparing the figures for the GBW of the shunt-peaked and common-emitter stages it is noted that the value for the common-emitter stage is less than that for the shunt-peaked stage by the factor $R_{Li}/(R_{Li} + r_{bb'})$. For low-frequency work this factor would be expected to approach unity since R_{Li} will be much larger than $r_{bb'}$. In high-frequency design, R_{Li} may be quite small. The shunt-peaked stage GBW figure may be twice that of the common-emitter stage in this instance.

It is incorrect to apply the bandwidth optimization theory to these stages because the GBW figure is not constant. Gain and bandwidth can be exchanged by varying R; however, Eq. (7.54) indicates that the D-factor change will affect GBW_i. For very small values of R, the D-factor approaches unity and, in this special case, the GBW is constant. We can now apply the results of the bandwidth optimization theory to the shunt-peaked stage so long as R is small enough to yield a value of unity for the D-factor. The midband gain must equal 1.65 and this aids in the determination of R. Setting Eq. (7.53) equal to 1.65 and solving for R gives

$$R = \frac{1.65}{\beta}(r_{bb'} + r_{b'e}) \approx 1.65 r_d. \tag{7.55}$$

In general, the value of R given by Eq. (7.55) is small and the D-factor will become

$$D = 1 + 1.65 r_d \omega_t C_{b'c} = 1 + \frac{1.65 C_{b'c}}{C_D} \approx 1.$$

Thus, overall bandwidth can be optimized for small load resistances. For high values of ω_t it is often difficult to use shunt-peaking with high bandwidths because the stray inductance may exceed the required value specified by Eq. (7.51). With the load resistance specified by Eq. (7.55), the required L becomes

$$L = \frac{1.65 r_{bb'}}{\omega_t}$$

which may fall in the nanohenry range for high-frequency transistors.

Shunt-peaked stages form the basis for many video amplifiers, but the design must often be accomplished by "cut-and-try" methods rather than specific design procedures. To realize a specified overall gain and bandwidth, one assumes a number of stages. Individual stage gain and bandwidth can then be calculated based on iterative stage theory. The necessary individual GBW can then be found. The required value of R can be found from Eq. (7.53) and then the actual GBW is found from Eq. (7.54). If this value equals or exceeds the required figure the design is acceptable. If not, a new value for n must be selected and the process repeated. While this stage yields a GBW figure that is higher than that of the base-compensated stage, the following section demonstrates that a viable design procedure for the base-compensated stage is possible.

We might also note at this point that the GBW figure can be increased further in a shunt-peaked stage by selecting R and L to create maximally flat magnitude response [5]. This procedure is more involved than the preceding method and will not be discussed here.

Base-compensated iterative stages [3]. Figure 7.26 shows the equivalent circuit for two base-compensated interstages of a multistage amplifier. The base-compensated stage is quite convenient to use in iterative stage amplifiers. The D-factor is determined by the effective collector load which consists of R_{Li} in parallel with an impedance

$$Z = \frac{R_1}{1 + j\omega R_1 C_1} + \frac{r_{b'e}}{1 + j\omega r_{b'e} DC_D} + r_{bb'}.$$

For proper compensation $R_1 C_1 = r_{b'e} DC_D = D/\omega_\beta$, allowing us to write

$$Z = \frac{R_1 + r_{b'e}}{1 + j\omega D/\omega_\beta} + r_{bb'}.$$

At frequencies below ω_β/D the impedance equals $R_1 + r_{b'e} + r_{bb'}$ which is typically one or two orders of magnitude larger than R_{Li}. The magnitude of Z decreases with

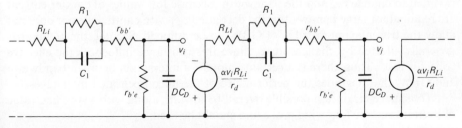

Fig. 7.26 Equivalent circuit of iterative stages.

frequency, but the frequency must exceed ω_β/D by one or two orders of magnitude before Z decreases to a value comparable to R_{Li}. Thus, for most amplifiers, the D-factor will be determined by R_{Li} and changes in R_1 will not affect D. The fact that loading of successive stages can be neglected allows the GBW figure to remain constant with adjustments in bandwidth. The midband gain for each stage is

$$A_{MBi} = \frac{-\beta R_{Li}}{R_1 + R_{Li} + r_{bb'} + r_{b'e}} \tag{7.56}$$

and the bandwidth is

$$\omega_{2i} = \frac{\omega_\beta}{D}\left[\frac{R_1 + R_{Li} + r_{bb'} + r_{b'e}}{R_{Li} + r_{bb'}}\right]. \tag{7.57}$$

The GBW is

$$\text{GBW}_i = \frac{\omega_t}{D}\frac{R_{Li}}{R_{Li} + r_{bb'}} \tag{7.58}$$

which is exactly the same as the common-emitter GBW given by Eq. (7.50). We again emphasize, though, that A_{MBi} and ω_{2i} can be controlled by R_1 (and C_1) while GBW_i remains constant.

An optimum value of R_{Li} can be found that maximizes GBW_i with respect to this variable. We find this value by differentiating Eq. (7.58) with respect to R_{Li} and equating the result to zero. This value of resistance, designated R_o, is found to be

$$R_o = \sqrt{\frac{r_{bb'}}{\omega_t C_{b'c}}}. \tag{7.59}$$

The D-factor corresponding to this value of load resistance is given by

$$D_o = 1 + \sqrt{r_{bb'}\omega_t C_{b'c}} \tag{7.60}$$

and the maximum value of GBW_i is

$$\text{GBW}_o = \frac{\omega_t}{(D_o)^2}. \tag{7.61}$$

While the value of optimum load resistance agrees with that calculated for uncompensated common-emitter stages [1], there are important differences between the two cases. In the uncompensated stage as previously discussed, emitter current is varied to change $r_{b'e}$ and the bandwidth. Because low values of emitter current are required for large bandwidths, the decrease in f_t with emitter current does not allow the true maximum value of GBW_i to be attained. Second, the D-factor is generally affected by changes in $r_{b'e}$ since this resistance influences the effective collector load. Thus, there is a certain amount of interaction between bandwidth and GBW_i that degrades the performance of the uncompensated interstages.

To achieve maximum possible overall bandwidth along with a specified value of overall gain, the following procedure should be used.

1. Select a bias point that maximizes ω_t with respect to emitter current.
2. Calculate the optimum value of load resistance to maximize GBW_i.
3. Find the value of R_1 from Eq. (7.56) that results in an individual stage midband gain of 1.65. This value will be

$$R_i = \frac{\beta R_{Li}}{1.65} - r_{bb'} - r_{b'e}. \tag{7.62}$$

The correct compensating capacitance should be chosen from Eq. (7.25).

4. The maximum overall bandwidth can now be calculated from

$$\omega_{2o\,\text{max}} = \frac{\omega_{2i}}{1.2\, n^{1/2}} = \frac{\text{GBW}_i}{1.98 n^{1/2}}. \tag{7.63}$$

D. Multistage Design with Constant GBW Stages

The issue of multistage design has been touched on previously, but not considered in detail. For stages with nonconstant GBW it is virtually impossible to develop practical analytic design procedures. It is for this reason that cut-and-try methods are suggested for shunt-peaked stages. In stages having constant GBW values, it is still rather difficult to develop a practical design procedure. A useful procedure should allow the designer to do the following:

1. Determine whether a specific device type can be used to satisfy the overall voltage gain and bandwidth requirements of the amplifier.
2. Include the effects of source and load resistances that make these end stages noniterative.
3. Calculate the minimum number of stages required for the design along with individual stage voltage gains.

Such a theory has been developed [3] applying to op amp and base-compensated stages, but will not be discussed here. The interested reader is referred to the literature.

It is a simple matter to determine if the individual stage GBW is sufficient to meet the overall gain and bandwidth of the amplifier. If ω_{or} is the required overall bandwidth and A_{or} is the required overall gain, then the individual stage bandwidth ω_2 which can be expressed as $\text{GBW}_i/1.65$ must satisfy the inequality

$$\frac{\text{GBW}_i}{1.65} \geqq (\omega_{or}\, 1.2 n^{1/2}).$$

Since 1.65^n must equal A_{or} then n can be written as

$$n = \frac{\ln A_{or}}{\ln 1.65} = 2 \ln A_{or}.$$

Substituting this value for n results in the requirement on GBW_i that

$$GBW_i \geq 2.8\omega_{or} \sqrt{\ln A_{or}}. \tag{7.64}$$

The value for GBW_i can be calculated for a given device and configuration. For example, Eq. (7.61) would be used for a base-compensated stage. If the actual GBW_i is smaller than $2.8\omega_{or} \sqrt{\ln A_{or}}$ then a device with a higher GBW_i must be used. If the equality sign holds, the required overall bandwidth is achieved by using n stages with an individual stage gain of 1.65. When the GBW_i is greater than the required value, either an excess overall bandwidth or an excess overall gain will result from using n stages. This indicates that fewer stages can be used to achieve the overall bandwidth and gain specifications. The procedure for reducing the number of stages can be found in reference [3].

E. Feedback Stages

Chapter 6 considered the op amp in relation to the exchange of gain for bandwidth. With this circuit, gain and bandwidth can be exchanged directly leading to a constant GBW figure. For a chain of iterative op amps as is shown in Fig. 7.27 we note that the end-stage effects can easily be included in the analysis. In general $R_1 \gg R_S$ and the output impedance with voltage feedback will be much smaller than R_L. Thus, the circuit approaches an iterative stage amplifier.

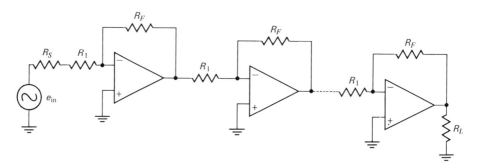

Fig. 7.27 A video amplifier using op amps.

Given a specified overall gain and overall bandwidth of an amplifier, we start the design process by applying Eq. (7.64):

$$GBW_i \geq 2.8\omega_{or} \sqrt{\ln A_{or}}.$$

If the individual op amp GBW_i is too small, a higher GBW op amp must be found. If the equality sign holds, then the individual stage gain is 1.65 and

$$n = 2 \ln A_{or}$$

stages must be used. If GBW_i exceeds the minimum required value, fewer than n stages can be used. Either a cut-and-try approach can be used to calculate the number of stages required or the procedure in reference [3] can again be used.

In transistor amplifiers, the input impedance and the feedback factor are almost always frequency-dependent at high frequency, making the analysis more difficult. If the feedback factor F becomes less at high frequencies, the bandwidth can sometimes be extended beyond that obtained with a constant feedback factor. As the frequency increases and the overall gain starts to decrease, F becomes less, tending to offset the decrease in gain. An inductor in series with the feedback network or a capacitor in shunt with the network will accomplish the desired effect. We have considered one example of this earlier where an external-emitter resistance was used to exchange gain and bandwidth. The GBW-figure was lowered when R_E was used, but was increased when a capacitor was shunted across R_E. Another example of a feedback network giving a value of F that decreases with frequency is the use of an inductor and series resistor for collector-to-base feedback.

When designing high-frequency feedback circuits, it is difficult to apply most of the "short-cut" techniques that we have previously considered. The reflected-impedance method cannot be applied easily, since the amplifier gain varies with frequency, resulting in a frequency-dependent reflected impedance. Loop or node equations form the basis for the most effective method of analyzing high-frequency feedback circuits; however, the results can be quite complex. Usually, simplifying assumptions can be made to obtain approximate results.

The following example demonstrates the application of the node method of analysis.

Example 7.5. For the circuit of Fig. 7.28 $f_t = 100 \text{ MHz}$, $C_{b'c} = 8 \text{ pF}$, $r_{bb'} = 200 \Omega$, $\beta = 80$, and $r_{b'e} = 2 \text{ k}\Omega$. (a) Evaluate the open loop gain and bandwidth (assuming proper bias). (b) Find the closed loop bandwidth.

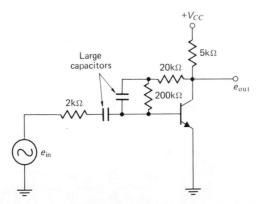

Fig. 7.28 Feedback stage.

Solution. (a) The equivalent circuit of Fig. 7.29 applies when feedback is not present. The midband gain is

$$A_{MB} = \frac{-\beta R_C}{R_S + r_{bb'} + r_{b'e}} = \frac{-80 \times 5 \times 10^3}{2 \times 10^3 + 2 \times 10^2 + 2 \times 10^3} = -95.2.$$

312 HIGH-FREQUENCY AMPLIFIERS

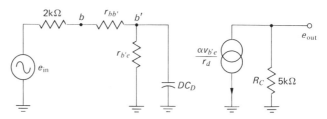

Fig. 7.29 Equivalent circuit for Example 7.5.

The bandwidth is

$$f_2 = \frac{B}{D}\frac{f_t}{\beta},$$

where

$$B = \frac{R_S + r_{bb'} + r_{b'e}}{r_{bb'} + R_S} = 1.91,$$

and

$$D = 1 + \alpha R_c \omega_t C_{b'c} = 1 + 5 \times 10^3 \times 6.28 \times 10^8 \times 8 \times 10^{-12} = 26.1.$$

Thus

$$f_2 = \frac{1.91 \times 10^8}{26.1 \times 80} = 91.5 \text{ kHz}.$$

(b) The circuit of Fig. 7.30 applies to the closed loop amplifier with the input voltage source converted to a current source. The appropriate node equations are as follows:

1. $\dfrac{e_{in}}{R_S} = \dfrac{v_b}{R_S} + \dfrac{v_b - v_{b'e}}{r_{bb'}} + \dfrac{v_b - e_{out}}{R_f},$

2. $0 = \dfrac{v_{b'e} - v_b}{r_{bb'}} + \dfrac{v_{b'e}(1 + j\omega r_{b'e} DC_D)}{r_{b'e}},$

3. $0 = \dfrac{\alpha v_{b'e}}{r_d} + \dfrac{e_{out} - v_b}{R_f} + \dfrac{e_{out}}{R_C}.$

Solving for $G = e_{out}/e_{in}$ results in

$$G = -8.77 \frac{(1 - j\omega 4.1 \times 10^{-10})}{(1 + j\omega 2 \times 10^{-7})}.$$

The midband gain is -8.77. The imaginary term in the numerator is negligible up to very high frequencies. The upper corner frequency is then determined by the denominator and is

$$f_2 = \frac{1}{2\pi} \times \frac{1}{2 \times 10^{-7}} = 796 \text{ kHz}.$$

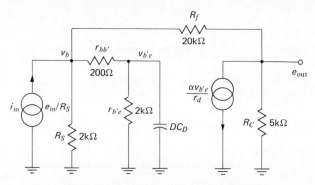

Fig. 7.30 Equivalent circuit of feedback amplifier.

Comparing the open-loop and closed-loop values (Table 7.1) shows that the GBW has been decreased by using feedback. Actually, the bandwidth of the closed loop amplifier will be slightly higher than calculated because the D-factor will be smaller when R_F is present. When the new D-factor is considered, the GBW figures compare very well for the two cases.

Table 7.1 Comparison of open loop and closed loop amplifier

	A_{MB}	f_2	GBW
Open loop values	−95.2	91.5 kHz	54.7×10^6 rad/s
Closed loop values	−8.77	796 kHz	43.8×10^6 rad/s

We should understand the significance of the ω-term appearing in the numerator. This term is negligible for frequencies up to and above the upper corner frequency. At very high frequencies, i.e., above

$$\omega_3 = \frac{1}{4.1 \times 10^{-10}} \text{ rad/s},$$

the gain expression approaches a constant value of

$$G_\infty = 0.018.$$

At very high frequencies, when C_D becomes essentially a short circuit, the amplifier itself has no gain. However, there is a direct transmission from input to output through R_F, resulting in the above value of G_∞. That this value is correct can be verified by assuming that C_D is a short circuit and analyzing the resulting network.

The design engineer is usually interested in simplifying a given problem if the results of the simplified analysis are reasonably accurate. This particular problem can be reduced from a three-node to a two-node circuit by neglecting $r_{bb'}$. We might expect this simplification to yield reasonably accurate results, since $r_{bb'}$ is much smaller than R_S, R_f, and $r_{b'e}$. The circuit now reduces to that shown in Fig. 7.31.

314 HIGH-FREQUENCY AMPLIFIERS

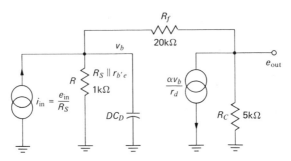

Fig. 7.31 Simplified equivalent circuit of feedback amplifier.

The two node equations are

$$1. \quad \frac{e_{in}}{R_S} = \frac{v_b}{R}(1 + j\omega RDC_D) + \frac{v_b}{R_f} - \frac{e_{out}}{R_f},$$

$$2. \quad 0 = \frac{\alpha v_b}{r_d} - \frac{v_b}{R_f} + \frac{e_{out}}{R_f} + \frac{e_{out}}{R_C}.$$

Solving for the gain results in

$$G = \frac{-\alpha R R_C R_f}{\alpha R R_C R_S + r_d R_S (R_C + R_f + R) + j(\omega/\omega_t) DR(R_f + R_C) R_S}.$$

Substituting values into this equation gives a gain of

$$G = \frac{-8.8}{1 + j\omega 1.85 \times 10^{-7}}.$$

The midband gain is -8.8, while the bandwidth is

$$f_2 = 861 \text{ kHz}.$$

Both gain and bandwidth are calculated to be slightly higher than the values obtained when using the more accurate circuit of Fig. 7.30. For most applications, the approximate results are within engineering accuracy.

If we note that the term $\alpha R R_C R_S$ is larger than the other real term in the denominator we can write an expression, admittedly inaccurate, but reflecting the general behavior of the circuit. This expression is

$$G = \frac{-(R_f/R_S)}{1 + j(\omega/\omega_t)D(R_f + R_C)/R_C}.$$

Note that the midband gain is decreased directly with R_f, but that the bandwidth does not increase directly, unless R_f is much larger than R_C. This neglects the effect of R_f on the D-factor, as mentioned previously.

There are many other configurations that can be employed in high-frequency feedback amplifiers. By now the reader is aware of many of the difficulties involved in these design problems. Multistage problems are very complex when the design is based on conventional circuit-analysis methods. More sophisticated methods based on pole-zero configurations or root-locus are often used for these problems. These techniques will not be dealt with here. The interested reader is referred to the references listed at the end of the chapter for more advanced discussions.

F. High-Frequency Integrated Amplifiers [6]

The inability to fabricate inductors and large coupling capacitors requires the use of special techniques in the design of monolithic wideband amplifiers. One circuit that can be used for an integrated amplifier is shown in Fig. 7.32.

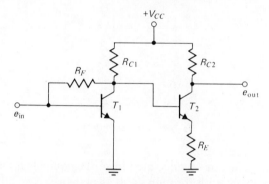

Fig. 7.32 Shunt-series cascade.

The first stage uses shunt feedback while the second uses series feedback. This configuration minimizes dc level problems since the collector of stage 1 can easily be set to a bias voltage that is compatible with the requirements of stage 2. Since both stages have been discussed previously, the analysis of this configuration will be left to the student.

Compound device stages can also be used in some high-frequency applications. The cascode connection of Fig. 7.33 is designed to minimize the Miller-effect capacitance, thereby lowering the D-factor.

The multiplication effect of the depletion-layer capacitance comes about because of the dependent voltage source connected to one side of this capacitance. An input voltage change results in a much greater voltage change across the capacitance. A large collector load causes a high voltage gain from base to collector and leads to a large D-factor. The cascode arrangement of Fig. 7.33 presents a low impedance at the collector of transistor T_1. The load at this point consists of the impedance to ground looking into the emitter of T_2. Since this impedance is small, the voltage gain and voltage swing at the collector of T_1 are relatively small. We, therefore, conclude that the Miller effect is minimized for transistor T_1. Although the voltage swing at the collector of T_1 is small, transistor T_2 acts as a common-base stage with high voltage gain. Thus, the overall gain can be high without a high D-factor.

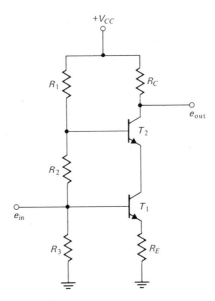

Fig. 7.33 Cascode connection.

The midband gain can be calculated from the equivalent circuit of Fig. 7.34, neglecting the feedback path through the bias network. A small capacitor across R_1 makes this assumption valid.

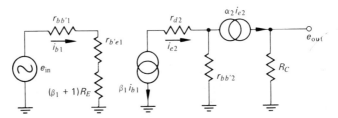

Fig. 7.34 Cascode equivalent circuit.

The overall gain is found to be

$$A_{MB} = -\frac{\beta_1 \alpha_2 R_C}{r_{bb'1} + (\beta_1 + 1)(r_{d2} + R_E)} \approx \frac{-R_C}{r_d + R_E}. \qquad (7.65)$$

A second upper corner frequency is introduced by the common-base stage, but this frequency is considerably higher than the corner frequency of the common-emitter stage. The D-factor of the first stage is

$$D \approx 1 + r_{d2}\omega_{t1}C_{b'c1}.$$

Since the emitter currents of both stages are nearly equal, the diode resistances are also of almost equal value. Using the definition of $\omega_t = 1/r_d C_D$ allows us to write

$$D \approx 1 + \frac{C_{b'c1}}{C_{D1}}$$

which will usually approach a value of unity.

The corner frequency of the cascode circuit will then have an approximate value of

$$\omega_2 = B\omega_\beta, \tag{7.66}$$

where B is the broadband factor.

The cascode configuration is obviously useful in applications requiring a large load resistance. In the simple common-emitter stage, the D-factor would limit the bandwidth excessively while the cascode circuit would overcome this problem.

A second compound device amplifier that is useful is the common-collector/common-emitter pair of Fig. 7.35. This configuration is appropriate in applications involving large source resistances. The common-emitter stage would lead to a poor broadband factor whereas the pair of Fig. 7.35 minimizes this effect.

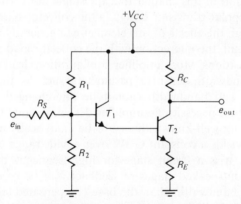

Fig. 7.35 Common-collector/common-emitter pair.

The differential stage is quite popular in monolithic amplifiers. In addition to the excellent temperature stability of this pair, the high-frequency response is also quite good with some minor modifications to the circuit. Figure 7.36 indicates this modified amplifier.

This amplifier is actually a common-collector/common-base pair. The emitter follower has good high-frequency properties since no collector load is present to introduce the Miller effect. The common-base stage also possesses good high-frequency response and the overall amplifier becomes a useful element in monolithic design work.

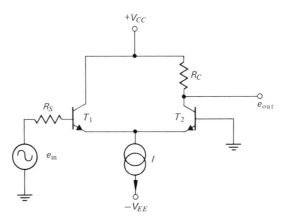

Fig. 7.36 Common-collector/common-base pair.

The symmetrical differential stage can also be used to neutralize the depletion-layer capacitance. The close matching of components required for neutralization can only be done with integrated circuits [6].

G. General Remarks on High-Frequency Design

It should be obvious from this chapter that the single most troublesome element in high-frequency bipolar transistor design is the collector-base depletion-region capacitance. Although the effects of this element can generally be reflected to the input side of the circuit, the interaction with the collector load resistance leads to nonlinear design equations. Most amplifier configurations lead to GBW products that depend on bandwidth sensitive parameters such as the load resistance. Consequently, gain and bandwidth cannot be interchanged directly. While a constant GBW is not a necessity in amplifier design, it simplifies a given design procedure by allowing well-known theory to be analytically applied. The base-compensated stage yields a constant GBW over a wide range of bandwidths and voltage gains; thus, it is a useful stage for high-frequency design. It must be emphasized that shunt-peaked stages or feedback can be used to create stages with greater overall bandwidths than the base-compensated stage. However, the design is not as analytical with cut-and-try methods being the major tool in the design process.

An important point that has not yet been considered is the dependence of design equations on transistor parameters. The base-compensated stage and the shunt-peaked stage both depend on element values that are determined from transistor parameters. This is a problem in production circuits that makes practical design more difficult. One of the best methods to overcome this difficulty is to base the design on feedback stages. The performance of the feedback amplifier can be made to depend primarily on external feedback networks that can be accurately determined. Production circuits depend heavily on feedback to make the behavior of several circuits more uniform.

7.4 TUNED STAGES

It has been noted that video stages are required to pass signals that contain a broad band of frequencies. A stage that has a lower corner frequency of 100 Hz and an upper frequency of 4 MHz is an example of a video amplifier. There are many other applications that require a narrowband amplifier. For example, the intermediate frequency stages of conventional radio receivers are designed to amplify a narrow band of frequencies centered about a frequency of 455 kHz. A signal containing only frequencies outside this band is not to be amplified.

Narrowband amplifiers are very important in fields that depend on the simultaneous transmission of several channels of information. The general field of communication in its present form actually owes its existence to the narrowband amplifier. In radio broadcasting, several stations in a geographic area simultaneously transmit signals with each station using a different carrier frequency. The ability of a receiver to select any one of these signals and amplify only the signal selected is the key factor in radio communication. The same general idea is applied in television receivers, two-way radio communication units, long-distance telephone facilities that carry several conversations simultaneously, and even in the field of satellite communications. Some of these fields use microwave equipment rather than transistor circuitry to achieve the required high frequencies, but tuned transistor amplifiers are involved in a large number of communication circuits. Solid-state radio and television receiving circuits account for a large portion of the total narrowband transistor amplifiers used.

For narrowband applications the transistor proves to be very troublesome. The inherently low impedances associated with the transistor and a tendency toward instability under certain conditions are the two major problems encountered for tuned stages. The tendency for unstable operation is due primarily to the presence of the collector-base capacitance and resistance. The circuit of Fig. 7.37 shows a tuned stage and its equivalent. We noted in Chapter 4 that r_v and r_s could be

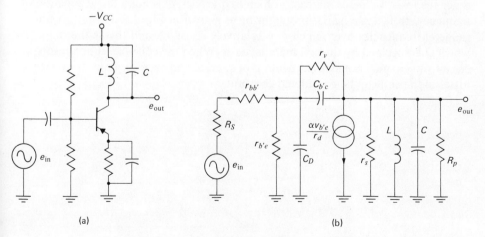

Fig. 7.37 Tuned stage: (a) actual circuit; (b) equivalent circuit.

neglected when the load impedance was small. The resonant circuit of the narrow-band stage does not always satisfy this requirement. At resonance, the impedance of the tuned circuit might range from several kilohms up to a few megohms, depending on the components used. Since the output impedance of the transistor can be less than 50 kΩ, it can lower the effective Q of the circuit and cause less selectivity. In order to examine this effect in more detail, it is necessary to analyze the parallel resonant circuit.

A. Inductors and Tuned Circuits

A pure inductance would have no loss of power when a current flows through the inductance. On the other hand, practical inductors do show a power loss due to several factors:

1. The coil wire has a finite resistance.
2. Eddy current and hysteresis effects in the core cause a power loss which is frequency-dependent and also dependent upon the core material.
3. Skin effect causes the coil wire resistance to increase at high frequencies.
4. Proximity effect causes radiation of power to the surroundings.

The external source feels the effect of these losses, but does not distinguish between the mechanisms responsible for the losses. In fact, so far as the external circuit is concerned, a resistance in series with a pure inductor will be equivalent to the practical inductor. The resistance, however, will be frequency-dependent to account for the frequency-dependent losses. The Q of a coil, which compares the energy stored to the energy dissipated per cycle, is

$$Q = \frac{\omega L}{R}.$$

Since the resistive losses increase at high frequencies, Q is not a linear function of frequency. In fact, the variation might be as shown in Fig. 7.38. There is quite a range of frequencies over which Q is relatively constant, and this is the range in which Q is specified by the coil manufacturer. When the coil is used in a resonant circuit, only a small range of frequency is of interest, and over this range the resistive losses of the coil can be considered constant.

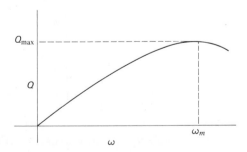

Fig. 7.38 Variation of Q with frequency.

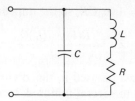

Fig. 7.39 Parallel-tuned circuit (tank circuit).

A parallel LC-circuit would be represented by the circuit of Fig. 7.39. The Q of this circuit is still equal to $\omega L/R$, but since we are interested in the resonant Q, we should use subscripts to indicate this:

$$Q_0 = \omega_0 L/R, \quad (7.67)$$

where

$$\omega_0 = 1/\sqrt{LC} \quad (7.68)$$

and is the resonant frequency. The impedance of the tank circuit is

$$Z = \frac{1}{j\omega C} \bigg\| (R + j\omega L) = \frac{R + j\omega L}{-\omega^2 CL + j\omega RC + 1}. \quad (7.69)$$

The magnitude of this function is plotted in Fig. 7.40 for various values of $Q_0 = \omega_0 L/R$. This plot demonstrates the selective properties of the tank circuit. If a constant current enters the tank at resonant frequency, a large voltage will be developed across the circuit. If the frequency of the input current is off resonance, considerably less voltage will result. From Eq. (7.69) the impedance at resonance can be found. For $\omega = \omega_0 = 1/\sqrt{LC}$, we have

$$Z_0 = \frac{R[1 + j(\omega_0 L/R)]}{j\omega_0 RC}.$$

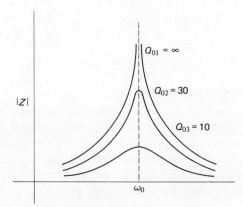

Fig. 7.40 Magnitude of Z as a function of frequency.

At resonance $Q_0 = \omega_0 L/R = 1/\omega_0 CR$, so the impedance is

$$Z_0 = \frac{R(1 + jQ_0)Q_0}{j}.$$

For the type of coils that must be used in tuned circuits, Q_0 must be quite high, ranging from a low of 10 up to several hundred. In this case the equation for Z_0 is approximately

$$Z_0 = RQ_0^2. \tag{7.70}$$

It is more convenient to work with an equivalent circuit, such as that of Fig. 7.41. The circuits are equivalent only if $R_p = Q_0^2 R$ and only in the region of resonance. In this case the Q_0 is figured from

$$Q_0 = \frac{R_p}{\omega_0 L}, \tag{7.71}$$

since

$$\frac{R_p}{\omega_0 L} = \frac{Q_0^2 R}{\omega_0 L} = \frac{Q_0^2}{Q_0} = Q_0.$$

It can be shown that the bandwidth between 3-db points for a high Q_0 circuit is

$$f_2 - f_1 = BW = f_0/Q_0, \tag{7.72}$$

where f_0 is the resonant frequency, f_2 is the upper 3-db point, and f_1 is the lower 3-db point. A higher Q_0 is seen to lead to a narrower bandwidth. Usually the manufacturer will specify the coil Q at a particular frequency along with the inductance L. This allows the calculation of the resistance R_p.

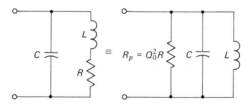

Fig. 7.41 Equivalent circuits.

B. Narrowband Transistor Amplifiers

Figure 7.42 shows a tuned transistor amplifier with a resistance R_1 shunting the tank circuit. It will be shown in the following section that the simple equivalent circuit used is not always appropriate for tuned stages, but we will use it here as a basis for discussing tuned amplifiers. When the tank circuit is added to the collector circuit, the output resistance of the transistor and R_1 appear in parallel with R_p. The equivalent resistance shunting the tank is

$$R_{sh} = R_p || R_1 || r_{out}.$$

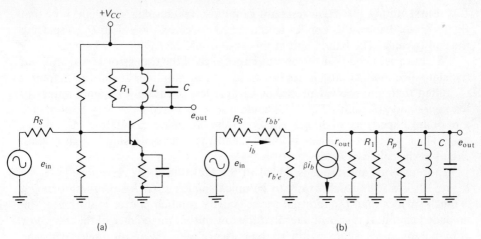

Fig. 7.42 (a) Tuned amplifier. (b) Simple equivalent circuit.

The "in circuit" or effective Q is lower than the coil Q as a result of these additional losses. The effective Q is given by

$$Q_{\text{eff}} = \frac{R_{sh}}{\omega_0 L} = \frac{R_{sh}}{R_p} Q_0. \tag{7.73}$$

The magnitude of the load impedance will exhibit a resonant peak at ω_0, but this peak will be broader than that for the LC circuit due to the lower effective Q. Voltage gain can be expressed as

$$A = \frac{-\beta Z_L}{r_{bb'} + r_{b'e} + R_S},$$

where Z_L is the collector load impedance, and the magnitude of the gain will be

$$|A| = \frac{\beta |Z_L|}{r_{bb'} + r_{b'e} + R_S}.$$

Since gain is proportional to collector impedance, gain exhibits a peak at the resonant frequency also. It is not uncommon for a properly designed tuned amplifier to exhibit a Q_{eff} of 50 to 100. The resonant gain is then

$$A_{\text{res}} = \frac{-\beta R_{sh}}{r_{bb'} + r_{b'e} + R_S}. \tag{7.74}$$

Using the relation given in Eq. (7.73) we can rewrite Eq. (7.74) as

$$A_{\text{res}} = \frac{-\beta \omega_0 L Q_{\text{eff}}}{r_{bb'} + r_{b'e} + R_S}. \tag{7.75}$$

A tuned amplifier might have a very high resonant gain if Q_{eff} is high. For example, if the sum of $r_{bb'} + r_{b'e} + R_S$ is 3 kΩ, $\beta = 100$, $\omega_0 = 6.28 \times 10^5$ rad/sec,

$L = 1$ mH, and $Q_{\text{eff}} = 50$, a resonant gain of -1000 results. We might note that Eq. (7.72) for bandwidth applies to the tuned collector stage with Q_{eff} replacing the coil Q value. The bandwidth in this case would be $f_0/Q_{\text{eff}} = 2$ kHz.

We have assumed that the output impedance of the transistor is real and that the simple equivalent circuit applies to the tuned amplifier stage. In Chapter 4 we noted that this equivalent circuit became less accurate for higher values of load resistance. Certainly Eq. (7.75) could not be expected to be very accurate for a resonant impedance of 30 kΩ as used in the preceding calculation of resonant gain. Thus, we must consider in more detail the effects of using a high Q tank circuit as the collector load.

If no external resistance is added to the tank the shunt resistance is $R_{sh} = R_p || r_{\text{out}}$. It is not unusual for r_{out} to be much smaller than R_p, meaning that Q_{eff} is much smaller than Q_0 and that the selectivity is much poorer than that of the unloaded circuit. It is shown later that the low output impedance of the transistor at high frequencies is due mainly to the presence of $C_{b'c}$. For good selectivity, the loading of the tank circuit by the transistor output impedance must be minimized.

The instability problem is again due to the presence of $C_{b'c}$. At frequencies just below resonance, the load impedance is quite high and slightly inductive. A portion of the output voltage is fed back through $C_{b'c}$ to the input. Under these conditions the feedback voltage can be greater than the initial applied voltage. As will be seen in Chapter 11 this will cause oscillations to occur; that is, an output voltage can be present even when there is no input signal.

Without going into further detail at present, it is sufficient to note that the presence of $C_{b'c}$ is responsible for most of the difficulties that occur in tuned amplifiers.

C. Neutralization of the Effects of $C_{b'c}$ and r_v

The feedback current through $C_{b'c}$ and r_v can be cancelled by adding an equal and opposite amount of current. Consider the circuit of Fig. 7.43. The transformer inverts the phase of the output voltage and since the neutralizing impedance equals the feedback impedance, the neutralizing current is equal in magnitude, but

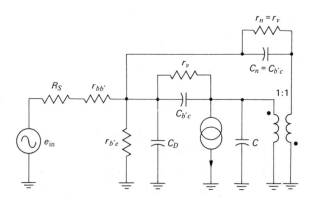

Fig. 7.43 Theoretical neutralization.

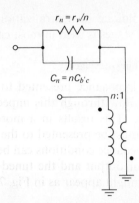

Fig. 7.44 Practical neutralizing network.

opposite in direction to the feedback current. The net current from output to input is now approximately zero. A scheme allowing more practical element values uses a stepdown transformer with turns ratio $n:1$, a neutralizing capacitance of $nC_{b'c}$, and a resistance of r_v/n, as shown in Fig. 7.44.

The above theory has assumed that the circuit designer has access to the b'-terminal, which of course is a false assumption. In practice, the neutralizing network must be connected to terminal b. The method of calculating r_n and C_n remains the same, but the neutralization is now a function of frequency and some adjustment in C_n and r_n may be required. It is also possible for the circuit to oscillate at some frequency where the neutralization is less effective.

An empirical approach to selecting r_n is necessary in cases where r_v is not known. If the y-parameters of the transistor are known, then both r_v and $C_{b'c}$ can be found, since

$$y_{12} = \frac{1}{r_v} + j\omega C_{b'c}.$$

It should be realized that the impedance of the neutralizing network will be reflected to the primary side of the transformer. The capacitance of the resonant circuit is $C + C_{b'c}/n$. Figure 7.45 shows the practical neutralized circuit.

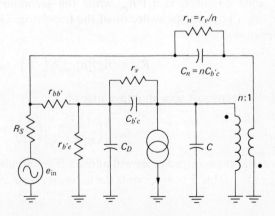

Fig. 7.45 Practical circuit.

The technique of neutralization is occasionally used in practice, but mismatching of stages is used in most cases. This method will be discussed next.

D. Mismatching of Stages

If the load impedance presented to the transistor is small, more of the output current will flow through this impedance and less will be fed back to the input through $C_{b'c}$. This results in a more stable circuit. On the other hand, a high impedance must be presented to the tank circuit to obtain a high Q circuit. Both of these impedance conditions can be met by using a step up transformer between the transistor output and the tuned circuit. In practice, a tapped transformer is used which might appear as in Fig. 7.46.

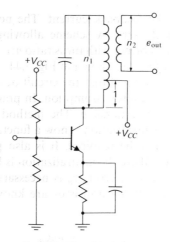

Fig. 7.46 Mismatched stage.

The load presented to the tuned transformer by the transistor is $n_1^2 r_{out}$, which can be quite high. The following stage will also cause an impedance of $(n_1/n_2)^2 r_{in\,2}$ to appear across the tuned circuit. The impedance presented to the transistor by the tuned circuit is $(1/n_1)^2 R_p$, while the secondary reflects an impedance of $(1/n_2)^2 r_{in\,2}$ back to the collector of the transistor. The tuned circuit has a shunt impedance of

$$R_{sh} = R_p \,||\, n_1^2 r_{out} \,||\, \left(\frac{n_1}{n_2}\right)^2 r_{in\,2},$$

while the transistor sees an impedance of

$$r = r_{out} \,\bigg|\bigg|\, \left(\frac{1}{n_1}\right)^2 R_p \,\bigg|\bigg|\, \left(\frac{1}{n_2}\right)^2 r_{in\,2}.$$

Obviously these impedances will allow a high-circuit Q, while causing the circuit to be more stable. For *rf*-circuits the necessary turns ratio ranges from 3:1 to 10:1.

The big advantage of mismatching is that stability is obtained for all frequencies, unlike neutralization which guarantees stability only for a certain band of frequencies. Mismatching also reduces the alignment problem when several tuned stages are used. Each tank circuit can be tuned with little or no effect on adjacent stages. When neutralization is used, the tuning often affects the neutralization of adjacent stages.

E. Combination of Mismatching and Neutralization

A combination of the two techniques can be used for critical applications. The circuit of Fig. 7.47 demonstrates one scheme of applying both methods.

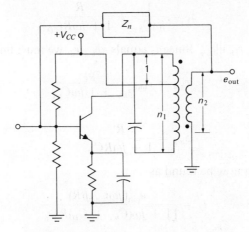

Fig. 7.47 Mismatching and neutralization.

The output impedance of the transistor will be higher due to neutralization. This allows a higher Q to be obtained by using both techniques, rather than either one alone.

*F. Effect of $C_{b'c}$ on the Output Circuit

The first sections of this chapter examined the effect of $C_{b'c}$ on the input circuit in terms of the D-factor. In this section, we are concerned with the output impedance of the stage and the influence of $C_{b'c}$ on this parameter. To find the frequency response of the transistor, the effects of $C_{b'c}$ can be reflected either to the input or the output. The analysis is simpler if $C_{b'c}$ is reflected to the input side as a Miller capacitance. However, this does not show the influence of $C_{b'c}$ on the output side of the circuit. To consider this we can use the circuit of Fig. 7.48. The ratio of $e_{\text{out}}/i_{\text{out}}$ will give the output impedance of the transistor (neglecting r_v and r_s). To find this impedance, the current $i_{\text{out}} = i_1 + i_2$ must be found. The component i_1 is seen to be simply

$$i_1 = e_{\text{out}}/z_1,$$

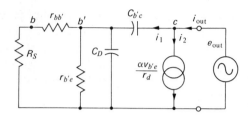

Fig. 7.48 Circuit used to calculate the effect of $C_{b'c}$ on the output circuit.

where z_1 is the impedance of $C_{b'c}$ in series with the impedance between b' and e. Thus

$$z_1 = \frac{1}{j\omega C_{b'c}} + \frac{R}{1 + j\omega RC_D}, \qquad (7.76)$$

where $R = (R_S + r_{bb'}) \| r_{b'e}$. Since i_2 equals $\alpha v_{b'e}/r_d$, we must find $v_{b'e}$. This value is

$$v_{b'e} = e_{\text{out}} \frac{z_{b'e}}{z_{b'e} + 1/j\omega C_{b'c}}$$

where

$$z_{b'e} = \frac{R}{1 + j\omega RC_D}.$$

The current i_2 can now be found as

$$i_2 = \frac{e_{\text{out}}(j\omega C_{b'c}\beta R)}{[1 + j\omega(C_{b'c} + C_D)R]r_{b'e}}. \qquad (7.77)$$

If we let $z_2 = e_{\text{out}}/i_2$, then

$$z_2 = \frac{[1 + j\omega(C_{b'c} + C_D)R]r_{b'e}}{j\omega C_{b'c}\beta R}.$$

This impedance represents a resistance of value $R_2 = r_{b'e}(C_{b'c} + C_D)/\beta C_{b'c}$ and a series capacitance $C_2 = C_{b'c}\beta R/r_{b'e}$. The output impedance of the transistor can then be represented by two parallel paths from collector to ground, one path of impedance z_1, the other of impedance z_2, as shown in Fig. 7.49. To review what we have done: When e_{out} is present, $C_{b'c}$ causes additional current to flow, which would

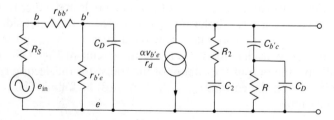

Fig. 7.49 Effect of $C_{b'c}$ reflected to output circuit.

be zero if $C_{b'c}$ were absent. The impedance at the output side of Fig. 7.49 causes this same amount of current to flow and hence can represent the effect of $C_{b'c}$.

It should be emphasized that the circuit of Fig. 7.49 can also be used to find the frequency response of the stage. While the same results are obtained, the analysis is more difficult in this case than that of reflecting $C_{b'c}$ to the input circuit.

Simplifications of Fig. 7.49. Since $C_{b'c}$ is much smaller than C_D, we can accurately represent z_1 by a single capacitor of value $C_{b'c}$. Using the values defined previously, we find R_2 and C_2 to be

$$R_2 = \frac{1}{\omega_t C_{b'c}}, \tag{7.78}$$

$$C_2 = \frac{\beta C_{b'c}}{B}, \tag{7.79}$$

where B is the broadband factor. The circuit now appears as shown in Fig. 7.50. We must remember that r_s also appears across the output terminals. At low frequencies, the capacitances C_2 and $C_{b'c}$ offer a high reactance to the output signal, and these paths can be neglected. However, at high frequencies, the path containing R_2 can offer a much lower impedance than $C_{b'c}$ or r_s. As frequency is increased, the output resistance approaches a value

$$R_2 = \frac{1}{\omega_t C_{b'c}}.$$

When tuned amplifiers are required to amplify frequencies above ω_β, the output impedance can often be approximated by this value. In order to deliver maximum power to the tuned circuit in this case, the optimum turns ratio for mismatching is

$$n_{1\,opt} = \sqrt{R_p/R_2} = \sqrt{R_p \omega_t C_{b'c}}. \tag{7.80}$$

As mentioned previously, the value of the turns ratio generally falls in the range of 3 to 10.

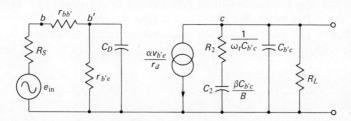

Fig. 7.50 Alternative equivalent of Fig. 7.49.

Relation of $C_{b'c}$ to tuned circuits. The capacitor $C_{b'c}$ can be added to the tank-circuit capacitance to calculate the resonant frequency. The remaining impedance becomes equal to R_2 at high frequencies. This value might range from a few hundred ohms up to a few kilohms.

At lower frequencies this impedance is unimportant, since the series capacitor will offer a high impedance. At high frequencies, however, the resistance R_2 will load the tuned circuit, causing a lower effective Q.

When the feedback effects of $C_{b'c}$ are neutralized, the resistance R_2 approaches infinity and C_2 approaches zero. It is this factor that causes the circuit Q to be higher after neutralization than before. There is another component of output resistance due to r_v, but this is greater than that due to $C_{b'c}$, except at low frequencies. This effect is also neutralized by the previous scheme to help increase the total output impedance.

When the effects of $C_{b'c}$ are neutralized, the simple equivalent circuit becomes more accurate to use for higher loads. Of course, neutralization will not be perfect and a finite value of output resistance will result.

G. The Tunnel Diode Amplifier

Amplification by the tunnel diode can be discussed in connection with the simple circuits of Fig. 7.51. Assuming that the appropriate values of R_S and R_L are selected, the output voltage can be written as

$$e_{out} = e_{in} \frac{R_L}{R_L + R_S + R_t - r_t}.$$

The midband voltage gain is then

$$A_{MB} = \frac{e_{out}}{e_{in}} = \frac{R_L}{R_L + R_S + R_t - r_t}. \tag{7.81}$$

Figure 7.52 shows the load-line for the amplifier circuit. If the sum of R_L, R_S, and R_t approaches the magnitude of r_t, the denominator of Eq. (7.81) approaches zero and the gain tends to infinity. However, a gain of infinity results in unstable operation; thus, the sum of external loop resistance must always be less than the magnitude of the negative resistance of the tunnel diode. The maximum possible sinusoidal output power can be evaluated by assuming that the negative resistance region of the characteristics can be represented by a straight line from the peak point to the valley point. We also assume that $R_S + R_t$ approaches zero and R_L approaches r_t. The greatest peak-to-peak voltage swing would then be $(V_v - V_p)$ while the corresponding peak-to-peak current swing is $(I_p - I_v)$. The maximum

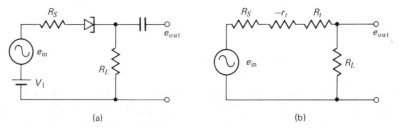

Fig. 7.51 Tunnel diode amplifier: (a) amplifier circuit; (b) equivalent circuit.

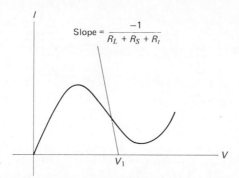

Fig. 7.52 Load-line for amplifier of Fig. 7.51.

output power is

$$P_{max} = \frac{(V_v - V_p)(I_p - I_v)}{8}. \qquad (7.82)$$

This is an upper limit that will never be reached because the negative resistance region is not entirely linear and because the load resistance must always be smaller than r_t when $R_t + R_S = 0$. In view of the rather modest voltage and current swings possible, it is obvious that the power output of the amplifier will be quite low, typically in the microwatt range.

It has been mentioned that the external series resistance must not exceed r_t. If this condition occurs the circuit becomes a potential switching circuit with more than one possible operating point. The use of the tunnel diode for switching purposes will be treated in Chapter 8. For the amplifier circuit we will consider the upper limit of external resistance to be

$$(R_S + R_L) < (r_t - R_t).$$

There is also a lower limit on the value of external resistance. The external resistance must not decrease to that value required to cause sinusoidal oscillations in the tunnel diode amplifier. We have not considered the general conditions leading to sinusoidal oscillations in an electronic circuit. These conditions are treated in a later chapter; however, it is appropriate that we consider the criterion for sinusoidal oscillations from the viewpoint of negative conductance or resistance.

It is well known that if the impedance or admittance of a circuit has poles and zeros in the left half of the Laplace or s-plane, the circuit response will be stable. For a high Q, parallel resonant circuit, the zeros of the admittance are very near the $j\omega$-axis of the s-plane as shown in Fig. 7.53(a).

If a very small current at the resonant frequency excites the circuit, a large voltage will appear. We note that the real part of the conductance is very small for this case and the susceptance is zero for resonance (and near zero for frequencies near resonance). If the real part of the admittance becomes zero or tends to a negative value, the zero positions move to the $j\omega$-axis or tend to move into the

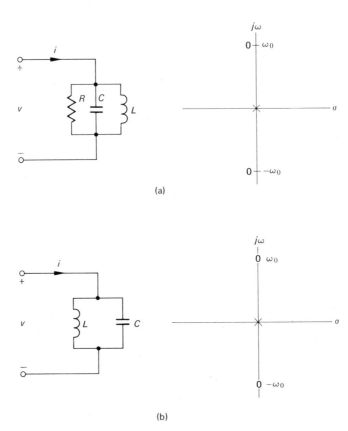

Fig. 7.53 Pole-zero diagram of admittance of parallel tuned circuit with (a) resistive losses; (b) no losses.

right half of the *s*-plane. Figure 7.53(b) shows the pole-zero diagram for the case of an admittance with zero real part at the resonant frequency. The circuit can now oscillate with no external energy supplied in the form of ac current. Thus, if the real part of an admittance (or impedance) becomes negative at the same time that susceptance approaches zero, the circuit can oscillate. These conditions can be developed in a much more rigorous manner, but the simple results just stated allow us to calculate the lower bound of external resistance in the tunnel diode amplifier.

The circuit of Fig. 7.54(a) shows the equivalent circuit of a tunnel diode amplifier with an external inductor that accounts for any wiring or added inductance. Combining the inductances and resistances to give $L_T = L + L_t$ and $R_T = R_L + R_S + R_t$ results in the equivalent circuit of Fig. 7.54(b). The real part of the terminal admittance is found to be

$$\text{Re}[Y] = \frac{\frac{1}{r_t}(R_T r_t - R_T^2 - \omega^2 L_T^2)}{R_T^2 + \omega^2 L_T^2}. \tag{7.83}$$

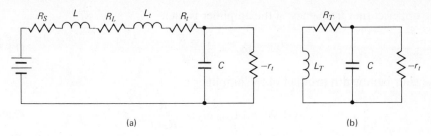

Fig. 7.54 Tunnel diode amplifier: (a) equivalent circuit; (b) reduced form of ac circuit.

Since the denominator is positive for all values of ω, we need only cause the numerator to remain positive to ensure a positive real part for the admittance. To avoid oscillations in the amplifier, the condition must hold that

$$R_T r_t - R_T^2 - \omega^2 L_T^2 > 0. \tag{7.84}$$

From this expression we see that $r_t R_T$ must exceed the value of $R_T^2 + \omega^2 L_T^2$. We have some control over ω since if the circuit oscillates it will do so at the frequency that causes the susceptance to approach zero. It is easy to show that this will be the resonant frequency

$$\omega_0 = \frac{1}{\sqrt{L_T C}}. \tag{7.85}$$

Generally, the resonant frequency will be quite large as a result of small values of L_T and C. When this is true the term $\omega^2 L_T^2$ is much larger than R_T^2 in Eq. (7.84). Thus, the condition reduces to

$$R_T r_t > \omega_0^2 L_T^2$$

or

$$R_T > \frac{\omega_0^2 L_T^2}{r_t}.$$

Using Eq. (7.85) we can write this inequality as

$$R_T > \frac{L_T}{C r_t}.$$

For stable amplification the conditions on external resistance can be stated by

$$\frac{L_T}{C r_t} - R_t < (R_S + R_L) < (r_t - R_t). \tag{7.86}$$

Once the appropriate value of $R_L + R_S$ is selected, the bandwidth of the circuit can be found. The capacitance C sees an equivalent resistance (neglecting the inductive effects) of

$$R_{eq} = -r_t \| (R_T) = \frac{R_T r_t}{r_t - R_T}.$$

The upper corner frequency of the amplifier is then

$$\omega_2 = \frac{r_t - R_T}{R_T r_t C}. \tag{7.87}$$

The gain-bandwidth product of the amplifier is

$$\text{GBW} = |A_{MB}|\,\omega_2 = \frac{R_L}{R_T r_t C}. \tag{7.88}$$

If $R_S + R_t$ is much smaller than R_L the GBW can be approximated by

$$\text{GBW} \approx \frac{1}{r_t C}.$$

This value can be quite large and is relatively constant while R_L is varied to exchange gain and bandwidth.

The configuration of the amplifier can also appear as shown in Fig. 7.55. Conditions on R_S and R_L can be derived for this case in a manner similar to that used for the configuration of Fig. 7.51.

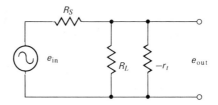

Fig. 7.55 Alternate configuration of a tunnel diode amplifier.

Tuned amplifiers can easily be constructed using the tunnel diode. Figure 7.56 shows the equivalent circuit of a parallel tuned amplifier which neglects the diode parameters L_t and R_t.

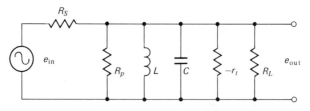

Fig. 7.56 Parallel tuned amplifier.

The inductance L is chosen to resonate with C at the appropriate frequency from

$$\omega_0 = \frac{1}{\sqrt{LC}}. \tag{7.89}$$

The resistance R_p represents the coil losses and is given in terms of coil Q as

$$R_p = Q_0 \omega_0 L. \tag{7.90}$$

At resonance, the voltage gain is

$$A_r = \frac{R}{R + R_S}, \tag{7.91}$$

where

$$R = \frac{r_t R_L R_p}{r_t R_p + r_t R_L - R_L R_p}. \tag{7.92}$$

The bandwidth can be calculated from

$$\text{BW} = f_0/Q_{\text{eff}} \tag{7.93}$$

where Q_{eff} is the loaded circuit Q. This value is

$$Q_{\text{eff}} = \frac{R_{sh}}{\omega_0 L} = \frac{R_{sh} Q_0}{R_p},$$

where R_{sh} is the parallel combination of R_S and R. The GBW product is found to be (in rad/sec)

$$\text{GBW} = \frac{1}{R_S C}. \tag{7.94}$$

A 455 kHz amplifier is shown in Fig. 7.57.

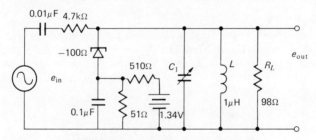

Fig. 7.57 Tunnel diode amplifier tuned to 455 kHz.

REFERENCES AND SUGGESTED READING

1. G. Bruun, "Common emitter transistor video amplifiers." *Proc. IRE* **44**, 1561–1572, November, 1956.
2. D. J. Comer, *Computer Analysis of Circuits*. Scranton, Pennsylvania: International, 1971, Chapter 8.
3. D. J. Comer and J. M. Griffith, "Designable video amplifiers using base-compensated stages." *IEEE Transactions on Circuit Theory* **CT 17**, 1:94–99, February, 1970.
4. D. J. Comer and J. M. Griffith, "Optimization of bandwidth in noniterative amplifiers." *Proc. IEE*, 384 (March, 1969).

5. M. S. Ghausi, *Electronic Circuits*. New York: Van Nostrand, 1971, Chapters 5 and 7.
6. A. B. Grebene, *Analog Integrated Circuit Design*. Cincinnati, Ohio: Van Nostrand-Reinhold, 1972, Chapter 8.
7. P. Lorrain and D. Corson, *Electromagnetic Fields and Waves*, 2nd ed. San Francisco: W. H. Freeman, 1970, Chapter 11.

PROBLEMS

Section 7.1A

7.1 The small-signal depletion-layer capacitance for an abrupt junction is given by

$$C = k/\sqrt{\phi - V},$$

where k is a constant, ϕ is the barrier voltage, and V is the applied junction voltage (V has a positive value for forward-bias and a negative value for reverse bias). Given that the collector-base capacitance is 10 pF for a reverse bias of 6 V, and that $\phi = 0.7$ V, find the capacitance for a 12-V reverse bias.

* **7.2** Repeat Problem 7.1 for the case of a graded junction where $C = k_1/(\phi - V)^{1/3}$.

7.3 Using the formula given in Problem 7.1, evaluate C at 6-V reverse bias, assuming that $C = 10$ pF at 18 V.

7.4 Using the formula given in Problem 7.2, evaluate C at 6-V reverse bias, assuming that $C = 10$ pF at 18 V.

7.5 Given that ω_β is 10^6 rad/s, find the diffusion capacitance C_D for emitter currents of (a) $I_E = 1$ mA, (b) $I_E = 5$ mA. Assume that $\beta = 50$.

* **7.6** Repeat Problem 7.5 for a transistor with $\omega_\beta = 2 \times 10^7$ rad/s.

Sections 7.1B–7.1F

7.7 The transistor shown in the figure is an abrupt junction transistor with $\omega_t = 6.28 \times 10^6$ rad/s, $\beta = 70$, and $C_{b'c} = 20$ pF at $V_{CB} = 6$ V. Assuming that $\phi = 0.8$ V, find the bandwidth of e_{out}/i_{in} when $I_E = 1$ mA.

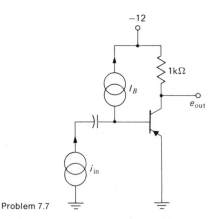

Problem 7.7

7.8 Repeat Problem 7.7 given that I_E is changed to 4 mA.

7.9 Assuming that f_β is defined as the frequency at which the short-circuit base-to-collector current gain is down 3 db from the midband value, and that f_t is defined as the frequency at which this gain is equal to unity, prove that $f_t = \beta f_\beta$.

7.10 When the diffusion capacitance decreases and can no longer be considered to be much greater than the depletion-layer capacitance, the formula for f_t is

$$f_t = \frac{1}{2\pi r_d(C_D + C_{b'e} + C_{b'c})}.$$

Explain why this quantity decreases at low-emitter bias currents.

*7.11 The transistor in the circuit shown in the figure has the following parameters: $\beta = 100$, $r_{bb'} = 100\ \Omega$, $V_{BE} = 0.6$ V, $f_t = 10^7$ Hz, and $C_{b'c} = 10$ pF at the bias point used. Calculate the midband voltage gain and the upper corner frequency of the circuit. What is the lower corner frequency? Neglect the loading effect of the bias resistors.

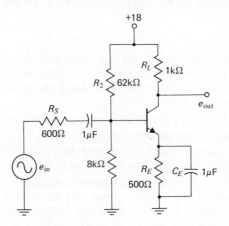

Problem 7.11

7.12 Repeat Problem 7.11 given that $f_t = 10^8$ Hz. Explain why the upper corner frequency does not increase by the same factor as f_t increases.

7.13 Repeat Problem 7.11 given that $f_t = 10^8$ Hz, $C_{b'c} = 5$ pF, and R_L is lowered to 200 Ω.

7.14 Repeat Problem 7.11 given that R_S is lowered to 200 Ω and R_L is lowered to 250 Ω.

7.15 Find the Miller-effect capacitances for Problems 7.11 and 7.12. Find C_D for both problems and compare to C_M. Calculate the D-factor for both cases.

7.16 Repeat Problem 7.11 for the case of $R_S = 3.6$ kΩ.

7.17 Repeat Problem 7.12 for the case of $R_S = 3.6$ kΩ.

*7.18 Repeat Problem 7.12 assuming that R_2 is changed to 100 kΩ and that $C_{b'c}$ remains constant at 10 pF.

Section 7.1G

7.19 An FET transistor has the following parameters: $g_m = 6000\ \mu$mhos, $r_{ds} = 10^5\ \Omega$, $C_{gs} = 3.5$ pF, $C_{ds} = 1.0$ pF, and $C_{gd} = 3.0$ pF. If this device is used to couple a 20-kΩ source to a 2-kΩ drain load, calculate the midband gain and the upper corner frequency.

*7.20 Repeat Problem 7.19 if the load resistance is changed to 1 kΩ and the source resistance is increased to 50 kΩ.

7.21 If the gain of the stage of Problem 7.19 is to be increased to -40 by increasing the load resistance, what is the new value of upper corner frequency?

7.22 If the corner frequency of the stage of Problem 7.19 is to be increased to 1.0 MHz by decreasing the load resistance, what is the new value of midband gain?

Section 7.2A

7.23 A midband voltage gain of -20 and a bandwidth of 560 kHz is required for an amplifier that couples a 200 Ω source resistance to a variable load. A transistor is used with $r_{bb'} = 100\ \Omega$, $\beta = 99$, $C_{b'c} = 8$ pF and $f_t = 10$ MHz (at the bias point used). What is the smallest value of load resistance that will allow the transistor to satisfy this design? Using this value of R_L, calculate the required value of $r_{b'e}$.

*7.24 A common-emitter amplifier using the transistor of Problem 7.23 is constructed. If $R_S = 200\ \Omega$ and $R_L = 200\ \Omega$, what is the gain that will result when the bandwidth is 1 MHz?

Sections 7.2B–7.2C

7.25 Given that a 3-kΩ resistor is inserted in the base lead of the circuit of Problem 7.11, calculate the required shunt capacitor for compensation. Evaluate the midband voltage gain and the upper corner frequency. Assume that the parameters of Problem 7.11 apply. Neglect the loading effect of bias network.

*7.26 Given that a 15-kΩ resistor is inserted in the base lead of the circuit of Problem 7.11, calculate the required shunt capacitor for compensation. Evaluate the midband voltage gain and the upper corner frequency. Assume that the parameters of Problem 7.11 apply. Calculate the lower corner frequency. Neglect the loading effect of bias network.

7.27 In the circuit of Fig.7.59, the emitter resistance is changed to 75 Ω. Assuming that the bias network is adjusted to give $I_E = 4$ mA, and that $C_E = 0$, calculate the midband voltage gain and the upper corner frequency of the amplifier. Assume that the parameters of Problem 7.11 apply and that $C_{b'c} = 10$ pF. Calculate the correct value of C_E to extend the bandwidth by emitter compensation, and calculate the new bandwidth. Use the same value of D' for both cases. Neglect the loading effect of bias network.

7.28 Repeat Problem 7.27 given that R_E is changed to 30 Ω ($I_E = 4$ mA).

7.29 Repeat Problem 7.27 assuming that R_E is changed to 30 Ω, $R_L = 100\ \Omega$, and $R_S = 200\ \Omega$.

7.30 Repeat Problem 7.27 given that $f_t = 10^8$ Hz.

7.31 An amplifier is to be designed to couple a 700-Ω source to a dc-coupled 2-kΩ load. The overall midband voltage gain is to be -10 and the upper corner frequency is to exceed 80 kHz. Design this stage, using base compensation, if required, and the transistor parameters given in Problem 7.11. Explain why the uncompensated stage is inappropriate for this case.

7.32 Repeat Problem 7.31 given that the required voltage gain is -80 and the upper corner frequency is to exceed 100 kHz.

7.33 Repeat Problem 7.31, using emitter compensation. Is it necessary to use C_E?

7.34 Repeat Problem 7.32, using emitter compensation. Assuming that the gain is still to be -80, what is the maximum upper corner frequency that can be obtained using C_E?

*7.35 Using base compensation, what is the maximum upper corner frequency that can be obtained with the transistor of Problem 7.11 when it couples a 3-kΩ load to a 500-Ω source with an overall gain of -15? Neglect the loading effect of bias network. Assume that $r_d = 26\ \Omega$.

7.36 Repeat Problem 7.35 for the case of emitter compensation (R_E and C_E).

Section 7.3A

*7.37 Given three nonidentical, noninteracting stages. The upper corner frequencies for the stages are $f_{21} = 1$ MHz, $f_{22} = 2$ MHz, and $f_{23} = 10$ MHz. If these stages are cascaded, what is the overall upper corner frequency?

7.38 Repeat Problem 7.37 given that $f_{21} = 1$ MHz, $f_{22} = 1$ MHz, and $f_{23} = 2$ MHz.

7.39 An amplifier requiring an overall bandwidth of 3.1 MHz is constructed using identical stages with constant values of GBW $= 1 \times 10^8$ rad/sec. Calculate the maximum possible gain for this amplifier.

Section 7.3B

7.40 Design an amplifier that has a midband gain of ± 100 and a bandwidth exceeding 250 kHz, using transistors with the parameters given in Problem 7.11. The amplifier is to couple a 1-kΩ source to a 2-kΩ load.

*7.41 Three stages are cascaded having the values $GBW_1 = 1 \times 10^8$ rad/sec, $GBW_2 = 2 \times 10^8$ rad/sec, and $GBW_3 = 8 \times 10^7$ rad/sec. The bandwidths of the stages are $\omega_{21} = 8 \times 10^6$ rad/sec, $\omega_{22} = 20 \times 10^6$ rad/sec, and $\omega_{23} = 40 \times 10^6$ rad/sec. Calculate the overall midband voltage gain and the overall bandwidth.

7.42 Assuming the values of GBW in Problem 7.41 remain constant as gain and bandwidth are exchanged, calculate the maximum possible overall midband voltage gain for an overall bandwidth of $\omega_{20} = 6.28 \times 10^6$ rad/sec.

Section 7.3C

7.43 Given a shunt-peaked interstage using a transistor with $f_t = 300$ MHz, $r_{bb'} = 75\ \Omega$, $\beta = 75$, $r_{b'e} = 1$ kΩ, and $C_{b'c} = 5$ pF. The gain per stage must be -10. Find the appropriate value of L and the bandwidth of the interstage.

*7.44 Repeat Problem 7.43 if the interstage gain is to be -4.

7.45 For the transistor of Problem 7.43, calculate the value of R required to achieve maximum overall bandwidth for an amplifier with an overall gain of 80. Assume that the emitter-bias current is 4 mA for this case.

Sections 7.3D–7.3E

7.46 Design an amplifier that has a midband gain of ± 600 and a bandwidth exceeding 1 MHz using transistors with the parameters of Problem 7.11. Use the minimum number of base-compensated stages. Assume source and load resistances must equal the interstage value.

7.47 Repeat Problem 7.46 if the required overall gain is 500 and the overall bandwidth must exceed 800 kHz.

7.48 Repeat Problem 7.46 using op amp stages with overall open loop gain of 100,000 and bandwidth of 1.1×10^3 rad/sec.

Section 7.4A

7.49 A coil has a Q_0 of 75 and an inductance of 40 μH. If the coil is to be used at a resonant frequency of 10 MHz, what is the required capacitance? Given that a resistance of 200 kΩ is placed in parallel with the tank circuit, calculate the Q of the circuit and the bandwidth.

*7.50 Repeat Problem 7.49, assuming that $Q_0 = 50$ and $f_0 = 1$ MHz.

7.51 A parallel tank circuit includes a capacitor of 100 pF and a resistor of 100 kΩ. Given that the resonant frequency is 8 MHz and the bandwidth is 80 kHz, find the circuit Q, the inductance, and the effective parallel resistance of the coil R_p.

Sections 7.4B–7.4D

7.52 In the circuit of Fig. 7.45, $r_v = 1$ MΩ, $C_{b'c} = 5$ pF, and the output impedance of the transistor after neutralization is 50 kΩ. If the inductance appearing at the transformer primary is 40 μH and a resonant frequency of 1.6 MHz along with a bandwidth of 40 kHz is required, what are suitable values for C, C_n, r_n, and the transformer turns ratio n? If $r_{bb'} = 100\ \Omega$, $r_{b'e} = 2$ kΩ, $R_S = 4$ kΩ, and $\beta = 80$, what is the voltage gain of the stage at resonance? If the transformer primary has an unloaded $Q_0 = 80$, what value of resistance must be shunted across the tank circuit? Neglect the high-frequency effects of the transistor.

*7.53 Repeat Problem 7.52, given that the secondary of the transformer is to drive a stage with 2-kΩ input impedance in addition to the neutralizing network.

7.54 In the circuit of Fig. 7.47, $n_1 = 10$, $n_2 = 1/2$, $r_v = 1$ MΩ, $C_{b'c} = 5$ pF, and the output impedance of the transistor after neutralization is 20 kΩ. The tank circuit has an unloaded Q_0 of 100 and an inductance of 60 μH. Given that the input impedance to the following stage is 1 kΩ, find proper values of r_n, C_n, and C to give a resonant frequency of 2 MHz. Calculate the bandwidth of the circuit.

Section 7.4G

7.55 Given that the tunnel diode is not included in the circuit of Fig. 7.51, calculate the output ac power in terms of e_{in}. Compare this to the power output when the tunnel diode is present by calculating the insertion power gain.

*7.56 Calculate the limiting values of $R_S + R_L$ for the amplifier of Fig. 7.51, given that $r_t = 50\ \Omega$, $L_t = 1 \times 10^{-9}$ H, $C = 1$ pF, and $R_t = 2\ \Omega$. What is the minimum value of additional series inductance that causes instability for all values of $R_L + R_S$?

LARGE-SIGNAL CIRCUITS

Large-signal circuits are those in which a device operating point makes such large excursions that the device parameters cannot be treated as constants. For example, the current gain of a transistor might decrease from 80 to zero as the transistor switches from the active region to cutoff. A transistor need not switch out of the active region to require large-signal analysis since r_d and β can change appreciably even when the transistor remains entirely within this region.

As device parameters vary, the analysis equations become nonlinear and are then extremely difficult to solve. Nonlinear analysis equations preclude the application of simple design procedures; thus, a great deal of effort is expended in developing linear analysis methods for large-signal circuits. This has led to the rather common usage of piecewise linear circuit models that represent a device by constant parameter networks within a given operating region. As the boundary of a region is crossed, certain parameters of the model may change immediately to new values. In general, the use of piecewise linear methods greatly simplifies nonlinear analysis and often allows the development of practical design procedures.

Digital or switching circuits make up the largest and most important class of large-signal circuits. This chapter is devoted strictly to this type of circuit. The transistor is considered first after which special switching devices are considered.

8.1 LOW-SPEED TRANSISTOR SWITCHING

The transistor very closely approximates certain switch configurations. For example, when the switch of Fig. 8.1(a) is open no current flow through the resistor is allowed and the output voltage is $+12$ V. Closing the switch causes the output voltage to drop to zero volts and a current of $12/R$ flows through the resistance. When the base voltage of the transistor of Fig. 8.1(b) is negative the transistor is cut off and no collector current flows. The output voltage is $+12$ V just as in the case of the open switch. If a large enough current is now driven into the base to saturate the transistor, the output voltage becomes very small, ranging from 20 mV to 500 mV depending on the transistor used. The saturated state corresponds closely to the closed switch. During the time that the transistor switches from cutoff to saturation, the active region equivalent circuit applies. For high-speed

342 LARGE-SIGNAL CIRCUITS

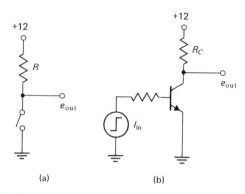

Fig. 8.1 Comparison of transistor to a switch: (a) switch; (b) transistor.

switching of this circuit, appropriate reactive effects must be considered. For low-speed switching these reactive effects can be neglected.

There are several possible equivalent circuits that can be used as dictated by the application at hand and the method of analysis used. The Ebers-Moll model may be appropriate for computer-aided design for all regions of operation, but is generally inappropriate for manual design. Section 4.4 introduced computer-aided design models for use in various regions, some of which can be used for manual analysis also. Figure 8.2 shows equivalent circuits that are useful in switching circuit analysis.

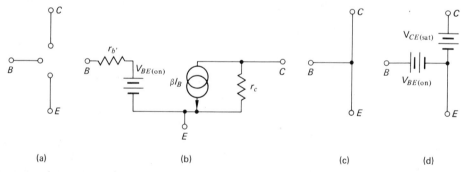

Fig. 8.2 Equivalent circuits: (a) cutoff region; (b) active region; (c) saturation region; (d) more accurate saturation region equivalent.

In the active region equivalent circuit, the quantity $r_{b'}$ includes the ohmic base resistance $r_{bb'}$ and the linearized value of $r_{b'e}$. A large number of low-speed switches will operate in either the cutoff or saturation region, making it unnecessary to use the active region equivalent.

Example 8.1. In the circuit shown in Fig. 8.1(b) the input current switches from 0 to 100 μA. Given that the current gain of the silicon transistor is 50 and R_C

is 800 Ω, find the output voltage after switching. Repeat for an input current of 500 μA. Assume that $V_{CE(\text{sat})} = 0.2$ V.

Solution. The collector current is

$$I_C = \beta_0 I_B = 50 \times 100 \ \mu A = 5 \ \text{mA}$$

after switching. The output voltage is

$$V_C = V_{CC} - I_C R_C = 12 - 5 \times 0.8 = 8 \ \text{V}.$$

The transistor is obviously not saturated. In the second case, an input current of 500 μA would cause an I_C of 25 mA if saturation did not occur. The collector saturation current is

$$I_{C(\text{sat})} = \frac{V_{CC} - V_{CE(\text{sat})}}{R_C} = \frac{11.8}{0.8} = 14.75 \ \text{mA};$$

hence, the transistor is saturated. The collector current is 14.75 mA, the base current is 500 μA, and the emitter current is 15.25 mA. Note that I_C no longer equals $\beta_0 I_B$ under saturation conditions. The collector current required for saturation is almost completely independent of the transistor used. If V_{CC} is much larger than the saturation voltage of the transistor, the maximum collector current is

$$I_{C(\text{sat})} = V_{CC}/R_C.$$

From the Ebers-Moll equations of Chapter 4, the collector leakage current that flows when emitter and base are shorted is I_{CS}. This quantity is of the same order of magnitude as I_{co}.

When a resistance is placed across the emitter and base leads, the leakage current flowing out of the base causes a voltage drop between base and emitter for a germanium or high-power silicon transistor. The junction is now slightly forward-biased and electrons are injected from emitter to base to cause additional collector current to flow. As R becomes higher, a greater forward bias of the base-emitter junction occurs and I_E increases further, resulting in a larger value of I_C. In the limit as R becomes infinite, giving zero base current, the simultaneous solution of the two current equations gives

$$I_C = \frac{I_{co}}{1 - \alpha} = (\beta + 1)I_{co}. \tag{8.1}$$

The collector current I_C varies from a low value of I_{CS} to $(\beta + 1)I_{co}$ as the resistance between base and emitter varies from zero to infinity. It is often useful to reverse-bias the base-emitter junction to minimize the collector leakage current. From Eq. (4.27) with both junctions reverse-biased the collector current is found to have a magnitude of

$$I_C = (I_{CS})(1 - \alpha_I) = \frac{I_{co}(1 - \alpha_I)}{1 - \alpha_N \alpha_I} \tag{8.2}$$

which is slightly less than I_{co}. The base current can also be found in this case by finding the emitter current and adding collector and emitter currents. The resulting base current magnitude is very near I_{co}. When the base is reverse-biased, the magnitudes of collector and base currents are almost equal while the emitter current is near zero. In situations where leakage current is important, an appropriate cutoff model would consist of an open circuit emitter and a current generator of value I_{co} from collector to base.

The circuit of Fig. 8.3 shows a transistor in the cutoff region. If $V_{BB} = 1$ V and $R = 100$ kΩ, a leakage current of $I_{co} = 10$ μA can reduce the emitter-base bias to zero volts. As discussed previously, it is desirable to ensure that a reverse bias appears across the base-emitter junction. In fact, many circuits will be designed so that this bias always exceeds some safe value. For example, the reverse bias may never be allowed to decrease below 0.2 V. In the circuit of Fig. 8.3 V_{BB} would have to be increased to satisfy this requirement. The maximum value of I_{co} occurring in the circuit is used to calculate the minimum value of reverse bias. The base-emitter loop equation gives

$$V_{BE(\text{min})} = RI_{co(\text{max})} - V_{BB}. \tag{8.3}$$

If I_{co} is 30 μA at the highest temperature to which the circuit will be exposed and $R = 100$ kΩ, a bias battery of $V_{BB} = 4$ V is necessary to assure a 1 V reverse bias.

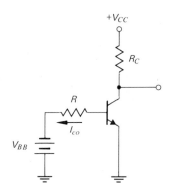

Fig. 8.3 Transistor switch.

Example 8.2. In the circuit of Fig. 8.4 both transistors have average values of $\beta_0 = 50$, $V_{BE(\text{on})} = 0.5$ V, $V_{CE(\text{sat})} = 0.3$ V, $r_{bb'} = 200$ Ω, and $r_c = 30$ kΩ. Sketch the output waveform, neglecting switching times of the transistors.

Solution. When the input signal is at the zero volt level the first stage will be cut off with a -3 V base voltage. If we assume that the second stage is in the active region, we can verify this assumption by analyzing the equivalent circuit of Fig. 8.5. A value for $r_{b'}$ could be assumed, but in many switching circuits the drop across this element is negligible. The voltage at the base then equals 0.5 V. The base current

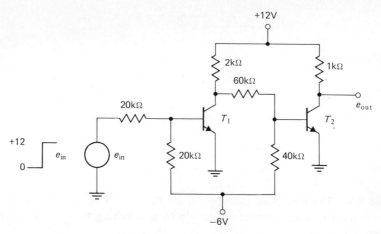

Fig. 8.4 Two-stage switching circuit.

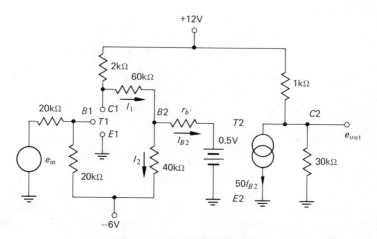

Fig. 8.5 Equivalent circuit of switching circuit for $e_{in} = 0$ V.

of T_2 can be found by noting that

$$I_{B2} = I_1 - I_2 = \frac{12 - 0.5}{62} - \frac{6.5}{40} = 23 \ \mu A.$$

The collector current source has a value of 1.15 mA. Converting the current source and 30 kΩ resistor to a voltage source allows the collector voltage to be calculated. This value is $e_{out} = 10.5$ V.

When the input voltage switches to $+12$ V, the first stage saturates. Using the appropriate equivalent circuit shows that the second stage is cut off and the output voltage is 12 V. The output voltage is shown in Fig. 8.6.

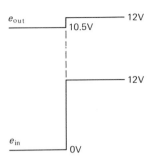

Fig. 8.6 Response of switching circuit.

Example 8.3. The transistor parameters are $\beta_0 = 50$, $V_{BE(on)} = 0.5$ V, and $r_c = 26$ kΩ. Sketch the output waveform, neglecting switching times of the transistor for the circuit of Fig. 8.7.

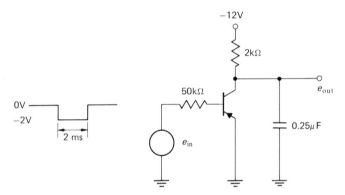

Fig. 8.7 RC switching circuit.

Solution. The transistor will be in the cutoff region when the input signal is zero, and the voltage on the capacitor will be 12 V. When the input voltage swings to -2 V, the active region equivalent of Fig. 8.8 is appropriate.

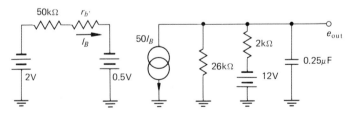

Fig. 8.8 Equivalent circuit in active region.

The value of $r_{b'}$ can be estimated as 1 kΩ although assuming this element is zero would lead to little error. The output voltage will move toward a target voltage that can be calculated by taking the Thévenin equivalent of the output circuit at the capacitor terminals. The time constant can be calculated from the equivalent resistance of this circuit. The target voltage is found to be $V_T = -8.4$ V and the time constant is 0.46 ms. The duration of the input pulse allows more than four time constants for the collector voltage to respond; thus, the target voltage will be reached before the pulse ends.

At the end of the pulse the transistor switches back to the cutoff region and the capacitor voltage approaches -12 V with a time constant of 0.5 ms. The charging and discharging time constants are almost equal in this case. If the input pulse height were increased to a value sufficient to cause saturation, a very small discharge time constant would result. Figure 8.9 shows the actual output signal and a waveform that assumes a much larger input signal which saturates the transistor.

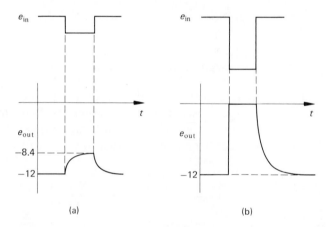

Fig. 8.9 Output waveforms: (a) actual output; (b) output for larger input signal.

8.2 HIGH-SPEED TRANSISTOR SWITCHING

There are three major effects that extend switching times in a transistor:

1. The depletion-region capacitances are responsible for delay time when the transistor is in the cutoff region.
2. The diffusion capacitance and the Miller-effect capacitance are responsible for the rise and fall times of the transistor as it switches through the active region.
3. The storage time constant accounts for the time taken to remove the excess charge from the base region before the transistor can switch from the saturation region to the active region.

There are other second-order effects that are generally negligible compared to the previously listed time lags.

A. Overall Transient Response

Before discussing the individual transistor switching times it is helpful to consider the response of a common-emitter switch to a rectangular waveform. Figure 8.10 shows a typical circuit using an npn transistor.

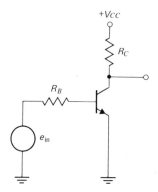

Fig. 8.10 Common-emitter switch.

A rectangular input pulse and the corresponding output are shown in Fig. 8.11. In many switching circuits, the transistor must switch from its off state to saturation and later return to the off state. In this case delay time, rise time, saturation storage time, and fall time must be considered in that order to find the overall switching time.

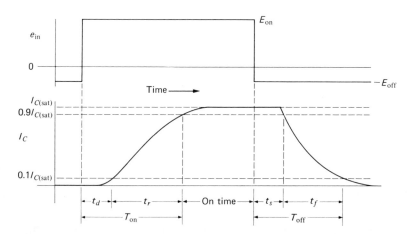

Fig. 8.11 Collector-current response of a common-emitter switch driven by a rectangular pulse.

The total waveform is made up of five sections: delay time, rise time, on time, storage time, and fall time. In working with exponentials it is convenient to consider 10- to 90-percent points. This causes some slight adjustments to be made in calculating the various switching times. For example, the total delay time t_d shown in Fig. 8.11 consists of the passive delay due to the depletion capacitances t'_d plus the time taken for the waveform to rise to 10 percent of its final value. The rise time t_r is the 10- to 90-percent rise time, and the total turn-on time T_{on} is the sum of t_d and t_r. The total storage time t_s is the sum of the saturation storage time t'_s and the time taken by the waveform to fall to the 90-percent point. The fall time t_f is the 90- to 10- percent fall time, and the sum of t_f and t_s is the total turn-off time T_{off}. The following list summarizes these points and serves as a guide for future reference:

t'_d = passive delay time; time interval between application of base drive and start of collector-current response.

t_d = total delay time; time interval between application of base drive and the point at which I_C has reached 10 percent of the final value.

t_r = rise time; 10- to 90-percent rise time of I_C waveform.

t'_s = saturation storage time; time interval between removal of forward base drive and start of I_C decrease.

t_s = total storage time; time interval between removal of forward base drive and point at which $I_C = 0.9 I_{C(\text{sat})}$.

t_f = fall time; 90- to 10-percent fall time of I_C waveform.

T_{on} = total turn-on time; time interval between application of base drive and point at which I_C has reached 90 percent of its final value. This value is given by the sum $(t_d + t_r)$.

T_{off} = total turn-off time; time interval between removal of forward base drive and point at which I_C has dropped to 10 percent of its value during on time. This value is given by $t_s + t_f$.

Not all applications will require evaluation of each of these switching times. For instance, if the base drive is insufficient to saturate the transistor, t'_s will be zero. We will now consider the detailed calculation of the various switching times.

The high-frequency Ebers-Moll model [1] can be used for computer-aided design in programs that accept nonlinear element values. Several programs such as SCEPTRE, NET 1, and ECAP II use this model for high-speed switching analysis. It is, of course, much too complex for manual analysis or design and simpler models are used in this instance. The following sections will develop simple transistor models for the various regions of operation.

B. Passive Delay Time

If the input voltage of the common-emitter switching circuit of Fig. 8.10 is negative or less than $V_{BE(on)}$, the transistor will be cut off. Figure 8.12 shows the applicable equivalent circuit.

In most switching circuits, the input voltages will be relatively large. The base resistance R_B will necessarily be large to limit base current to reasonable levels.

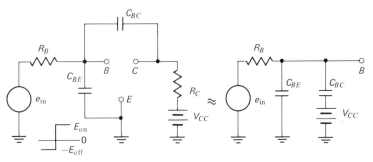

Fig. 8.12 Equivalent circuit in cutoff region.

The ohmic base resistance is then negligible compared to R_B. Reverse-biased junctions exhibit very large resistances, normally in the megohm range, and the only elements required between terminals of the equivalent circuit are the depletion-region capacitances. To simplify succeeding calculations with little loss of accuracy, we will assume that R_C can be neglected in delay-time calculations since it is small compared to R_B, and the cutoff equivalent circuit becomes even less complex.

The base terminal will assume the level of the negative input signal $-E_{\text{off}}$ if this signal is present long enough to charge the capacitances C_{BE} and C_{BC} to this level. If the input voltage then switches to a positive value to turn the transistor on, the base voltage cannot change instantaneously and will move toward a target value of E_{on} at a rate determined by the capacitors C_{BE} and C_{BC} and the resistance R_B. The collector current will not start to flow until the base voltage reaches the value $V_{BE(\text{on})}$. The time interval between the positive transition of the input signal and the point where V_B reaches $V_{BE(\text{on})}$ is the passive delay time t'_d.

Calculation of t'_d would be very simple if it were not for the fact that both the depletion-region capacitances depend on the voltages appearing across each of these elements. As the base voltage changes, so also does each capacitor voltage and a nonlinear problem results. In small-signal circuits the dc voltages across these capacitances determined the small-signal values which were taken as constant at the given operating point. In Chapter 2 it was shown that the small-signal depletion-region capacitance depends on the junction voltage and varies as

$$C_{ss} = \frac{dQ}{dV} = \frac{k_1}{(\phi - V)^{1/2}} \tag{8.4}$$

for a uniformly-doped, abrupt junction transistor and as

$$C_{ss} = \frac{k_2}{(\phi - V)^{1/3}} \tag{8.5}$$

for most high-frequency transistors. The quantity ϕ is the barrier voltage, k_1 and k_2 are constants, and V is the applied junction voltage. The voltage V has a positive value in these equations for a forward bias and a negative value for reverse bias, regardless of the type of transistor used.

The problem of calculating the delay time can be simplified to a practical extent by using average or piecewise-linear values for the depletion-region capacitances. There have been several proposals for the determination of appropriate piecewise-linear capacitance values. Perhaps the most reasonable choice is to define the average value of capacitance as the total change in charge on the capacitance as voltage varies from one value to another, divided by the total voltage change; that is,

$$C_{AV} = \frac{Q(V_2) - Q(V_1)}{V_2 - V_1}. \tag{8.6}$$

The total change of charge can be found from the small-signal capacitance since

$$Q = \int_{Q_1}^{Q_2} dQ = \int_{V_1}^{V_2} \frac{dQ}{dV} dV = \int_{V_1}^{V_2} C_{SS} \, dV. \tag{8.7}$$

The average capacitance can then be written as

$$C_{AV} = \frac{\int_{V_1}^{V_2} C_{SS} \, dV}{V_2 - V_1}. \tag{8.8}$$

For high-frequency transistors C_{SS} is given by Eq. (8.5) and the average value of capacitance is

$$C_{AV} = \frac{1.5 k_2}{(V_1 - V_2)} [(\phi - V_2)^{2/3} - (\phi - V_1)^{2/3}]. \tag{8.9}$$

It must be remembered that V_1 and V_2 take on negative values for reverse bias and positive values for forward bias.

Once the average values of C_{BE} and C_{BC} have been found, the time constant of the base circuit will be

$$\tau_d = (C_{BE} + C_{BC}) R_B. \tag{8.10}$$

If the initial voltage at the base is $-E_{off}$ and the target voltage is E_{on}, the time required for the base voltage to reach $V_{BE(on)}$ can be calculated from the general charging equation,

$$v(t) = v_i + (v_t - v_i)(1 - e^{-t/\tau}), \tag{8.11}$$

where v_i is the initial voltage and v_t is the target voltage. For the base circuit, the passive delay time is then obtained by solving

$$V_{BE(on)} = -E_{off} + (E_{on} + E_{off})(1 - e^{-t'_d/\tau_d})$$

for t'_d. This gives

$$t'_d = \tau_d \ln \left(\frac{E_{on} + E_{off}}{E_{on} - V_{BE(on)}} \right). \tag{8.12}$$

Obviously, the base voltage never reaches the target value of E_{on}. Because of the emitter-base diode the base voltage is clamped to the value $V_{BE(on)}$.

An example will demonstrate the method of calculating the passive delay time.

Example 8.4. The common-emitter switch of Fig. 8.10 uses the element values $V_{CC} = 12$ V, $R_C = 2$ kΩ, and $R_B = 40$ kΩ. The small-signal collector-base capacitance is 10 pF at a junction voltage of 6 V while the emitter-base capacitance is 6 pF at 6 V. If the input signal switches from -6 V to $+6$ V at $t = 0$, find the passive delay time assuming $V_{BE(on)} = 0.5$ V.

Solution. The average values of capacitance must be found, but the appropriate constant for use in Eq. (8.9) is required for each capacitor. The specification of small-signal capacitances allows these constants to be evaluated from Eq. (8.5). If we assume $\phi = 1.0$ V, then k_2 for the collector-base capacitance is

$$k_2 = C_{b'c}(\phi - V)^{1/3} = 10^{-11}(6.6)^{1/3} = 1.82 \times 10^{-11}.$$

The corresponding value for the emitter-base capacitance is

$$k_2 = 6 \times 10^{-12}(6.6)^{1/3} = 1.09 \times 10^{-11}.$$

Knowing these values allows us to calculate C_{BC} and C_{BE}. The values of V_2 and V_1 from Eq. (8.9) for each capacitance must also be evaluated. For C_{BC} the initial voltage is $V_1 = -E_{off} - V_{CC} = -18$ V. The voltage at the edge of the active region is $V_2 = -V_{CC} + V_{BE(on)} = -11.5$ V. These values give

$$C_{BC} = \frac{1.5 \times 1.82 \times 10^{-11}}{-18 + 11.5}[(12.5)^{2/3} - (19)^{2/3}] = 6.4 \text{ pF}.$$

For C_{BE} the values are $V_1 = -E_{off} = -6$ V and $V_2 = V_{BE(on)} = 0.5$ V. These voltages result in

$$C_{BE} = \frac{1.5 \times 1.09 \times 10^{-11}}{-6 - 0.5}[(0.5)^{2/3} - (7)^{2/3}] = 7.6 \text{ pF}.$$

The time constant is

$$\tau_d = 14 \times 10^{-12} \times 4 \times 10^4 = 0.56 \text{ } \mu s.$$

The passive delay time is

$$t'_d = (0.56 \text{ } \mu s) \ln\left(\frac{12}{5.5}\right) = 0.44 \text{ } \mu s.$$

C. Active-Region Switching Time

When the transistor reaches the active region, the high-frequency equivalent circuit developed in Chapter 7 can be applied. There are two large-signal effects that must be considered in switching-circuit design, however, that are not present in the small-signal circuit. In switching through the active region, the B-factor can change due to changes in r_d, and the D-factor will change due to changes in $C_{b'c}$. We have demonstrated in the previous section that an average value of depletion-layer capacitance can be calculated. The calculation of the D-factor, based on the average value of $C_{b'c}$, will be sufficiently accurate for most purposes. The average value of $C_{b'c}$ as the transistor switches in the active region will be different from the average value in the cutoff or delay region due to the different voltages appearing across the element in these two regions.

Fortunately, the B-factor is rarely needed in high-speed switching. As mentioned previously, the voltage swings normally used in switching circuits are somewhat high, requiring an external base resistance to limit the base current. This resistance is usually compensated with a shunt capacitance or a "speed-up" capacitor. The value of this resistance is much greater than $r_{b'e}$, and hence changes in $r_{b'e}$ become insignificant. A third effect that can sometimes add to the switching time is the decrease in f_t at very-low emitter currents. This departure from ideal need only be considered in very critical design.

The circuit of Fig. 8.10 demonstrates a practical switching circuit that uses no speed-up capacitor. There are two cases in which we are interested as the transistor switches into its active region. Case 1 concerns situations where the base current is limited to values that do not cause saturation of the transistor. Case 2 covers situations where saturation does occur.

Turn-on-times

CASE 1. *Turning on; no saturation.* In the following discussions, β_0 will be used to represent the large-signal current gain of the transistor. Manufacturers often denote this value as h_{FE} to differentiate it from the small-signal value of h_{fe}. The base current required to saturate the transistor of Fig. 8.10 is

$$I_{B(\text{sat})} = \frac{I_{C(\text{sat})}}{\beta_0} = \frac{V_{CC}}{\beta_0 R_C}. \tag{8.13}$$

Values of base drive less than that given by Eq. (8.13) will not cause saturation. Once the transistor is in its active region, the collector current will approach its final value exponentially with a time constant equal to $\bar{D}/\omega_\beta$. The broadband factor B will be approximately unity for values of R_B much larger than $r_{b'e}$. In this case, the transistor can be considered to be driven by a current source. However, an average value of B can be used if needed. If the collector voltage change is small, the initial value of $C_{b'c}$ can be used for the average value of collector depletion-layer capacitance C_{BC}. For the case in which the active region swing is larger, the average value of capacitance is calculated from Eq. (8.9). The average value of the D-factor is then

$$\bar{D} = 1 + \alpha R_C \omega_t C_{BC}.$$

Perhaps it is worth emphasizing that the emitter-base depletion-region capacitance is insignificant in the active region. The diffusion capacitance is approximately zero in the cutoff region, but rapidly increases to a value far in excess of the depletion-region capacitance when the active region is entered.

It is very important to note that while the emitter current may change appreciably during the switching time, the time constant of the input circuit remains constant. The parameter ω_β is given by

$$\omega_\beta = \frac{1}{r_{b'e} C_D}.$$

The diffusion capacitance C_D varies directly with emitter current while r_d varies inversely with emitter current. Except for nonideal behavior of β, the beta cutoff frequency is constant over the active region.

Example 8.5. The circuit of Fig. 8.10 has the element values $V_{CC} = 18$ V, $R_C = 2$ kΩ, and $R_B = 100$ kΩ. The transistor parameters are $f_t = 10$ MHz, $\beta_0 = 50$, $V_{BE(on)} = 0.6$ V, and $C_{b'c} = 12$ pF at 6 V. The resistance r_c is large enough to neglect. If e_{in} switches from 2 V to 10 V, sketch the output waveform.

Solution. The initial and final voltages are calculated from the equivalent circuit of Fig. 8.13. The 100 kΩ input resistance is large enough to approximate an input current source. The initial base current is

$$I_{B1} = \frac{2 - 0.6}{100} = 0.014 \text{ mA}$$

and the final current is

$$I_{B2} = \frac{10 - 0.6}{100} = 0.094 \text{ mA}.$$

The initial and final values of collector current are $I_{C1} = 0.7$ mA and $I_{C2} = 4.7$ mA. The voltage levels are

$$V_{C1} = 18 - 2 \times 0.7 = 16.6 \text{ V}$$

and

$$V_{C2} = 18 - 2 \times 4.7 = 8.6 \text{ V}.$$

The voltage waveform switches between these levels exponentially with a time constant of $\bar{D}/\omega_\beta$. We must now evaluate C_{BC} to find $\bar{D}$. Using Eq. (8.9) with $V_1 = -16.6 + 0.6 = -16$ V and $V_2 = -8.6 + 0.6 = -8$ V, C_{BC} is found to be $C_{BC} = 9.85$ pF. The D-factor is

$$\bar{D} = 1 + 0.98 \times 2 \times 10^3 \times 2\pi \times 10^7 \times 9.85 \times 10^{-12} = 2.21$$

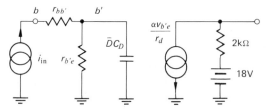

Fig. 8.13 Equivalent circuit.

and the time constant of switching is

$$\tau = \frac{2.21}{2\pi \times 2 \times 10^5} = 1.76 \ \mu s.$$

The equations for collector current and voltage are

$$I_C = I_{C1} + (I_{C2} - I_{C1})(1 - e^{-t\omega_\beta/\bar{D}})$$
$$= 0.7 + 4(1 - e^{-t/1.76 \ \mu s})$$

and

$$V_C = V_{C1} + (V_{C2} - V_{C1})(1 - e^{-t\omega_\beta/\bar{D}})$$
$$= 16.6 - 8(1 - e^{-t/1.76 \ \mu s}).$$

These waveforms are shown in Fig. 8.14. The rise time of the collector current is 2.2τ or 3.87 μs.

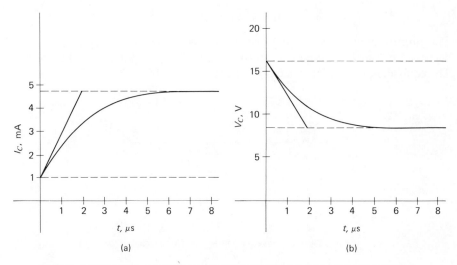

Fig. 8.14 Collector current and voltage waveforms: (a) collector current; (b) collector voltage.

The active region time constant can be decreased by using a compensating capacitor in parallel with R_B. In Chapter 7 this element was used to extend upper corner frequency of the high-frequency amplifier. The value of C_B for compensation is again given by

$$C_B = \frac{\bar{D}}{R_B \omega_\beta}. \tag{8.14}$$

In Chapter 7 it was also shown that for the compensated base circuit the upper 3-db radian frequency equals the reciprocal of the time constant of the rise time.

The time constant with compensation is then (assuming the presence of a source resistance R_S)

$$\tau_c = \left(\frac{R_S + r_{bb'}}{R_S + R_B + r_{bb'} + r_{b'e}}\right)\frac{\bar{D}}{\omega_\beta} \approx \left(\frac{R_S + r_{bb'}}{R_B}\right)\frac{\bar{D}}{\omega_\beta}. \quad (8.15)$$

The time constant decreases with source resistance; however, practical transistor stages will often have a source resistance that is also the load resistance of the preceding stage. There is then some lower limit imposed on the value of R_S.

In the preceding example, if we assume values of 3 kΩ for R_S, 200 Ω for $r_{bb'}$, and 800 Ω for $r_{b'e}$ the time constant of the circuit decreases from 1.8 μs to a compensated value of

$$\tau_c = \frac{3.2 \times 10^3}{1.04 \times 10^5} 1.8 \ \mu s = 0.055 \ \mu s.$$

The capacitance required for compensation is

$$C_B = \frac{2.17}{10^5 \times 4\pi \times 10^5} = 17 \ \text{pF}.$$

The improvement in switching speed with compensation is quite marked. A decrease in switching time of approximately 30 was achieved in this example. The equivalent circuit for this case appears in Fig. 8.15.

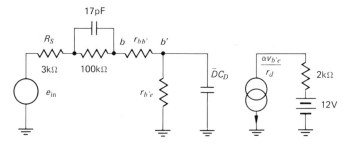

Fig. 8.15 Base-compensated switching circuit.

The voltage gain of a switching stage is typically quite low as in Example 8.5. An 8 V input transition resulted in an output transition of 8 V. For logic circuits all logic level signals are large and low gain is desirable. In applications where a small pulse is to be amplified, a smaller base resistance can be used. If the sum of R_S and R_B is not much greater than $r_{b'e}$, then an average value of $r_{b'e}$ must be used to evaluate the broadband factor. A suitable value of $r_{b'e}$ [2] is

$$\bar{r}_{b'e} = \frac{(\beta + 1)(0.026)}{I_{E(ave)}}.$$

The broadband factor is then

$$\bar{B} = \frac{R_S + r_{bb'} + R_B + \bar{r}_{b'e}}{R_S + r_{bb'}}. \quad (8.16)$$

If a stage switches from the cutoff region to the active region, the active region switching time begins at t'_d with the initial collector current taken as $I_{C1} = 0$. Thus, the total switching time can be broken into two distinct components. The passive delay time starts at the leading edge of the input signal and ends when the base voltage reaches $V_{BE(on)}$. At this point the collector current becomes nonzero and switches toward the final current value to complete the active region switching.

CASE 2. *Turning on; saturation.* The majority of switching circuits will allow the transistor to saturate for a variety of reasons. It is easier to design saturating circuits, in addition to the fact that the temperature drift becomes less critical if the transistor is completely saturated. Equation (8.13) gives the minimum base current required for saturation, but no maximum base drive has been determined. The transistor determines the maximum currents that can be allowed. In numerous applications, the base current that flows for saturation lies between the minimum value given by Eq. (8.13) and ten times this value. We will see later that the saturation delay time is increased as the forward base drive increases. To minimize this delay time, I_B is often limited to values slightly above $I_{B(sat)}$.

If a base current step I_B, which is larger than $I_{B(sat)}$, is applied to the circuit of Fig. 8.10, the transistor will saturate. However, the collector current does not proceed from zero to $I_{C(sat)}$ instantaneously. The collector current switches through the active region with the same time constant that it previously had, that is,

$$\tau = \bar{D}/\omega_\beta.$$

The difference in this case is that the target collector current is higher than the limiting value of collector current $I_{C(sat)}$. The waveform appears in Fig. 8.16. The time taken to switch through the active region can be decreased considerably by increasing the target current. For circuits where saturation storage time is unimportant, overdriving the stage is a very effective method of reducing turn-on time.

When I_{B2} is greater than $I_{B(sat)}$, the circuit is said to be overdriven. The degree of overdrive is given by an overdrive factor K, which is defined by

$$K = \frac{I_{B2}}{I_{B(sat)}} = \frac{\beta_0 I_{B2}}{V_{CC}/R_C}. \qquad (8.17)$$

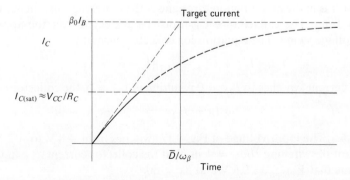

Fig. 8.16 Collector-current waveform for an overdriven switch.

The collector current is described by an exponential up to the time of saturation, with $\beta_0 I_{B2}$ as the target current, rather than I_{C2}. Hence

$$I_C = I_{C1} + (\beta_0 I_{B2} - I_{C1})(1 - e^{-t\omega_\beta/\bar{D}}).$$

If $I_{C1} = 0$, the time required for the collector current to cross the active region can be calculated from the above equation to be

$$t_{on} = \frac{\bar{D}}{\omega_\beta} \ln\left[\frac{I_{B2}}{I_{B2} - I_{C(sat)}/\beta_0}\right]. \tag{8.18}$$

The time to cross the active region should not be confused with the total turn-on time T_{on} which includes passive delay time. In terms of K this can be expressed as

$$t_{on} = \frac{\bar{D}}{\omega_\beta} \ln \frac{K}{K-1}. \tag{8.19}$$

If K is larger than five, the collector current rise will be almost linear, and the turn-on time becomes

$$t_{on} \approx \bar{D}/K\omega_\beta.$$

As K increases, raising the collector target current, the turn-on time decreases.

Turn-off times

There are two cases to consider when calculating turn-off or fall time of a non-saturated transistor switch. The first occurs when switching takes place between two values of collector current that are both within the active region. We shall call the initial base and collector current level I_{B2} and I_{C2} and the final levels I_{B3} and I_{C3}. The second case of practical interest occurs when the input voltage source switches to a negative value (assuming an npn transistor). The base current I_{B3} will now be negative until the capacitance $\bar{D}C_D$ discharges to allow the base-emitter junction to become reverse-biased. The collector current will switch from the initial value of I_{C2} to a final value of zero.

CASE 1. *Fall time of nonsaturated switch with $I_{B3} \geq 0$.* When the base current drive of the switch is suddenly lowered, the collector current will respond exponentially. The transistor is in the active region throughout the switching time; hence the time constant is $\bar{D}/\omega_\beta$. The capacitance C_{BC} must be evaluated over the appropriate collector voltage swing. The equation for collector current is

$$I_C = I_{C2} + (I_{C3} - I_{C2})(1 - e^{-t\omega_\beta/\bar{D}}). \tag{8.20}$$

If the base current switches to $I_{B3} = 0$ resulting in $I_{C3} = 0$, the collector current is

$$I_C = I_{C2} e^{-t\omega_\beta/\bar{D}}. \tag{8.21}$$

Example 8.6. The input voltage of Fig. 8.17 switches from $+8$ V to $+2$ V. If the time constant of switching $\bar{D}/\omega_\beta = 1$ μs, find the collector current as a function of time. Assume that $V_{BE(on)} = 0.5$ V and $\beta_0 = 50$.

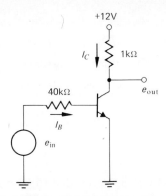

Fig. 8.17 Circuit for Example 8.6.

Solution. The initial base current is

$$I_{B2} = \frac{8 - 0.5}{40} = 0.188 \text{ mA.}$$

Initial collector current is then 9.4 mA. The base current switches to a value of

$$I_{B3} = \frac{2 - 0.5}{40} = 0.0375 \text{ mA.}$$

The final value of collector current is 1.87 mA. Collector current can then be expressed as

$$I_C = 9.40 - 7.53(1 - e^{-t/1 \, \mu s}).$$

CASE 2. *Fall time of a nonsaturated switch with $I_{B3} < 0$.* If the input voltage switches to a negative value the base-emitter junction will ultimately become reverse-biased and I_{B3} will become zero. Immediately upon switching, however, the capacitance $\bar{D}C_D$ will not allow the transistor base voltage to change instantaneously; thus, the base current will be negative. The negative base current will exist until the capacitance is discharged which occurs simultaneously with the collector current reaching zero value.

The negative base current removes charge from the base region very rapidly and the collector current decreases to zero much sooner than the case wherein I_{B3} is positive or zero. The collector target current is $\beta_0 I_{B3}$ which now has a negative value. The collector current moves from the initial value of I_{C2} toward the negative target value of $\beta_0 I_{B3}$, but will stop charging when zero current is reached. The equation for collector current in this instance is

$$I_C = I_{C2} + (\beta_0 I_{B3} - I_{C2})(1 - e^{-t\omega_\beta/\bar{D}}).$$

The time required to switch from I_{C2} to zero current is called turn-off time t_{off} and should not be confused with the total turn-off time T_{off} which includes storage time. Turn-off time can be found by equating I_C to zero in the preceding equation and solving for t_{off}, giving

$$t_{off} = \frac{\bar{D}}{\omega_\beta} \ln\left(1 - \frac{I_{B2}}{I_{B3}}\right). \tag{8.22}$$

Noting that I_{B3} will have a negative value for this case we can also write

$$t_{off} = \frac{\bar{D}}{\omega_\beta} \ln\left(1 + \left|\frac{I_{B2}}{I_{B3}}\right|\right). \tag{8.23}$$

We again emphasize the fact that $\bar{D}$ must be evaluated over the appropriate voltage swing.

The improvement in switching speed due to base compensation will result for turn-off time as well as for turn-on time. The compensated time constant is again given by Eq. (8.15).

Example 8.7. Rework the problem of Example 8.6 with the input voltage swinging from $+8$ V to -6 V. Note that $\bar{D}$ will be different in this case; thus, let us assume a time constant of $\bar{D}/\omega_\beta = 1.2$ μs.

Solution. The initial base and collector currents are the same as those in Example 8.6: $I_{B2} = 0.188$ mA and $I_{C2} = 9.4$ mA. Upon switching I_{B3} becomes

$$I_{B3} = \frac{-6 - 0.5}{40} = -0.163 \text{ mA}.$$

The collector target current is then

$$\beta_0 I_{B3} = -8.15 \text{ mA}.$$

Turn-off time is

$$t_{off} = 1.2 \ln\left(1 + \frac{0.188}{0.163}\right) = 0.86 \text{ μs}.$$

The collector current waveform is sketched in Fig. 8.18.

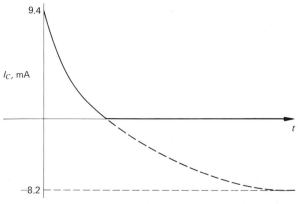

Fig. 8.18 Collector-current waveform.

D. Saturation Storage Time

When the base current drive is greater than that required for saturation, a decrease in base current will not result in an instantaneous decrease in collector current. For example, if I_B is switched from an overdriven value to zero, the collector current will remain at $I_{C(\text{sat})}$ for a finite time before it starts its exponential decay to zero current. This finite delay is called the saturation storage time. The origin of this storage time is easily seen in terms of the gradient of minority carriers in the base region. Consider the schematic of Fig. 8.19.

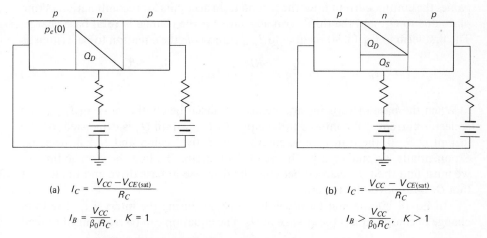

(a) $\quad I_C = \dfrac{V_{CC} - V_{CE(\text{sat})}}{R_C}$

$\quad\quad I_B = \dfrac{V_{CC}}{\beta_0 R_C}, \quad K = 1$

(b) $\quad I_C = \dfrac{V_{CC} - V_{CE(\text{sat})}}{R_C}$

$\quad\quad I_B > \dfrac{V_{CC}}{\beta_0 R_C}, \quad K > 1$

Fig. 8.19 Hole distribution in base region for $K = 1$ and $K > 1$.

The collector current can never exceed a value of V_{CC}/R_C. At the edge of saturation, the base-to-collector voltage is approximately zero and the concentration of holes at the collector junction is also approximately zero. As the number of holes injected from emitter to base is increased, driving the transistor further into saturation, a very slight increase in I_C causes the collector-base junction to become forward-biased. The concentration of holes at the collector junction then increases, and the gradient of holes across the base is negligibly changed. The value of $V_{CE(\text{sat})}$ will be slightly lower as the transistor is saturated harder. The hole gradients in both cases of Fig. 8.19 are the same, but there is excess charge in the base that does not contribute to the gradient for the overdriven case. When the base current is interrupted, the collector current remains constant until the excess charge is depleted. That is, the excess charge supplies carriers to keep the collector current at its saturated value until Q_S is removed from the base region. In addition, I_{B3} can be negative and Q_S also supplies charge for this current. The storage time can most easily be derived by what is referred to as "charge control" considerations [2]. We can write the base current in terms of the components of charge in the base plus the derivatives of these components. When the transistor is saturated as shown in Fig. 8.19(b), the base current is made up of recombination current that is proportional to the total charge present. In addition, the base current must account for any change of total charge carriers that occurs in the base. If the charge is divided into two components, the first consisting of Q_D, or the charge contributing to diffusion current, and the second Q_S, or the excess charge stored in the base, the base current can be written as

$$I_B = \frac{Q_D}{\tau_D} + \frac{Q_S}{\tau_S} + \frac{dQ_T}{dt} = \frac{qD_p p_e(0)A}{(\beta + 1)W} + \frac{Q_S}{\tau_S} + \frac{dQ_T}{dt} \quad (8.24)$$

The first term in the above equation is the recombination base current that accompanies the emitter current flow, the second term accounts for recombination taking place due to Q_S, and the third term accounts for the change in total base charge. The first term in Eq. (8.24) is equal to $I_{B(\text{sat})}$, allowing the equation to be written as

$$I_B = \frac{V_{CC}}{\beta_0 R_C} + \frac{Q_S}{\tau_S} + \frac{dQ_T}{dt} = I_{B(\text{sat})} + \frac{Q_S}{\tau_S} + \frac{dQ_T}{dt}. \quad (8.25)$$

When the base drive is instantaneously reduced below the value of $I_{B(\text{sat})}$, the collector current will continue to be equal to $I_{C(\text{sat})}$ until Q_S is exhausted. At the instant $Q_S = 0$, the transistor is entering the active region and I_C will decrease exponentially, as predicted by the equivalent circuit. To find the storage time t'_s, we must find the time elapsed between the decrease in base drive and the instant that Q_S becomes equal to zero.

In Eq. (8.25), the first term will be constant during the interval t'_s. The total change in charge equals the change in Q_S. The equation can be rearranged to read

$$I_B - I_{B(\text{sat})} = \frac{Q_S}{\tau_S} + \frac{dQ_S}{dt}. \quad (8.26)$$

Under steady-state conditions before switching, $I_B = I_{B2}$ and $dQ_S/dt = 0$; hence

$$Q_{SO} = \tau_S(I_{B2} - I_{B(\text{sat})}), \quad (8.27)$$

where Q_{SO} is the excess stored charge before switching. Solving Eq. (8.26) for Q_S gives

$$Q_S = Ce^{-t/\tau_S} + \tau_S(I_B - I_{B(\text{sat})}).$$

The above equation is good only for the case where I_B is constant; therefore let us apply this equation at $t = 0^+$, just after switching. I_B will be constant at a value I_{B3}, and the equation is

$$Q_S = Ce^{-t/\tau_S} + \tau_S(I_{B3} - I_{B(\text{sat})}).$$

At $t = 0^+$, $Q_S = Q_{SO}$ and C is

$$C = Q_{SO} - \tau_S(I_{B3} - I_{B(\text{sat})})$$
$$= \tau_S(I_{B2} - I_{B(\text{sat})}) - \tau_S(I_{B3} - I_{B(\text{sat})}) = \tau_S(I_{B2} - I_{B3}).$$

Now Q_S is

$$Q_S = \tau_S(I_{B2} - I_{B3})e^{-t/\tau_S} + \tau_S(I_{B3} - I_{B(\text{sat})}).$$

The time taken for Q_S to become zero can be found from this equation by setting $Q_S = 0$ and solving for t'_s, giving

$$t'_s = \tau_S \ln \frac{(I_{B2} - I_{B3})}{(I_{B(\text{sat})} - I_{B3})}. \tag{8.28}$$

The storage time depends on the initial value of I_B, since this determines how hard the transistor is saturated; it depends on the final value of I_B since this determines the rate at which charge is removed from the base; and it depends on $I_{B(\text{sat})}$. In addition, the recombination time τ_S appears in the equation for t'_s. This constant can be related to transistor parameters, but it is easier to obtain its value from the manufacturer's specifications. The time constant often is not given directly, but the storage time is given along with the circuit used to measure t'_s. We can evaluate I_{B2}, I_{B3}, and $I_{B(\text{sat})}$ for the measuring circuit, and τ_S can be found. This allows us to apply τ_S to any circuit, regardless of the values of I_{B2}, I_{B3}, and $I_{B(\text{sat})}$, to calculate t'_s. Note that I_{B3} must be less than $I_{B(\text{sat})}$ to bring the stage out of saturation. In many instances I_{B3} will have a negative value (opposite to the direction of I_{B2}) in which case t'_s will become much shorter. As I_{B3} approaches an infinite negative value, t'_s approaches $\tau_S \ln 1$ or zero. If I_{B3} is zero, corresponding to opening the base lead, the formula is

$$t'_s = \tau_S \ln \frac{I_{B2}}{I_{B(\text{sat})}} = \tau_S \ln K. \tag{8.29}$$

Of course, Eq. (8.28) does not apply if I_{B2} is less than $I_{B(\text{sat})}$.

E. Relating Switching Times to Overall Response

The switching times indicated in Fig. 8.11 can be calculated from the relations developed in the preceding sections. There are a few interesting relationships that will aid in these calculations. For example, the total delay time t_d equals t'_d plus the 0–10% rise time. It can be shown that the 0–10% rise time can be expressed as

$$t_{0-10} = \frac{\bar{D}}{\omega_\beta} \ln \left(\frac{K}{K - 0.1} \right). \tag{8.30}$$

The 10–90% rise time of a saturating switch can be expressed as

$$t_r = \frac{\bar{D}}{\omega_\beta} \ln \left(\frac{K - 0.1}{K - 0.9} \right). \tag{8.31}$$

These equations can be used in calculating total turn-on time. Corresponding equations can be developed for the turn-off times.

Any of the above equations can be modified when driven by a low impedance source. This means that in all equations in which $\bar{D}/\omega_\beta$ appears, $\bar{D}/\bar{B}\omega_\beta$ should replace this quantity.

In considering the waveform of Fig. 8.11 and the related equations, several observations can be made. For example, t_r can be decreased considerably by using high values of I_{B2} and K. However, as the transistor is saturated harder the saturation storage time increases and, in some instances, this increase in t'_s is greater than the decrease in t_r. We also note that the fall time t_f can be decreased by switching to a large negative value of base current I_{B3}. To accomplish this a large value of E_{off} must be used. This leads to a greater passive delay time t'_d when the circuit again switches to the on state.

Changing base-drive conditions appears to increase some of the switching times while decreasing others. The use of the speed-up capacitor overcomes this difficulty. We choose R_B to keep the saturation current low, thus minimizing t'_s. The proper value of speed-up capacitor bypasses R_B during transients so that the apparent value of base drive is much higher than the steady-state value. This results in a decreased rise time. The same is true for switching the transistor to the off state. A small value of E_{off} resulting in low values of delay time is used, but due to the fact that R_B is bypassed as the input source switches, a large initial negative value of base drive is accomplished. The vast improvement in switching times brought about by using the speed-up capacitor explains why it is used in the majority of high-speed switching circuits.

Example 8.8. The transistor switch of Fig. 8.10 is used with values of $R_B = 100$ kΩ, $R_C = 4$ kΩ, $V_{CC} = 12$ V, and the input voltage switches from -6 V to $+12$ V. The transistor specifications are as follows: $\beta_0 = 60$, $\omega_t = 10^8$ rad/sec, $C_{b'c} = 10$ pF at $V_{BC} = -12$ V, $C_{b'e} = 20$ pF at $V_{BE} = -6$ V, $V_{BE(on)} = 0.6$ V, $r_{bb'} = 100$ Ω and $\tau_S = 2$ μs. Determine the passive delay time and the 0- to 100-percent rise time when the input voltage switches to $+12$ V. Determine the saturation storage time and the 100- to 0-percent fall time when the input switches to -6 V.

Solution. To calculate the passive delay time the average value of $C_{b'e}$ and $C_{b'c}$ as base voltage charges from -6 V to 0.6 V must be found. The average value of $C_{b'e}$ is found from Eq. (8.9) to be $C_{BE} = 26$ pF. The collector-base capacitance is $C_{BC} = 9$ pF. The total capacitance that must be charged through R_B is 35 pF. Equation (8.12) gives a passive delay time of

$$t'_d = 3.5 \ln \left[\frac{12 + 6}{12 - 0.6} \right] = 1.6 \ \mu s.$$

The overdrive factor K is

$$K = \frac{\beta_0 I_{B2}}{V_{CC}/R_C} = \frac{60 \times 0.114}{12/4} = 2.3.$$

From Eq. (8.19) the time required to switch through the active region is

$$t_{on} = \frac{\bar{D}}{\omega_\beta} \ln \frac{K}{K - 1}.$$

To use this equation the D-factor must be found. The average collector-base depletion capacitance is evaluated from Eq. (8.9) with $V_1 = -11.4$ V and $V_2 = 0$ V. This value is found to be $C_{BC} = 13.4$ pF and $\bar{D}$ is 6.25. This gives

$$t_{on} = \frac{6.25}{1.667 \times 10^6} \ln \frac{2.3}{1.3} = 2.14 \ \mu s.$$

The saturation storage time can be found from Eq. (8.28) with $I_{B2} = 0.114$ mA, $I_{B3} = -0.066$ mA, and $I_{B(sat)} = 0.05$ mA. This time is

$$t_s' = 2 \ln \left[\frac{0.114 + 0.066}{0.05 + 0.066} \right] = 0.88 \ \mu s.$$

To find the 100- to 0-percent fall time, Eq. (8.23) is applied. The base current that flows as the transistor starts to shut off will be $I_{B(sat)}$. The fall time is then

$$t_{off} = \frac{6.25}{1.667 \times 10^6} \ln \left(1 + \frac{0.050}{0.066} \right) = 2.12 \ \mu s.$$

The D-factor for turn-off is the same as that for turn-on since the same voltage swings apply.

A speed-up capacitor can be added to the circuit to improve switching times. This element is often chosen arbitrarily to be one to two times the value given by Eq. (8.14) to account for the additional charge present in the base region when the transistor is saturated and the charge on the delay capacitances C_{BE} and C_{BC}.

8.3 TUNNEL DIODE SWITCHING

A. The Basic Tunnel Diode Switch

The circuit of Fig. 8.20 shows a simple switching circuit. The series resistance R must be greater than r_t in order for the circuit to function as a switching circuit. The load-line for $V = V_1$ shows three intersections with the characteristic curve.

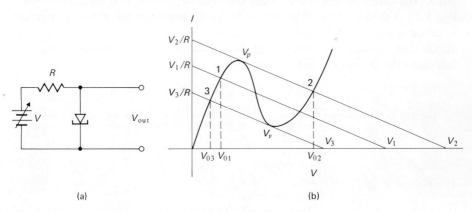

Fig. 8.20 Tunnel diode switch: (a) switching circuit; (b) load-line.

Let us assume that point 1 is the initial operating point corresponding to an output voltage of V_{01}. As V is increased the operating point moves along the characteristic curve toward the peak voltage V_p. When V is increased to V_2 and tends to exceed this value, the load-line intersects the V–I curve only at point 2. The output voltage switches from V_{01} to V_{02} very rapidly with a possible transition speed of the order of 1 ns. If V is now reduced to V_3 and then to a lower value, the circuit switches rapidly to operating point 3 and an output voltage of V_{03}. The input and output voltage waveforms are shown in Fig. 8.21.

It should be noted that the output voltage is quite small, measuring only a few tenths of a volt. It is possible to use the tunnel diode in transistor switching circuits to produce relatively high-speed switching at higher voltage levels.

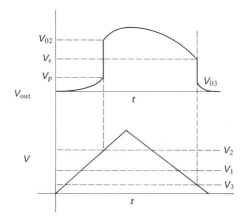

Fig. 8.21 Input and output waveforms for tunnel diode switch.

8.4 THE SILICON-CONTROLLED RECTIFIER

The SCR is a three-terminal, three-junction device capable of very high current gains. The three terminals are called the cathode, anode, and gate, as shown in Fig. 8.22. A device allowing access to the fourth region of the pnpn-arrangement is called a silicon-controlled switch. Another device, with only a cathode and anode lead, is called the Shockley diode. All devices with this pnpn-arrangement belong to the thyristor family.

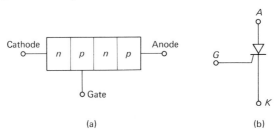

Fig. 8.22 Silicon-controlled rectifier: (a) physical layout of SCR; (b) symbol for SCR.

The characteristics of the SCR which make it important are: (1) when the gate is not positive with respect to the cathode, a large voltage can be applied to the anode with no accompanying current conduction; (2) raising the gate voltage a small amount to the triggering level causes a very high current flow with very little voltage drop from anode to cathode; and (3) the gate cannot be used to cause the anode-to-cathode current to stop flowing; only the combination of a gate voltage below the trigger level and the temporary interruption of anode current will restore the SCR to the nonconducting state.

The anode-to-cathode voltages that can be applied to the device without conduction taking place range from 50 V to 1000 V for different types of SCR. Continuous anode current ratings up to 250 A with surge current ratings of 5000 A are available. Generally, the maximum gate trigger voltage and maximum gate current are specified by the manufacturer. We shall see later that the gate trigger voltage varies with temperature; thus it must be specified at a particular temperature.

A. Theory of Operation

The operation of the SCR can be described by considering the device to consist of an interconnected arrangement of a pnp- and an npn-transistor, as shown in Fig. 8.23. Part (a) shows two separate transistors, but it is easily seen that if the two devices were brought into contact with each other, they would form a pnpn-device.

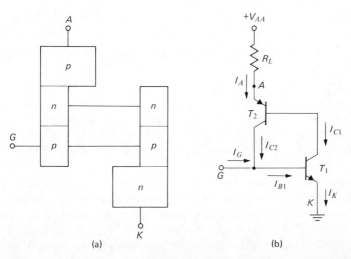

Fig. 8.23 An SCR equivalent consisting of two transistors: (a) physical layout; (b) schematic representation.

We can see from Fig. 8.23(b) that the two transistors form a feedback circuit. The current I_{B1} is amplified by T_1, again by T_2, and returned to the base of T_1. If an attempt is made to operate the SCR such that both transistors are in the active region, the loop gain is much greater than unity. The result is the saturation of both transistors, which lowers the loop gain to less than unity. This corresponds to the high-conductance, low-voltage state of the SCR. On the other hand, the loop gain can be limited to values less than unity, when both transistors are off, by controlling the gate voltage. If the gate voltage is negative or lower than $V_{BE(on)}$, the collector current of T_1 consists only of leakage current I_{co}. Then β_1 is very low, since the transistor is cut off. A rough plot of I_C as a function of V_{BE} and I_B for a single transistor is shown in Fig. 8.24. When V_{BE} is less than $V_{BE(on)}$, I_C is very nearly I_{co}, and I_B is approximately $-I_{co}$. As V_{BE} increases, I_B decreases to zero, and I_C becomes βI_{co}. Since β depends on I_C, dropping off at low values of I_C, the value of I_C corresponding to $I_B = 0$ can be quite small.

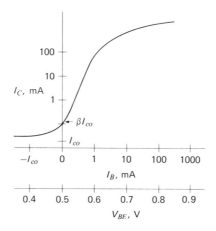

Fig. 8.24 Collector-current variation with V_{BE} and I_B for a silicon transistor.

The SCR will not conduct appreciable anode current so long as T_1 is cut off and $\beta_1 \beta_2$ is less than unity. Referring to Fig. 8.23(b), we see that the voltage appearing at the anode would very nearly equal V_{AA}, the power supply voltage. For low values of gate voltage, the product $\beta_1 \beta_2$ can easily be less than unity. As V_G increases, β_1 increases, as demonstrated by the graph of Fig. 8.24. Furthermore, since more current flows in the collector of T_2, β_2 also increases. That gate voltage which causes $\beta_1 \beta_2$ to approach unity is referred to as the gate trigger voltage. At this point, as the loop current increases, β_1 and β_2 also tend to increase. In Chapter 6, it was shown that a loop gain of greater than unity cannot exist when positive feedback is present. The two transistors would conduct infinite current if $\beta_1 \beta_2 \geq 1$. As this point is approached, the current increases rapidly, saturating both transistors. Under saturated conditions, the product $\beta_1 \beta_2$ is limited to a value less than unity.

Once the transistors saturate, the anode voltage drops to a very low value, conducting maximum current through the load. Reducing the gate voltage below the trigger level will not cause $\beta_1 \beta_2$ to decrease, since I_{C1} and I_{C2} are both quite large. If a very-low-impedance, high-current gate source is available to shunt I_{C2}

to ground, the SCR can be turned off, using the gate. However, a source of this type is almost never available; thus for all practical purposes, the gate has no control over the SCR after breakdown has occurred. In order to restore the SCR to the nonconducting state, the gate voltage must be decreased below the trigger level, and the anode current must be temporarily interrupted. Actually, as the anode current becomes less than a value referred to as the holding current, the product of $\beta_1 \beta_2$ will decrease below unity. The significance of the holding current is that (1) in order to sustain conduction when the SCR is turned on, the anode current must equal or exceed the holding current; and (2) in order to restore the SCR to the nonconducting state, the anode current must become less than the holding current. In rectifier applications where an ac voltage serves as the power supply voltage, the SCR will shut off as the anode voltage becomes negative. In dc applications the problem of turning the SCR off is a very troublesome one.

Example 8.9. The SCR of Fig. 8.25 has a holding current $I_H = 100$ mA, a maximum gate trigger voltage of 0.75 V, and a maximum required gate trigger current of 10 mA. Calculate the maximum value of V_{in} that will cause the SCR to break down. Given that V_{in} is zero, calculate the value to which V_{AA} must be reduced to turn the SCR off.

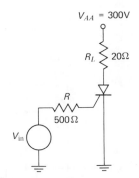

Fig. 8.25 The SCR circuit for Example 8.9.

Solution. In order to supply 10 mA and 0.75 V to the gate, V_{in} must be

$$V_{in} = V_G + I_G R$$
$$= 0.75 + 0.5 \times 10 = 5.75 \text{ V}.$$

When the SCR turns on, the anode voltage will drop to a low voltage, for example, 2 V. The anode current is

$$I_A = \frac{300 - 2}{20 \; \Omega} = 14.9 \text{ A}.$$

In order to drop I_A below the 100 mA holding current, the voltage across R_L must be less than 2 V. At $I_A = 100$ mA, assuming an SCR voltage of 1 V, V_{AA} must be lowered to 3 V to turn the SCR off.

Turn-on time. The turn-on time of the SCR depends on the same parameters that determine the transistor turn-on time. There will be a passive delay time due to the nonlinear gate capacitance associated with the reverse-biased junctions. The time to switch through the active region is determined by ω_t, but since regeneration is present, the switching time is minimized. Rather than giving parameters from which the turn-on time can be calculated, most manufacturers specify turn-on times under given conditions.

The passive delay time will depend directly on the gate source and impedance present. On the other hand, for normal operation, the time to switch through the active region is determined mainly by the SCR, rather than the source. This is due to the fact that the gate usually supplies only enough drive to start the regeneration process of the SCR. Once regeneration starts, the switching time is a function of the internal parameters of the device.

Turn-off time. The turn-off time of the SCR is often extended over the expected value, due to rate effect. The collector-base depletion-layer capacitances of both transistors appear in parallel and constitute a feedback path from the base of T_2 to the base of T_1 (gate). Figure 8.26 shows this path.

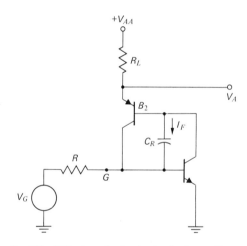

Fig. 8.26 SCR schematic showing feedback capacitance.

When the anode current is interrupted and V_G is zero, the SCR shuts off and V_A starts rising from approximately ground toward V_{AA}. As this change occurs, however, the point B_2 follows the anode voltage. This means that a current flows in the capacitive feedback path equal to

$$I_F = C_R \frac{dV_A}{dt}. \tag{8.32}$$

If this current exceeds the minimum gate current required to turn the SCR on, the SCR starts turning on again. When this occurs, I_F drops due to decreased dV_A/dt and the device does not reach the on state. Thus dV_A/dt is reduced by the rate effect and the maximum allowable value of dV_A/dt to avoid this effect is

$$\frac{dV_A}{dt} = \frac{I_{G(\min)}}{C_R}.$$

Using a value of $I_{G(\min)} = 1$ mA and $C_R = 50$ pF we can write a typical value of dV_A/dt as

$$\frac{dV_A}{dt} = 2 \times 10^7.$$

The time constant of the anode voltage rise would then be

$$\tau = \frac{V_{AA}}{2 \times 10^7}.$$

Using a power supply voltage of 200 V gives a time constant of

$$\tau = 1 \times 10^{-5} = 10 \ \mu s.$$

In low-frequency applications, this time constant might be acceptable, but to extend the frequency performance of the SCR, rate effect must be overcome.

If we use a gate source with low impedance which can swing negative during turn-off time, rate effect will be minimized. This is a result of the feedback current being shunted through the source, rather than flowing into the gate-cathode junction. Often the input source will not be able to swing negatively or will be capacitor-coupled to the SCR. The circuit of Fig. 8.27 overcomes this difficulty. The cathode will now be positive with respect to the gate, due to the drop across the forward-biased diodes. The gate resistor R_G can shunt the feedback current to ground,

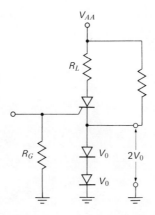

Fig. 8.27 A circuit for minimizing rate effect.

keeping the gate reverse-biased. The amount of current that can be allowed to flow through R_G before the gate becomes forward-biased is

$$I_G = \frac{3V_0}{R_G}.$$

The factor 3 accounts for the two diode drops plus the gate-cathode drop required for forward bias. There are other methods of reducing rate effect which will not be considered here.

Turn-off methods. Often the most difficult design problem relating to the SCR is that of turn-off. It has been mentioned that the gate voltage must be less than the minimum turn-on voltage and the anode current must be temporarily interrupted to turn the SCR off. Interrupting or diverting the anode current can present a serious problem. In Chapter 11, it is shown that for an ac power supply voltage, the anode current stops when the voltage becomes negative. For a dc supply, additional circuitry must be used to turn the SCR off.

One method of diverting the anode current is to use a transistor, as shown in Fig. 8.28(a). During the on time of the SCR, the transistor must be reverse-biased. When the SCR is to be turned off, the transistor is saturated, diverting the anode current through the shunt path. While this circuit is a simple one, it has several disadvantages. The transistor must have a maximum collector-to-emitter breakdown voltage equal to or exceeding the power supply voltage. Furthermore, the current rating of the transistor must be such that the entire anode current can be diverted through the transistor. Of course, the surge current rating of the transistor can be much higher than the continuous current rating, but this rating still may not be adequate.

The capacitor-commutation scheme of Fig. 8.28(b) uses a second SCR to turn off the first. When SCR_1 is turned on, it automatically diverts the anode

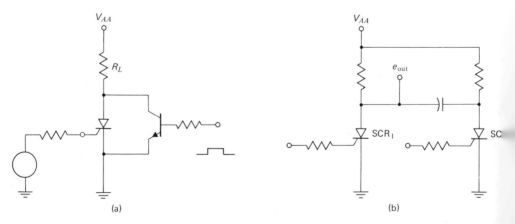

Fig. 8.28 Turn-off methods: (a) shunt transistor; (b) capacitor commutation.

current of SCR_2, thereby ensuring that SCR_2 is off. When SCR_1 is to be turned off, a gate triggering pulse is applied to SCR_2. As SCR_2 turns on, the anode current of SCR_1 is interrupted, causing turn-off of SCR_1. While it is easy to satisfy the voltage and current ratings of the second SCR, the repetition frequency is lowered, due to the recovery time of the capacitor. Since one SCR is always on, the current drain from the source can be quite high also. There are other methods of turn-off that can be found in the literature.

B. Applications of the SCR

The SCR is extremely useful in high-current applications such as rectification or driving electric motors or other high-current mechanisms. However, it can also be a useful element in low-current switching applications. Since the device has two stable states rather than one, it can serve as a memory element in circuits such as binary counters, shift registers, or multivibrators. A few of these applications will be demonstrated.

Incandescent lamp driver. The circuit of Fig. 8.29 allows any one of several lamps to be turned on by the application of a logic-level signal to the appropriate gate. So long as V_G remains above the trigger level, the lamp will remain on. If the gate voltage is removed, the SCR will shut off when the rectified voltage approaches zero, dropping the anode current below I_H. Lamps rated as high as several hundred watts can be driven in this manner.

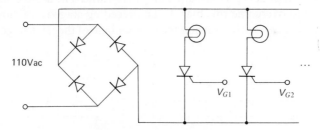

Fig. 8.29 Incandescent lamp driver.

Reversible high dc current driver. There are occasions when it is required to drive current through a load in either direction. One example of this might be a dc servomotor. The bridge circuit of Fig. 8.30 satisfies the requirement of current flow in either direction. If a positive voltage is applied to the inputs marked V_{INF}, then SCR_1 and SCR_4 will turn on, while SCR_2 and SCR_3 will be forced off. Current will then travel through the load via SCR_1 and SCR_4. When a pulse is applied to the inputs marked V_{INR}, then SCR_2 and SCR_3 turn on, and SCR_1 and SCR_4 turn off. Current now flows through the load in the opposite direction. Pulse-width modulated inputs can drive the bridge circuit, giving proportional speed control. The turn-off transistors must have a sufficiently high surge-current rating.

374 LARGE-SIGNAL CIRCUITS

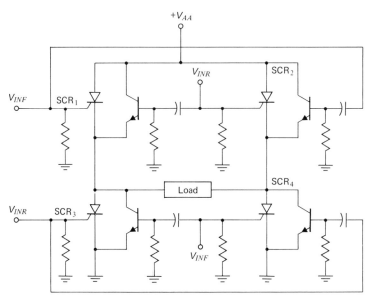

Fig. 8.30 Reversible current driver.

Ring counter. The circuit of Fig. 8.31 shows a ring counter which uses only one active element per stage.

The gate of each SCR is prebiased to a negative value. The input pulses are not sufficiently large to overcome this bias. If the preceding stage is on, however, the cathode of that stage is positive, raising the gate voltage of the next stage. An input

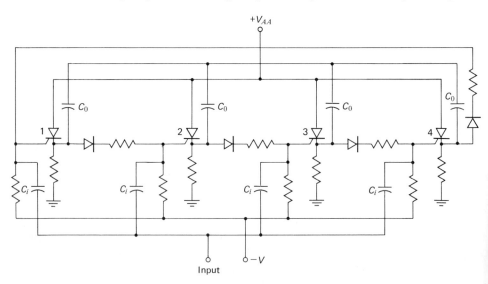

Fig. 8.31 Four-position ring counter.

pulse will now cause this stage to turn on. As the stage turns on, a positive pulse is coupled through the capacitors C_0 to all other stages, turning the preceding stage off. Thus as each input pulse is applied, the stage following the only on stage turns on and extinguishes the preceding stage.

REFERENCES AND SUGGESTED READING

1. D. J. Comer, *Computer Analysis of Circuits*. Scranton, Pennsylvania: International Textbook Co., 1971, Chapter 7.
2. D. T. Comer, *Large Signal Transistor Circuits*. Englewood Cliffs, N.J.: Prentice-Hall, 1967, Chapter 6.
3. T. Kohonen, *Digital Circuits and Devices*. Englewood Cliffs, N.J.: Prentice-Hall, 1972, Chapter 6.

PROBLEMS

Section 8.1

8.1 In the circuit shown in the figure R_B is infinite (open circuit) and $R_C = 10$ kΩ. If $\beta_0 = 100$, $I_{co} = 3$ μA, and $V_{CC} = 12$ V, calculate the collector voltage.

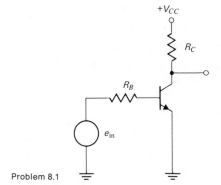

Problem 8.1

* **8.2** Calculate the negative value of input voltage that guarantees a reverse bias of 2 V on the emitter-base junction of the transistor shown in Problem 8.1. Use values of $R_B = 50$ kΩ and $I_{co(max)} = 5$ μA.

8.3 The circuit of Problem 8.1 has a 4 kΩ load resistance, a power supply of 20 V and a transistor with $\beta_0 = 90$ and $V_{BE(on)} = 0.6$ V. Calculate the value of R_B required to drive the stage to the edge of the saturation region when the input voltage is $+6$ V.

8.4 Repeat Problem 8.3 if $R_C = 2$ kΩ and $V_{CC} = 12$ V.

* **8.5** The transistor of Problem 8.1 has a $\beta_0 = 90$ and $V_{BE(on)} = 0.6$ V. If $R_B = 40$ kΩ, $e_{in} = +6$ V, and $V_{CC} = 20$ V, what value of R_C is required to put the stage at the edge of saturation?

8.6 Repeat Problem 8.5 if $V_{CC} = 12$ V.

8.7 The transistor of Problem 8.1 has a $\beta_0 = 90$ and $V_{BE(on)} = 0.6$ V. If $R_B = 40$ kΩ, $R_C = 1$ kΩ, and $V_{CC} = 20$ V, what minimum input voltage level is required to saturate the transistor?

* **8.8** Sketch the output waveform of the circuit in the figure showing all voltage levels if e_{in} switches from $+6$ V to 0 V. Assume that $V_{BE(on)} = 0.5$ V and $\beta_0 = 50$. Neglect transistor capacitances.

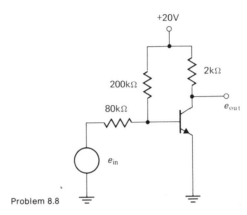

Problem 8.8

8.9 Repeat Problem 8.8 if e_{in} switches from 0 to -6 V.

8.10 Rework Example 8.2 if the 60-kΩ resistance is decreased to a value of 20 kΩ.

***8.11** Rework Example 8.3 if the input signal levels are 0 and -8 V.

8.12 Rework Example 8.3 if the collector load resistance is changed from 2 kΩ to 8 kΩ.

Section 8.2B

8.13 Calculate the passive delay time of the circuit shown in Fig. 8.10 if e_{in} switches from -10 V to $+4$ V at $t = 0$. The circuit element values are $R_C = 3$ kΩ, $R_B = 40$ kΩ, and $V_{CC} = 12$ V. The small-signal collector-base capacitance is 8 pF at a junction voltage of 6 V while the emitter-base capacitance is 10 pF at 6 V. Assume that $V_{BE(on)} = 0.6$ V.

*** 8.14** Repeat Problem 8.13 if the power supply voltage is changed to $V_{CC} = 20$ V.

8.15 Repeat Problem 8.13 if the input signal switches from -2 V to $+6$ V at $t = 0$.

Section 8.2C

8.16 Rework Example 8.5 if e_{in} switches from $+4$ V to $+6$ V. Does the switching time constant of this problem differ from that of the example? If so, explain why.

8.17 Rework Example 8.5 if e_{in} switches from $+1$ V to $+16$ V.

*** 8.18** Calculate the appropriate value for speed-up capacitance in Problem 8.16. Calculate the new time constant if $R_S = 2$ kΩ and $r_{bb'} = 100$ Ω.

8.19 Find the time required for the transistor of Example 8.5 to reach the edge of the saturation region when an input signal that swings from -2 V to $+24$ V is applied. Assume that $C_{BE} + C_{BC} = 20$ pF in the delay region and $C_{BC} = 10$ pF over the active region swing.

*** 8.20** Repeat Problem 8.19, given that R_B is lowered from 100 kΩ to 60 kΩ.

8.21 The transistor shown in the figure has $\beta_0 = 60$, $V_{BE(on)} = 0.6$ V, and $f_t = 10$ MHz. Calculate the fall time if $C_{BC} = 8$ pF averaged over the swing of interest. The input signal switches from 2 V to 0 V.

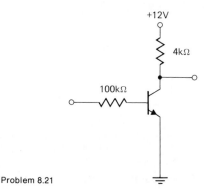

Problem 8.21

8.22 If e_{in} switches from $+3$ V to -10 V in Problem 8.21 find t_{off}. Assume that $C_{BC} = 9$ pF over the swing of interest.

*8.23 Repeat Problem 8.22 if e_{in} switches from $+2$ V to -6 V.

Sections 8.2D–8.2E

8.24 Assuming that the storage time constant $\tau_S = 1$ μs for the transistor of Problem 8.21, calculate the saturation storage time when e_{in} switches from $+20$ V to 0 V.

8.25 Repeat Problem 8.24 given that e_{in} swings from $+20$ V to $+3$ V.

*8.26 Repeat Problem 8.24 given that e_{in} swings from $+20$ V to -10 V.

8.27 If the transistor of Problem 8.21 is switched from 10 V to 2 V giving a saturation storage time of 2 μs, find the storage time constant τ_S.

8.28 If $\tau_S = 3$ μs, $f_t = 20$ MHz, $\beta_0 = 90$, $V_{BE(on)} = 0.6$ V, and $C_{BC} = 10$ pF over the swing of interest, find the total time for I_C to become zero in the circuit of Problem 8.21 following a change of input voltage from $+10$ V to -2 V.

8.29 Repeat Problem 8.28 for an input voltage that changes from $+4$ V to -2 V.

*8.30 Repeat Problem 8.28 for an input voltage that changes from $+10$ V to -10 V.

8.31 Sketch the collector voltage waveform for the circuit of Problem 8.21 if a pulse swinging from -6 V to $+10$ V and back to -6 V is applied to the input. Use $\beta_0 = 60$, $V_{BE(on)} = 0.6$ V, $f_t = 10$ MHz, and $\tau_S = 2$ μs. Assume that $C_{BC} = 10$ pF over the active region and $C_{BE} + C_{BC} = 20$ pF during delay time.

8.32 Repeat Problem 8.31 if the lower level of the pulse is -10 V and the upper level is $+4$ V.

*8.33 Repeat Problem 8.31 if the lower level of the pulse is $+2$ V and the upper level is $+20$ V.

Section 8.3

8.34 Assume the tunnel diode in the switching circuit of Fig. 8.20 obeys the ideal characteristics shown. Sketch the output voltage as a function of time for the input shown in the figure on the following page, assuming that $R = 50$ Ω.

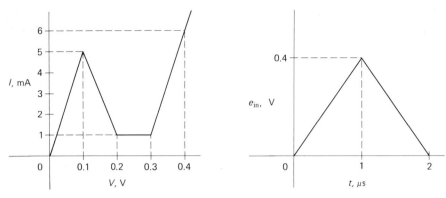

Problem 8.34

8.35 Repeat Problem 8.34 given that the positions of the resistor R and the tunnel diode are interchanged; V_{out} now appears across R.

Section 8.4

8.36 The input gate voltage to an SCR is $10 \sin(377t - 10°)$. The supply voltage V_{AA} applied to the SCR is $150 \sin 377t$. The load resistance is $100 \, \Omega$ and the gate trigger voltage is 1 V. Sketch the current waveform of the load.

8.37 Repeat Problem 8.36 assuming that the input gate voltage is $10 \sin(377t - 150°)$.

*8.38 The SCR of Example 8.9 is used in the circuit shown in the figure. Calculate R_2 so that a 0.5-V input pulse will reliably trigger the SCR.

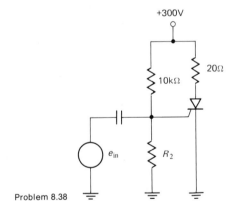

Problem 8.38

LOGIC COMPONENTS AND FAMILIES

The most useful logic element is one having two controllable stable states. These two states can be defined to correspond to the two digits of the binary number system, zero and one. We will see in Chapter 10 how several of these binary circuits can be interconnected to manipulate large numbers expressed in binary form.

As an example of voltage level definition, a circuit may be said to be in the "one" state for all output voltages above 1.2 V while all output voltages less than 0.2 V are identified as the "zero" state. One or both of the voltage levels corresponding to the binary states may be negative and in some cases, the more negative voltage may be defined as the "one" state. Systems having the most positive voltage level defined as the "one" state are called positive logic systems. On some occasions it is convenient to use a negative logic system wherein the "zero" state corresponds to the most positive of the two voltage levels. For example, any voltage over $+2$ V would perhaps be interpreted as binary zero while any voltage less than -4 V is identified as binary one.

This chapter considers the various types of integrated logic circuit that are used in digital system design. The integrated circuit has had a tremendous impact on the digital systems field. The marked improvements over discrete circuits in size, cost, and reliability have resulted in many new and interesting design possibilities. There are three classifications of integration today depending on the level of complexity of the circuit or system. Small-scale integration (SSI) results in an integrated circuit that performs a basic logic function, such as the OR function or the AND function. An OR gate is a circuit that may have two or more inputs and a single output. If a voltage level corresponding to binary one is applied to one or more inputs, the output signal is also a binary one. If all zeros are applied at the input, the output also corresponds to the zero state. In the AND gate, the output state is zero for all input conditions except the situation wherein all inputs are in the one state. Using SSI techniques a silicon chip may be designed to include two or three logic gates. Such a chip might contain 20 to 40 actual components.

The next step up in complexity is medium-scale integration (MSI). This level of integration may represent an order of magnitude increase in complexity over the SSI circuit. Logic functions equivalent to the capability of 10 or 20 logic gates may be integrated on a single chip in this case. Large-scale integration (LSI) requires the most complex fabrication techniques available. A single chip may be fabricated to perform the logic for an entire digital system. The capability of 100 to 1000 logic gates may be included within a chip constructed by LSI techniques. This device may contain 10,000 to 20,000 components and could be designed to include the entire logic for a sophisticated electronic calculator.

All three levels of integration are widely used in the digital field. LSI is directed toward finished products requiring compactness or very complex functions while SSI is used more in preliminary design work or applications not placing a premium on miniaturization.

Most digital design is done without the necessity of circuit design. All circuits required to perform the conventional logic functions are available as integrated circuits; thus, our attention will be directed toward gaining a knowledge of the building blocks available.

9.1 LOGIC GATES

The two basic gates used in digital systems are the OR gate and the AND gate. We will first consider the simplest implementations of these gates and later consider alternate, more practical realizations of the same functions.

A. OR Gates

The OR gate has one output and two or more inputs. If all inputs are in the 0 state, the output is also in the 0 state. The presence of a 1 bit on one or more of the inputs will lead to a 1 bit at the output. A simple two-input OR gate using positive logic is shown in Fig. 9.1. Typical logic levels might be 0.5 volts or less for the 0 state and 2.5 volts or more for the 1 state.

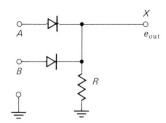

Fig. 9.1 Two-input positive logic OR gate.

Low impedance input sources will result in an output signal X that will

1. equal 0 if $A = B = 0$,
2. equal 1 if $A = 0$ and $B = 1$,
3. equal 1 if $A = 1$ and $B = 0$, and
4. equal 1 if $A = B = 1$.

When either input assumes the positive level, the corresponding diode will become forward-biased. The output voltage then equals the input level minus a small diode drop. If the input swings to +4 V, the output would reach +3.5 V. This same information can be expressed in the truth tables of Fig. 9.2.

The truth table of Fig. 9.2(a) is valid for the OR gate of Fig. 9.1, assuming positive logic is used. The function table is more general and can be used to determine the output signal whether positive or negative logic is applied. We will discuss the use of this truth table in more detail shortly.

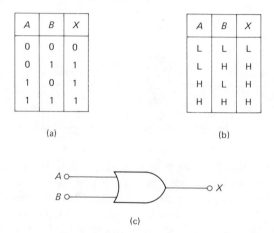

Fig. 9.2 Truth tables: (a) truth table for positive logic; (b) function table; (c) symbol for OR gate.

B. AND Gates

Figure 9.3 indicates a simple AND gate. This circuit has one output and may have two or more inputs. The output equals 0 for the input conditions $A = B = 0$, $A = 0$ and $B = 1$, or $A = 1$ and $B = 0$. The only condition leading to a 1 output is $A = B = 1$.

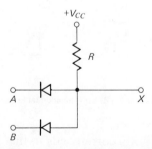

Fig. 9.3 Two-input positive logic AND gate.

The truth tables are shown in Fig. 9.4.

If both input sources generate a high voltage the diodes will allow the output to reach this same voltage level plus the drop across the diodes. A +6 V signal applied to both inputs would result in a +6.5 V output. If either one or both

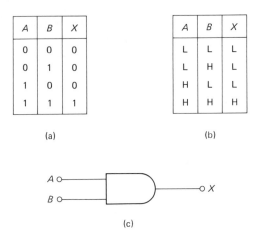

Fig. 9.4 Truth tables: (a) truth table for positive logic; (b) function table; (c) symbol for AND gate.

sources are at zero volts, the output level is at 0.5 V. These characteristics satisfy the AND gate definitions for a positive logic circuit.

There is an interesting and useful relationship between the OR gate and the AND gate. If we change from positive logic to negative logic, the OR gate becomes an AND gate and the AND gate becomes an OR gate. For the circuit of Fig. 9.1 this can be seen from the function table of Fig. 9.2(b). If the low level voltage is defined as binary 1 while the high level is binary 2 the negative logic truth table of Fig. 9.5 results.

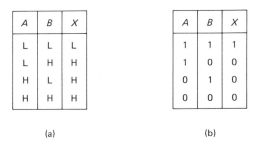

Fig. 9.5 Truth tables for the gate of Fig. 9.1: (a) function table; (b) truth table for negátive logic.

The truth table of Fig. 9.5(b) corresponds to that for an AND gate. From a voltage viewpoint we note that if either or both inputs are at a high voltage or at binary 0, at least one diode is forward-biased, ensuring that the output is at a high level also. Only if both inputs are low will the output be low, indicating the 1 state.

Figure 9.6 shows the truth tables for the circuit of Fig. 9.3. Using negative logic, this gate becomes an OR gate.

A	B	X
L	L	L
L	H	L
H	L	L
H	H	H

(a)

A	B	X
1	1	1
1	0	1
0	1	1
0	0	0

(b)

Fig. 9.6 Truth tables for the gate of Fig. 9.3: (a) function table; (b) truth table for negative logic.

C. NOR Gates

The simple gates of Figs. 9.1 and 9.3 can be used in some applications, but suffer from impedance loading problems. If such a gate is to drive succeeding circuits, the input impedance of these following stages must be much greater than R in order to negligibly affect the operation of the gate. The output impedance of any stage that precedes the gate must be much less than R to minimize attenuation. Thus, if three gates are arranged in cascade, as often required in logic systems, serious loading problems result. The common method of overcoming this situation is to add one or two stages of amplification to the gate.

If a single inverting amplifier is used the OR gate becomes a NOR gate as shown in Fig. 9.7. If a noninverting amplifier is used the OR gate characteristics are preserved.

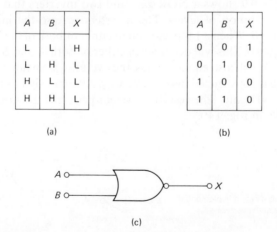

Fig. 9.7 Positive logic NOR gate: (a) function table; (b) truth table for positive logic; (c) symbol for NOR gate.

The small circle following the OR gate represents an inversion and is added to the OR gate to construct the NOR gate symbol.

D. NAND Gates

A NAND gate is an AND gate followed by an inversion as indicated in Fig. 9.8. It can easily be shown that if negative logic is used, the NOR gate of Fig. 9.7 becomes a NAND gate and the NAND gate of Fig. 9.8 becomes a NOR gate.

A	B	X
L	L	H
L	H	H
H	L	H
H	H	L

(a)

A	B	X
0	0	1
0	1	1
1	0	1
1	1	0

(b)

(c)

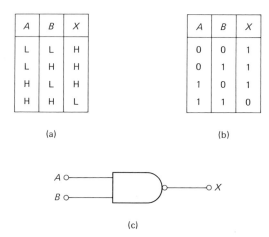

Fig. 9.8 Positive logic NAND gate: (a) function table; (b) truth table for positive logic; (c) symbol for NAND gate.

The effect of redefining levels on the logic function suggests a method of using the same gate along with logic level inverters to perform more than one function. The system of Fig. 9.9 shows a NOR gate and two inverters that act as an AND gate with no redefinition of levels. The inverters reverse the input logic levels presented at A and B. When a 1 is presented to either or both of the NOR inputs the output X will equal 0. This will occur when either or both A and B equal 0. If both A and B equal 1, X will also equal 1. Thus, the overall system functions as an AND gate. An OR gate is constructed in the same way. All input signals are inverted and applied to a NAND gate. The resulting system performs the OR function. Such a system is shown in Fig. 9.10.

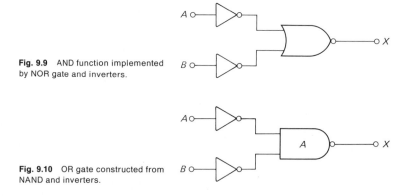

Fig. 9.9 AND function implemented by NOR gate and inverters.

Fig. 9.10 OR gate constructed from NAND and inverters.

The utility of NOR and NAND gates in constructing OR and AND gates leads to the popularity of these circuits. All integrated circuit logic families provide inverters that can easily be used with other gates to provide the logic functions of interest. Several inverters may be included on a single chip, limited only by the number of pins available for external connection. NOR and NAND gates are often used as simple inverters; thus, a given gate serves a variety of functions.

9.2 LOGIC FAMILIES

The actual implementation of gates and other logic elements can be accomplished with several different configurations, each emphasizing certain devices. Those gates based on a particular element configuration are said to belong to a logic family. All circuits of a given family must be compatible in terms of operating characteristics. In general, the electronics of an entire computer or digital system will use only a single family of logic circuits. Hundreds of gates and other basic circuits will be interconnected to form the various subsystems of the digital system.

When we speak of a logic family being compatible we imply compatibility with respect to voltage levels, impedance characteristics, and switching times. The normal output voltage swing of any gate must serve as a suitable input signal to any circuit of the family. Output impedances must be low compared to input impedances to allow several gate inputs to be driven by a single output. The switching speed of all circuits must be such that other circuits can be activated by the transition times of an output signal.

The minimum voltage swing and voltage levels are specified in terms of a maximum lower level and a minimum upper level corresponding to the two possible states. For example, the lower level may never exceed 0.5 V while the upper level will never be less than 2.4 V. The minimum voltage transition will then be $2.4 - 0.5 = 1.9$ V.

There are several factors leading to the specification of allowable loading. In determining the number of circuits that can be driven by a given stage (fan-out specification), one must consider the dc current available from the driving circuit and the input current required by each driven stage. Noise margin also affects the fan-out of a circuit. Noise margin specifies the maximum amplitude noise pulse that will not change the state of the driven stage, assuming the driving stage presents a worst-case logic level to the driven stage. This figure generally decreases with increased loading and consequently affects the fan-out of the driving stage. A third factor affecting fan-out is the input capacitance of the driven stages. Too many stages connected to a single output will present more capacitance at this terminal, lowering switching times to an unacceptable level. The fan-out figure is specified by the manufacturer and indicates the number of similar circuits that can be driven by the given circuit without sacrificing reliability. Some circuits within a family may require more input current than others or may have more input capacitance. This circuit may then cause more than one unit of equivalent loading. If so, the equivalent loading may be specified as four units, for example, indicating that the required input current is equal to the total required by four normal circuits.

The fan-in figure specifies the number of circuits that can be connected to the input of the gate. Quite often fan-in or fan-out figures can exceed the maximum specified values at the expense of switching times.

For monolithic logic circuits the graph of Fig. 9.11 is useful in defining switching times for an inverting gate. There is a finite delay between the application of the input pulse and the output response. A quantitative measure of this delay is the difference in time between the point where e_{in} rises to 50% of its final value and the time when e_{out} falls to its 50% point. This quantity is called leading-edge delay t_{d1}. The trailing-edge delay t_{d2} is the time difference between 50% points of the trailing edges of the input and output signals. The propagation delay is defined as the average of t_{d1} and t_{d2}, or

$$t_{pd} = \frac{t_{d1} + t_{d2}}{2}.$$

Fall and rise times are defined by 10- to 90-percent values as the output voltage swings between lower and upper voltage levels.

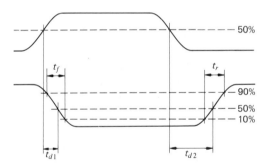

Fig. 9.11 Definition of switching times.

Propagation delay time of an integrated circuit is a function of passive delay time, rise and fall times, and saturation storage time of individual transistors of the circuit. However, since input and output capacitance will influence the integrated circuit switching times, the fan-in and fan-out will affect these values. Switching times are most often specified by graphs showing the various times as functions of fan-out with specified driving conditions. Figure 9.12 shows a typical graph.

There will always be a certain amount of stray capacitance at the input and output terminals; thus, the careful designer will allow for the effect on switching time of these stray values.

In the next few paragraphs we will consider some of the more popular logic families.

9.2 LOGIC FAMILIES

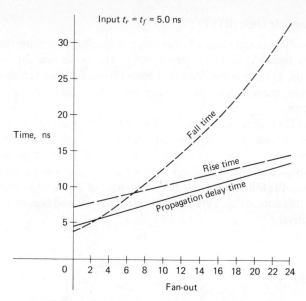

Fig. 9.12 Switching times as a function of loading.

A. Resistor-Transistor Logic (RTL)

A four-input RTL gate is shown in Fig. 9.13. For positive logic the gate functions as a NOR gate. The RTL family is constructed in relatively simple configurations and is consequently one of the cheapest lines of logic. On the other hand, this family is more inflexible than other popular families. The fan-out specification is usually small with a typical value of four or five for gates. Buffer or current amplifier stages are available to increase the number of parallel inputs that can be driven. The noise margin for RTL is low, as is the switching speed. Because of the disadvantages of RTL it finds limited application in digital system design.

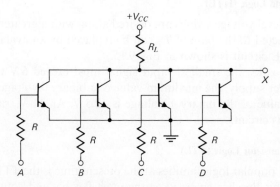

Fig. 9.13 A four-input RTL gate.

B. Diode-Transistor Logic (DTL)

The DTL family generally has a better fan-out specification than the RTL family, although this figure is still not outstanding. The noise margin is better and a variation of the DTL line can lead to a high-threshold logic family (HTL) with a very good noise margin. Switching times are of the same order of magnitude as the RTL circuit.

A typical DTL gate is depicted in Fig. 9.14. This gate functions as a NAND gate for positive logic since all three inputs must be high in order to cause T_2 to turn on. If any input is low the base of T_1 is clamped to a voltage near ground by the forward-biased diode. The drop across the emitter-base junction of T_1 plus the drop across the following diode keeps the base of T_2 negative and ensures that T_2 is off for this condition. Typical levels for binary 0 and binary 1 are 0.4 V and 2.4 V, respectively.

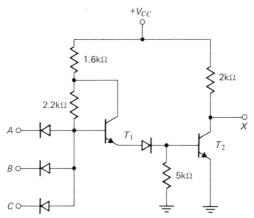

Fig. 9.14 A three-input DTL gate.

C. High-Threshold Logic (HTL)

If greater logic level voltage swings are desired along with a greater noise margin, the diode connected to the base of T_2 can be replaced by an avalanche or Zener diode. This HTL circuit is shown in Fig. 9.15.

If V_Z is six volts, the smallest input signal must exceed 6 V to turn T_2 on. For a 15-V power supply, the maximum value of a binary 0 voltage is 1.5 V while the minimum value of the binary 1 voltage is 12.5 V. An 11-V transition would be present in this circuit.

D. Transistor-Transistor Logic (TTL)

One of the most popular logic families at the present time is the TTL family. Since the late 1960s this line has emerged as the most flexible and continues to be in great demand after the mid-1970s. This family possesses good fan-out figures and

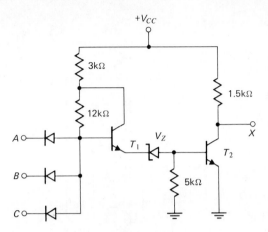

Fig. 9.15 A three-input HTL gate.

relatively high-speed switching. The Schottky-clamped TTL lowers switching times even further with propagation delay of gates in the area of two nanoseconds. The basic TTL gate is shown in Fig. 9.16.

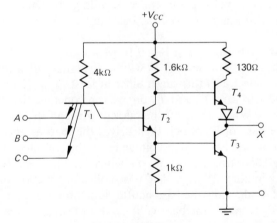

Fig. 9.16 Basic TTL gate.

The TTL family is based on the multiemitter construction of transistors made possible by integrated circuit technology. The operation of the input transistor can be visualized by the circuit of Fig. 9.17, which shows the bases of the three transistors connected in parallel as are the collectors while the emitters are separate.

If all emitters are at ground level the transistors will be saturated due to the large base drive. The collector voltage will be only a few tenths of a volt above ground. The base voltage will equal $V_{BE(on)}$ which may be 0.5 V. If one or two of the emitter voltages are raised, the corresponding transistors will shut off. The transistor with an emitter voltage of zero volts will still be saturated, however, and

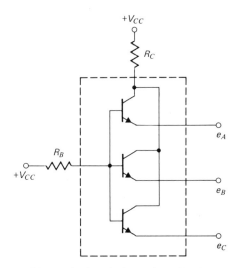

Fig. 9.17 Discrete circuit equivalent to the multiemitter transistor.

this will force the base voltage and collector voltage to remain low. If all three emitters are raised to a higher level, the base and collector voltages will tend to follow this signal.

Returning to the basic gate of Fig. 9.16 we see that when the low logic level appears at one or more of the inputs, T_1 will be saturated with a very small voltage appearing at the collector of this stage. Since at least $2V_{BE(on)}$ must appear at the base of T_2 in order to turn T_2 and T_3 on, we can conclude that these transistors are off at this time. When T_2 is off the current through the 1.6 kΩ resistance is diverted into the base of T_4 which then drives the load as an emitter follower.

When all inputs are at the high voltage level, the collector of T_1 attempts to rise to this level. This turns T_2 and T_3 on, which clamps the collector of T_1 to a voltage of approximately $2V_{BE(on)}$. The base-collector junction of T_1 appears as a forward-biased diode while the base-emitter junctions are reverse-biased diodes in this case. As T_2 turns on the base voltage of T_4 drops, decreasing the current through the load. The load current tends to decrease even faster than if only T_4 were present, due to the fact that T_3 is turning on to divert more current from the load. At the end of the transition T_4 is off with T_2 and T_3 on. For positive logic the circuit behaves as a NAND gate.

This arrangement of the output transistors is called a totem pole. As we discussed the emitter follower in Chapter 5 we found that the output impedance is asymmetrical with respect to emitter current. As the emitter follower turns on, the output impedance decreases. Turning the stage off increases the output impedance and can lead to distortion of the load voltage especially for capacitive loads. The totem pole output stage overcomes this problem as discussed in the preceding paragraph.

There are two standard methods of improving the high-speed switching characteristics of TTL. The first is to add clamping diodes to the input emitters of the gate to reduce transmission line effects by providing more symmetrical impedances. This improvement is shown in Fig. 9.18 along with smaller resistors and a Darlington connection at the output. These gates exhibit a typical propagation delay time of 6ns.

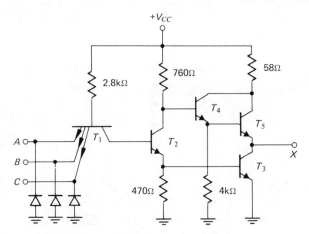

Fig. 9.18 High-speed TTL gate.

A very significant improvement in TTL switching speed results from using Schottky barrier diodes to clamp the base-collector junctions of all transistors to avoid heavy saturation. Figure 9.19 shows the arrangement of the clamping diode.

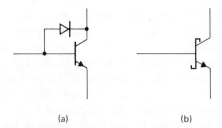

Fig. 9.19 (a) Schottky-clamped transistor. (b) Symbol for clamp.

The low forward voltage across the Schottky diode causes the diode to divert most of the excess base current around the base-collector junction. The transistor current can then decrease rapidly without the delay associated with excess base charge. The Schottky-clamped TTL gates exhibit propagation delay times of two to three nanoseconds.

A very wide choice of logic circuits is available in the TTL family, making this line the most versatile of all families.

E. Emitter-Coupled Logic (ECL)

One of the older families is the ECL family. For many years this configuration was unrivalled in high-speed switching applications, but now competes with the Schottky-clamped TTL family in the high-speed area. The good switching characteristics of this family again result from the avoidance of heavy saturation of any transistors within the gate. Figure 9.20 shows an ECL gate with two separate outputs. For positive logic X is the OR output while Y is the NOR output.

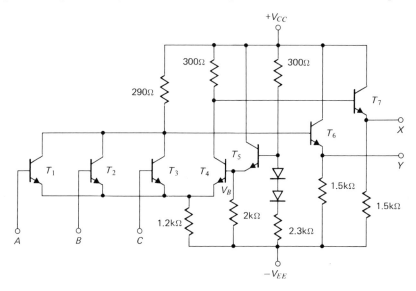

Fig. 9.20 An ECL gate.

Often the positive supply voltage is taken as zero volts and V_{EE} as -5 V. The diodes and emitter follower T_5 establish a base reference voltage for T_4. When inputs A, B, and C are less than the voltage V_B, T_4 conducts while T_1, T_2, and T_3 are cut off. If any one of the inputs is switched to the 1 level which exceeds V_B, the transistor turns on and pulls the emitter of T_4 positive enough to cut this transistor off. Under this condition output Y goes negative while X goes positive. The relatively large resistor common to the emitters of T_1, T_2, T_3, and T_4 prevents these transistors from saturating. In fact, with nominal logic levels of -1.9 V and -1.1 V, the current through the emitter resistance is approximately equal before and after switching takes place. Thus, only the current path changes as the circuit switches. This type of operation is sometimes called current mode switching. Although the output stages are emitter followers they conduct reasonable currents for both logic level outputs and, therefore, minimize the asymmetrical output impedance problem.

The ECL family has a disadvantage of requiring more input power than the TTL line. Furthermore, the great variety of logic circuits realizable with TTL cannot be duplicated with ECL.

F. Complementary-Symmetry MOS (CMOS)

Although p-channel (p-MOS) and n-channel (n-MOS) devices offer advantages in terms of packing density over the CMOS family, the latter has become very popular in recent years. Two major reasons for the growing popularity of CMOS are the extremely low power dissipation of CMOS gates and other logic circuits and the compatibility of this family with TTL logic circuits. While p-MOS and n-MOS logic families continue to grow in importance, we will here consider only the well-established CMOS family. Because switching speeds are lower than in many bipolar logic families, CMOS is used in low- to medium-frequency applications.

The inverter of Fig. 9.21 is the basic building block for the CMOS gate. Both p-channel and n-channel devices are enhancement-type MOS transistors. When the input voltage is near ground potential the n-channel device T_1 is off. The voltage from gate to source of T_2 is approximately $-V_{DD}$ and, therefore, T_2 is on. With T_2 in the high conductance state and T_1 in the low conductance state, the power supply voltage drops across T_1 and appears at the output. When e_{in} increases and approaches V_{DD}, T_2 turns off while T_1 turns on, dropping the power supply voltage across T_2 and leading to zero output voltage. The high impedance of the off transistor results in negligible current drain on the power supply. The threshold voltage can be controlled during fabrication and is generally designed so that switching occurs at an input voltage of approximately $V_{DD}/2$.

The output voltage levels of a gate are typically less than 0.01 V for the 0 state and 4.99 V to 5.00 V for the 1 state ($V_{DD} = 5$ V). The required input voltages for this case might be 0.0 V to 1.5 V for a 0 and 3.5 V to 5.0 V for a 1. The noise margins in either state are then approximately 1.5 V.

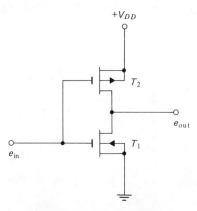

Fig. 9.21 Basic CMOS inverter.

Figure 9.22 shows a two-input NOR gate using CMOS. If both A and B are held at the 0 logic level, both n-channel devices are off while both p-channel transistors are on. The output is near V_{DD}. If input A moves to the 1 state the upper p-channel device turns off while the corresponding n-channel turns on. This leads

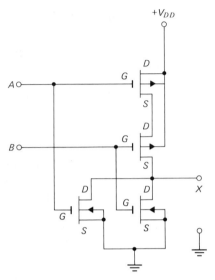

Fig. 9.22 A CMOS NOR gate for positive logic.

to an output voltage near ground. If input B is at the upper logic level while A is at the 0 level, the lower p-channel device shuts off while the corresponding n-channel device turns on again, resulting in an output 0. When both A and B equal 1, the output is obviously 0.

A positive logic NAND gate is shown in Fig. 9.23. In this case at least one of the series n-channel devices will be off and one of the parallel p-channel devices will be on if A, B, or both A and B are at the low logic level. The output will then be

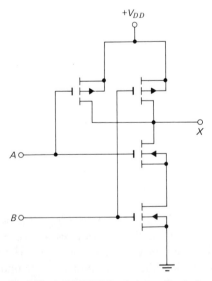

Fig. 9.23 A CMOS NAND gate for positive logic.

high. Only when both A and B are raised to the high logic level will X equal the low level. For this input combination, both p-channels are off while both n-channels will be on.

There are several other logic circuits that can be fabricated in the CMOS family [3]. Various types of flip-flops, memories, and shift registers represent only a few of the useful CMOS circuits.

9.3 MULTIVIBRATORS

The linear circuits considered in Chapters 1 through 7 and the switching circuits considered previously exhibit certain common characteristics. For example, the level of the output voltage is always determined by the level of the input signal and, when no input signal is present, there is only one state that can exist at the output. Multivibrators are switching circuits that do not share these characteristics with the above-mentioned circuits. In general, the output can appear in more than one state, the output level is not always determined by the input signal level, and in some cases an input signal is not required to cause an output.

A multivibrator can be further classified into one of three types: bistable, astable, or monostable. Over the years, other names have developed which apply to these three types of circuit. The flip-flop or regenerative trigger is another name used in connection with bistable multivibrators, although these circuits are only subclasses of the bistable. Astables are often called free-running multivibrators, and monostables are referred to as one-shot or single-shot multivibrators. The basic difference among the three types of multivibrator is the number of stable states associated with each type. As the names imply, the bistable has two stable states, the astable has no stable states, and the monostable has one stable state. The waveforms of Fig. 9.24 show typical input and output signals that might be expected from the three types.

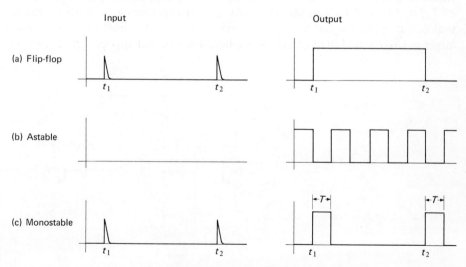

Fig. 9.24 Multivibrator waveforms.

The output of the flip-flop circuit changes from one stable state to another upon application of an input signal. It will remain in this state until excited to its original state by a second input signal. The astable requires no input, and the output switches between two quasi-stable states. The length of time spent in each state is determined by certain components of the circuit. The output of the one-shot circuit remains in a stable state until excited by the input signal. The second state is quasi-stable and the output reverts back to the original stable state after a time T determined by circuit components. In all cases the output signals are independent of the amplitudes of the input signals, assuming that these amplitudes are large enough to initiate switching.

Multivibrators make up a very important class of circuits. They find application in almost all types of waveform generation. Digital computers, digital data-transmission systems, and timing systems are examples of systems which require the use of a great number of multivibrator circuits.

We will consider both discrete-element and integrated circuit multivibrators. For conventional digital systems, the IC multivibrator is popular; however, many special applications involving variable periods require discrete-element astables or one-shots. Furthermore, a thorough understanding of the basic operation of discrete-element multivibrators will simplify the following discussion on the IC versions.

A. Monostable Multivibrators

Figure 9.25 shows a conventional one-shot multivibrator circuit. For proper values of R_B, the transistor T_2 will be saturated; R_1 and R_2 form a voltage divider between the collector voltage of T_2, which will be approximately zero, and the supply voltage $-V_{BB}$. The base of T_1 will then be negative with respect to the emitter; hence T_1 will be off. The output voltage is zero and this corresponds to the one stable state of the circuit. If an external positive trigger pulse is applied to the base of T_1, it will tend to turn this transistor on. At this point regeneration occurs; that is, the applied signal is amplified and returned to the base of T_1 as a much larger positive signal than originally applied. The original applied signal is ampli-

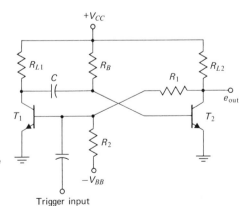

Fig. 9.25 Conventional transistor monostable multivibrator.

fied and inverted by T_1. The negative-going signal at the collector of T_1 is coupled through C to the base of T_2. Again amplification and inversion take place, resulting in a positive-going signal at the collector of T_2. This signal is coupled back to the base of T_1 to cause saturation of T_1. The circuit is now in the quasi-stable state. Base drive to keep T_1 saturated is furnished through R_{L2} and R_1 so long as T_2 remains off. However, as the capacitor charges toward a positive voltage, the base of T_2 becomes less negative, finally allowing T_2 to turn on again. Base drive to T_1 is interrupted and T_1 turns off, completing the cycle. Regeneration can only occur if the amplification around the loop is sufficient. This means that both transistors must be in their active regions for regeneration to take place. The amplitude and duration of the trigger signal must be such that this condition is met.

For all but the most critical application, the equivalent circuits of Figs. 8.2(a) and (c) can be used to represent the transistor. We will first consider an analysis based on these simple models and then extend the results to include nonideal effects. The equivalent circuit of Fig. 9.26 applies to the one-shot before the trigger pulse is applied. The capacitor has a voltage of V_{CC} across it with the polarity shown. When the positive trigger pulse is applied, transistor T_1 will saturate and T_2 will be turned off. Neglecting switching times, the equivalent circuit is as shown in Fig. 9.27 instantaneously after switching.

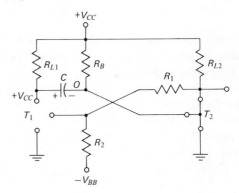

Fig. 9.26 Equivalent circuit for the stable state of one-shot.

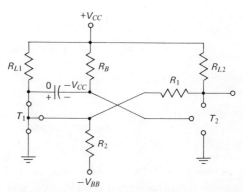

Fig. 9.27 Equivalent circuit of one-shot instantaneously after application of trigger pulse.

The charge on C cannot change instantaneously; therefore, as the collector of T_1 switches from $+V_{CC}$ to ground, the base of T_2 is forced by the capacitor to switch from ground to $-V_{CC}$. Then T_2 shuts off causing base drive to be supplied through R_{L2} and R_1 to T_1. The capacitor will charge from $-V_{CC}$ toward a voltage of $+V_{CC}$ through R_B, but when it reaches zero volts T_2 will turn on. This forces T_1 off and is the point in time at which the circuit reverts to the original state. The length of time that T_2 remains in the off state is referred to as the period of the one-shot. The period can be found by calculating the time required for the capacitor to charge from $-V_{CC}$ V to zero volts. The charging circuit for the capacitor is shown in Fig. 9.28. The voltage on the capacitor can be expressed as

$$v_c = v_i + (v_t - v_i)(1 - e^{-t/R_B C}),$$

where the initial voltage is $-V_{CC}$ and the target voltage is $+V_{CC}$. Thus

$$v_c = -V_{CC} + 2V_{CC}(1 - e^{-t/R_B C}).$$

Setting $v_c = 0$ and solving for the period T gives

$$T = R_B C \ln 2 = 0.69 R_B C. \tag{9.1}$$

The output voltage during the period T is calculated as

$$V_{C2} = \frac{R_1}{R_1 + R_{L2}} V_{CC}. \tag{9.2}$$

At the end of the period, the capacitor has zero volts appearing across it. As T_2 switches on, this effectively ties one side of the capacitor to ground. The other side charges through R_{L1} to $+V_{CC}$ V, since the collector of T_1 now presents an open circuit to the capacitor. The collector voltage of T_1 then charges to $+V_{CC}$ governed by the recovery time constant $R_{L1}C$. From three to five of these time constants are allowed for recovery time, that is, before another trigger pulse is applied to the circuit. Waveforms at various points in the circuit are shown in Fig. 9.29.

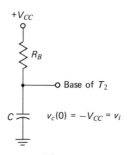

Fig. 9.28 Charging circuit of timing capacitor.

Nonideal effects. In the preceding analysis, the simple equivalent circuits of Figs. 8.2(a) and (c) were used. These circuits do not reflect several aspects of tran-

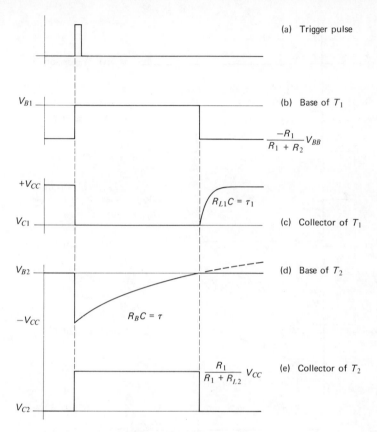

Fig. 9.29 One-shot waveforms.

sistor behavior which occasionally must be considered. The following items can cause slight modifications in the design of the one-shot circuit:

1. $V_{BE(\text{on})}$; nonzero values of base-to-emitter voltage drop when the transistor is on.
2. $V_{CE(\text{sat})}$; nonzero values of collector-to-emitter saturation voltage.
3. $r_{bb'}$; nonzero values of ohmic base resistance.
4. I_{co}; leakage current in germanium transistors.
5. Switching times of transistors.
6. Breakdown voltage of reverse-biased emitter-base diode.

Changes in $V_{BE(\text{on})}$ and I_{co} with temperature can cause the period of the one-shot to change.

The presence of $V_{BE(\text{on})}$ requires that the capacitor charge to a value of $V_{BE(\text{on})}$, rather than to zero volts, before T_2 will turn on to end the period. The finite saturation voltage affects the output voltage, since the lower level will now be

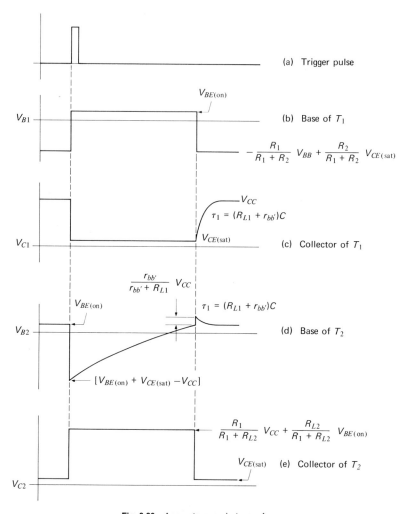

Fig. 9.30 Accurate one-shot waveforms.

$V_{CE(\text{sat})}$ rather than zero volts. Furthermore, the initial charge on the capacitor, instantaneously upon switching, will be $V_{CC} - V_{BE(\text{on})}$. The waveforms of Fig. 9.30 reflect these changes. The presence of $V_{CE(\text{sat})}$ and $V_{BE(\text{on})}$ modify both collector and base waveforms slightly. Only rarely will these quantities need to be considered, since their effects are small. The discontinuity in the voltage V_{B2} at the end of the period occurs as T_1 switches off, giving the equivalent circuit shown in Fig. 9.31. The capacitor appears as a short circuit and V_{B2} has an instantaneous value of

$$V_{CC} \frac{r_{bb'}}{R_{L1} + r_{bb'}} + V_{BE(\text{on})},$$

decaying exponentially to $V_{BE(\text{on})}$.

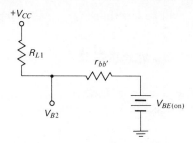

Fig. 9.31 Equivalent circuit instantaneously after T_1 shuts off.

The presence of I_{co} generally does not affect voltage levels, unless R_{L1} and R_{L2} are quite large. However, I_{co} can modify the period of the one-shot as shown next.

The equivalent circuit of the capacitor while charging is shown in Fig. 9.32. The circuit from collector to base of T_2 can be represented by a current source equal to leakage current I_{co} when T_2 is turned off. The source I_{co} and R_B can be converted to a voltage source as shown. The target voltage is now $V_{CC} + R_B I_{co}$, rather than V_{CC}. The period T is calculated by finding the time necessary to charge C to a value of $V_{BE(on)}$. The governing equation is

$$V_{BE(on)} = v_i + (v_t - v_i)(1 - e^{-T/R_B C})$$
$$= -[V_{CC} - V_{BE(on)} - V_{CE(sat)}]$$
$$+ (V_{CC} + R_B I_{co} + V_{CC} - V_{BE(on)} - V_{CE(sat)})(1 - e^{-T/R_B C}).$$

Solving for T gives

$$T = R_B C \ln \frac{2V_{CC} + R_B I_{co} - V_{BE(on)} - V_{CE(sat)}}{V_{CC} + R_B I_{co} - V_{BE(on)}}. \tag{9.3}$$

Equation (9.3) reduces to

$$T = R_B C \ln 2 = 0.69 R_B C$$

when leakage current, $V_{BE(on)}$, and $V_{CE(sat)}$ are neglected. For silicon transistors I_{co} can be neglected, but changes in $V_{BE(on)}$ and $V_{CE(sat)}$ with temperature can still

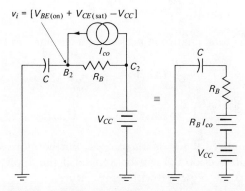

Fig. 9.32 Equivalent charging circuit including I_{co} effects.

affect T. Usually, $V_{CE(sat)}$ will be less than $V_{BE(on)}$, allowing Eq. (9.3) to be written as

$$T = R_B C \ln \frac{2V_{CC} - V_{BE(on)}}{V_{CC} - V_{BE(on)}} \tag{9.4}$$

for silicon transistors. The change in T when $V_{BE(on)}$ changes is found by differentiating the period with respect to $V_{BE(on)}$. This gives

$$\frac{dT}{dV_{BE(on)}} = R_B C \frac{V_{CC}}{(2V_{CC} - V_{BE(on)})(V_{CC} - V_{BE(on)})}$$

$$\approx \frac{R_B C}{2V_{CC}}.$$

We can relate the change in period to the change in temperature by noting that $V_{BE(on)}$ decreases by 2 mV/°C for silicon. If we use the symbol K for temperature (to avoid confusing temperature with period T), then

$$\frac{dT}{dK} = \frac{dT}{dV_{BE(on)}} \frac{dV_{BE(on)}}{dK} = -0.001 \frac{R_B C}{V_{CC}}.$$

Since dT/dK is constant with temperature K, the total change in period for a given change in temperature is

$$T_2 - T_1 = \Delta T = -0.001 \frac{R_B C}{V_{CC}} [K_2 - K_1]. \tag{9.5}$$

From Eq. (9.5) the period is seen to decrease as the temperature increases. This effect is minimized for larger values of V_{CC}. The percentage change in period is

$$\frac{\Delta T}{T_1} \times 100 = \frac{-0.1[K_2 - K_1]}{V_{CC} \ln 2} = \frac{-0.145[K_2 - K_1]}{V_{CC}}.$$

For germanium transistors a change in period occurs due to both $V_{BE(on)}$ changes and I_{co} changes. We find $\partial T/\partial I_{co}$ to be

$$\frac{\partial T}{\partial I_{co}} = -\frac{R_B}{2V_{CC}} R_B C$$

for $V_{CC} \gg I_{co} R_B$. The total change in period for a germanium transistor is

$$T_2 - T_1 = \Delta T = \frac{\partial T}{\partial I_{co}} (\Delta I_{co}) + \frac{\partial T}{\partial V_{BE(on)}} (\Delta V_{BE(on)})$$

$$= \frac{R_B C}{2V_{CC}} [-R_B(I_{co2} - I_{co1}) + (V_{BE(on)2} - V_{BE(on)1})]. \tag{9.6}$$

Both terms are minimized if V_{CC} is maximized. Furthermore, the compensation techniques studied in Section 3.8 can be applied if the application calls for an extremely small variation of period with temperature.

Use of external emitter resistance. The source V_{BB} can be replaced by a common-emitter resistor which is usually bypassed by a capacitor. The resulting circuit is shown in Fig. 9.33. When T_2 is on, the voltage at the output is (neglecting $V_{CE(sat)}$ and $V_{BE(on)}$)

$$V_{min} = \frac{R_E}{R_E + R_{L2}} V_{CC}. \tag{9.7}$$

The output voltage during T is (assuming that $R_{L1} = R_{L2}$)

$$\begin{aligned} V_{max} &= \frac{R_E}{R_E + R_{L2}} V_{CC} + \frac{R_1}{R_1 + R_{L2}} V_{CC}\left(1 - \frac{R_E}{R_E + R_{L2}}\right) \\ &= \frac{V_{CC}}{R_E + R_{L2}}\left[R_E + \frac{R_1 R_{L2}}{R_1 + R_{L2}}\right]. \end{aligned} \tag{9.8}$$

The reverse bias on T_1 when T_2 is on is

$$V_{BE(off)} = \frac{V_{CC} R_E}{R_E + R_{L2}}\left(1 - \frac{R_2}{R_1 + R_2}\right). \tag{9.9}$$

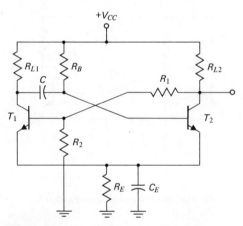

Fig. 9.33 One-shot multivibrator requiring single source.

Design procedure. There are many degrees of freedom in designing a one-shot, but the following design procedure can be used in most cases.

1. Select V_{CC} and R_L.
2. Select R_E.
3. Select R_B and C.
4. Select R_1 and R_2.

The selection of V_{CC} and R_L is based on the required output voltage swing and output impedance requirements; R_E is determined by the output voltage swing and the required reverse bias on T_1. Then R_B is selected to give T_2 the necessary base current, and C is selected to give the proper period T. It can easily be shown that Eq. (9.1) applies to this circuit also. Finally, R_1 and R_2 are selected to give T_1 the proper reverse bias when T_1 is off and to establish the necessary base drive when T_1 is on. A design example will demonstrate this procedure.

Example 9.1. Design a one-shot multivibrator with a period of 1 ms and an output impedance of 5 kΩ or less. The voltage swing should be at least 9 V. Assume that $\beta_{0(min)} = 50$, $I_{co} = 0$, $V_{BE(on)} = 0$, and $V_{CE(sat)} = 0$.

Solution.
1. Select V_{CC} and R_L.
   To ensure a 9-V swing, V_{CC} will be chosen as 12V. A value of $R_L = 5$ kΩ will result in an output impedance of approximately 5 kΩ. The circuit configuration is shown in Fig. 9.34.

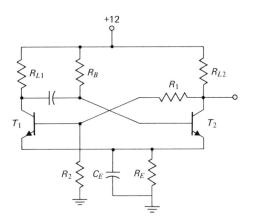

Fig. 9.34 The one-shot circuit of Example 9.1.

2. Select R_E.
   If 2 V are dropped across R_E, a sufficient reverse bias on T_1 can be obtained, and a 9-V transition of the output voltage can be obtained. Note that if the voltage drop across R_E is too large, for example 4 V, the output swing could not be 9 V. The voltage appearing at the emitter is

$$V_E = \frac{R_E}{R_E + R_L} V_{CC} = \frac{R_E}{R_E + 5K} V_{CC},$$

so

$$R_E = 1 \text{ k}\Omega.$$

3. Select R_B and C.

Since R_B must cause saturation of T_2, we will pick the current through R_B to be $1.5 I_{B(sat)}$, giving a slight safety factor. The collector current for saturation is

$$I_{C(sat)} = \frac{V_{CC}}{R_E + R_L} = \frac{12}{6 \text{ k}\Omega} = 2 \text{ mA};$$

therefore

$$I_{B(sat)} = \frac{I_{C(sat)}}{\beta_{0(min)}} = \frac{2 \text{ mA}}{50} = 40 \text{ }\mu\text{A}.$$

We will allow 60 μA to flow through R_B. Since

$$I_B = \frac{V_{CC} - V_B}{R_B} = \frac{V_{CC} - V_E}{R_B},$$

then

$$R_B = \frac{12 - 2}{60 \text{ }\mu\text{A}} = \frac{10 \text{ V}}{60 \text{ }\mu\text{A}} = 167 \text{ k}\Omega.$$

The period T must be 1 ms, so

$$C = \frac{T}{0.69 R_B} = \frac{10^{-3}}{0.69 \times 167 \times 10^3} = 0.0087 \text{ }\mu\text{F}.$$

4. Select R_1 and R_2.

We will assume that a reverse bias of 1 V is required on T_1. When T_1 is off, the emitter voltage is 2 V; therefore R_1 and R_2 must be selected to set the base of T_1 at 1 V. Since the collector of T_2 is at 2 V, the resistors R_1 and R_2 form a voltage divider, as shown in Fig. 9.35.

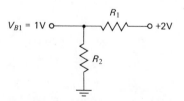

Fig. 9.35 Bias circuit of T_1 when T_2 is on.

It is easy to see that $R_1 = R_2$ is required to achieve the assumed reverse bias. This can be verified by Eq. (9.9). The other requirement on R_1 and R_2 is that this network furnish the required base drive to saturate T_1 during the one-shot period. Assuming that a base current of 60 μA is required (same as I_{B2}), we can write

$$I_{B1} = I_1 - I_2 = 60 \text{ }\mu\text{A},$$

where I_1 is the current through R_1, and I_2 is the current through R_2. By inspection

$$I_1 = \frac{V_{CC} - V_{B1}}{R_{L2} + R_1} = \frac{V_{CC} - V_E}{R_{L2} + R_1} = \frac{10}{5 + R_1}$$

and

$$I_2 = \frac{V_{B1}}{R_2} = \frac{V_E}{R_2} = \frac{2}{R_2}.$$

Since $R_1 = R_2$, then

$$I_{B1} = \frac{10}{5 + R_1} - \frac{2}{R_1} = 60\ \mu\text{A}$$

gives

$$R_1 = 125\ \text{k}\Omega = R_2.$$

In practice, the requirements on I_{B1} and the reverse bias on T_1 are not so stringent and a trial-and-error approach is often most efficient in selecting R_1 and R_2. Before we finish the design we must check to see if the requirement of a 9-V swing will be met. The lower level of the output voltage has been found to be

$$V_{min} = 2\ \text{V}.$$

The upper level from Eq. (9.8) is

$$V_{max} = \frac{12}{6}\left[1 + \frac{125 \times 5}{125 + 5}\right] = 2[1 + 4.8] = 11.6\ \text{V}.$$

That is, when T_2 is shut off, the collector voltage swings to 11.6 V. The total net swing is

$$V_{swing} = V_{max} - V_{min} = 11.6\ \text{V} - 2\ \text{V} = 9.6\ \text{V}.$$

The list of values is as follows:

$$V_{CC} = 12\ \text{V},$$
$$R_{L1} = R_{L2} = 5\ \text{k}\Omega,$$
$$R_E = 1\ \text{k}\Omega,$$
$$R_B = 167\ \text{k}\Omega,$$
$$C = 0.0087\ \mu\text{F},$$
$$R_1 = 125\ \text{k}\Omega,$$
$$R_2 = 125\ \text{k}\Omega.$$

In general, R_{L1} and R_{L2} are chosen to be equal. This guarantees that V_E will have the same value, regardless of the state of the circuit. The capacitor C_E bypasses R_E to give the circuit a higher loop gain, thereby reducing the switching times of the transistors. In many applications C_E will not be necessary. Reasonable values of C_E range from 1 to 100 μF, although for shorter periods the value can be much smaller.

Some logic families provide monostable multivibrators along with the other logic elements of the families. These multivibrators often require an external timing capacitor and resistor. An equation for the period in terms of the external elements will be given by the manufacturer. For example, the TI SN74121 [5] has a period that is typically 30 to 35 ns without external elements. This period can be varied from 40 ns to 28 seconds by choosing the correct combination of timing components. Capacitance can be varied from 10 pF to 1000 μF while the timing resistance can take on values from 1.4 kΩ to 40 kΩ. The period in this case is defined by the same equation that applies to the discrete element case, namely,

$$T = RC \ln 2 = 0.69 \, RC.$$

The input and output signals for this IC one-shot are TTL-compatible and would obviously be useful in TTL logic system design.

There are two types of one-shot that find heavy application in the digital area. One is called a retriggerable monostable and the other is the nonretriggerable type. In the nonretriggerable circuit a second trigger pulse, occurring during the period, is ignored by the one-shot. No effect on the period is noticeable and the circuit cannot be triggered again until the period and recovery time are completed. The discrete circuit monostable discussed earlier in this section is nonretriggerable.

A retriggerable monostable references the conclusion of the period to the input trigger pulse. For example, if the circuit is triggered to start the period and a second trigger pulse is applied at $t = 0.6T$, the output of the one-shot will last $1.6T$ rather than T. When a trigger pulse occurs the period will extend a full period from that point even if an earlier period has not been completed.

B. Astable Multivibrators

Figure 9.36 shows a conventional astable multivibrator. It can be seen that both T_1 and T_2 can be saturated simultaneously with the base drive furnished by R_1 and R_2. Of course, this is not the mode of operation desired for the astable. The normal mode of operation is present when only one transistor is on at a given time. When this transistor switches off, the other transistor is forced on. We will see later that

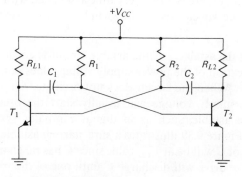

Fig. 9.36 Conventional astable multivibrator.

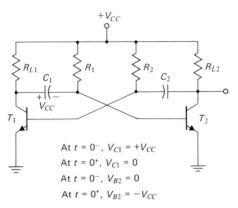

Fig. 9.37 Instantaneous voltage on C_1 as T_1 turns on.

special circuits can guarantee the existence of this mode, but for the present time let us assume that T_1 is off and T_2 is on. Let us further assume that at time $t = 0$, T_1 switches on. It is at this point that we will start the analysis. Prior to the turning on of T_1, the capacitor C_1 has a voltage of V_{CC} across it, as shown in Fig. 9.37. Just as in the case of the one-shot, the transition of V_{C1} from V_{CC} to zero volts causes the base voltage of T_2 to swing from zero to $-V_{CC}$ to preserve the charge on C_1. When T_2 is turned off, V_{C2} rises to $+V_{CC}$ V; T_2 remains off until C_1 charges to zero volts. This time is

$$t_1 = 0.69 R_1 C_1.$$

At the end of t_1, T_2 turns on, dropping V_{C2} from $+V_{CC}$ to zero volts. This forces T_1 off until C_2 can charge from $-V_{CC}$ to zero volts. Transistor T_1 will be off for a time

$$t_2 = 0.69 R_2 C_2.$$

At the end of time t_2, T_1 again turns on and starts the cycle once more. The cycle is seen to continue indefinitely once initiated. The total period of the astable is

$$T = t_1 + t_2 = 0.69(R_1 C_1 + R_2 C_2). \tag{9.10}$$

It should be noted that there is a recovery time associated with both collectors for the astable circuit. The waveforms are shown in Fig. 9.38 with exaggerated recovery times.

Methods of starting the astable. If the astable circuit is constructed as shown in Fig. 9.36, the application of the power supply voltage does not guarantee that the circuit will oscillate. When the circuit does not oscillate, removal and reapplication of the power supply voltage will usually start the oscillation. There are applications of the astable, such as in digital computer systems, where sure-starting is critical. Figure 9.39 illustrates a sure-starting astable. When V_{CC} is first applied, both sides of C will be at V_{CC} volts, since C has no charge. If both T_1 and T_2 turn on, the base drive will discharge C until one of the transistors turns off. So long as one transistor is off, the charge on C will be restored through one of the

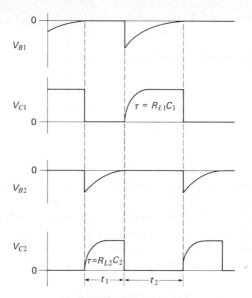

Fig. 9.38 Astable waveforms.

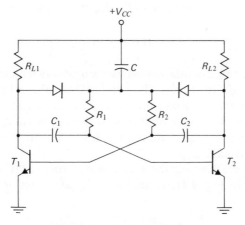

Fig. 9.39 Sure-starting multivibrator.

diodes. Basically, the addition of C automatically causes removal of base drive if the transistors turn on simultaneously.

Design procedure. The procedure for designing an astable multivibrator is simpler than that for a one-shot. The steps are outlined below:

1. Choose V_{CC} and R_L.
2. Pick R_1 and R_2.
3. Pick C_1 and C_2.

The extension to the nonideal case, considering $V_{CE(sat)}$, $V_{BE(on)}$, $r_{bb'}$, and I_{co}, should be obvious after examining the one-shot case.

Modifications of the astable multivibrator. When the astable changes state, a reverse-bias voltage of $-V_{CC}$ appears at the base of the transistor that turns off. In some instances, the reverse breakdown voltage of the emitter-base junction is less than $-V_{CC}$. While the breakdown is nondestructive if the circuit limits the base current to a reasonable level, the current flow through this junction can very drastically affect the period of the astable (or one-shot). A simple method of overcoming this difficulty is to add a diode in series with the emitter. The reverse-bias voltage now appears across both the diode and the emitter-base junction. The total breakdown voltage is equal to the sum of the breakdown voltage of the diode and that of the junction.

There are applications of the astable or one-shot wherein it is desirable to vary the period in accordance with an input voltage. A popular approach to this problem is shown in Fig. 9.40. By connecting R_1 and R_2 to a variable voltage, the target voltage for the capacitors can be controlled. The total period becomes

$$T = (R_1 C_1 + R_2 C_2) \ln \left(1 + \frac{V_{CC}}{v_t}\right). \tag{9.11}$$

Since $\ln(1 + x) \approx x$ for small values of x, Eq. (9.11) can be written as

$$T = (R_1 C_1 + R_2 C_2) \frac{V_{CC}}{v_t}, \tag{9.12}$$

when $v_t \gg V_{CC}$. For this case, the period varies inversely with the control voltage. A circuit whose period varies directly with control voltage has been developed, but is considered later. The lower limit on v_t for the circuit of Fig. 9.40 is that voltage which will cause enough base current through R_1 and R_2 to saturate the transistors. For values of v_t that are not much larger than V_{CC}, Eq. (9.11) can be applied to find the change in period.

One method of synthesizing the astable waveform with IC monostable multivibrators is shown in Fig. 9.41. When the start gate signal is low, the circuit is

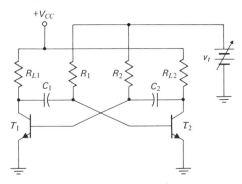

Fig. 9.40 Variable-period astable.

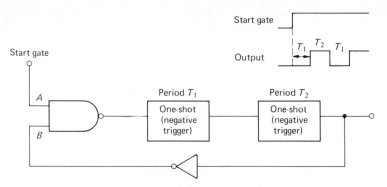

Fig. 9.41 Generation of repetitive waveform with one-shots.

inactive. Gate input B will be high under this condition. When input A swings positive to open the start gate, the NAND gate output goes negative initiating the period T_1. At the end of T_1, the negative transition occurring at the input of the second one-shot initiates the period T_2. During T_2 the NAND gate is closed and has a high level output. At the end of T_2 the NAND gate output swings negative, initiating the period T_1 to continue the repetitive process until the start gate signal drops to the low level.

C. Bistable Multivibrators

While there are various forms of the bistable multivibrator, the most popular is the flip-flop which we will consider in this section. Although discrete-element astable and one-shot circuits continue to find much practical application, it is a rare occasion when a discrete-element flip-flop is used. There is such a broad range of monolithic flip-flop types that an IC version can be found for almost any application. We will, however, start our discussion of bistable multivibrators by considering the discrete-element circuit. This will provide a better understanding of the fundamental behavior of flip-flops and will aid the subsequent study of gating principles and IC flip-flops.

There are several types of flip-flop circuits used in logic design which will be separately considered.

The RS flip-flop. The simplest form of the bistable circuit has two input lines: one for setting the flip-flop to the 1 state and another for resetting the circuit to the 0 state. An RS flip-flop is indicated in Fig. 9.42 with Q as the output variable. The S input is the set line and the R input is the reset line. The capacitors are not necessary for low-speed operation, but serve as speed-up capacitors in high-speed switching applications.

The proper value of C_S along with other element values can cause the regeneration times of the transistors to equal the recovery times of the collectors [2]. The collector recovery times can then be related to the upper limit of the pulse repetition rate or the maximum frequency at which the bistable can be triggered

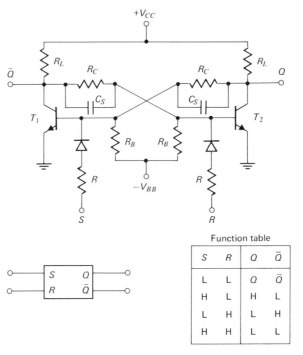

Fig. 9.42 Discrete-element *RS* flip-flop.

without failing to respond. If three time constants are allowed for completion of both regeneration time and collector recovery time, the maximum pulse repetition frequency is

$$\text{PRF}_{\max} = \frac{1}{6R_L C_S}. \tag{9.13}$$

Since the speed-up capacitor is given by (assuming that $D = 1$) $C_S = 1/R_C \omega_\beta$, Eq. (9.13) becomes

$$\text{PRF}_{\max} = \frac{\omega_\beta R_C}{6R_L}.$$

Neglecting current through R_B, the maximum value of R_C that will bring the transistor to the edge of saturation is

$$R_{C(\max)} = \beta R_L.$$

Therefore,

$$\text{PRF}_{\max} = \frac{\omega_t}{6}. \tag{9.14}$$

This expression neglects delay times and the *D*-factor, but points out that to obtain a maximum PRF, transistors with a large ω_t should be used.

In designing the RS flip-flop one must simply ensure that when a given transistor is on, the other transistor is forced to be off. Furthermore, the off transistor must supply enough base drive to keep the on transistor saturated after the input signal returns to the low level. The circuit must, of course, function at higher temperatures also.

To eliminate the necessity of the bias source an emitter resistance with capacitor bypass can be used as in the case of the one-shot circuit. With this configuration, as shown in Fig. 9.43, the following design procedure can be used [1].

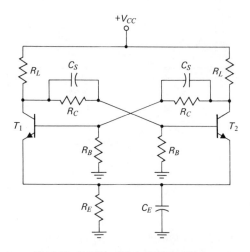

Fig. 9.43 Bistable with single dc supply.

1. Select V_{CC} and R_L from output-voltage swing and impedance requirements.
2. Choose R_E to set the emitters sufficiently positive to provide proper bias for the off condition. Often a value of R_E equal to 10 percent of R_L will satisfy the bias condition.
3. The coupling networks, R_B and R_C, are chosen to meet the on and off bias requirements of the transistors.
   a) The ratio of the voltage divider $R_B/(R_B + R_C)$ is chosen to ensure that $V_{BE(off)}$ is sufficiently negative.
   b) The combination is chosen to provide sufficient base current to the on transistor.
4. After assumed values have been checked and adjusted as necessary, the speed-up capacitors may be calculated.

An example will demonstrate the use of this procedure.

Example 9.2. Design a flip-flop employing the configuration of Fig. 9.43 to meet the following specifications. The transistors to be used have values of $\beta_{0(min)} = 50$, $f_t = 8$ MHz, and $C_{b'c}(AV) = 20$ pF.

1. The collector swing should be 10 V.
2. The output impedance should be 2 kΩ or less.
3. The on transistor is to be driven by $I_B \approx 1.5 I_{B(sat)}$.

Solution. The circuit of Fig. 9.44 can be used for the design calculations.

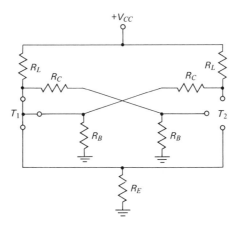

Fig. 9.44 Equivalent circuit for Example 9.2.

1. Select V_{CC} and R_L.
   To obtain the proper output impedance, R_L is chosen to be 2 kΩ. Since a 10-V swing is required, and approximately 10 percent of this drop will be absorbed by R_E, V_{CC} is chosen to be +12 V.
2. Select R_E.
   We choose R_E to be 10 percent of R_L, or $R_E = 200$ Ω. This will set the emitter voltage at $V_E = 1.09$ V.
3. Select R_C and R_B.
   The base current required for saturation is

$$I_{B(sat)} = \frac{V_{CC}}{R_E + R_L} \frac{1}{\beta_{0(min)}} = 109 \ \mu A.$$

Then $I_{B(on)}$ should be approximately $1.5 \times 109 = 165 \ \mu A$. Choose $R_B = 80$ kΩ and $R_C = 80$ kΩ. The base drive will be

$$I_B = I_1 - I_2 = \frac{(V_{CC} - V_E)}{R_L + R_C} - \frac{V_E}{R_B} = \frac{10.9}{82} - \frac{1.09}{80}$$
$$= 133 \ \mu A - 14 \ \mu A = 119 \ \mu A.$$

This value of base drive is not sufficient, so R_C and R_B must be chosen again. Pick $R_C = 60$ kΩ and $R_B = 80$ kΩ. This gives

$$I_{B(on)} = \frac{10.9}{62} - \frac{1.09}{80} = 162 \ \mu A,$$

which is sufficient. The off condition must now be checked. If we assume that a 0.2-V reverse bias is the minimum allowed bias, we can check the off condition. The equation for $V_{BE(\text{off})}$ is

$$V_{BE(\text{off})} = V_{B(\text{off})} - V_E.$$

Since

$$V_{B(\text{off})} = V_E \frac{R_B}{R_B + R_C},$$

this equation is

$$V_{BE(\text{off})} = V_E \left(\frac{R_B}{R_B + R_C} - 1 \right)$$

$$V_{BE(\text{off})} = 1.09 \left(\frac{80}{140} - 1 \right) = -0.46 \text{ V}.$$

The reverse-bias voltage satisfies our assumption.

4. The value of the speed-up capacitor is calculated from Eq. (7.25) as

$$C_S = \frac{D}{R_C \omega_\beta} = \frac{3 \times 50}{6 \times 10^4 \times 6.28 \times 8 \times 10^6} = 50 \text{ pF}.$$

This capacitor could be chosen somewhat larger to account for excess stored charge when the transistor is saturated, and charge on the delay capacitors when the transistor is off.

It is sometimes appropriate to change states in the RS flip-flop by turning one of the transistors off rather than turning one on. In this case the circuit might appear as indicated in Fig. 9.45 and is called an $\bar{R}\,\bar{S}$ flip-flop.

The RS or $\bar{R}\,\bar{S}$ flip-flop is often called a latch. These simple bistables can be constructed from two inverting gates. Figure 9.46 shows an $\bar{R}\,\bar{S}$ flip-flop using

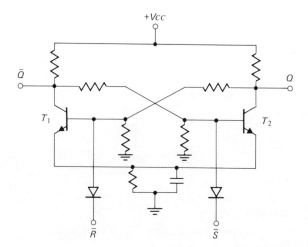

Function table

$\bar{S}$	$\bar{R}$	Q	$\bar{Q}$
L	L	H	H
H	L	L	H
L	H	H	L
H	H	Q	$\bar{Q}$

Fig. 9.45 An $\bar{R}\bar{S}$ flip-flop.

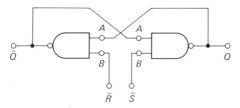

Fig. 9.46 An $\overline{RS}$ latch using NAND gates.

two NAND gates. Integrated circuit versions may include from two to four latches on a 14- to 16-pin chip.

The T flip-flop. A multivibrator that changes state or toggles with each successive input is called a T flip-flop. Figure 9.47 shows an example of a discrete-element T flip-flop. The two networks comprised of R_g, C_T, and D are called diode-steering networks and ensure reliable triggering. These gating circuits added to the basic RS flip-flop result in the T flip-flop. When the circuit is in a relaxed condition, one transistor will be on and one will be off. Let us assume that T_1 is off and T_2 is on. The collector potential of the off transistor will be near V_{CC} volts, while the on transistor will have a collector voltage level near zero volts. The collector voltages of the transistors will charge the trigger capacitors to these same levels through the gating resistors R_g. Thus, the diode associated with the off transistor will be reverse-biased while the diode connected to the on transistor will have approximately zero bias. When a negative transition appears at the input (called the clock input), the diode associated with T_1 will not pass this transition, while the other diode allows the base of T_2 to be pulled to a negative voltage. The transistor T_2 is forced off which will turn T_1 on, changing the bistable to the alternate state. After the trigger capacitors have been charged, another negative

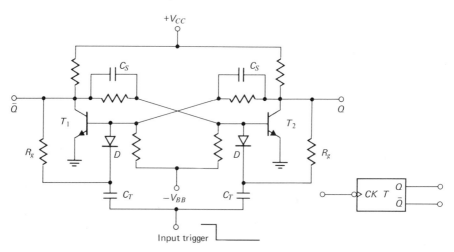

Fig. 9.47 Discrete-element T flip-flop.

transition can be applied, but this time will turn T_1 off and T_2 on. Each negative input transition is steered by the gating networks to the on transistor, guaranteeing a change of stage. Positive transitions of the input waveform are ignored by this circuit as both diodes block transmission to the transistor bases. The small circle shown on the CK input of the symbol reflects the fact that a negative transition is required to change states. The trigger capacitors must be quite small if high pulse repetition rates are required. A lower limit on C_T is imposed by the charge requirements of the transistor. Application of the trigger pulse should remove all base charge and stored charge to shut the transistor off. A minimum value for C_T is

$$C_{T(\min)} = \frac{V_{cc}}{V_t} C_S, \tag{9.15}$$

where V_t is the amplitude of the trigger pulse.

The trigger recovery time is that duration required for C_T to charge to the power supply voltage after switching. This recovery time imposes a limit on R_g in terms of the pulse repetition rate of the circuit. Diodes in parallel with the gating resistors can be used to reduce the trigger recovery time [2].

The T flip-flop can be used as a frequency divider for rectangular waveforms. Figure 9.48 shows the input and output signals when the T flip-flop is used as a divider. This circuit is very useful in counting applications.

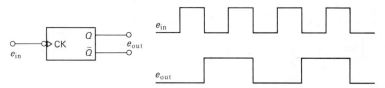

Fig. 9.48 A T flip-flop used as a frequency divider.

Addition of diodes to each base of the circuit of Fig. 9.47 for set and reset lines combines the features of a T flip-flop and an RS flip-flop. It comes as no surprise that this circuit, shown in Fig. 9.49, is called an RST flip-flop.

The JK flip-flop. This bistable circuit differs from the other flip-flop types in that there are two gating inputs along with the toggle or clock input. The level of the gate inputs determines to which state the clock input will shift the flip-flop. Figure 9.50 shows a discrete-element realization of this circuit. For the case of positive logic, if J and K are both at the 1 level, the clock pulse will be ignored. If J and K are at the 0 level, the circuit will change state or complement on each negative clock transition. Some JK flip-flops are designed so that both gates must be set to the 1 level in order to complement the circuit.

If $J = 1$ and $K = 0$ prior to the negative transition of the clock or trigger pulse, the flip-flop will switch to the $Q = 1$ state or remain in this state if Q were equal to 1 prior to the trigger pulse. The state $Q = 0$ can be guaranteed by applying gating levels of $J = 0$ and $K = 1$ followed by the trigger input. The gates must

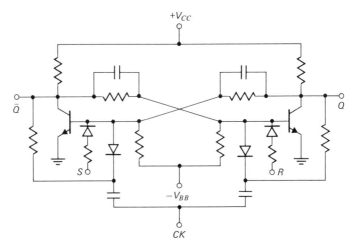

Fig. 9.49 An *RST* flip-flop.

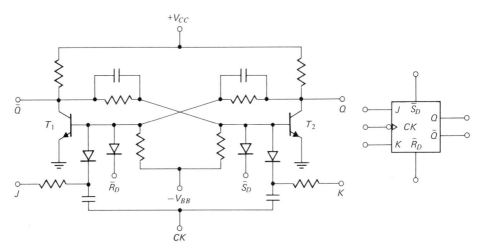

Fig. 9.50 *JK* flip-flop with direct set and reset.

be set for some reasonable time before the trigger pulse is applied if reliable operation is to be achieved.

It is often convenient to have direct set and reset lines to set information into the flip-flop without applying a trigger pulse. If $\bar{S}_D = 0$ and $\bar{R}_D = 1$, then $Q = 1$; if $\bar{S}_D = 1$ and $\bar{R}_D = 0$, then $Q = 0$. We will later see the reasons for having both direct and gated inputs when registers are considered.

Although a typical integrated JK flip-flop may have 20 transistors rather than two, the operating characteristics are the same as those of the flip-flop of Fig. 9.50. Trigger capacitors are not used in several of the IC logic families, but the circuits behave in the manner described with the transition of the clock signal leading to the change of state.

This particular type of circuit is called an edge-triggered JK flip-flop because the change of state occurs as a result of the negative clock transition. Another popular JK flip-flop is the master-slave version requiring a two-phase clock to set information into the circuit. In this type of circuit the master stage is an intermediate storage element that accepts the input signal when the first clock makes a transition. The stored information is transferred to the slave stage when the second clock signal changes.

A function table for the edge-triggered JK flip-flop of Fig. 9.50 is shown in Fig. 9.51. When both gates are at the low level (and $\bar{S}_D$ and $\bar{R}_D$ are high), a clock transition will be passed to both transistor bases. However, the speed-up capacitors serve as "memory" capacitors in this case, ensuring that the circuit will toggle or change to the complementary state [1].

Inputs					Outputs	
$\bar{S}_D$	$\bar{R}_D$	Clock	J	K	Q	$\bar{Q}$
L	H	x	x	x	H	L
H	L	x	x	x	L	H
L*	L*	x	x	x	H	H
H	H	⌐	L	L	$\bar{Q}_o$	Q_o
H	H	⌐	H	L	H	L
H	H	⌐	L	H	L	H
H	H	⌐	H	H	Q_o	$\bar{Q}_o$

Fig. 9.51 Function table for an edge-triggered JK flip-flop. Q_o is defined as the level of Q before application of the negative clock transition. (*This output condition will not persist when $\bar{S}_D$ and $\bar{R}_D$ are returned to the high state.)

The D flip-flop is closely related to the JK and has become more prominent since the development of the integrated circuit. As the complexity of a monolithic circuit increases or as more circuits are fabricated on a single chip, the limiting factor on flexibility is the number of pins available for external connections. If the package allows 16 connections, then the circuit must be fabricated in such a way that no more than 16 connections are required. The D flip-flop eliminates the K gate input by including an inverter on the chip from the J input to the K input. This forces K to always equal $\bar{J}$. Thus, if $J = 1$, a negative clock transition will result in $Q = 1$. If $J = 0$, then $Q = 0$ after the clock transition. The elimination of this input is critical in some cases, allowing an extra flip-flop to be fabricated on the chip.

REFERENCES AND SUGGESTED READING

1. D. J. Comer, *Introduction to Semiconductor Circuit Design*. Reading, Mass.: Addison-Wesley, 1968, Chapter 9.
2. D. T. Comer, *Large Signal Transistor Circuits*. Englewood Cliffs, N. J.: Prentice-Hall, 1967, Chapter 6.

3. E. R. Hnatek, *A User's Handbook of Integrated Circuits*. New York: Wiley, 1973, Chapters 6–8.
4. T. Kohonen, *Digital Circuits and Devices*. Englewood Cliffs, N. J.: Prentice-Hall, 1972, Chapter 7.
5. Texas Instruments Inc., *The TTL Data Book for Design Engineers*. Dallas: Texas Instruments Incorporated, 1973.

PROBLEMS

Section 9.1

9.1 Assume the OR gate inputs of Fig. 9.1 are driven by sources that swing from 0.5 V to 3.0 V. The source impedance is 300 Ω. If $R = 2$ kΩ calculate the output voltage for positive logic when (a) $A = B = 0$; (b) $A = 1, B = 0$; (c) $A = B = 1$. Assume the forward diode drop is 0.6 V.

* **9.2** Repeat Problem 9.1 for negative logic. Construct a truth table for the output and identify the function of the gate for this case.

9.3 Repeat Problem 9.1 for the AND gate of Fig. 9.3 if $V_{CC} = 10$ V and $R = 2$ kΩ.

9.4 Repeat Problem 9.1 for the AND gate of Fig. 9.3 using negative logic. Construct a truth table for the output and identify the function of the gate for this case.

* **9.5** To demonstrate the loading problems of the simple OR gate with no amplification calculate the output voltages Y and Z when $A = B = 1$. Assume $e_{in} = 3.0$ V for binary 1, and ideal diodes. (See the figure for Problem 9.5)

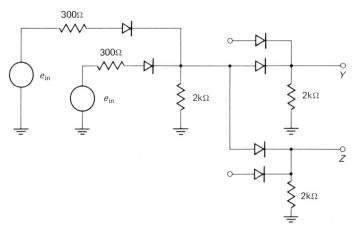

Problem 9.5

9.6 Using inverters and a three-input NAND gate, show how to construct a three-input OR gate.

Section 9.2

* **9.7** If $V_{CC} = 5$ V in Fig. 9.16 and all three inputs are tied together, calculate the input voltage at which the output voltage begins to drop. Assume $V_{BE(on)} = 0.6$ V for all transistors and state any other necessary assumptions.

9.8 Calculate the voltage appearing at the emitters of T_1, T_2, T_3, and T_4 in Fig. 9.20 when $A = B = C = 0$ (positive logic). Assume $V_{CC} = 0$ and $V_{EE} = 5.0$.

Section 9.3A

* **9.9** The one-shot multivibrator of Fig. 9.25 uses matched transistors with $\beta_0 = 60$, $V_{CE(sat)} = 0.2$ V, and $V_{BE(on)} = 0.5$ V. The following element values are employed: $V_{CC} = +12$ V, $V_{BB} = 6$ V, $R_{L1} = R_{L2} = 3$ kΩ and $R_B = 150$ kΩ. Choose C, R_1, and R_2 to cause a 1-ms period. The output voltage swing should be at least 11 V and T_1 should be reverse-biased by 2 to 4 V when T_2 is on.

9.10 Derive Eq. (9.9).

9.11 Prove that the period of a one-shot circuit using an external emitter resistance is given approximately by $T = 0.69 R_B C$.

9.12 Design a one-shot circuit using a single 18-V power supply and matched transistors with $\beta_0 = 60$, $V_{CE(sat)} = 0$, and $V_{BE(on)} = 0.5$ V. The period should be 50 μs and T_1 requires a reverse-bias of at least $\frac{1}{2}$ V. The output impedance should be less than 1 kΩ. (Neglect switching times of the transistors.) What is the recovery time constant of the circuit?

9.13 Repeat Problem 9.12 for a 12-V power supply and a required period of 100 μs.

9.14 Repeat Problem 9.12, given that the supply voltage is 12 V, but that the timing resistor R_B is returned to 24 V.

Section 9.3B

* **9.15** Silicon transistors are used for an astable multivibrator. Given that $t_1 = t_2 = 50$ μs at 25°C, calculate the periods at 75°C. A 20-V power supply is used.

9.16 Repeat Problem 9.15, assuming that the temperature changes from 25°C to -30°C.

9.17 Using the transistors and power supply of Problem 9.12, design an astable multivibrator with $t_1 = 40$ μs and $t_2 = 60$ μs. The output impedance should be 4 kΩ or less.

9.18 Design a one-shot circuit using the transistors and power supply of Problem 9.12. The timing resistor should be returned to the wiper of a potentiometer, as shown in the figure, to give a period that can be varied from 20 μs to 40 μs.

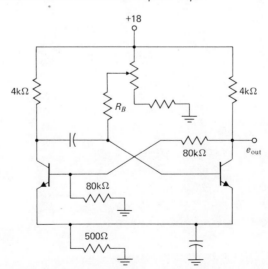

Problem 9.18

9.19 In the astable circuit of Fig. 9.40, $V_{CC} = 12$ V, $R_{L1} = R_{L2} = 4$ kΩ, $R_1 = R_2 = 80$ kΩ, and $C_1 = C_2 = 0.01$ μF. Plot the total period as a function of v_t as v_t varies from 6 V to 24 V.

9.20 Repeat Problem 9.19, given that $R_1 = R_2 = 400$ kΩ and v_t varies from 60 V to 80 V. Use both Eqs. (9.11) and (9.12) to calculate T and plot the results on the same graph.

*9.21 The astable circuit of Fig. 9.40 uses silicon transistors with $V_{BE(on)} = 0.6$ and $\beta_{0(min)} = 80$. Given that $v_t = 30$ V and $V_{CC} = 18$ V, select reasonable element values to give a symmetrical total period of 1 ms and an output impedance of 3 kΩ or less.

9.22 Repeat Problem 9.21, assuming that $v_t = 6$ V and $V_{CC} = 18$ V.

Section 9.3C

9.23 Work Example 9.2 using silicon transistors with $\beta_0 = 80$, $f_t = 10$ MHz, $C_{B'C} = 10$ pF, $V_{BE(on)} = 0.5$ V, and $V_{CE(sat)} = 0.3$ V.

9.24 Derive Eq. (9.15) for the minimum value of trigger capacitor.

9.25 Figure 9.51 shows the truth table for the JK flip-flop of Fig. 9.50. Sketch the output Q for the gating and clock signals shown in the figure for Problem 9.25. Assume $\bar{S}_D = \bar{R}_D =$ high.

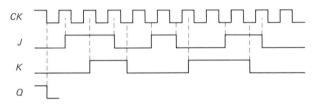

Problem 9.25

DIGITAL SYSTEMS

The greatest user of logic circuits is the digital computer. It naturally follows that when logic circuits are mentioned one immediately thinks of the digital computer. It must be emphasized that there are numerous other important applications of the digital or logic circuit. Automatic control of processes is often accomplished by digital methods. Digital communications is an extremely important field that depends heavily on logic circuits. Digital methods are applied routinely in instrumentation. While this chapter discusses logic circuits in connection with the digital computer it should be recognized that the same circuits are often applied directly to other important fields. Furthermore, techniques discussed in this chapter allow the logic designer to understand the operation of the IC microprocessors that have recently become so prominent.

10.1 THE DIGITAL COMPUTER

Most logic functions and electronic calculations are performed by binary circuits, that is, circuits that have only two possible operating points as discussed in Chapter 9. The output voltage then corresponds to either a predetermined lower or upper level. One of these states can be used to represent a binary 1 while the alternate state represents binary 0. Binary numbers form the basis of operation of the digital computer. A register consisting of several binary circuits can store large binary numbers. For example, a register containing 10 binary circuits can store any number from 0 up to 1023. Twenty circuits would be required to store numbers reaching one million. Ferrite cores also exhibit two distinct magnetic states and can be used in storing binary numbers. Dynamic registers for transporting binary numbers from one location to another are based on the simple two-state circuit. Thus, as we consider the digital computer we must learn to think in terms of simple binary operations. We must recognize that the basic circuits treated in Chapters 8 and 9, when appropriately interconnected, make up one of the most powerful scientific tools developed by man: the digital computer.

A. Computer Organization

We will now consider the overall organization of the computer in order to provide motivation for further study of digital circuits and systems. The typical digital computer is generally composed of five basic sections. As shown in block diagram form in Fig. 10.1 these are the input and output units, the memory or storage section, the arithmetic unit, and the control unit. The arithmetic unit plus the control unit make up the central processing unit (CPU) of the computer.

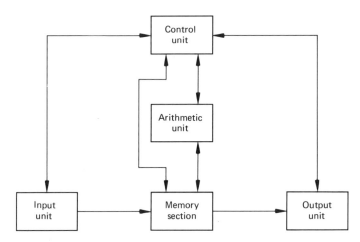

Fig. 10.1 Five basic sections of a digital computer.

The input unit provides access to the computer by an input device such as a typewriter, card reader, paper tape reader, teletype terminal, or special-purpose device. The input unit must convert the signals generated by the input device into appropriately coded electrical signals. A typewriter input requires conversion from mechanical movement of the keys to the required digital code. Card readers or paper tape readers sense hole positions on the cards or paper tape and convert the resulting signals into the required code. The input unit must be able to indicate to the control unit that information is ready to be transmitted. It must also receive messages from the control unit signifying that the computer is ready to receive input information.

For digital process control systems the input signal may be the analog voltage of measurement circuitry. This analog signal, which may represent the error between a present position and a desired position of some mechanical apparatus, is then converted to digital form for use by the computer. The input unit must contain an analog-to-digital (A/D) converter in this instance.

There are several possible input devices and each input unit is different. Some units might accept two or three different input devices while others are designed to operate with only one device. Common among all units is the requirement that the control signals and input information must be presented to the memory or control unit of the computer in the proper form, namely, binary-coded electrical signals.

There are also several possible output devices that can be used. Among these are the typewriter, card punch, paper tape punch, teletype terminal, and line printer. Much faster than the other devices, the line printer is easily capable of printing 30 lines of 130 characters in one second. The output unit operates under the direction of the control unit. Often a device will serve both input and output functions. The teletype terminal is a prime example of such an I/O device.

For general-purpose computers it is more efficient to use paper tape or punched card input and a line printer output. Most calculations in a program are done at electronic speeds; thus, program run time is often limited by the speeds of the input and output devices. Large time-sharing computers are required to consider inputs from many unrelated, usually remote input devices and transmit results back to the same number of output devices. These remote stations are referred to as I/O terminals and will often communicate with the computer over telephone lines. Since a terminal can transmit or accept information very slowly compared to central processor speeds, the control unit switches very rapidly from one I/O device to another. The information sent by each terminal is stored in a particular section of the memory. While information is being accepted from one terminal, an altogether different program may be in the process of being executed. Time-sharing computers such as the IBM System/360 model 67 can service 256 terminals when operating in the time-shared mode.

The rapid access memory of the computer is usually composed of magnetic cores. Often a bulk storage memory will be available, although access to this memory section is comparatively slow. The rapid access memory is often called the main memory while the bulk storage section is called the file.

There have been some tremendous advances in magnetic core storages in the last decade. While some small special-purpose computers may have only a 32,000-bit capacity, the main memory of some computers exceeds 10 million-bit capacity. Main memory cores of 10 to 50 mils in diameter are commonly used at the present time. Wires threaded through the ferrite cores are used for reading and writing information. As the ferrite cores decrease in diameter the cycle time for reading or writing information decreases. Since the minimum diameter of the core has a practical limit which is now being approached, it appears that further dramatic improvement in main memories will result from new memory technologies. It is difficult, however, to visualize the ferrite core being replaced in the near future as a versatile main memory component.

Information that must be available in very short times and data sets that are required for rapid calculations are stored in the main memory. An entire section of this memory is set aside for permanent instructions. The permanent instruction section includes the instructions that control the main program plus the built-in subprograms or functions.

The file memory might store as many as a billion bits of information, but access to this information is comparatively slow. In most cases mechanical motion is involved in reading from or writing into the file. Magnetic discs or drums can be used as files with systems that require somewhat random access to the stored information. Bits may be stored serially, but the magnetic heads can be quickly moved to new blocks of information. Magnetic tape, on the other hand, offers only serial access and is very inefficient if used in other than a serial mode.

It is possible to use other types of memories in special-purpose machines. Integrated circuits offer attractive features in certain high-speed, low-capacity memories. Recirculating delay line memory units can be used in conjunction with ferrite cores to offer advantages in special cases.

The arithmetic unit performs the basic operations of addition and subtraction. Adders, subtractors, and registers including the accumulator make up this unit. More complex operations such as multiplication and division are performed by algorithms calling for repeated addition or subtraction. Utilizing stored instructions from the memory, the arithmetic unit carries out these and other more complex operations under the direction of the control unit.

While it is beyond the scope of this text to discuss each section of the digital computer in detail, the basic circuits and subsystems will be treated in succeeding sections of this chapter.

B. Binary Numbers

We are so used to measuring quantities and counting with decimal numbers that we rarely stop to examine the significance of the number system. Before proceeding to the binary number system let us review some basic ideas concerning the decimal system. In this system each column position corresponds to a different power of ten to be multiplied by the number occupying this position. For example, the number 124 means $1 \times 10^2 + 2 \times 10^1 + 4 \times 10^0$. The least significant number is multiplied by 10^0, the next least significant figure is multiplied by 10^1, and each figure of higher significance is multiplied by an appropriate power of ten. The decimal number 16,052 corresponds to

$$1 \times 10^4 + 6 \times 10^3 + 0 \times 10^2 + 5 \times 10^1 + 2 \times 10^0.$$

The number 10 serves as the base for the decimal number system. Decimal numbers smaller than unity conform to the same pattern as the integer numbers. The decimal number 0.1368 corresponds to

$$1 \times 10^{-1} + 3 \times 10^{-2} + 6 \times 10^{-3} + 8 \times 10^{-4}.$$

The base ten system is very convenient for manual calculations. In fact, the main argument for conversion to the metric system of weights and measures consists of the fact that the various units are related to others by multiples of ten. That is, 1 kilogram = 1000 grams = 10^6 milligrams. It is possible to use other bases for number systems. Three rather popular ones for the digital computer are the binary-coded decimal system, the octal system, and the binary system. The binary system uses the base 2 while the octal system uses 8 as the base. The binary system is particularly well suited to digital circuits because only two symbols are required in this system. Normally, the symbols used are 0 and 1 corresponding to the two possible states of most digital circuits such as flip-flops and astable or monostable multivibrators.

The same scheme that is used for decimal numbers is applied to binary numbers, except that each column position is multiplied by some power of 2. The binary number 101011 represents

$$1 \times 2^5 + 0 \times 2^4 + 1 \times 2^3 + 0 \times 2^2 + 1 \times 2^1 + 1 \times 2^0 = \text{decimal } 43.$$

Binary numbers less than one follow the same pattern. The fraction 13/16 can be represented by $0.1101 = 1 \times 2^{-1} + 1 \times 2^{-2} + 0 \times 2^{-3} + 1 \times 2^{-4}$. Other examples of binary numbers are:

Binary	Decimal
101111	47
10	2
110000	48
110001	49
1010	10
110010	50
1	1

A great deal of arithmetic is performed in the computer by using binary numbers. Addition can easily be accomplished by noting that when the sum of binary digits reaches two a carry-bit is added to the next column. Some examples of addition are indicated below.

Binary	Decimal
11111	31
+ 00001	+ 1
100000	32
100101	37
+001001	+ 9
101110	46
100111	39
+ 100010	+34
1001001	73

The implementation of a digital adder will be considered in a later section of this chapter. We will also consider a subtractor at that time.

Binary-coded decimal is used for input-output operations since the normal input or output devices are based on decimal operation. A keyboard or punched card input requires a conversion from decimal to binary-coded decimal. This step is accomplished by an encoder which will be discussed later in this chapter. Conversion from binary-coded decimal to binary can be accomplished before arithmetic operations are performed.

In the binary-coded decimal scheme, each decimal position is represented by a four-bit binary number. For example, the number 346 would become

$$\begin{array}{ccc} 0011 & 0100 & 0110 \\ 3 & 4 & 6 \end{array}$$

This type of representation is rather inefficient since 12 binary bits are required to represent three-place decimal numbers up to 999. In simple binary systems 10 stages can reach decimal numbers up to $2^{10} - 1 = 1023$.

Octal systems combine the features of a binary system and the binary-coded decimal system to some extent. We will not consider this system in detail; however, we will note that the base is 8 and that the number of stages required to represent a given number in octal equals the number required for binary. In an octal system, the number 010 111 101 011 represents

$$2 \times 8^3 + 7 \times 8^2 + 5 \times 8^1 + 3 \times 8^0 = \text{decimal } 1515.$$

10.2 BOOLEAN ALGEBRA

For simple logic functions one can easily deduce by inspection the type and number of gates required, the most efficient configuration, and the number of inputs required per gate. For example, to OR the three signals A, B, and C, a three-input OR gate can be used. Alternatively, if only two-input OR gates are available, two such gates can realize the OR condition.

In a digital computer or related equipment, logic functions that may be quite complex must be performed. There may be constraints imposed on the system used to synthesize the desired function. Perhaps due to cost or size requirements only gates with four inputs or fewer can be used. Perhaps a given circuit output can drive only 10 gates before fan-out is exceeded. There are several factors that tend to influence the physical configuration of a logic system.

Boolean algebra offers a systematic method of relating the operation of hardware to the mathematics of the logic function. It is possible to minimize required hardware using Boolean algebra or to apply constraints such as those mentioned earlier. This approach to logic design is important enough to justify the following discussion of the fundamental principles of Boolean algebra.

It should be emphasized that the two voltage levels assumed by a circuit in a binary system must be defined in terms of the bits they are to represent. In positive logic the 1 is represented by the most positive voltage level, while the 0 is represented by the remaining voltage level. The two levels spoken of may be $+10$ V and 0 V, $+12$ V and -6 V, or -2 V and -12 V. Note that the 0 bit need not correspond to a voltage level of zero volts. Negative logic applies if the 1 is defined as the lower voltage of two levels. The levels for the 1 and 0, respectively, could be -12 V and -6 V, -12 V and 0 V, 0 V and $+10$ V, or 0.5 V and 1.0 V. Once the bit levels have been defined, it is customary to maintain these levels throughout the system. While a redefinition of levels in one section of a system is sometimes undertaken, it is an exception to the rule.

A. Truth Tables and Identities

In logic systems it is necessary to label all input and output leads. The OR gate of Fig. 10.2 has inputs A and B while the output is labeled X. The variables A, B, and X can take on either of the two values 1 and 0.

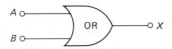

Fig. 10.2 A two-input OR gate.

In logic design it is often useful to invert the signal. That is, if the two levels being used are $+2.4$ V and 0.5 V, an inverter circuit will present a 0.5-V output when $+2.4$ V appears at the input while the 0.5-V input results in a $+2.4$-V output. Basically, the two levels are interchanged. The Boolean expression indicating that the signal has been inverted is a dash above the signal name. The invert of A is designated $\bar{A}$ and is sometimes called NOT A. Whatever value A takes on, $\bar{A}$ will take on the opposite bit or complement of A.

As discussed in Chapter 9 the truth table is very useful in logic design. This table lists the various possible input combinations that can occur, along with the outputs corresponding to each of these combinations. The truth table for the OR gate of Fig. 10.2 is shown in Fig. 10.3. The same information given by the truth table is represented by the Boolean expression.

$$A + B = X.$$

The plus sign represents the OR operation rather than an addition. The signal X is created by combining A and B in the OR operation.

Inputs		Output
A	B	X
0	0	0
0	1	1
1	0	1
1	1	1

Fig. 10.3 OR gate truth table.

There are several Boolean identities associated with the OR gate that are useful. They are shown in Fig. 10.4. These identities, which can easily be verified by truth tables, are listed on the following page for later reference.

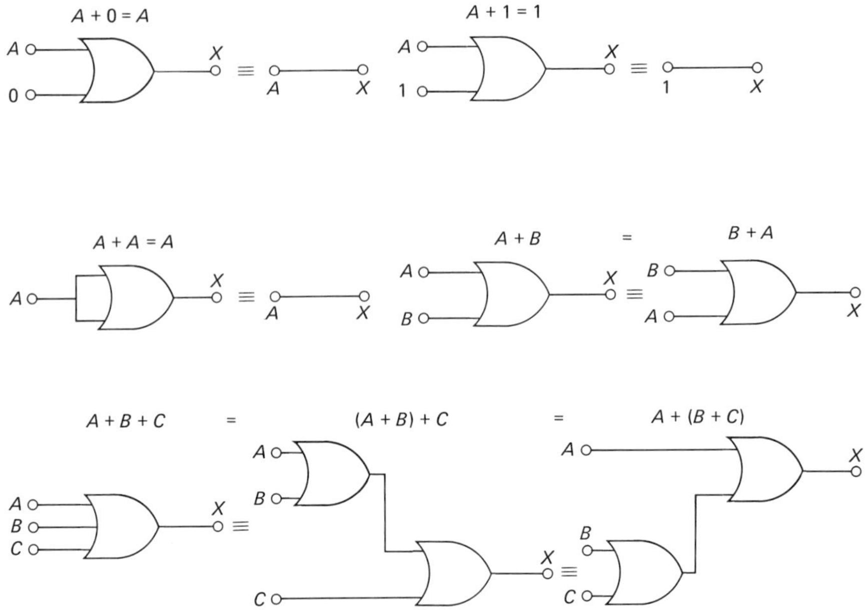

Fig. 10.4 Boolean OR identities showing implementation.

$$A + 0 = A \tag{10.1}$$

$$A + 1 = 1 \tag{10.2}$$

$$A + A = A \tag{10.3}$$

$$A + B = B + A \tag{10.4}$$

$$A + B + C = (A + B) + C = A + (B + C). \tag{10.5}$$

From the design viewpoint Eq. (10.1) states that signal X can be obtained by either connecting a lead directly to signal A or performing the OR operation on the signals A and the 0 bit. Of course, the former method is more efficient since it does not require a gate. Similarly, Eqs. (10.2) and (10.3) can be implemented without the use of a gate.

In reducing a complex logic expression, terms such as $A + 0$, $A + 1$, or $A + A$ may occur. These can be replaced by the equivalents A, 1, or A, respectively, and not only simplify the Boolean expression, but also minimize the number of gates required for implementation. This fact will be demonstrated in a later section.

A two-input AND gate is shown in Fig. 10.5 along with the truth table for this circuit. The Boolean expression meaning the AND operation of A and B is

$$A \cdot B = X$$

or

$$AB = X.$$

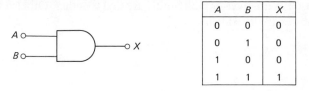

Fig. 10.5 Two-input AND gate and truth table.

Some important Boolean AND identities are listed below and demonstrated in Fig. 10.6.

$$A0 = 0 \tag{10.6}$$

$$A1 = A \tag{10.7}$$

$$AA = A \tag{10.8}$$

$$AB = BA \tag{10.9}$$

$$ABC = (AB)C = A(BC). \tag{10.10}$$

Again, these identities are useful in simplifying complex logic functions and decreasing the required number of circuits or gates for implementation.

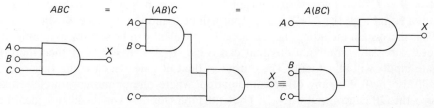

Fig. 10.6 Boolean AND identities showing implementation.

An inverting OR gate or NOR gate can function as an inverter if only one input is used or if both inputs are tied together. An inverting AND gate or NAND gate can be used as an inverter if all inputs are tied together. Other useful Boolean inverter identities are listed below.

$$\bar{\bar{A}} = A \tag{10.11}$$

$$\bar{A} + A = 1 \tag{10.12}$$

$$\bar{A}A = 0 \tag{10.13}$$

$$A + \bar{A}B = A + B \tag{10.14}$$

B. DeMorgan's Laws

There are two Boolean relationships that allow reduction of complex expressions. These are referred to as DeMorgan's laws and are given by

$$\overline{ABC \cdots N} = \bar{A} + \bar{B} + \bar{C} + \cdots + \bar{N} \tag{10.15}$$

and

$$\overline{A + B + C + \cdots + N} = \bar{A}\bar{B}\bar{C} \cdots \bar{N}. \tag{10.16}$$

Before discussing the simple proofs of these relations we should relate their significance to practical circuits. Let us consider the system of Fig. 10.7. We will consider only three variables, but the discussion can easily be extended to any number of variables.

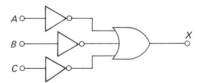

Fig. 10.7 Logic systems demonstrating DeMorgan's first law.

For the AND gate the only condition that will result in an output 1 is when $A = B = C = 1$. The inverted output X will then be 0. If any of the three variables is not equal to 1, the AND gate output will be 0 and the variable X will equal 1. That the system including the OR gate is equivalent can be seen by assuming the several possible combinations that can occur. If $A = B = C = 1$ then the OR gate inputs will all equal 0. The output of the OR gate, which is the variable X, will also be 0. For any other combination, where one or more variables equal zero, the OR gate output will be 1. The truth tables for the two schemes are identical and therefore, the systems are logically equivalent.

The second law can be demonstrated for three variables by the circuits given in Fig. 10.8. The only condition that will cause an output 1 is $A = B = C = 0$. All other combinations result in a 0 output.

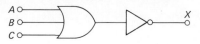

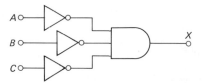

Fig. 10.8 Logic systems demonstrating DeMorgan's second law.

Application of these two laws along with the Boolean identities allows a designer to implement a given expression in terms of the gates available. For example, only NAND gates or NOR gates and inverters may be available in a particular logic family. Obviously, if an AND gate is required, one can follow a NAND by an inverter to produce this function. It can be more efficient in many such cases, however, to use DeMorgan's laws to synthesize the desired expression.

Let us demonstrate the key points in applying this method. Suppose we have a function $X = A(B + CD)$ which we must implement with NOR gates and inverters. The form of the function we should obtain is one that can be realized directly with NOR gates. An expression such as

$$X = \overline{(\overline{R + S}) + (\overline{P + Q})}$$

results from two stages of NOR gates as shown in Fig. 10.9. We can approach this form for the function X by applying DeMorgan's laws to obtain

$$X = \overline{\overline{A} + \overline{B + CD}} = \overline{\overline{A} + \overline{B} \cdot \overline{CD}} = \overline{\overline{A} + \overline{B}(\overline{C} + \overline{D})}$$
$$= \overline{\overline{A} + \overline{B}\overline{C} + \overline{B}\overline{D}} = \overline{\overline{A} + \overline{\overline{B} + C} + \overline{B + D}}.$$

The final expression can be realized as shown in Fig. 10.10. The exclusive OR gate (designated OE) is an example of a function that can be realized in many different ways. A two-input OE will have a 1 output if either input equals 1, but a zero output results if both inputs are 1 or 0. Problem 10.11 requires several realizations of this function, which can be expressed as

$$X = A\overline{B} + \overline{A}B. \tag{10.17}$$

Figures 10.11 and 10.12 show two implementations of the OE.

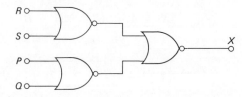

Fig. 10.9 Two-stage gating.

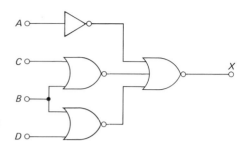

Fig. 10.10 Implementation of $X = A(B + CD)$ using NOR gates.

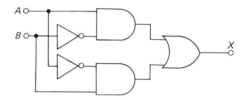

Fig. 10.11 Exclusive OR with $X = A\bar{B} + \bar{A}B$.

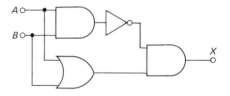

Fig. 10.12 Exclusive OR with $X = \overline{AB}(A + B)$.

C. Minimization of Boolean Functions

The use of Boolean algebra allows the investigation of many different schemes. It must be pointed out that applying DeMorgan's laws and the identities results in a rather random generation of the various possibilities. Systematic methods for obtaining the least number of gates, eliminating inverters, or obtaining some other desired goal are not always available. There are certain cases such as encoders and decoders in which certain key goals can be obtained. The number of gates or the number of inputs can be minimized by application of Boolean principles.

As an example of a useful reduction of an expression, let us assume that we must realize the function

$$X = A + AB. \tag{10.18}$$

This can be realized directly by the system of Fig. 10.13. We can reduce the expression as follows:

$$X = A + AB = A(1 + B) = A1 = A.$$

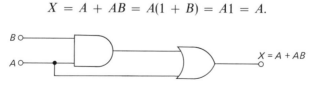

Fig. 10.13 System realizing $X = A + AB$.

The logical equivalent of $A + AB$ is simply the variable A. Consequently, an alternate realization of the function X is a wire connected from A to X. This consists of two fewer gates and would certainly be the preferred method of implementation. Expressions of this type occur often in the reduction of a more complex expression and it is almost impossible to minimize gates without applying Boolean algebra.

There are various specific methods of minimizing Boolean functions in order to implement the function with a minimum amount of circuitry. Karnaugh maps or Vietch diagrams provide graphical means of minimization while tabulation methods can also be used. These methods follow an orderly step-by-step procedure to obtain the final result, but become exceedingly complex as the number of Boolean variables increases.

Most schemes require that the original expression to be minimized be expressed in terms of a sum-of-products form rather than a product-of-sums form. Several of the relations developed earlier can be used to obtain the desired starting form. For example, the expression

$$(A + B + C)(\bar{A} + D)$$

can be expressed as

$$A\bar{A} + \bar{A}B + \bar{A}C + AD + BD + DC$$

to give the required starting form.

The Karnaugh map is perhaps the most straightforward method available to reduce a sum-of-products expression to simplest form. Unfortunately, the technique is much more difficult to apply to expressions involving five or more variables. Let us first look at the simple two variable expression

$$X = A\bar{B} + AB. \tag{10.19}$$

This expression could be represented by a map that includes exactly four locations corresponding to the four possible combinations of the two variables as shown in Fig. 10.14.

Fig. 10.14 A map of all possible combinations of a two-variable expression.

The second map of Fig. 10.14 represents the same information as the first one, but is more convenient to use. The expression for X takes on a value of one for two of the four possible combinations of the variables A and B. We can represent this function by placing a 1 in the locations corresponding to the combinations $A\bar{B}$ and AB. Zeros are placed in the remaining locations. Figure 10.15 reflects this representation.

To reduce the expression X to simplest terms we look for adjacent locations containing values of one. If these locations correspond to a term involving the

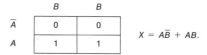

Fig. 10.15 A simple Karnaugh map.

value of a variable and another term involving the complemented value of that variable, a couple is formed. In Fig. 10.15, for example, the map shows a one at locations $A\bar{B}$ and at AB. From our experience with Boolean algebra we recognize that the expression X can immediately be reduced to $X = A(B + \bar{B}) = A$.

For expressions involving more terms the Karnaugh map provides an orderly means of locating all terms that can be reduced in the manner described. Let us consider the four-variable function

$$F = ABCD + A\bar{B}CD + A\bar{B}\bar{C}D + \bar{A}BCD + \bar{A}B\bar{C}\bar{D} + \bar{A}\bar{B}CD = \bar{A}\bar{B}C\bar{D}.$$

With four variables there are 16 possible combinations; thus, the Karnaugh map must contain 16 locations. Figure 10.16 shows a standard arrangement of the four-variable Karnaugh map. We can note several adjacent locations containing binary one that dictate a simpler expression. For example, the column containing four ones tells us that a one should occur whenever $C = D = 1$ or $CD = 1$, regardless of the values of A or B. The second row containing three ones indicates that a one should also occur in the expression whenever $\bar{A}B\bar{D} = 1$. The fourth row indicates the occurrence of a one in the expression when $A\bar{B}D = 1$. Thus, the entire expression can be reduced to

$$F = CD + \bar{A}B\bar{D} + A\bar{B}D. \tag{10.20}$$

	$\bar{C}$	$\bar{C}$	C	C	
$\bar{A}$	0	0	1	0	$\bar{B}$
$\bar{A}$	1	0	1	1	B
A	0	0	1	0	B
A	0	1	1	0	$\bar{B}$
	$\bar{D}$	D	D	$\bar{D}$	

$F = ABCD + A\bar{B}CD + A\bar{B}\bar{C}D + \bar{A}BCD + \bar{A}B\bar{C}\bar{D} + \bar{A}BCD + \bar{A}BCD$

Fig. 10.16 A four-variable Karnaugh map.

If F is implemented by hardware, seven four-input AND gates plus a seven-input OR would be required for the original expression. Alternatively the reduced expression could be implemented using two three-input AND gates plus a two-input AND gate along with a three-input OR gate. Obviously the reduced expression is much more economical.

The procedure for reduction consists of first drawing the Karnaugh map, and then locating all adjacent squares that contain a one. Adjacent squares containing a one indicate the presence of a couple which will allow a reduction of at least two terms. A square containing a one with no adjacent squares containing a one indicates a term in the original expression that must be included with no further reduction. We must note that the two squares at the ends of any row or at the ends of any column must be considered as adjacent locations. Some simple examples of these ideas are demonstrated in Fig. 10.17.

$F = A\bar{B}C + \bar{A}\bar{B}\bar{C}$

	$\bar{C}$	C	
$\bar{A}$	0	(1)	$\bar{B}$
$\bar{A}$	0	0	B
A	0	0	B
A	0	(1)	$\bar{B}$

$F_R = BC$

$F = AB\bar{C}D + ABCD + \bar{A}B\bar{C}D + \bar{A}BCD$

	$\bar{C}$	$\bar{C}$	C	C	
$\bar{A}$	0	0	0	0	$\bar{B}$
$\bar{A}$	0	1	1	0	B
A	0	1	1	0	B
A	0	0	0	0	$\bar{B}$
	$\bar{D}$	D	D	$\bar{D}$	

$F_R = BD$

$F = AB\bar{C}D + ABC\bar{D} + \bar{A}B\bar{C}\bar{D} + \bar{A}BC\bar{D} + \bar{A}\bar{B}\bar{C}\bar{D}$

	$\bar{C}$	$\bar{C}$	C	C	
$\bar{A}$	1	0	0	1	$\bar{B}$
$\bar{A}$	0	0	0	0	B
A	0	1	0	0	B
A	1	0	0	1	$\bar{B}$
	$\bar{D}$	D	D	$\bar{D}$	

$F_R = AB\bar{C}D + \bar{B}\bar{D}$

$F = \bar{A}\bar{B}\bar{C}\bar{D} + \bar{A}\bar{B}\bar{C}D + \bar{A}BCD$

	$\bar{C}$	$\bar{C}$	C	C	
$\bar{A}$	1	1	0	0	$\bar{B}$
$\bar{A}$	0	1	0	0	B
A	0	0	0	0	B
A	0	0	0	0	$\bar{B}$
	$\bar{D}$	D	D	$\bar{D}$	

$F_R = \bar{A}\bar{B}\bar{C} + \bar{A}C\bar{D}$

Fig. 10.17 Some examples of reduction.

In the first map, the couple is formed from locations at opposite ends of a column. The second map shows a quadruple that can be represented by a single two-variable term. The third map also contains a quadruple although it is not so obvious as the last. In addition, this map contains a four-variable term that cannot be reduced. The last map demonstrates that one square containing a one can be used to form more than one couple. With these basic ideas, a great deal of reduction can be performed.

There are also expressions that imply "don't-care" conditions such as

$$F = AB + AC + B\bar{C}.$$

When $A = B = 1$ occurs, the expression must equal 1 regardless of the value of C. When $A = C = 1$ the expression must equal 1 regardless of the value of B. The map for this expression is shown in Fig. 10.18. We note that the two couples represented by the reduced expression account for all occurrences of ones on the map. The third couple is not required in this case.

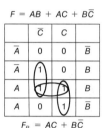

Fig. 10.18 A simple map.

For expressions involving five or more variables these methods can be extended although they become considerably more complex [2].

Example 10.1. Four variables are used to represent a four-bit binary number. When certain numbers occur, a logical 1 should be generated. The numbers for which the output should indicate logical 1 are 1, 4, 5, 9, 11, and 12. Assume that each variable is also present in complemented form.

Solution. Logical 1 is to be generated for any of the following combinations.

Decimal	Binary	Four-variable term
1	0001	$\bar{A}\bar{B}\bar{C}D$
4	0100	$\bar{A}B\bar{C}\bar{D}$
5	0101	$\bar{A}B\bar{C}D$
9	1001	$A\bar{B}\bar{C}D$
11	1011	$A\bar{B}CD$
12	1100	$AB\bar{C}\bar{D}$

Therefore, the expression is

$$F = \bar{A}\bar{B}\bar{C}D + \bar{A}B\bar{C}\bar{D} + \bar{A}B\bar{C}D + A\bar{B}\bar{C}D + A\bar{B}CD + AB\bar{C}\bar{D}.$$

This expression could be realized directly with six four-input AND gates plus one six-input OR.

On the other hand, the Karnaugh map of Fig. 10.19 leads to a simpler expression. The three couples circled in the figure are the only ones necessary to represent the expression giving

$$F_R = B\bar{C}\bar{D} + \bar{A}CD + A\bar{B}D.$$

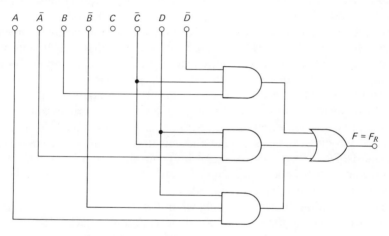

Fig. 10.19 Karnaugh map for Example 10.1.

This reduced expression is implemented in Fig. 10.20 and requires only 4 total gates and 12 total inputs, compared to 7 total gates and 30 total inputs to implement the original expression. It is left as an exercise for the student to realize this expression with NAND–NOR gate logic.

Fig. 10.20 Implementation of the reduced expression.

10.3 ENCODING AND DECODING

When a given key of a computer keyboard is depressed, a unique code must be transmitted to the computer to identify the character corresponding to that key. This is accomplished by the encoding process in which a distinct binary code is generated for each key depressed. The opposite process in which a code is converted into a signal that activates the particular key corresponding to the code is called decoding. When the computer has made a calculation and the program calls for a printout, the binary or binary-related code must be decoded to perhaps set a latch which activates a solenoid driver and prints the appropriate character.

There is no particular code used by all computers, although many remote terminals communicate with the computer over telephone lines with ASCII (American Standard Code for Information Interchange) [2]. The code used internally by the computer depends on the design of the CPU. Let us first consider the decoding problem.

A. Decoding

The basic problem in decoding generally consists of converting the binary code presented by some fixed number of stages in a register or set of latches to one of several outputs. We will consider the case of decoding four binary-coded decimal lines into one of ten decimal lines. The same techniques used will apply directly to a larger number of stages. Figure 10.21 shows a binary-coded decimal-to-decimal decoder.

The print stobe controls the time at which conversion takes place. The gates simply examine the outputs of the BCD register. When a given code is present, for example 0101, the correct AND gate which is number 5 in this case will open when strobed. No other gate will open at this time. AND gate 5 will open because $\overline{Q4} = Q3 = \overline{Q2} = Q1 = 1$ for this particular number.

It is obvious that the number of gates required equals the number of output characters, while the maximum number of inputs to each gate equals the number of binary positions used in the register plus one input for the print or gating pulse. Note that not all inputs of Fig. 10.21 are required for BCD-to-decimal conversion. For example, $Q4$ and $Q1$ could be used to activate the ninth line. If 16 output lines had been required with 16 possible input codes (4-bit code), four input lines plus a strobe input would be required in each case. For larger binary codes gate expanders can often be used. Most integrated circuit gates have compatible gate expanders that allow many inputs to be added to a given gate. In the rare instance in which expanders are unavailable and the number of binary stages exceeds the number of inputs, tree decoding can be used. Let us assume that we want to decode the binary characters $ABCD$, $A\overline{B}C\overline{D}$, $\overline{A}BC\overline{D}$, and $\overline{A}\overline{B}C\overline{D}$, and we have only two-input AND gates. One possible decoder is shown in Fig. 10.22. If a gating strobe is necessary for control an additional gate is required.

Multistage gating similar to that shown in Fig. 10.22 can be used to minimize the number of inputs to the decoder. Single-level decoding such as that shown in Fig. 10.21 minimizes the total number of gates, but does not minimize the total number of inputs. With certain types of decoder (for example, diode decoders or IC decoders), it is more efficient to minimize the total number of inputs, thus minimizing the total number of diodes or pin connections required. In this situation, multistage gating can be used.

To calculate the total numbers of gates and inputs required, let us assume an n-bit code is to be decoded. There are 2^n possible output characters with an n-bit code. Therefore, the total number of gates required for single-stage gating is

$$G_n = 2^n, \qquad (10.21)$$

and each gate requires n inputs (neglecting gate strobe input). The total number of inputs is then

$$I_n = n \times 2^n. \qquad (10.22)$$

For a 4-bit binary code the number of gates is $G_n = 2^4 = 16$ and the number of inputs is $I_n = 4 \times 16 = 64$.

If the binary code can be represented by $ABCD$ then we can decode two lines

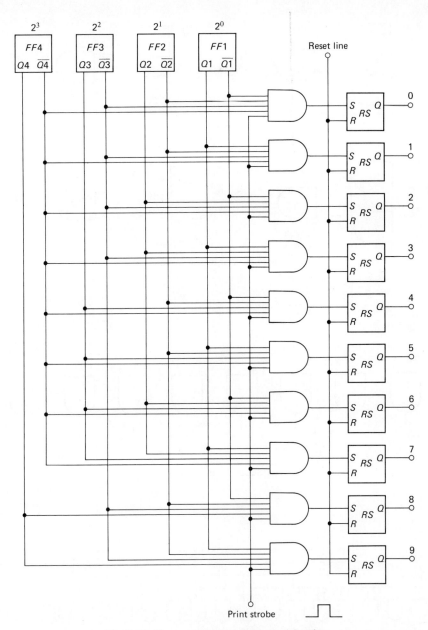

Fig. 10.21 Binary-coded decimal-to-decimal decoder.

at a time, as shown in Fig. 10.23. Obviously more gates are required in two-stage decoding. In this instance 24 total gates are required. On the other hand, a smaller number of inputs is necessary. Sixteen inputs for the first stage of decoding are required while 32 inputs for the second stage are necessary. A total of 48 inputs are used, compared to the single-stage figure of 64.

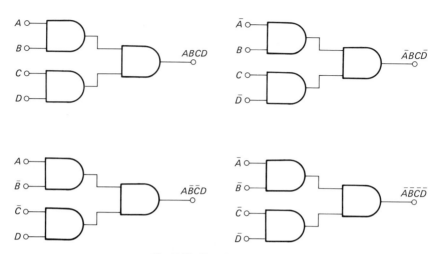

Fig. 10.22 Tree decoding matrix.

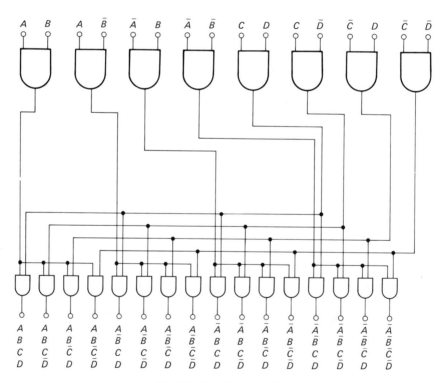

Fig. 10.23 Two-stage decoding.

In the two-stage decoder of Fig. 10.23 the decision to subdivide the four binary variables into two groups of two each was arbitrary. Perhaps it would be more efficient to decode three variables and combine these results with the fourth variable at the second stage of decoding. We must consider the general problem of subdivision of variables in order to determine the most efficient method of subdividing the binary variables.

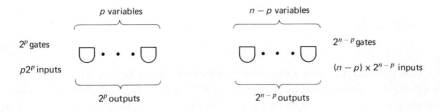

Fig. 10.24 General case of two-stage gating.

Figure 10.24 indicates the general case of two-level gating, assuming an n-bit code. The scheme of Fig. 10.24 assumes that the original n variables are subdivided into a group of p variables and the remaining group of $(n - p)$ variables. To decode p variables requires 2^p gates and $p2^p$ inputs. The $(n - p)$ variables are decoded by 2^{n-p} gates and $(n - p)2^{n-p}$ inputs. The second stage of decoding consists of 2^n two-input gates with a total of 2^{n+1} inputs. The total number of gates required is

$$G_n = 2^n + 2^p + 2^{n-p}. \tag{10.23}$$

The total number of inputs is

$$I_n = p2^p + (n - p)2^{n-p} + 2^{n+1}. \tag{10.24}$$

To minimize the number of inputs, Eq. (10.24) can be differentiated with respect to p and this result equated to zero. The differentiation is simplified by noting that $e^{0.693} = 2$ and writing Eq. (10.24) as

$$I_n = pe^{0.693p} + (n - p)e^{0.693(n-p)} + 2^{n+1}.$$

Equating the derivative to zero gives

$$\frac{dI_n}{dp} = 0.693p2^p + 2^p - 0.693(n - p)2^{n-p} - 2^{n-p} = 0. \tag{10.25}$$

Solving Eq. (10.25) for p results in a value of

$$p = n/2. \tag{10.26}$$

Subdividing the variables into two equal groups minimizes the total number of inputs in two-stage gating. Returning to Eq. (10.24) with $p = n/2$ gives a minimum number of total inputs of

$$I_{n(\min)} = n2^{n/2} + 2^{n+1}. \tag{10.27}$$

The total number of gates required for this case is

$$G_{n0} = 2^n + 2^{(n+2)/2}. \tag{10.28}$$

It is possible to reduce the total number of inputs further by using three or more stages of decoding. The two groups of $2^{n/2}$ gates involved in the first stage of decoding in Fig. 10.24 can be subdivided to form an additional stage of decoding. Subdivision of the variables can continue to reduce the total number of inputs until the number of variables in a given group equals two or three.

Consider the decoding of an 8-bit binary code given by

$$(A_1 A_2 A_3 A_4 A_5 A_6 A_7 A_8).$$

If single-stage gating is used, $2^8 = 256$ gates and $8 \times 2^8 = 2048$ inputs are required. Two-stage decoding can be accomplished by subdividing the code to give

$$[(A_1 A_2 A_3 A_4)(A_5 A_6 A_7 A_8)].$$

The input stage decoding requires two groups of $2^4 = 16$ gates or 32 gates. Each gate has four inputs giving 128 inputs. The output stage decoding requires 256 two-input gates. Two-stage decoding results in 288 gates and 640 inputs.

Obviously it is possible to subdivide the code further giving three stages of decoding. Each group of four bits can use two stage decoding before the outputs of each group are decoded at the last stage. The code is subdivided to give

$$\{[(A_1 A_2)(A_3 A_4)][(A_5 A_6)(A_7 A_8)]\}.$$

Note that there are three levels of brackets with this subdivision suggesting the three stages of decoding to be used. The input stage consists of four groups of four gates each with two inputs. The next stage consists of two groups of 16 two-input gates. The last stage consists of 256 two-input gates. The total number of gates is 304 and the total number of inputs is 608.

It is not always possible to subdivide the binary bits into groups of two. For example, if a 5-bit code is decoded the logical choices for subdivision are

$$[(A_1 A_2)(A_3 A_4 A_5)] \quad \text{or} \quad [(A_1 A_2)(A_3 A_4)\, A_5].$$

It can be shown that the minimum subdivision should consist of two variables rather than one. Thus, the former of the two choices is appropriate. In minimizing the number of input gates, the binary bits are subdivided as near the center of the bits as possible. Each subdivision is then subdivided further until only groups of two or three bits remain.

10.3 ENCODING AND DECODING

To calculate the total number of gates and inputs required to decode $[(A_1A_2)(A_3A_4A_5)]$ we note that the input stage decoding of $[A_1A_2]$ requires four two-input gates or eight inputs. The term $[A_3A_4A_5]$ requires eight three-input gates or 24 inputs. The total number of output gates is 32 and the output gates each require two inputs. The total number of gates is 44 and the total number of inputs is 96.

B. Encoding

In the encoding process we are generally given several single lines or latches, each corresponding to a specific character. When a given line is activated the appropriate binary code must appear at the output of the encoder. Conversion of decimal numbers to BCD will serve as an appropriate example. Figure 10.25 shows this decimal-to-BCD encoder.

Diode encoding can be used rather than active OR gates, but the basic operation is the same in either case. Activation of a single decimal line sets all four latches to the corresponding binary code.

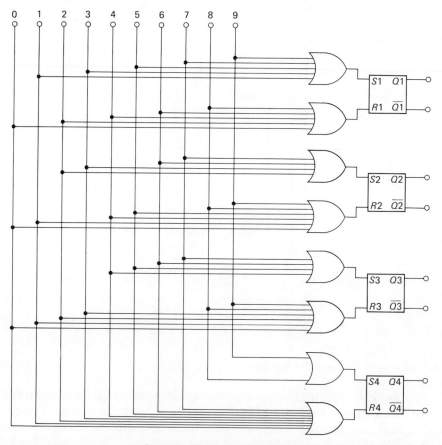

Fig. 10.25 Decimal-to-BCD encoder.

There are methods of encoding that reduce the number of inputs required at the expense of an increased number of gates. Using three stages of encoding for the encoder of Fig. 10.25 results in a reduction from 40 to 33 inputs and an increase from 8 to 13 gates. This scheme is indicated in Fig. 10.26. In both Figs. 10.25 and 10.26 the number of inputs can be approximately halved by encoding only the set lines and inverting these lines to generate the corresponding reset signal.

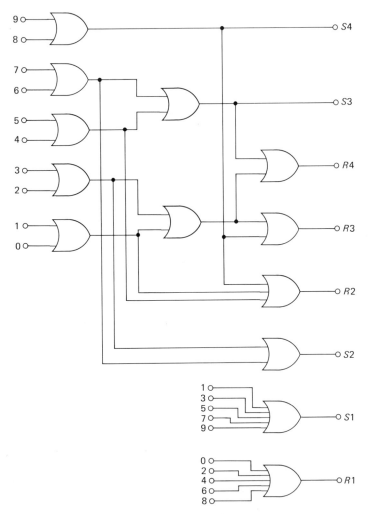

Fig. 10.26 Three-stage encoding.

For systems requiring an encode strobe it is convenient to use type D flip-flops as the binary register. Figure 10.27 shows a version of the decimal-to-BCD encoder that must be strobed with a negative going transition after the decimal line has been activated. The time delay required between activation of a decimal line and application of a strobe must be sufficient to allow the flip-flop gate to be set. Only 15 inputs and four gates are required, although the D flip-flop is more complex than the RS type.

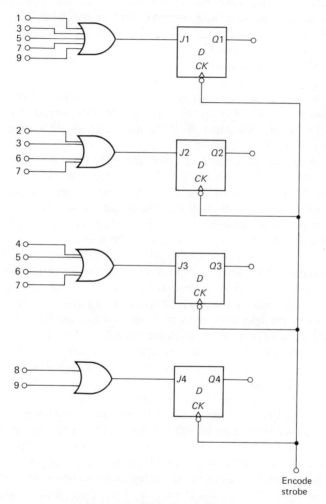

Fig. 10.27 Strobed decimal-to-BCD encoder.

C. Code Conversion

It is often necessary to convert from one code to another within the computer. For example, binary code may require conversion to BCD. The most straightforward method of accomplishing such a conversion is to decode the first code so that each code activates an individual line. Encoding of these lines can then be done to produce the desired output code.

10.4 REGISTERS AND TIMING CIRCUITS

A. Clock Signals and Synchronization

The digital computer may carry out several hundred thousand operations in one second; thus, a very accurate time reference must be available. This reference is established by a digital master clock circuit. All other clocks within a system must be referenced to the master clock.

In a simple system an astable multivibrator might be sufficient to generate the clock signal. Most systems require a higher degree of accuracy which can be provided by more precise timing circuits such as the crystal-controlled oscillator. The slave clocks or subclocks may then be multivibrator circuits that can periodically be corrected in accordance with the master clock. This process of clock correction is called synchronization.

Within a given system it may be a simple matter to drive the subclocks directly from the master clock. A subclock may be used, for example, to generate eight negative transitions as required by some circuit of the system. In this case, a gate referenced to the master clock may activate an astable for the required time to generate eight pulses. If the astable starts in synchronism with the master clock, it cannot generate an appreciable error in eight bit times.

There are, however, many modern applications of systems that involve transmission of digital information over long distances without the accompanying transmission of a clock signal. A prominent example of this situation occurs in the time-sharing systems that link terminals to the computer by leased telephone lines. The cost of the telephone lines prohibits sending anything but the digital signal; thus the clock signal is not available at both the terminal location and the computer. In this case, the transmitted signal is referenced to the clock of the sending unit and the receiving unit is required to generate a time reference that is synchronized to the sending clock.

To demonstrate a method of synchronization, let us consider the 4-bit binary word of Fig. 10.28. This code represents the binary number 1101 (decimal 13) plus a start bit. The least significant bit follows the start bit. The word of Fig. 10.28(a) is represented by nonreturn to zero or NRZ code. Each period of the clock corresponds to one bit time and if successive bits of the word contain binary ones, the signal does not return to the zero level during these bit times. The return to zero or RZ code indicates the bit value during one-half of the clock period and maintains the zero level during the remaining half of the period. Both codes are popular in digital systems and we will later see how to generate each one or convert from one to the other.

To read the transmitted signal into the receiving circuitry, a shift pulse must be created during each bit time to shift each individual bit into a register. If the clock signal of Fig. 10.28(c) can be accurately generated at the receiver, based only on the presence of the start bit, the shift pulses can then be produced.

At first thought one might conclude that a receiving clock with the exact period of the transmitting clock would suffice to control the shift pulse timing. It is, unfortunately, impossible to match the periods of two free-running clocks

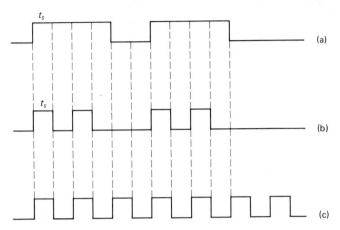

Fig. 10.28 Serial transmission of a binary word: (a) NRZ code; (b) RZ code; (c) clock signal.

to the required precision. If the periods of 1-MHz clocks are matched to one part in 100,000 (0.001 percent), there can be an error of 10 bit times in just one second. The error obviously is cumulative and after 1,000,000 cycles can be very large.

There are two basic methods of generating the required receiving clock signal. The first is to start a receiver clock upon reception of the start bit and allow exactly five clock periods to occur before shutting this clock off. If this receiving clock period is approximately the same as that of the transmitting clock, not much error can accumulate in five bit times. The second method allows the receiving clock to run continuously with a period as nearly equal to the transmitting clock as possible and then synchronize the receiving clock each time a binary one is received. The start bit will always be a one; hence the clock is always synced prior to reception of the actual word.

The first method can easily be implemented using the clock-signal generation method of Fig. 9.41 plus a third one-shot to generate the five-bit time gate. Figure 10.29 shows this scheme. The start bit triggers a one shot creating a $4\frac{3}{4}$ bit time

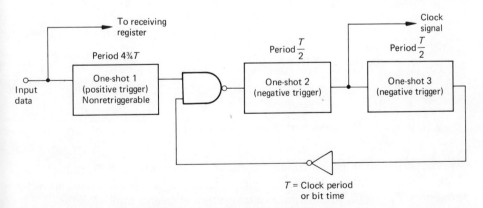

Fig. 10.29 Receiving clock signal generation.

gate. This gate initiates the period of the second one-shot and the third one-shot works, as explained in connection with Fig. 9.41, to generate a continuous clock signal until the gate is closed. This occurs when the first one-shot times out at $t = 4\frac{3}{4}T$. Figure 10.30 shows the pertinent timing chart.

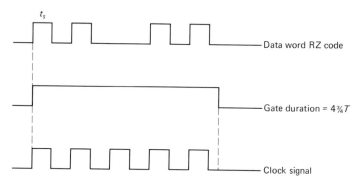

Fig. 10.30 Timing chart for the system of Fig. 10.29.

The system of Fig. 10.29 will operate with either NRZ or RZ code. Figure 10.31 shows a circuit that can be used only for RZ code to synchronize the receiving clock although, with some modification, it can be used for either code to start the clock at the correct time. In some systems it is desirable to have a receiving clock that operates continually, but is corrected or synced each time a word is received.

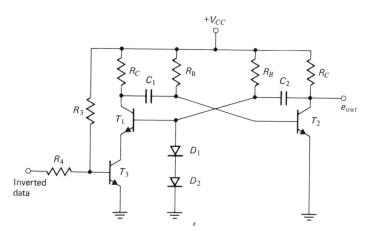

Fig. 10.31 Synchronizable astable clock.

The period of the astable circuit is chosen to be equal to that of the transmitting clock. When no data is being transmitted, transistor T_3 is saturated, allowing transistors T_1 and T_2 to function as a conventional astable multivibrator.

When a start bit arrives to signal the beginning of the transmitted binary code, the astable must be forced to the beginning of the state with T_2 on and T_1 off. As the leading edge of the start bit arrives, there are only two possible states in which the astable can be. Let us assume first that the astable is such that T_1 is on and T_2 is off. No matter what part of the timing cycle the astable has reached, the start bit turns T_3 off which forces T_1 off and turns T_2 on. When this occurs the capacitor C_2 will be charged to the appropriate voltage to start the astable period at this point. The start bit duration must be less than or equal to the half-period of the astable so that normal operation of the astable will now continue. If T_1 is already off when the start bit arrives, we desire this condition to remain until the trailing edge of the bit, whereupon the normal timing cycle of the astable should resume. Transistor T_3 will remain off until the trailing edge of the start bit forcing T_1 to remain off and T_2 on, even though the base of T_1 has reached a positive voltage. Since T_1 cannot saturate when the base voltage reaches zero or is slightly positive, the capacitor C_2 could charge to a higher voltage than normal, thus affecting the succeeding half-period. The diodes prevent this from occurring by clamping the base voltage to a slightly positive level. When the trailing edge of the start bit occurs, T_3 turns on as does T_1, and the normal timing cycle continues. We note also that each succeeding nonzero bit of the binary code has the same syncing effect as the start bit.

Obviously, the circuit of Fig. 10.31 requires the RZ code to function properly since the half-period is forced to equal the duration of the start bit. If a gate rather than the inverted data signal is used to drive T_3, the circuit functions in a manner similar to that of the system of Fig. 10.29. In this case, transistor T_3 is held off until the gate goes positive (in response to the leading edge of the start bit). The gate then lasts for $4\frac{3}{4}$ bit times, allowing five clock pulses to be generated. Again both RZ and NRZ codes will operate the circuit in this configuration.

Before leaving the subject of binary clocks we should consider the split-phase clock that is often useful in digital system design. In order to shift binary information into a JK flip-flop, the gates must be properly set before the shift pulse is applied to the flip-flop. If the shift pulse and gating signals are applied at the same time, the gating capacitors have not had time to charge to the levels necessary to effectively steer the shift pulse to the proper transistor. It is useful to use the normal clock pulses to control the information presented to the gates and a delayed clock signal to shift the information into the flip-flop.

One method that can be used to create a delayed clock signal is to first generate a rectangular waveform of frequency equal to twice that of the desired clock frequency. This signal is designated $2C$ while the clock signal is designated C. It is quite easy to obtain the signal C by applying $2C$ to the trigger input of a T flip-flop.

A delayed clock signal DC that lags C by one-quarter cycle can be created by inverting $2C$ and applying this signal to the trigger input of a T flip-flop. A complete clock system is shown in Fig. 10.32, along with the timing chart for the clock and delayed clock outputs.

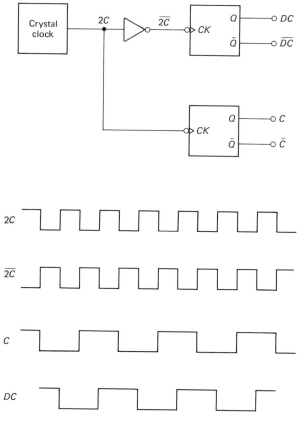

Fig. 10.32 Clock system with delayed clock signal.

B. Binary Counters

A binary pulse counter can be constructed using the T flip-flop circuit as the basic element. One flip-flop is required for each column of the maximum binary number to be counted. For example, if the counter must count from 0 to decimal 1000, ten stages are required since nine stages can reach a maximum decimal number of $2^9 - 1 = 511$. Ten stages can reach binary 1111111111 which corresponds to decimal $2^{10} - 1 = 1023$.

Figure 10.33 shows a four-position binary counter that will proceed from 0 up to decimal $2^4 - 1 = 15$. The first negative transition of the counter input line will cause $Q1$ to assume the positive level of logic. All other stages are still in the 0 bit position. As $Q1$ makes this transition from 0 to 1, a positive transition is presented to $CK2$ with no corresponding change in state of $FF2$. The counter now reads 0001. A second negative transition causes $Q1$ to switch from 1 back to 0. This negative transition is presented to the CK input of $FF2$ resulting in a change in state to $Q2 = 1$. The counter now reads 0010 corresponding to decimal 2. As additional negative input transitions occur the count advances to consecutive

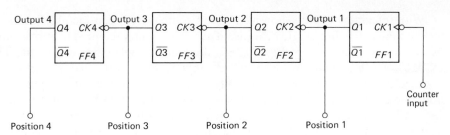

Fig. 10.33 Four-position binary counter.

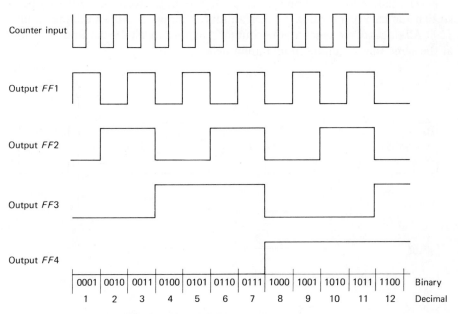

Fig. 10.34 Waveforms of outputs of counter stages.

binary numbers. Figure 10.34 shows a timing chart for the counter input and all outputs.

The simple counter indicated in Fig. 10.33 presents speed problems if several stages are required. If 12 stages are used and all stages reach the binary count corresponding to a 1 bit in each stage, the next bit will cause all stages to revert to the 0 bit. However, there will be a small delay in each stage which accumulates as we progress to the higher order stages. After the transition of the input pulse, the output of $FF1$ makes a slightly delayed transition. This delayed transition causes $FF2$ to make a transition that is delayed still further. Each flip-flop stage adds to the delay so that the transition of the last stage is delayed from the input transition by twelve times the average propagation delay time per stage. It is possible to decrease this delay in critical applications at the expense of circuit simplicity.

With integrated circuits, however, circuit complexity is considerably less important than in the discrete element case. Thus, the circuitry required to implement the synchronous counter of Fig. 10.35 is done on a single chip such as the TI SN74163, which actually includes considerably more circuitry than shown here in a medium-scale integrated TTL family. The synchronous counter applies the input line to all stages through AND gates. The negative transitions of the input signal will not reach the T input of a given stage unless the appropriate conditions are present to open the gate. For example, after six negative transitions, the counter contains 0110. None of the AND gates is open; thus on the seventh transition, only $Q1$ changes to binary one giving a count of 0111. Now all gates are opened so that the eighth transition reaches all four T inputs and the resulting count is 1000. All stages of the synchronous counter that must toggle see the input transition at the same time resulting in minimum delay.

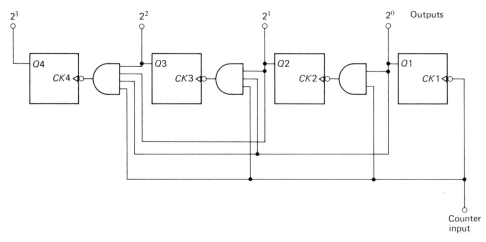

Fig. 10.35 Synchronous counter.

There are occasions in digital systems that require a backward or down counter. A binary number is set into the counter which then counts toward zero as input pulses occur. Figure 10.36 shows a simple method of constructing a down counter. The timing chart assumes that the original count was 1011 or decimal 11. After 11 negative transitions of the input signal the counter contains a count of 0000.

A counter that can count in either direction depending on which of two input lines is activated is shown in Fig. 10.37. The forward count opens the AND gate connecting $Q1$ to $T2$ through an OR gate. Unless the forward count goes negative, none of the AND gates connected to the Q lines can be opened. Thus, the inactive input line should have a positive voltage level while the active line will transmit the pulses to be counted. If pulses are applied to the backward line while the forward line has a positive level, the count will decrease with each input pulse. The

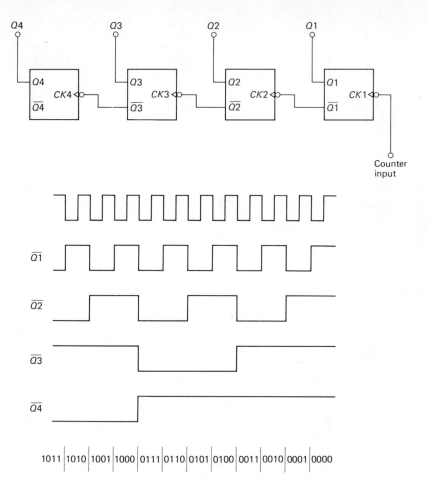

Fig. 10.36 Backward counter and timing chart.

delay circuits, represented by a square labeled D, can be made from OR gates or AND gates with a capacitor in shunt with the input. This delay is necessary to prevent slivers from being passed by the AND gates as one input goes negative simultaneously to the other input going positive. This is referred to as a race problem and will be discussed in more detail shortly. Let us consider the inputs to the AND gate connected to $Q1$. If a negative transition of the forward count line occurs, $FF1$ will change from $Q1 = 0$ to $Q1 = 1$. Ideally this transition would occur simultaneously with the input transition. In the actual circuit there will be finite rise and fall times associated with the input signal and the output of the flip-flop. The resulting signal may appear as shown in Fig. 10.38.

The condition leading to the possible sliver is termed a race condition. If the input pulse falls fast enough or wins the race, no sliver will occur. If $Q1$ rises while

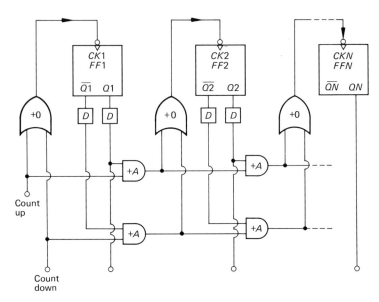

Fig. 10.37 Up-down counter.

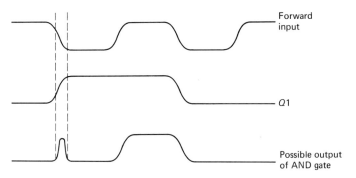

Fig. 10.38 Possible undesired AND gate sliver.

the input pulse is still falling, the sliver can occur. Thus, the necessity of delaying the outputs of each flip-flop, thereby avoiding a possible race condition. Race conditions should always be avoided even if the circuit is constructed and the sliver is found not to exist. Different loading conditions or temperature changes at a later time may lead to the production of this sliver which could complement the succeeding stage and cause an incorrect count to register.

The requirement for delay in the circuit of Fig. 10.37 is satisfied quite readily by using the master-slave flip-flop rather than an edge-triggered circuit. This flip-flop consists of two latches, the first of which accepts input data which is later transferred to the second latch. Thus, output data does not change simultaneously with new input information. A typical master-slave unit would respond to a clock

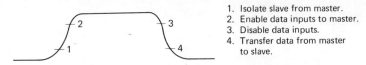

Fig. 10.39 Master-slave timing chart relative to clock.

pulse as indicated in Fig. 10.39 [4]. Replacing the flip-flops of Fig. 10.37 with master-slave units eliminates the necessity of delay circuits.

An example of an IC up-down counter is the TI SN74163. This is a synchronous, 4-bit, up-down counter with a clear input, a borrow and a carry output. Data can be loaded into the counter in parallel if a preset count is desired. Figure 10.40 shows the block diagram for this circuit which has a complexity equivalent to approximately 55 gates.

This counter can be cascaded with other stages increasing the bit capacity by four per stage. The carry and borrow outputs connect directly to the up-count and down-count inputs, respectively, of the adjacent stage. The clear input will reset all bits to zero at any time this input is activated.

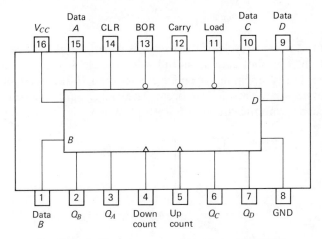

Fig. 10.40 An IC synchronous up-down counter (TI SN74193).

C. Timed Clock Gates

It is often necessary to generate a clock signal containing a precise number of negative or positive transitions. We will see in later sections that a shift register will receive one bit of a data word each time a negative clock transition occurs. If the word contains eight bits, then exactly eight transitions are required to fill the register.

There are several methods of generating a fixed number of clock pulses. The simplest is to use a one-shot of proper period duration to gate a continuous clock. This scheme is shown in Fig. 10.41. The trigger pulse must occur at the proper

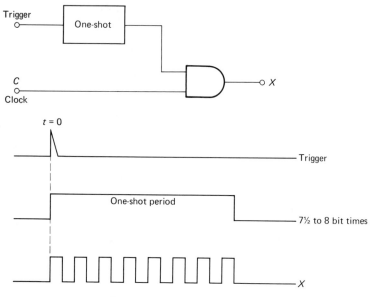

Fig. 10.41 One-shot clock gating.

time relative to the clock cycle. If the trigger is generated at the beginning of the clock period, the one-shot gate opens, allowing the clock signal to be passed by the AND gate. The gate remains open for greater than $7\frac{1}{2}$ but less than 8 bit times, allowing eight negative transitions to occur. The trigger signal can be synchronized to the clock signal with the circuit of Fig. 10.42. When it is desired to generate the

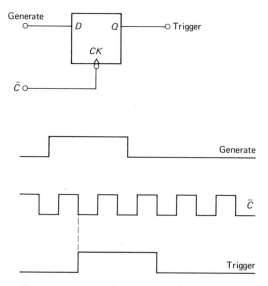

Fig. 10.42 Trigger synchronization using a D flip-flop.

eight clock pulses, an asynchronous generate signal occurs. This may occur at any point of the clock cycle. After the appropriate set-up time of the D input, a negative signal at the CK input will cause the D flip-flop to change state. The signal $\bar{C}$ is the inverted clock signal; thus the flip-flop sets $Q = 1$ at the start of a clock cycle.

One-shot circuits are considered to be less reliable than most other logic elements. They are sometimes triggered by extraneous noise (although bistables may also suffer this problem) and are often avoided in critical timing applications. An alternate method of generating eight clock pulses without the one-shot is shown in Fig. 10.43.

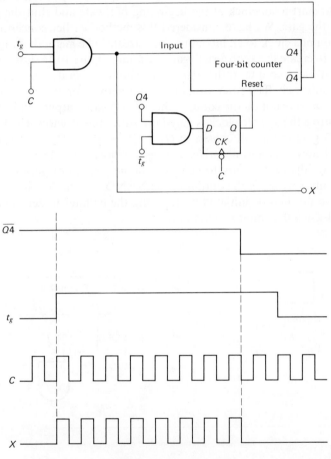

Fig. 10.43 Counter clock gating.

Again the trigger that opens the gate must be synced with the beginning of the clock period. The counter is at zero count prior to this time; thus $\overline{Q4} = 1$. The gated clock pulses are applied to the counter which responds to the negative transitions. The eighth negative transition switches the fourth flip-flop of the counter to the one state. This closes the gate since $\overline{Q4} = 0$ after the eighth negative transition. The counter must be reset before it can be used to again gate the clock. The D-input of the flip-flop is gated to binary one when the trigger strobe falls, since $Q4 = 1$ at this point. The first negative clock transition after t_g resumes a zero value raises the reset line to reset the counter to zero. Since $Q4$ and the D-input now equal zero, the following negative clock transition returns the flip-flop to the zero state, deactivating the reset line. The circuit is now ready to generate another series of clock pulses.

Both of the preceding methods gate the pulses of a continuous clock to produce the desired number of clock pulses at the gate output. It is possible and sometimes desirable to start a subclock at the beginning of a gate and shut the clock off at the end of the gate. We have considered this method earlier concerning the synchronization of a clock to an incoming data stream. The method of Fig. 10.29 uses a one-shot to create the appropriate gate duration. Figure 10.44 shows an alternate method that uses a counter rather than a one-shot to generate nine negative transitions of a subclock. The reset circuitry is not shown for the sake of simplicity. Assuming a zero count on the counter, the NAND gate output will be high. When t_g goes positive the astable clock is gated on. After nine negative clock transitions the NAND gate input will consist of $Q_1 = \overline{Q}_2 = \overline{Q}_3 = Q_4 = 1$. The output of the NAND now becomes negative, closing the clock gate. The counter must be reset after t_g falls to be ready to generate the next series of pulses. This scheme can be modified by using the output of the NAND to gate the astable. The series of pulses would then be initiated by resetting the counter to zero and would be terminated when the count reaches nine.

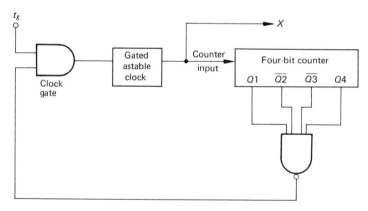

Fig. 10.44 Alternate method of clock gating.

D. Registers

There are two basic types of register used in the computer or other digital systems: static registers in which all positions are generally filled simultaneously (parallel operation), and dynamic shift registers that can be filled or emptied serially from one end of the register. In several instances a dynamic register will combine parallel and serial operation. For example, the register may be filled serially from a receiving line and shifted into a memory in a parallel shift operation.

We will first consider a static register that is filled and emptied in parallel. Although the practical computer will normally include larger registers to hold larger binary words, the eight-bit register of Fig. 10.45 demonstrates the fundamental operation of static registers.

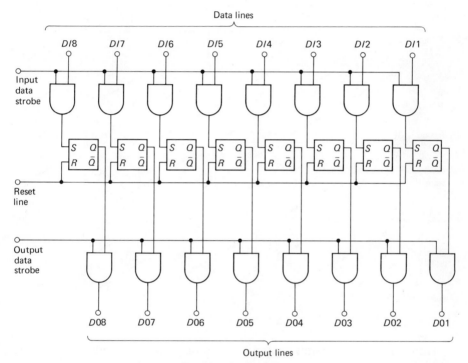

Fig. 10.45 Static register.

The reset line is activated to clear the register before any data are entered. When the input data lines are set a strobe pulse appearing on the input data strobe line will set the data into the register. After the data word has been entered it can be shifted on to another section of the computer. Activating the output data strobe causes the data to appear on the output lines during the strobe. Several sets of output gates can be used to shift the data into one of several different locations, depending on which output data strobe is activated.

A shift register can be used to receive serial information and transmit this serial information to some other location. It can also be used to receive parallel information and transmit serially or to receive serially and transmit in parallel. This register is often used to receive or transmit data over a communication link. If it is desired to shift a six-bit code onto a transmission line, the output of the flip-flop storing the least significant bit is connected to the line through a gate. When the data is to be transmitted, the gate is opened and six evenly spaced clock pulses are applied to the shift register. The first pulse causes all information to shift toward the least significant position with the bit that was previously in this position shifting onto the line. Let us assume that the code to be transmitted is 110010. If this information has been entered into the six-position shift register, each clock pulse will cause the data to move as shown in Fig. 10.46.

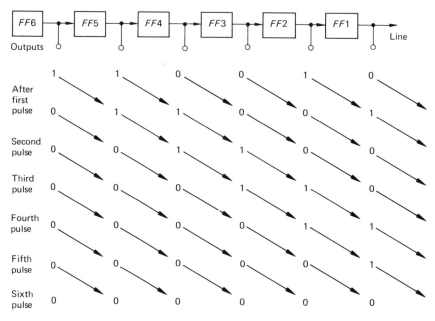

Fig. 10.46 Shift register.

The operation of the shift register is based on the fact that a given stage will appropriately gate the next stage so that the clock pulse will set this stage to the same state as that of the preceding stage. Figure 10.47 shows a shift register based on JK flip-flops. The gates of $FF8$ must be connected such that $J8 = 0$ and $K8 = 1$ to cause zeros to fill the register as the code shifts onto the line. The register can also be filled by connecting $J8$ to the incoming data line and $K8$ to the inverted data signal. Parallel information can be entered through the direct

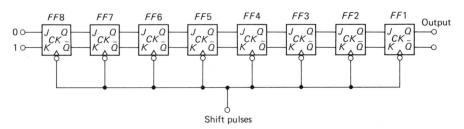

Fig. 10.47 Eight-position shift register.

set inputs if appropriate. The state of $Q8$ and $\overline{Q8}$ prior to a shift pulse gates $FF7$ so that the state of $Q7$ after the pulse corresponds to the state of $Q8$ prior to the pulse.

When receiving input data, the gates $J8$ and $K8$ must be set prior to the application of the shift pulse. The delayed clock signal generated by the system of Fig. 10.32 is useful in this instance. The incoming data is synchronized to the clock signal while the shift pulses are derived from the delayed clock.

IC shift registers are available from several manufacturers. Several hundred bit registers can be fabricated on a single chip for serial in–serial out registers. Obviously there cannot be lines available to allow direct sets to each stage. There are various types of eight-bit registers available in monolithic circuits. The TI SN74164 is an eight-bit, parallel out, serial shift register, while the SN74165 is a parallel load shift register. These two units could be used in a digital communication system with the SN74165 as the sending register and the SN74164 as the receiving register. The sending register might be filled in parallel, then transmit serially to the receiver. When this register is filled, a parallel transfer can empty its contents into the appropriate unit.

With some additions to a shift register an input/output or I/O register can be constructed. Let us consider the register of Fig. 10.48 which is capable of receiving an eight-bit binary code plus the start bit. The register is designed to

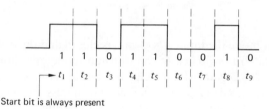

Start bit is always present

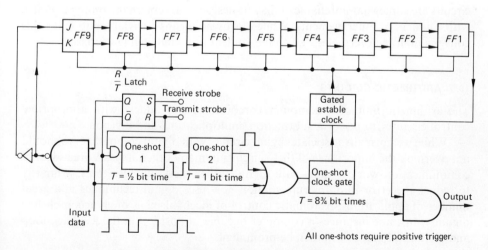

Fig. 10.48 I/O register and input code.

receive and transmit an NRZ (non-return to zero) code. As discussed earlier, when two successive ones are present in the code the signal does not return to the zero level between bits for the NRZ code. The mode of operation is selected by a switch. The code to be shifted is also shown in Fig. 10.48. When the latch is in the receive mode the output AND gate is closed, allowing nothing to shift onto the output line. The input AND gate is open at this time, allowing the incoming code to be presented to the shift register. The start bit will pass through the OR gate to trigger the clock gate one-shot, allowing the astable clock to generate nine shift pulses. Figure 10.49 shows the time relation between the clock pulses and the input code. The shift pulses correspond to the negative-going edge of the clock pulses. This allows each input bit to be presented to the gates for one-half bit time before it is shifted into the register. At the end of $8\frac{1}{2}$ bit times the register will be full including the start bit. The binary code is now present to be used as required by the overall system.

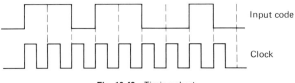

Fig. 10.49 Timing chart.

In the transmit mode the input AND gate is closed and the output AND gate is opened. As the latch is switched to the transmit mode one-shot 1 is triggered, opening the output AND gate. After a delay of $\frac{1}{2}$ bit time, one-shot 2 and the clock gate one-shot are triggered. This $\frac{1}{2}$ bit time delay is necessary to create a start bit lasting a full bit time. The information is then serially shifted onto the output line. Each time the transmit strobe occurs the transmit cycle is initiated. Since one-shot circuits are somewhat unreliable, it may be desirable to replace the timing circuitry with counters or other elements. While it is by no means a trivial job to make this conversion, it is nevertheless left to the student as a challenging exercise.

10.5 ARITHMETIC CIRCUITS

The arithmetic unit of a computer combines binary adders with appropriate control circuitry to perform subtraction, multiplication, and division.

While we normally associate digital circuits with digital computers, we should not overlook the importance of these circuits in many other areas. A great deal of communication work is done with digital rather than analog signals, primarily to reduce the error rate. Instrumentation now uses digital techniques in a great number of applications. Automatic control of mechanisms is often accomplished digitally. In some instances a computer occupies a position in the feedback loop and digital techniques will then be prominent.

A. Adders

Many principles that apply to these areas are common to the digital computer field. Hence, we can discuss these principles in connection with digital computer systems, realizing that they apply equally well to other areas that utilize digital circuits. The adder is an example of a system that is used in several fields.

A. Adders

Let us consider the problem of adding a single column binary number A to a second single column number B. The various possible combinations of A and B are

$$
\begin{array}{ccccc}
A & 0 & 1 & 0 & 1 \\
B & \underline{0} & \underline{0} & \underline{1} & \underline{1} \\
  & 00 & 01 & 01 & 10
\end{array}
$$

For the case where both A and B are equal to 1, a bit must be carried to the column of higher significance. Thus, output signals for both columns must be present. We will designate the carry bit by C and the sum bit by S. We can now construct a truth table as shown in Fig. 10.50.

A	B	S	C
0	0	0	0
1	0	1	0
0	1	1	0
1	1	0	1

Fig. 10.50 Truth table for adder.

Expressions for S and C can easily be written as

$$S = A\bar{B} + \bar{A}B$$

and

$$C = AB.$$

The sum S can be generated by an exclusive OR function while C can be generated by an AND gate. Any of the schemes discussed previously to implement the OE gate can be used. The half-adder is shown in Fig. 10.51.

The half-adder forms the basis for a full adder. In adding two binary numbers with more than one column, a single half-adder can be used only for the least

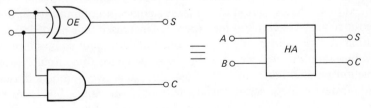

Fig. 10.51 A half-adder.

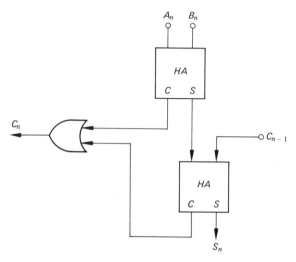

Fig. 10.52 A full adder.

significant column. The next most significant column must consider three inputs rather than two. The two numbers being added in that column and the carry-bit from the adjacent column must be added together. An additional OR gate and half-adder can be used to create a full adder as in Fig. 10.52. If we are concerned with the addition of the nth column of the two binary numbers, the previous carry-bit C_{n-1} will be considered by the full adder. The sum S_n will be presented as output while the carry-bit C_n is available to be used in forming the column of higher significance. A four-position binary parallel adder is shown in Fig. 10.53. The four columns of the binary number A are represented by $A_3 A_2 A_1 A_0$, while B is represented by $B_3 B_2 B_1 B_0$. The sum S will be given by $S_4 S_3 S_2 S_1 S_0$.

It is assumed that the numbers A and B are stored in registers or storage units to be presented to the adder inputs. Neglecting delays in circuit operation, the sum S appears immediately upon presentation of the binary numbers A and B to the input. For example, if the numbers A and B are $A = 1001$ and $B = 1101$, then the various outputs will immediately assume the values $S_4 = 1$, $S_3 = 0$, $S_2 = 1$, $S_1 = 1$, and $S_0 = 0$. The carry lines will contain the information $C_3 = 1$, $C_2 = 0$, $C_1 = 0$, and $C_0 = 1$.

The parallel adder requires a full adder for each column of the binary numbers to be added (with the exception of the least significant column). It is possible to decrease the required circuitry at the expense of increased time needed to complete the addition. A serial adder only requires one full adder plus some simple control circuitry, but addition time increases by the same factor that circuitry decreases. A 10-position parallel adder uses 10 stages and adds two numbers in one clock period (or faster) while a serial register uses only one stage, but takes roughly 10 clock periods to complete the addition. Figure 10.54 shows a serial adder. One possible configuration showing input registers and accumulator is shown in

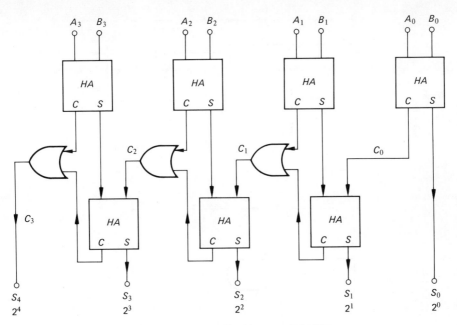

Fig. 10.53 A four-position binary parallel adder.

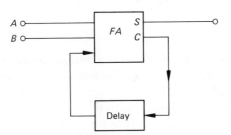

Fig. 10.54 Serial adding stage.

Fig. 10.55. If we assume that the numbers to be added, A and B, have been set into their respective shift registers, the addition will start when the add gate is opened. The full adder will consider the least significant column of the numbers A and B and create the sum at output S during the first clock period. At the start of the second clock period, this sum is shifted into register A which doubles as the accumulator. The second least significant column of the numbers A and B is presented to the full adder during this second clock period. If a carry signal was generated during the first clock period it was delayed one bit time and will appear at the full adder input during the second clock period. As the sum of each column is formed it is stored in the accumulator at the end of the clock period as each previously stored bit shifts further along the register. If the numbers contain 16 bits, the addition will be completed and stored in the accumulator 17 clock periods after opening the add gate.

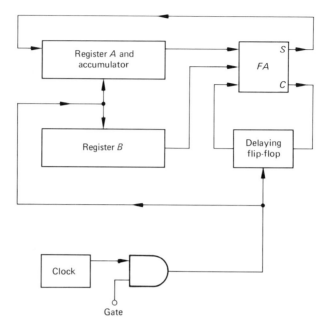

Fig. 10.55 Serial adder.

B. Subtractors

Subtractors can be constructed by creating a borrow function in a manner similar to the creation of the carry function in the full adder. A system can be designed based on the full adder that will add or subtract depending on an input control signal. Another method that is used effectively involves the formation of the complement of the number to be subtracted. Subtraction can then be accomplished by adding the two numbers. There are some additional considerations that must be made to apply this method, but we will not discuss these points in further detail.

C. Multiplication

Multiplication is performed by applying an algorithm that is based on repeated addition. Suppose it were required that 52 be multiplied by 83. A very straightforward method would start with an accumulator that initially contains a zero value. The number 83 could then be added to this accumulator 52 times, each time storing the intermediate sum in the accumulator. A reversible counter could be set to 52 and decremented each time an addition is made. At the end of 52 additions, this counter would reach zero and the number in the accumulator would represent the product of 52 and 83. This method is quite inefficient especially if large numbers are to be multiplied. Multiplication and division are required so often in a typical computer program that it is essential to minimize the time required in these processes.

A procedure that decreases the total time required for multiplication applies repeated addition and column shifting, much as we do in longhand multiplication. To multiply 52 and 83 in this way, the product of 2 and 83 would be formed by repeated addition, then the product of 5 and 83 would be formed. The latter product would be shifted one position toward the most significant column and added to the former product. In effect, we would create the product 50 × 83 by forming the product 5 × 83 and shifting the result one column to the left. In a digital computer, the multiplication is carried out in binary or some closely related number system. This same principle of repeated addition is generally used in computer multiplication of binary numbers. Division can be done using a subtractor along with the appropriate algorithm. We can therefore conclude that adders and subtractors form the basis of the important arithmetic unit of a computer. Of course, any arithmetic unit is fairly complex, but the individual components of the unit include little more than registers, gates, triggers, adders, and subtractors.

*10.6 MEMORIES

There are several aspects of computer operation that can only be discussed briefly at this time. While we have not considered the control unit in detail we have considered the various circuits that compose this section of the CPU. One important area that has not been treated is that of memory and storage devices. We will now summarize the operating characteristics of magnetic memories and core storage.

The main memory of a computer is generally made up of magnetic cores while the file or bulk memory can be a magnetic tape, disc, or drum. Smaller special-purpose computers may have no need of a file memory; however, most general-purpose units will have one or more types of magnetic file memory.

A. Magnetic Recording

Magnetic tapes and discs have a thin layer of magnetic oxide coating on their surfaces. This coating is capable of being magnetized to create flux in either direction at localized points on the surface. The oxide can also be demagnetized if necessary. The tape or disc is magnetized by a head that has a slight gap as shown in Fig. 10.56. Flux is created in the head and the gap by running current through the winding. The flux fringes in the vicinity of the gap and passes through the oxide coating, magnetizing the region of tape under the gap.

If a pulse of current is applied the flux ϕ will vary as shown. As the tape moves, a small area of the tape will be magnetized by this flux. Magnetic lines of flux will emanate from this spot until the spot is demagnetized at some later time. If a second head is placed in contact or very near the moving tape, this flux will produce a changing magnetic field in the gap of the read winding. As the center of the spot nears the center of the gap the flux in the read head increases; as the center of the spot moves past this point, the read head flux decreases. The read

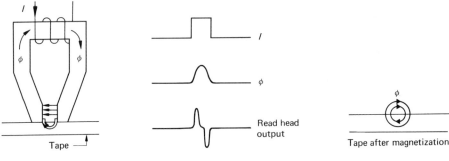

Fig. 10.56 Magnetic recording head.

head output voltage is also shown in Fig. 10.56. Certainly the speed of the tape influences the output voltage since the change in flux with time depends on tape speed.

If a constant current is impressed on the write windings the record flux will be constant and a longer area of tape will become magnetized. Except for the points at which this long path of magnetized material starts and stops, no voltage will be produced when this path moves under the read head. No change in flux will be present except at the end points of the path.

One type of recording, referred to as return-to-zero (RZ), applies a negative current to the write windings at all times except when a 1 bit is to be recorded. At this time the current switches positive for a short time. The negative current simply magnetizes the tape in the opposite direction to override any residual magnetization that may have been present. The waveforms resulting from RZ code along with a necessary clock signal are shown in Fig. 10.57. The clock signal can be generated by a parallel recording track or from an astable that is synchronized

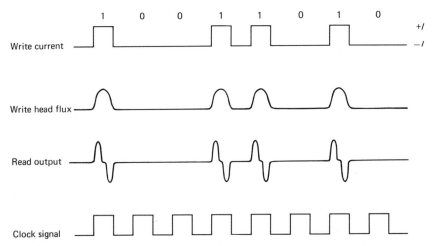

Fig. 10.57 RZ recording. (*Note:* Read output signal appears some time later than write current and flux.)

from the read winding signal. The read output pulses will be amplified and shaped into logic-level signals.

A second method of recording binary data is the nonreturn-to-zero (NRZ) technique where the write current does not return to zero between successive 1 bits. This code is shown in Fig. 10.58.

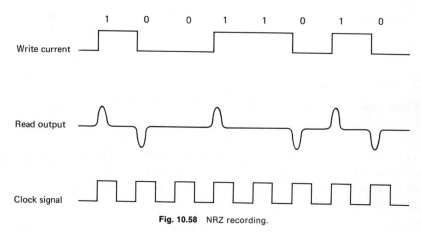

Fig. 10.58 NRZ recording.

We note that in NRZ recording there is no output signal for the case of a 1 bit succeeding another 1 bit. Thus, when a positive transition is made from 0 to 1 an RS flip-flop can be set to the $Q = 1$ state. When a negative pulse appears at the read output signifying a transition from 1 to 0, the latch can be reset to $Q = 0$. Figure 10.59 shows a scheme for generating the logic-level code.

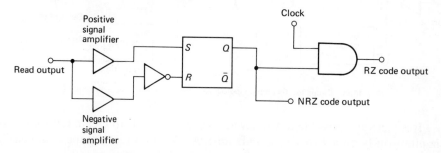

Fig. 10.59 Circuit for developing logic-level codes from NRZ recorded signal.

If an NRZ logic level code is desired, the output is taken at the latch output. If the RZ logic level code is desired, the latch output is combined with the clock at an AND gate. The NRZ method is capable of producing higher frequency operation than the RZ method since the time between transitions is at least halved.

A third method of magnetic recording is the nonreturn-to-zero-inverted (NRZI) technique. In this method any flux transition is interpreted as a 1 while

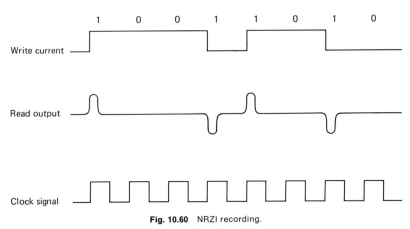

Fig. 10.60 NRZI recording.

no transition during a bit time is interpreted as a zero. Both positive and negative transitions must lead to a logic-level signal corresponding to the 1 bit. Figure 10.60 shows the appropriate waveforms while Fig. 10.61 indicates a circuit for conversion of the recording signals to logic-level signals.

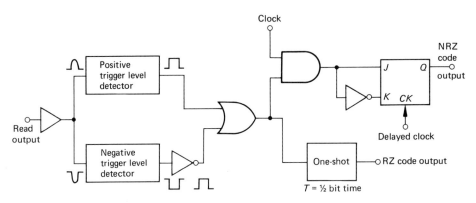

Fig. 10.61 Circuit for developing logic-level codes from NRZI recorded signal.

It is generally possible to use several recording heads to create several tracks simultaneously on a $\frac{1}{2}$ to 1-inch wide tape. A 2400-foot reel of tape could easily store 20 million 36-bit characters. Access to this information is rather slow compared to random access discs or magnetic drums. Magnetic tapes are generally used in applications wherein the information to be stored will be presented to the computer in serial fashion.

The gap width of a recording head might typically be 10^{-4} inches or less. For most magnetic recording the mechanical wear of the head and the magnetic medium can be minimized by using air heads. A very thin cushion of pressurized air keeps the head a small distance away from the magnetic surface.

A typical bit density for digital computer work is 1000 to 2000 bits per inch although densities of 10 times these figures have been reported in research. The writing frequency is typically in the 1-MHz region. The speeds of the logic units are considerably faster than bulk memory speeds; consequently, a buffer register is used in transmitting information between this memory and the main memory or CPU.

At higher recording frequencies the flux strength of the recorded bits decreases. To induce a reasonably large output voltage in the read windings, the relative speed between head and tape can be increased. This increase in speed can be accomplished more easily for a drum or disc than for tape. At very high speeds the tape is difficult to handle since it can be stretched, flexed, or broken. One method to reach relative head-to-tape speeds of greater than 120 inches per second is to use movable heads that travel in a direction opposite to that of the tape. Three heads might be mounted on a rotating cylinder. As the tape moves in one direction the cylinder rotates in the opposite direction. The output of each head is used at the correct time to reconstruct the signal. The axis of the rotating cylinder can also be changed to record diagonal tracks on the tape. As the tape moves, each head records a diagonal track adjacent to the track of the preceding head.

Not only are discs or drums easier to use at higher speeds, but also access to any sector or address on these mediums can take place in milliseconds giving a more rapid, random-access capability. For many programs requiring a file, the disc or drum is much more reasonable than the tape due to this random-access feature. It should be realized, however, that the main memory of the computer will have access times in the microsecond range which is a considerable improvement over the disc or drum memory.

B. Magnetic Core Storage

The main memory of a digital computer in most cases consists of magnetic cores. In the early 1960s some outstanding advances in core memories were made. Core size was reduced to the point that memories of several million bits could be produced with overall dimensions commensurate with the remaining components of the digital computer. It is felt by many that the core is now reaching its ultimate potential in terms of size and cycle time. There is a minimum inner diameter of the core because of the wires that must pass through the core. Cycle time is related to the core size in general and further improvement in core performance is expected to be minimal.

Large-scale integration (LSI) has made it possible to use integrated registers for relatively large memories at a cost that is competitive with core storage. Integrated circuit memories can have lower cycle times than cores and LSI memories of over 100,000 bits have been used in commercial computers.

A magnetic core is constructed from some type of magnetic material such as magnetite and magnesium or manganese. A powder is pressed into the desired shape and fired to form the ferrite core. Figure 10.62 shows a magnetic core with

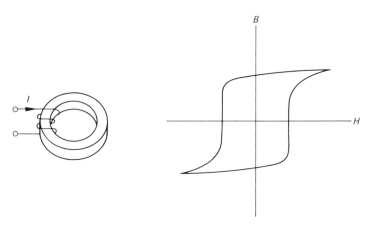

Fig. 10.62 Ferrite core and hysteresis curve.

a single winding. Ferrite cores might typically have a 30 mil inside and 50 mil outside diameter.

A curve showing magnetic flux of the core as a function of winding current is sketched in Fig. 10.63. With no current through the winding the core flux will have a value given by point 1 or point 4 depending on the magnitude and direction of the current last flowing in the winding. Let us assume that the flux corresponds initially to that of point 1. As current increases, the operating point moves to points 2 and 3, but always returns to point 1 when the current ceases. If current becomes negative the operating point moves along the curve toward point 6. As I_m is approached the core flux switches rapidly from positive to negative. Once the value of current moves to $-I_m$ or more negative, removal of the current results in a core flux corresponding to point 4. A current increase to or beyond the value

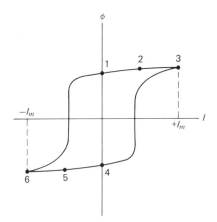

Fig. 10.63 Hysteresis curve in terms of core flux and winding current.

$+I_m$ with subsequent removal of current returns the operating point to point 1. Thus, the core has two stable states. The particular state of the core is determined by the winding current. We might observe that although the core has two externally controlled states we must consider some method of determining the state of the core at any specified time. The flux could be measured to determine direction, but this would involve complex equipment for each core. A far more efficient method utilizes the fact that as flux linkages change with time, a voltage is induced in the winding. Hence, a sense winding is used to detect flux change by registering a voltage as the change occurs.

If we arbitrarily define the 1 state to correspond to point 1 and the 0 state to correspond to point 4, we can consider the specific method of data readout from the core. Assuming the core rests in the 1 state, a negative current $-I_m$ applied to the input winding will change the flux from a positive to a negative value. A relatively large voltage will develop across the sense winding as the flux switches. If the core is in the 0 state, a negative current applied to the input winding changes the flux very little from point 4 to point 6 and back to point 4 as the current pulse drops to zero. The output voltage of the sense winding is three to four times as great when the core was in the 1 state prior to the current pulse. The duration of the output pulse is also longer when the core was in the 1 state prior to the input pulse. Suitable detection methods can discriminate between the outputs of the two cases and convert the larger output to a logic level 1 and the smaller output to a logic level 0.

The detection of the state of the core is accomplished at the expense of resetting the core to the 0 state. After the negative current pulse has been applied, the core will be in the 0 state regardless of the state prior to switching. This type of memory element in which data is erased or destroyed upon readout is called a destructive readout memory (DRO). A great deal of information stored in a memory must be retained after the initial readout; thus, nondestructive readout (NDRO) methods have been developed to be used with core memories. Before considering NDRO memories we will look more closely at DRO memories.

DRO core memory. Cores are generally arranged in several planes of square matrices. A given plane is threaded by four wires as shown in Fig. 10.64. The 1 bit can be written into a given core by applying currents of $\frac{1}{2}I_m$ to the appropriate X-select and Y-select lines. Only the core through which both activated lines pass will be set to the 1 state. Other cores will contain a wire with a current of $\frac{1}{2}I_m$, but this current is insufficient to set a core to the 1 state. Upon removal of the current the coincident core has been set, hence the name coincident-current memory. Any one of the nine cores can be selected by activating two of the six select wires.

To read the contents of a given core, a current of $-\frac{1}{2}I_m$ is passed through the appropriate two select lines. If a 1 is present the output appears on the sense winding. It is possible to wind the sense winding so that the polarity of the output voltage is the same for all cores; however, noise considerations may prevent this.

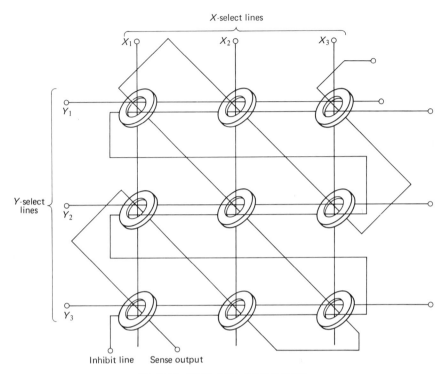

Fig. 10.64 Single plane of core memory.

The inhibit line would not be necessary for a single plane, but is required for a multiplane memory. At this point it is sufficient to note that if a current of $\frac{1}{2}I_m$ is applied to this winding when all cores are in the 0 state, it is impossible to write a 1 bit in any core. Let us assume that the inhibit line and the X_3 and Y_2 lines each carry a current of $\frac{1}{2}I_m$. The net current through the selected core is only $\frac{1}{2}I_m$ since the inhibit current opposes the select currents. Activation of this line inhibits the writing of a 1 in the selected core.

A complete coincident-current memory consists of several planes of cores as shown in Fig. 10.65. To store an n-bit word, n planes are required. Each Y-select line passes through a corresponding row on all planes and each X-select line passes through a corresponding row on all planes. For example, the Y_6 line threads the cores of row 6 on each plane. A given word is stored vertically in the planes with one bit stored in each plane. If a word contains 16 bits, and is stored in the $X_5 Y_8$ position, the least significant bit may be stored in the upper plane, the next least significant bit will be stored in the next plane, and so on up to the most significant bit which is stored in the sixteenth core plane. Each bit of this word will be stored in the $X_5 Y_8$ position of a plane.

The word capacity of the memory equals the number of cores on a single plane. A plane with 32X- and 32Y-select lines contains 1024 cores and the entire memory including all planes will store up to 1024 words.

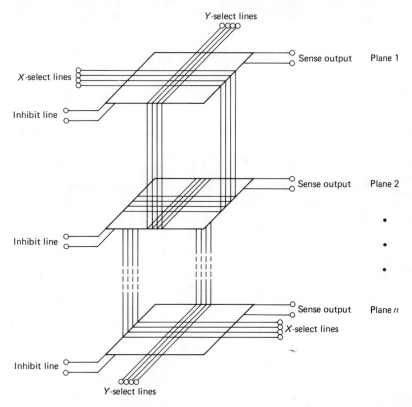

Fig. 10.65 Coincident-current memory.

To demonstrate the transmission of data between a CPU and a core memory we will consider the 64 word, 4-bit/word memory of Fig. 10.66. While the memory system of this figure is impractical because of the small number of bits per word and the fact that it is a DRO unit, it nevertheless demonstrates several points concerning practical NDRO memories.

When a data word is to be read into the memory, an appropriate address is determined by the computer. Both X and Y coordinates will be specified in 3-bit binary and these numbers will be entered into the X and Y address registers. The data word is entered into the data-in register. We will consider the case of the word 1110 (14) to be read into the X_5(101), Y_2(010) position. The X and Y decoders will convert the binary addresses to select the X_5 and Y_2 lines. When the write gate is opened the X_5 and Y_2 current drivers produce currents of $+\frac{1}{2}I_m$ through the X_5Y_2 core of all four planes. However, the inhibit driver of the LSB plane produces a current of $\frac{1}{2}I_m$ through all cores of this plane. The direction is such to cause a net current of $\frac{1}{2}I_m$ through the X_5Y_2 core of this plane. No other inhibit driver is activated as a result of the data set into the data-in register. The X_5Y_2 core of the upper three planes contain a 1 bit after opening the write gate. The

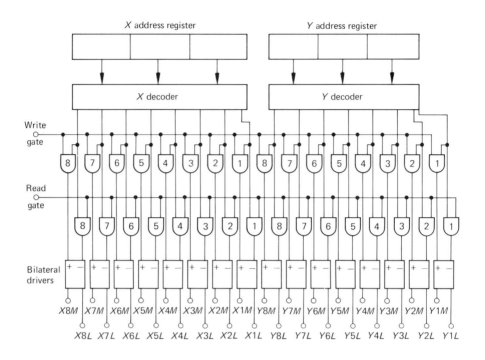

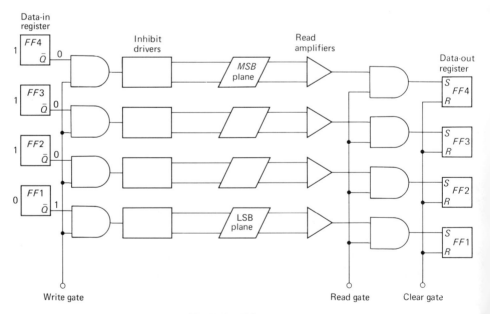

Fig. 10.66 DRO memory.

data word 1110 is now stored in the memory. It is seen that the purpose of the inhibit windings is to offset the effects of the select currents in all core planes that are to store the 0 bit.

When it is desired to read the data stored in this particular position, the two address registers must be filled with the appropriate address codes. The decoders again select the X_5 and Y_2 lines. The read gate is opened and the bilateral drivers now drive currents of $-\frac{1}{2}I_m$ through the X_5 and Y_2 lines. Those planes which contain the 1 bit in the X_5Y_2 core will exhibit an output as the cores switch to the 0 state. The core with a 0 bit will have an output that will not be passed by the read amplifier. The amplifier outputs are gated to flip-flops of the data-out register. All planes containing the 1 bit will cause the corresponding latch of this register to be set as the read gate is opened. We note that the data in the X_5Y_2 position is destroyed upon readout with all four cores reset to the zero position.

NDRO core memory. An NDRO memory can be obtained from the DRO system with little additional circuitry. When a word is read into the data-out register, this information is immediately rewritten in the appropriate cores. Figure 10.67 shows one plane of the control circuitry required for a complete NDRO system. The address registers, decoders, and line-drive circuitry are not shown for the sake of clarity.

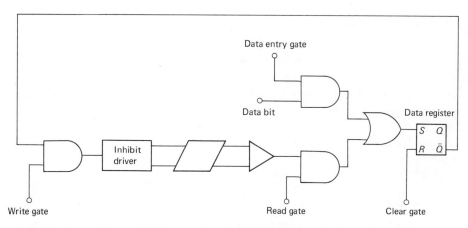

Fig. 10.67 One plane of an NDRO memory.

To enter data, the data word bits are presented to one input of each data entry AND gate. When the data entry gate is opened the bits are stored in the flip-flops of the data register. A write gate will result in the data word being entered into the memory at the position indicated by the address registers. The appropriate inhibit drivers will be activated from the flip-flops of the data register to write zeros in the planes requiring a 0 bit.

When the data is read out, a clear pulse resets the data register to zero. The read gate then opens (after the address registers have been set) and the core data fills the data register. The read gate is then closed and the write gate opens, rewriting the word into the same core position from which it was just derived. The data remains in the data register for use by the computer until the read-write cycle is again initiated or until new data is entered into the memory. The important waveforms of the read-write cycle are shown in Fig. 10.68.

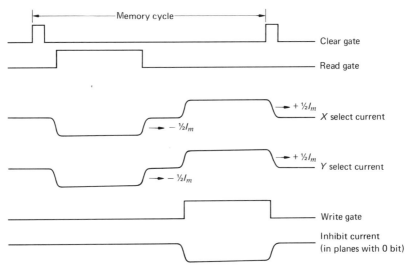

Fig. 10.68 Waveforms of NDRO memory.

The memory cycle consists of a clear gate, a read gate, and a write gate. This cycle time determines the speed of the memory since NDRO is required in general-purpose computers. Cycle times for present-day digital computers range from a few hundred nanoseconds up to several microseconds. Access times of the core memory are much shorter than access times of the disc or drum by factors of 1000 to 10,000.

C. Integrated Circuit Memories

There are several types of binary memory circuit included in a typical computer. We have already discussed the static register and the shift register, both of which are constructed from flip-flop circuits. These memories exhibit the characteristics of high speed and low storage capacity. In recent years, IC techniques have been improved to the point that large memories are available that, in many cases, compete favorably with core storage memories. As far back as the beginning of the 1970s, some computers were being designed with mainframes that were completely solid-state (IBM 370/135 and 370/145), while others combined semiconductor and core memory to advantage (IBM 370/155 and 370/165).

Both bipolar and MOS transistors can be used in flip-flop circuits to provide static storage of binary information. The word static in this case means that, once set, each memory cell maintains its state so long as power is supplied continuously to the circuit. The very high gate impedance of the MOS device has made the dynamic register a practical storage unit. In the dynamic register, the binary information is stored as charge on the inherent capacitances of the device gates. Ideally, these stored bits would remain as long as necessary with no further input power required; however, junction leakage slowly drains the charge. Dynamic memories then require the capacitances to be recharged at certain intervals to maintain the stored information. Since power consumption only takes place during recharging of the capacitors, the big advantage of the dynamic memory is low power consumption.

Although it is not the purpose of this section to discuss IC memories in detail, some of the more popular memory units will be considered.

The Texas Instrument SN7484A is a good example of a 16-bit memory. This IC can be used in the construction of a larger memory using the same organization as the magnetic core memory. In this case, however, nondestructive readout is possible with no additional circuitry. Figure 10.69 shows the logic diagram of this TTL memory.

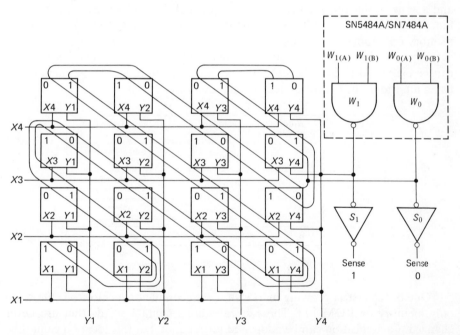

Fig. 10.69 16-bit active-element memory (TI SN7484A). (Courtesy of Texas Instruments Incorporated.)

The flip-flops are arranged in a four-by-four matrix with the X address lines determining the row and the Y address lines determining the column of the flip-flop to be addressed. Once the appropriate X and Y address lines have been raised to the binary one level, a bit can be written into the corresponding memory cell by opening the write gates by setting $W_1(A) = W_0(A) = 1$. If a zero is to be entered, then $W_0(B) = 1$ and $W_1(B) = 0$. To enter a one, $W_1(B) = 1$ and $W_0(B) = 0$. The memory can be read from by noting the level of the sense outputs after the X and Y address lines have been set. This memory requires from 20–30 ns to write new information and about the same time for readout. Thus, it is considerably faster than the core memory unit and does not require a read-write cycle each time data is read out.

Another interesting feature of this unit is that it can be expanded horizontally in the X or Y directions and several planes of this memory can be constructed. For example, if 64 8-bit words are to be stored, four ICs will make up each of eight planes. The organization is exactly the same as the core memory with each plane storing bits of equal significance for 64 different words.

A second method of IC memory organization is demonstrated by the Texas Instrument SN7489 64-bit read/write memory [5]. This chip features nondestructive readout of the 16 words of 4-bit length each. There are four select lines which address any of the 16 locations in binary fashion. There are four input data lines and four sense output lines to enter and read out of the memory. A memory enable and a write enable line are provided for control of the read/write operations. The function table of Fig. 10.70 indicates the conditions required in controlling the memory operations. This table assumes that the binary address lines have been activated. The 64 memory cells are again made from flip-flop circuits. The memory can be expanded to include more bits per word or more words per memory by using additional ICs. Typical access time for this circuit is 33 ns.

Memory enable	Write enable	Operation	Sense output lines
L	L	Write	Complement of input data
L	H	Read	Complement of selected word
H	L	Inhibit storage	Complement of input data
H	H	Do nothing	High

Fig. 10.70 Function table of the TI SN7489 memory.

One of the fastest growing markets for semiconductor memories is the read-only memory or ROM [1]. These storage units contain words that are never changed. Each location can be addressed for readout, but this location cannot be written into. There are several small ROMs available in IC form at present. These vary in size from units of perhaps 256-bit capacity up to larger MOS units of 4096 bits. The random-access memory or RAM can also be realized with bipolar transistor ICs or with MOS ICs, although for high-speed operation, the MOS unit is not so appropriate as the bipolar memory. If the MOS RAM is dynamic, periodic refreshing of the stored information is necessary to prevent complete discharge of the gate capacitors. Although this requirement for refreshing leads to slightly more complex operation, the packing density of MOS circuitry is quite good, leading to a high storage capacity for the MOS memory.

REFERENCES AND SUGGESTED READING

1. E. R. Hnatek, *A User's Handbook of Integrated Circuits.* New York: Wiley, 1973.
2. T. Kohonen, *Digital Circuits and Devices.* Englewood Cliffs, N.J.: Prentice-Hall, 1972.
3. M. M. Mano, *Computer Logic Design.* Englewood Cliffs, N.J.: Prentice-Hall, 1972.
4. Texas Instrument Staff, *Designing with TTL Integrated Circuits.* New York: McGraw-Hill, 1971.
5. Texas Instrument Staff, *The TTL Data Book for Design Engineers.* Dallas: Texas Instruments Inc., 1973.

PROBLEMS

Section 10.1

10.1 Convert the following decimal numbers to binary: (a) 17; (b) 238; (c) 1024; (d) 1023.

* **10.2** Convert the following binary numbers to decimal: (a) 11111111; (b) 10010110; (c) 11100011; (d) 1101010101.

10.3 Represent the following decimal numbers with binary-coded decimal numbers: (a) 146; (b) 23; (c) 8974; (d) 27981.

10.4 Convert the following decimal numbers to binary numbers with at least 1/1000 accuracy: (a) 0.123; (b) 0.279; (c) 0.999; (d) 0.111.

* **10.5** Convert the following binary numbers to decimal numbers with at least 1/1000 accuracy: (a) 0.11110; (b) 0.1010101; (c) 0.000001; (d) 0.100100.

Sections 10.2A–10.2B

Using Boolean algebra determine whether the given equations are true or false.

10.6 $\overline{AB} + C\bar{A} + \bar{B}C = \overline{AB}$

10.7 $\bar{A}\bar{B}C + \bar{A}BC = \bar{A}$

* **10.8** $\overline{\bar{A}\bar{B}\bar{C}\bar{D}} + D = \overline{A + B + C} + D$

10.9 $AB\bar{C} + A\bar{B}C + \bar{A}BC = \overline{\overline{AB} + \overline{AC} + \overline{BC} + ABC}$

10.10 $AB + CD + \bar{A}\bar{C} = \bar{A} + B + \bar{C} + D$

*__10.11__ Implement the exclusive OR gate using only inverters and (a) NOR gates; (b) NAND gates.

Section 10.2C

10.12 Minimize the expression $f = ABC + \bar{A}BC + \bar{A}\bar{B}C$.

10.13 Minimize the expression $f = \bar{A}\bar{B}\bar{C}\bar{D} + \bar{A}B\bar{C}D + \bar{A}BC\bar{D} + A\bar{B}CD + AB\bar{C}D$.

*__10.14__ Minimize the expression $f = (AB + CD)(A\bar{C} + BD)$.

10.15 Minimize the expression $f = (A + BC)(A + B + CD + \bar{A}C)$.

Section 10.3A

10.16 Consider the BCD-to-decimal decoder of Fig. 10.21. Only the ten decimal numbers are required at the output. Is it possible to reduce the total number of inputs to the gate and still achieve the desired decoding function? Can a reduction in inputs be effected if 16 distinct output lines are required? If so, calculate these numbers.

*10.17 Calculate the number of gates and inputs required to decode a 6-bit binary code with single-stage gating.

10.18 Minimize the number of inputs required to decode a 6-bit binary code. Calculate the total number of gates and inputs required.

10.19 Minimize the number of inputs required to decode an 11-bit code. Calculate the total number of gates and inputs required.

*10.20 Calculate the number of gates and inputs required for two-stage decoding of a 14-bit code.

10.21 Minimize the number of inputs required to decode a 14-bit code. Calculate the total number of gates and inputs required.

Section 10.3B

10.22 Design an encoder that has three binary output stages with the eight input lines corresponding to 0, 1, 2, 3, 4, 5, 6, and 7. Use single-stage gating.

10.23 Can the number of inputs to the encoder of Problem 10.22 be minimized with additional stages of gating? If so, show such a system.

Section 10.3C

10.24 Design a code converter that converts a 6-bit binary input to a BCD output. Use single-stage gating.

Section 10.4A

*10.25 Explain the difficulties that occur if the first one-shot of Fig. 10.29 is retriggerable.

10.26 Assuming an inverted data signal swinging between 3.5 V and 0.2 V, select all element values for the circuit of Fig. 10.31 to lead to half periods of 0.5 ms and an output impedance of less than 500 Ω. Use a 5-V power supply.

Section 10.4B

10.27 Sketch the waveforms appearing at point A, B, and C in the figure.

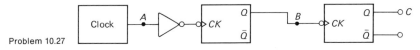

Problem 10.27

*10.28 Sketch the output waveform of the positive logic AND gate shown in the figure. Assume that the flip-flop changes state on the negative clock transitions.

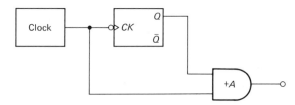

Problem 10.28

10.29 Draw a timing chart showing signals $Q1-Q4$ and $CK1-CK4$ for the synchronous counter of Fig. 10.35, assuming a clock signal is applied to the input.

Section 10.4C

10.30 Design a logic system that allows 14 negative transitions of a clock signal to appear on a line after which the line is gated off. Assume that a start strobe is available that is synced with the positive transition of the clock.

10.31 Repeat Problem 10.30 for 17 pulses and a start strobe that is asynchronous relative to the clock.

Section 10.4D

10.32 Design a 4-bit static register using only NAND gates, NOR gates, and inverters.

10.33 Redesign the I/O register of Fig. 10.48 eliminating all one-shots from the design. Show pertinent timing charts when the word 110110010 is transmitted or received. State any reasonable assumptions made.

10.34 Design a transmission system to transfer data in parallel from a static register to shift register 1 when a 0.5 ms transfer strobe occurs. When a 0.5-ms shift strobe occurs data from shift register 1 must be transferred to shift register 2 (remote location) over a two-wire system (1 ground line and a signal line). Assume that a 1 kHz clock signal is available plus a delayed clock signal ($\frac{1}{4}$ bit delay). Assume that the transfer and shift strobes occur in sync with the clock. A start bit must precede the eight-bit word when transferred to register 2. The static register is filled by a data input circuit which has already been designed. Explain the operation of your system in detail using a timing chart. Economize in design. Assume the following costs.

Circuit	Cost
two-input NAND or NOR	$ .10
three-input NAND or NOR	.12
Hex inverter	.30
RS flip-flop	.23
JK flip-flop	.36
Monostable	1.20
Astable	1.40

Section 10.5

10.35 Design a 4-bit parallel adder with associated registers with provisions for filling the registers serially, but performing the addition in parallel. A register should serve as the accumulator. Assume the availability of a clock signal and any control signals required. State and explain any additional assumptions.

10.36 Design a 4-bit serial adder using a single full adder. Use one shift register to store one of the addends and also to store the result. Assume the availability of a clock signal, a delayed clock signal ($\frac{1}{4}$ bit delay), and an add gate in sync with the clock signal to initiate the operation.

Section 10.6A

10.37 Show a general system for converting logic-level NRZ code to RZ code.

10.38 Show a general system for converting logic-level RZ code to NRZ code.

Section 10.6B

10.39 How many cores are required for a 4096-word coincident-current memory if there are 16 bits/word? How many core planes will be used? How many X-select and Y-select lines are required? How many bits are required for the address?

*__10.40__ A core storage unit contains a total of 8192 cores. If the X and Y address registers each have four binary output lines, what is the minimum number of inhibitor drivers that can be used if all positions of the memory are to be addressable?

Section 10.6C

10.41 Using the SN7484A IC shown in Fig. 10.69 design a 28-word, 4 bit/word memory.

11
WAVEFORM GENERATION

Power is supplied to electronic circuits by either a dc voltage source or a fixed-frequency ac source. If the power is supplied by an ac waveform it is converted to a dc signal by means of a power supply. This dc power is then transformed into the waveform required by various wave generation or wave shaping techniques. Examples of important signal types are sinusoids, ramps, rectangular waveforms, and pulses.

One of the most important signal generators is the sinusoidal oscillator. This chapter will first consider the design of the oscillator and then proceed to a consideration of various nonsinusoidal signal generation schemes. The multivibrator circuits of Chapter 9 are quite important in the production of rectangular waveforms, and will be considered further in this chapter.

11.1 SINUSOIDAL OSCILLATORS

A. General Design Considerations

When several stages of amplification are used in a negative feedback amplifier, reactive effects often cause an additional 180° phase shift around the loop changing negative feedback to positive feedback. It is not unusual that compensating networks must be added to avoid oscillation problems. One might conclude that oscillator design is very simple in view of the fact that instability occurs so often in amplifier design. It is, however, quite a different matter to construct a circuit that oscillates at a prescribed frequency with a prescribed stable amplitude than to simply achieve oscillations. A useful design procedure must lead to a circuit that meets specific requirements.

As was noted in Chapter 6, the gain expression for a closed loop circuit applying positive feedback is

$$G = \frac{A}{1 - AF}. \qquad (6.34)$$

As AF approaches unity the overall gain approaches infinity and a finite output voltage can result when no input signal is present. It is not quite as easy to apply Eq. (6.34) as it might first appear. Other factors tend to complicate the direct application of this equation. For example, both A and F are generally frequency-dependent and A varies with the operating point of the active devices used in the amplifier. Temperature or aging effects can also modify values of A and F to further complicate reliable design.

For sinusoidal oscillators, the term AF must have a magnitude of unity with a phase shift of $0°$ or some multiple of $360°$ at a single frequency. When the return ratio AF has unity magnitude and the correct phase angle, the output signal feeds back exactly the correct input signal to reproduce the output. Random noise or transients can cause the oscillations to begin, but they are sustained by recirculating the appropriate signal. It may appear that the circuit is manufacturing its own energy since the input signal is amplified to produce an output signal which in turn produces the input signal. There is dc power entering the oscillator being converted to ac power in the process of amplification, and thus there is a net power flow into the system.

If the return ratio is increased drastically the output signal will tend to follow this increase. In general, at least one device will be driven out of its active region and the return ratio will then drop to unity or lower. In relaxation oscillators such as the astable multivibrator, the return ratio tends to a large value, forcing the active devices to switch rapidly across the operating region. The gain around the loop is then very low until a charging capacitor forces a device into the active region once more. At this point the astable switches rapidly to an alternate state again dropping the value of the loop gain or return ratio.

The basic design goals for the sinusoidal oscillator are (1) to create a return ratio magnitude of unity or slightly larger at the desired oscillation frequency, (2) to create a phase shift in the return ratio of $0°$ or $360°$ at this same frequency, (3) to ensure that the first two conditions are not satisfied at other frequencies, and (4) to ensure that the first two conditions will continue to be satisfied as parameter values change due to aging, temperature change, or device replacement. The first three conditions can be met in a variety of ways while the fourth condition often requires special consideration.

Example 11.1. Given that the gain of an ideal amplifier with infinite input and zero output impedances is 100, find a suitable feedback network to convert this amplifier to a sinusoidal oscillator.

Solution. The circuit of Fig. 11.1 will lead to a magnitude for AF of unity and a zero-degree phase shift when the tank circuit is resonant. We can express the feedback factor as

$$F = \frac{Z_t}{Z_t + R_1},$$

where Z_t is the tank circuit impedance. This quantity is

$$Z_t = \frac{j\omega L}{1 - \omega^2 CL + j\omega L/R},$$

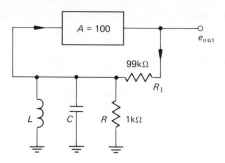

Fig. 11.1 Idealized oscillator circuit.

giving a return ratio of

$$AF = \frac{j\omega LA}{R_1(1 - \omega^2 CL) + j\omega L(1 + R_1/R)}.$$

Only at resonance will zero-degree phase shift occur and this frequency is given by

$$\omega_0 = 1/\sqrt{LC}.$$

At this frequency the magnitude of AF becomes

$$|AF(\omega_0)| = \frac{AR}{R_1 + R} = \frac{100R}{R_1 + R}.$$

The values $R = 1$ kΩ and $R_1 = 99$ kΩ lead to a return ratio of unity at the resonant frequency. An examination of the expression for AF shows that there is only one possible oscillation frequency since neither the phase shift nor magnitude conditions can be satisfied at any point other than resonance.

In practice R might be chosen slightly larger than 1 kΩ resulting in an open-loop value for $|AF|$ greater than unity. When the loop is closed, the gain A would readjust to a value such that $|AF| = 1$. We shall next examine the mechanics of this readjustment of gain.

B. Amplitude Stability

If the magnitude of the return ratio dropped below unity the circuit would cease oscillations. Since temperature changes, age, and operating-point changes can affect gain, the magnitude of AF might drop below unity as gain changes. It is standard procedure to design the circuit to lead to a value of $|AF|$ that is slightly greater than unity at the oscillation frequency. The greater the value of $|AF|$, the greater will be the amplitude of the output signal and the amount of distortion. This distortion will affect A and, in many cases, will automatically lower the gain to the required value. Because the variation of A with operating point is a complex nonlinear relationship, it is difficult to derive quantitative design equations. In order to demonstrate the nature of amplitude limiting, the following simplified example is used.

Example 11.2. In the circuit of Fig. 11.1, the gain of the amplifier is given by

$$A = B/r_d,$$

where B is a constant and r_d is the small-signal emitter-base diode resistance of the last transistor stage. We will also assume that the dc distortion present in the final-stage collector current is directly proportional to the amplitude of the output voltage swing; that is,

$$I_C = I_{CQ} - Ke_{\text{out}},$$

where I_{CQ} is the no-signal quiescent current, K is a constant, and I_C is the dc collector current that flows in the presence of the distorted signal. Find the amplitude of the output voltage.

Solution. The waveforms of Fig. 11.2 show the collector current before and after the distortion. As the distortion increases, I_C decreases which, in turn, increases r_d. The net result is a decreased gain. If A is designed to be 105 and $F = \frac{1}{100}$ at the oscillation frequency, the quantity AF would tend toward a value of 1.05. As AF increases toward unity, however, the overall gain as given by Eq. (6.34) approaches infinity. The output voltage could then reach a very large value. Since the distortion in I_C increases with output signal, the gain will automatically limit at a value of A which is approximately 100. In other words, e_{out} will stabilize at that value which limits AF to approximately unity. If we know the value of K and I_{CQ}, the amplitude of e_{out} can be found.

Since

$$A = \frac{B}{r_d} = \frac{BI_E}{0.026} = \frac{BI_C}{0.026\alpha}$$

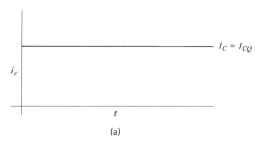

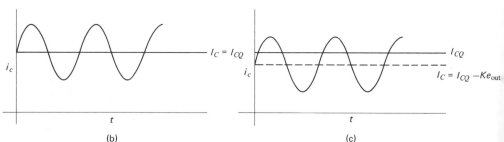

Fig. 11.2 Collector current: (a) no output signal; (b) nondistorted output signal; (c) distorted signal.

and $A = 105$ when $I_C = I_{CQ}$, then the gain can be written as

$$A = \frac{B}{0.026\alpha}(I_{CQ} - Ke_{out}) = 105 - \frac{BKe_{out}}{0.026\alpha}.$$

Since A must equal 100 (or slightly less) to limit AF to unity, the amplitude of the output voltage can be found as

$$e_{out} = \frac{0.026 \times \alpha \times 5}{BK} = \frac{0.130\alpha}{BK}.$$

If K is small, meaning that the gain variation with I_C is weak, the output voltage will be large. If K is large, the output voltage will be small. The amplitude stability is much greater for a large value of K. If a temperature change causes I_{CQ} to increase, Ke_{out} must increase to keep A constant. If K is large, then the change in e_{out} can be small, but if K is small then e_{out} must change a great deal to offset the change in I_{CQ}.

For good stability the change in gain with output voltage amplitude should be large and an increase in amplitude must result in decreased gain. Expressed mathematically, the derivative dA/de_{out} must be a large negative number for good stability.

In a practical amplifier the gain A depends on many parameters that are dependent on operating point. Transistor current gain and FET transconductance are functions of operating point. Input and output impedances may vary with bias points and the feedback factor F may be affected by impedance changes. In some cases, the gain may tend to increase rather than decrease, with output signal. In this case, the operating point can be affected drastically as the amplifier either leaves the operating region or moves to a region that causes the gain to decrease with further distortion.

For oscillators requiring stable, predictable operation, a dc signal proportional to output amplitude is generated and used to control the amplifier gain. In this way the gain A can be made to vary appropriately with e_{out} to create an amplitude-stabilized oscillator. Such an oscillator will be discussed in a later section.

C. Frequency Stability

The oscillation frequency of the ideal oscillator of Fig. 11.1 is a function only of L and C. These elements can change with temperature and the frequency at which zero-degree phase shift occurs will then change. In a less ideal oscillator parasitic capacitances will exist which may have a slight influence on the phase shift of the circuit. For example, the collector depletion-layer capacitance of a bipolar transistor or the gate capacitance of an FET may affect the oscillation frequency to some degree. As operating point changes, temperature changes, or device replacement modify these parasitic elements, the frequency of oscillation is changed.

The key to good frequency stability is to make the phase shift a rather strong function of frequency near the point of zero-degree phase shift. If this is the case, a change in value of a capacitance will lead to a phase shift resulting in a shift of frequency to restore the return ratio to zero-degree phase shift. If $|d\theta/d\omega|$ is large only a slight change in ω is required to correct the phase shift.

Let us return to the idealized oscillator of Fig. 11.1. The phase shift of AF can be expressed as

$$\theta = 90° - \arctan \frac{\omega L(R + R_1)/RR_1}{(1 - \omega^2 CL)}.$$

Since R_1 is much larger than R we can write

$$\theta = 90° - \arctan \frac{1}{Q(1 - \omega^2/\omega_0^2)},$$

where $Q = R/\omega L$ and $\omega_0^2 = 1/LC$. Differentiating this phase angle with respect to radian frequency, assuming Q is approximately constant near resonance, gives

$$\frac{d\theta}{d\omega} = \frac{-2\omega Q}{1 + Q^2\left(1 - \dfrac{\omega^2}{\omega_0^2}\right)^2}.$$

At the resonant frequency $\omega = \omega_0$ and this derivative becomes

$$\left.\frac{d\theta}{d\omega}\right|_{\omega=\omega_0} = -2\omega_0 Q.$$

The phase shift will change rapidly with frequency if the resonant circuit has a large value of Q. We conclude that the frequency stability is better for high Q circuits because the variation of phase shift with frequency near resonance is greater than that for a low Q circuit.

For extremely good frequency stability, crystal-controlled oscillators are used. Very high values of Q apply to the equivalent electrical circuit of the crystal leading to correspondingly high values of $d\theta/d\omega$ for these oscillators. A sometimes serious problem of the quartz crystal is the variation of electrical parameters with temperature. If temperature stability is important, the crystal can be placed in a heat oven maintaining a constant crystal temperature.

D. The RC Oscillator

A phase shift or RC oscillator is shown in Fig. 11.3. The circuit can be constructed without using an emitter follower, but there will be a much narrower range of element values that can satisfy the loop-gain requirements. Some approximations can be made in the circuit of Fig. 11.3 which allow several simplifications without involving a great loss of accuracy. For example, the load of stage 1 is R_L in parallel with the input impedance of the emitter follower, but this is approximately R_L. The gain of the emitter follower is assumed to be unity and the output impedance is assumed to be zero.

If only two RC stages were used, the phase shift around the loop could not equal 0°. The amplifier stage shifts the phase 180° so that each RC stage would

11.1 SINUSOIDAL OSCILLATORS

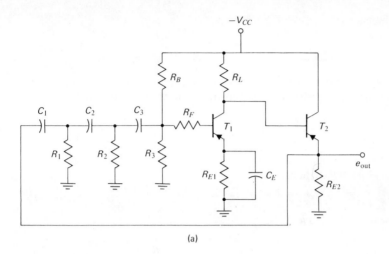

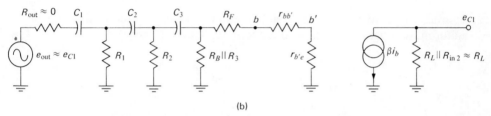

Fig. 11.3 RC oscillator: (a) actual circuit; (b) ac equivalent circuit.

then be required to shift the phase by 90°. Since a single RC stage cannot pass a finite signal with 90° phase shift, three stages are required as a minimum to produce oscillations. In order to simplify the design, the capacitors are chosen to be equal and both R_1 and R_2 are chosen to equal the resistance R, which is defined as

$$R = (R_B||R_3)||(R_F + r_{bb'} + r_{b'e}). \tag{11.1}$$

The equivalent circuit then appears as shown in Fig. 11.4. To analyze the circuit the gain $A = e_{C1}/e_A$ is found. The feedback factor F, which is a measure of the fraction of e_{C1} that reaches point A, is then found as $F = e_A/e_{C1}$ around the feedback loop.

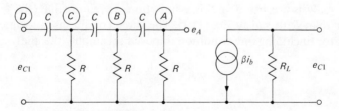

Fig. 11.4 Equivalent circuit of RC oscillator.

The gain is

$$A = \frac{e_{C1}}{e_A} = -\frac{\beta R_L}{R_F + r_{bb'} + r_{b'e}}. \quad (11.2)$$

When e_{C1} is applied to the feedback loop (point D), the voltage e_A is found by solving the appropriate loop equations. The result of this step is

$$F = \frac{e_A}{e_{C1}} = \frac{1}{1 - \dfrac{5}{\omega^2 C^2 R^2} + j\dfrac{1}{\omega^3 R^3 C^3} - j\dfrac{6}{\omega CR}}. \quad (11.3)$$

The product AF is then

$$AF = \frac{\dfrac{\beta R_L}{R_F + r_{bb'} + r_{b'e}}}{\dfrac{5}{\omega^2 C^2 R^2} - 1 + j\dfrac{6}{\omega CR} - j\dfrac{1}{\omega^3 C^3 R^3}}. \quad (11.4)$$

In order for AF to have no phase shift, the sum of imaginary terms in the denominator must equal zero at the frequency of oscillation. Hence

$$\frac{1}{\omega_0^3 C^3 R^3} = \frac{6}{\omega_0 CR},$$

giving

$$\omega_0 = \frac{1}{\sqrt{6}\,CR}. \quad (11.5)$$

Thus the phase-shift condition has been imposed on AF. The magnitude condition must now be satisfied; that is,

$$AF = \frac{\beta R_L/(R_F + r_{bb'} + r_{b'e})}{5/\omega_0^2 C^2 R^2 - 1} = \frac{\beta R_L}{29(R_F + r_{bb'} + r_{b'e})} = 1. \quad (11.6)$$

At the oscillation frequency, F is seen to be equal to $-\frac{1}{29}$. We now select R_L and $(R_F + r_{bb'} + r_{b'e})$ to satisfy Eq. (11.6). However, we must impose one constraint on the term $(R_F + r_{bb'} + r_{b'e})$ as a result of the simplifying steps that were taken previously. In Eq. (11.1), for convenience, we set

$$R = (R_B \| R_3) \| (R_F + r_{bb'} + r_{b'e}). \quad (11.1)$$

Since our analysis was based on this assumption, we require that $(R_F + r_{bb'} + r_{b'e})$ be greater than R to satisfy Eq. (11.1). The three conditions associated with the RC oscillator (including one condition imposed to simplify the analysis) are

$$\omega_0 = \frac{1}{\sqrt{6}\,CR}, \quad (11.5)$$

$$\frac{\beta R_L}{29(R_F + r_{bb'} + r_{b'e})} = 1 \quad \text{(or slightly larger)} \quad (11.6)$$

and
$$(R_F + r_{bb'} + r_{b'e}) > R.$$

Example 11.3. Assume that the silicon transistors in Fig. 11.3(a) have current gains of $\beta = 100$, $r_{bb'} = 200\ \Omega$, and T_1 is to be biased with an emitter current of 2 mA, giving $r_{be'} = 1313\ \Omega$. Select R, C, R_L, R_3, R_B, R_{E1}, C_E, R_{E2}, and R_F to give an oscillation frequency of $f_0 = 100$ kHz. Use a power supply of $V_{CC} = -12$ V.

Solution. Of course, there are many combinations of values that will satisfy this design. One starting point is to select R as a reasonable value, say 2 kΩ. Then C is calculated from Eq. (11.5) as

$$C = \frac{1}{2\pi \times 10^5 \times \sqrt{6} \times 2 \times 10^3} = 325 \text{ pF}.$$

We then choose R_{E1} to be 0.5 kΩ and C_E to be 1 μF to ensure complete bypassing of R_{E1} at the frequency of oscillation. We choose R_3 to be 10 kΩ, and R_B can be found, based on the bias requirement of $I_E = 2$ mA. The appropriate equations are (neglecting the small drop across R_F):

$$\frac{V_B}{10\text{ k}\Omega} + I_B = \frac{V_{CC} - V_B}{R_B}$$

and

$$V_B = I_E R_{E1} + V_{BE} = 1.0 + 0.5 = 1.5 \text{ V}.$$

We find R_B to be equal to 62 kΩ.

We can select R_{E2} as 5 kΩ in order to obtain a high input impedance to T_2. Next R_F is selected from Eq. (11.1). This value is found to be $R_F = 1.1$ kΩ. The minimum value of load resistance is found from Eq. (11.6) to be

$$R_L = \frac{29(R_F + r_{bb'} + r_{b'e})}{\beta} = \frac{29 \times 2.6}{100} = 758\ \Omega.$$

This value would be chosen slightly larger to cause the open-loop value of AF to be larger than unity.

The amplitude of the RC oscillator can be stabilized by deriving a signal which is proportional to the output amplitude and using this signal to control the gain. The arrangement shown in Fig. 11.5 would stabilize the output amplitude. The negative portion of the output waveform is amplified and rectified by the combination of T_3 and D_1. The capacitor C_F charges to the peak value of this amplified signal and is selected so that it will hold this charge for several cycles of the output waveform. The dc signal on the capacitor is proportional to the amplitude of the output signal. This voltage is applied to the bias network in such a way as to lower the base current when the signal on C_F becomes larger. Hence $r_{b'e}$ will increase and cause the voltage gain of T_1 to decrease. This circuit can be analyzed by the method used in Example 11.2.

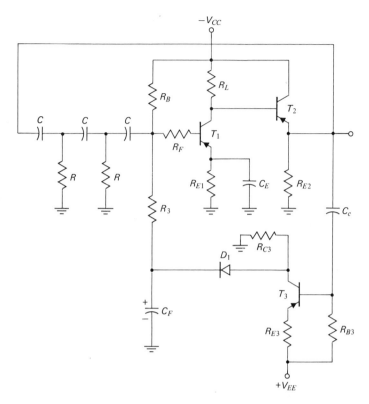

Fig. 11.5 Amplitude-stabilized *RC* oscillator.

Figure 11.6 shows an *RC* oscillator using the FET as the amplifying element. If $r_{ds} \| R_L$ is much smaller than R, Eq. (11.5) gives the oscillation frequency while the voltage gain from gate to drain must again equal -29; thus, $g_m(r_{ds} \| R_L) = 29$.

The range of element values generally used with the phase-shift oscillator dictates relatively low-frequency applications up to a few hundreds of kilohertz. The frequency can be varied without changing the amplitude of oscillations by tuning all three capacitors simultaneously. When this type of tuning is used, the resonant value of F remains at $\frac{1}{29}$.

E. Active Phase-Shift Oscillator

It is possible to use active phase shifters to achieve the required phase relation for the return ratio. Consider the circuits of Fig. 11.7.

It will be shown in Chapter 13 that the magnitude of the gain of the op amp circuit is approximately unity, while the phase shift is

$$\theta = -2 \tan^{-1} \omega CR.$$

11.1 SINUSOIDAL OSCILLATORS 497

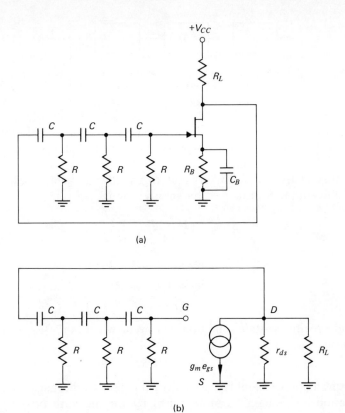

Fig. 11.6 (a) RC oscillator using the FET. (b) Equivalent circuit.

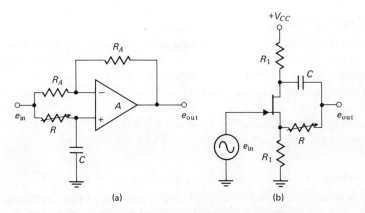

Fig. 11.7 Active phase shifters; (a) op amp circuit, (b) FET circuit.

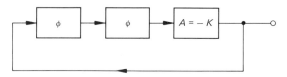

Fig. 11.8 Active phase-shift oscillator.

It can be shown that the same gain and phase relationships hold for the FET circuit, provided that $R_1 \ll R$ and $g_m r_{ds} \gg 2$. Figure 11.8 shows a block diagram of an active phase-shift oscillator.

With these active phase shifters and an amplifier of gain $-K$ having high-input and low-output impedances, the return ratio is

$$AF = -K\left(\frac{1-sCR}{1+sCR}\right)^2. \tag{11.7}$$

Letting $s = j\omega$ and rationalizing the denominator gives

$$AF = -K\frac{(1-j\omega CR)^4}{(1+\omega^2 C^2 R^2)^2}. \tag{11.8}$$

The zero-degree phase-shift condition will occur when $\tan^{-1} \omega CR = 1$ or at a frequency of

$$\omega_0 = 1/CR. \tag{11.9}$$

The inverting amplifier should have a unity gain to satisfy the magnitude condition for oscillations. This circuit can be tuned by varying both of the phase-shift resistors. Although tuning can be accomplished by varying only one resistor, Eqs. (11.8) and (11.9) will not apply. Note that neither variation of R nor of C affects the magnitude of AF; thus, the output signal remains constant as frequency changes.

F. The Tuned-Circuit Oscillator

The bipolar transistor or FET can be used as the active device in the tuned-circuit oscillator. These oscillators can operate in the class-A or class-C modes. We will first consider the tuned-collector oscillator in the class-A mode.

The tuned-collector oscillator is analyzed quite simply when it can be assumed that the transformer is ideal. The equivalent circuit for this stage is shown in Fig. 11.9. The equivalent circuit does not include $C_{b'c}$, but for high-frequency oscillators, this element should be included. The resonant frequency of the oscillator is

$$\omega_0 = 1/\sqrt{LC}.$$

At this frequency, the collector load is real and consists of the shunt combination of r_{out}, R_p, and the impedance reflected by the transformer. This load is

$$R = r_{out} \| R_p \| n^2(r_{bb'} + r_{b'e}),$$

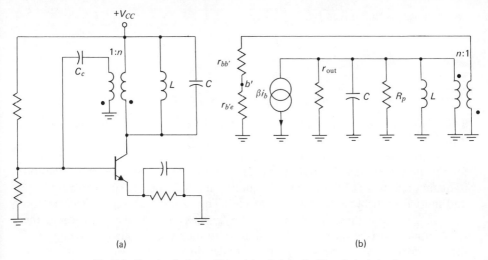

Fig. 11.9 Tuned-collector oscillator: (a) actual circuit; (b) equivalent circuit.

where R_p accounts for the losses of the inductor or transformer. The voltage gain from base to collector is

$$A = -\frac{\beta R}{r_{bb'} + r_{b'e}},$$

and the voltage fed back from collector to base is

$$F = -1/n.$$

The value of AF is then

$$AF = \frac{\beta R}{n(r_{bb'} + r_{b'e})}.$$

The value of n required for oscillations is

$$n = \frac{\beta R}{r_{bb'} + r_{b'e}}.$$

The term R is a function of n^2; thus the above equation is not too useful. An upper limit on n can be established by assuming that

$$n^2(r_{bb'} + r_{b'e}) \gg r_{out} \| R_p.$$

This gives a value for n of

$$n_{max} = \frac{\beta [r_{out} \| R_p]}{(r_{bb'} + r_{b'e})}. \tag{11.10}$$

Equation (11.10) predicts a fairly high value for n. This value can be lowered by including a resistance R_F in series with $r_{bb'}$, to lower the gain of the stage. The value of n_{max} is then given by

$$n_{max} = \frac{\beta[r_{out}||R_p]}{R_F + r_{bb'} + r_{b'e}}. \tag{11.11}$$

The use of R_F strengthens the assumption that

$$n^2(R_F + r_{bb'} + r_{b'e}) \gg r_{out}||R_p.$$

The turns ratio required is now given by Eq. (11.11), since loading effects of the input impedance can be ignored.

In practice, an ideal transformer will not be used. The inductance of the transformer may serve as the resonating inductance. The equation for oscillation frequency and Eqs. (11.10) and (11.11) can be put into terms of primary, secondary, and mutual inductances of the transformer. This is not a very practical method of design because these inductances must be measured. If the elements of the transformer must be measured, it is simpler to measure the input inductance of the transformer when the secondary is loaded in the same way that it will be loaded in the circuit. The circuit of Fig. 11.10 demonstrates this. The input inductance and the Q of the transformer can now easily be measured, and the previously developed theory can be applied directly.

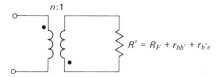

Fig. 11.10 Transformer measurement.

In order that the oscillator possess good frequency stability the effective Q of the tank circuit must be quite high. Inclusion of R_F in the base lead improves the frequency stability of the tuned-collector oscillator by leading to a higher value of effective Q. It has been mentioned that the tuned-circuit oscillator can be operated in the class-C mode and, in fact, the amplitude stability can be better for class-C than for class-A operation. We will now consider the tuned-drain, class-C, FET oscillator of Fig. 11.11. Again we will assume that the actual transformer can be represented by an ideal transformer and a resonating inductance L.

We can calculate a minimum value of transformer turns ratio based on the value of g_m that occurs near a bias voltage of $V_{GS} = 0$. Prediction of the amplitude of output voltage is extremely difficult because of the complex variation of g_m with bias voltage and because the mode of operation changes abruptly from class-A to class-C as oscillations initially occur.

Prior to applying power there is no voltage between gate and source. When power is applied the gate-bias voltage cannot change instantaneously; thus the stage operates at zero bias voltage. The gain from gate to drain at resonance is

$$A = -g_m r_{ds}||R_p.$$

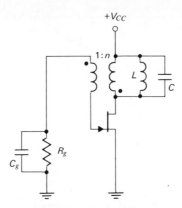

Fig. 11.11 Tuned-drain oscillator.

The turns ratio must be less than n_{max}, which is

$$n_{max} = g_m r_{ds} || R_p. \tag{11.12}$$

A smaller value of n will be used than that given by Eq. (11.12) to guarantee that oscillations not only will occur, but also will reach a reasonably high amplitude. As the ac signal is fed back to the gate, the gate-source diode along with R_g and C_g acts as a clamping circuit. As gate goes positive with respect to source this diode conducts heavily, but on the negative portion of the gate swing this diode appears as an open circuit. Thus, a negative dc voltage appears on the capacitor and biases the gate more negative. This voltage is approximately equal to the peak-to-peak value of the feedback voltage. The gate conducts over only a small portion of a cycle due to the negative gate voltage. During this part of the cycle the drain conducts heavily, resulting in a periodic surge of current. This periodic drain-current waveform contains a fundamental component which develops a voltage across the resonant circuit. This sinusoidal voltage is fed back to the input side of the circuit through the transformer. As voltage becomes larger, two things act to limit the amplitude. Gate-bias voltage becomes more negative causing the value of g_m to decrease. The peak drain current that flows during the time that the gate tries to swing positive depends on the gate-bias voltage. As this peak signal becomes larger, the gate-bias voltage becomes more negative and the amplitude of oscillations must limit due to this effect also. Obviously, a thorough mathematical treatment of this subject should be left to the student!

G. Hartley and Colpitts Oscillators

The Hartley oscillator is very similar to the tuned-collector oscillator, replacing the transformer with a tapped coil, as shown in Fig. 11.12. This arrangement does not require mutual coupling between inductors. Two separate inductors can be used in this circuit.

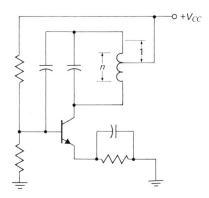

Fig. 11.12 Hartley oscillator.

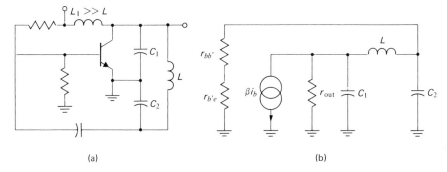

(a) (b)

Fig. 11.13 Colpitts oscillator: (a) actual circuit; (b) equivalent circuit.

The Colpitts oscillator uses only one inductor and two capacitors to obtain the required phase shift through the feedback loop. Figure 11.13 shows a Colpitts oscillator. The impedance between collector and ground of the Colpitts circuit is

$$Z_C = r_{out} \| X_{C1} \| \{X_L + [X_{C2} \| (r_{bb'} + r_{b'e})]\},$$

which can be written as

$$Z_C = \frac{r_{out}[R_{in} - \omega^2 C_2 L R_{in} + j\omega L]}{r_{out} - r_{out}\omega^2 C_1 L + R_{in} - R_{in}\omega^2 C_2 L + j\omega(C_2 + C_1)R_{in}r_{out} + j\omega L - j\omega^3 C_1 C_2 L R_{in} r_{out}},$$

where $R_{in} = r_{bb'} + r_{b'e}$. The gain is given by

$$A = -\beta Z_C / R_{in}.$$

The feedback factor is

$$F = \frac{X_{C2} \| R_{in}}{X_L + X_{C2} \| R_{in}} = \frac{R_{in}}{R_{in} - \omega^2 C_2 L R_{in} + j\omega L}.$$

The return ratio AF is

$$AF = \frac{-\beta r_{out}}{r_{out} - r_{out}\omega^2 C_1 L + R_{in} - R_{in}\omega^2 C_2 L + j\omega(C_2 + C_1)R_{in}r_{out} + j\omega L - j\omega^3 C_1 C_2 L R_{in} r_{out}} \qquad (11.13)$$

The phase angle of AF will be zero only if

$$j\omega(C_2 + C_1)R_{in}r_{out} + j\omega L = j\omega C_1 C_2 L R_{in} r_{out},$$

or if

$$\omega_0^2 = \frac{(C_2 + C_1)}{C_1 C_2 L} + \frac{1}{C_1 C_2 R_{in} r_{out}} \approx \frac{(C_2 + C_1)}{C_1 C_2 L} = \frac{1}{LC_{eq}}, \qquad (11.14)$$

where

$$C_{eq} = \frac{C_1 C_2}{C_1 + C_2}.$$

At the oscillation frequency, the magnitude of AF is

$$AF = -\frac{\beta r_{out}}{r_{out}(1 - \omega^2 C_1 L) + R_{in}(1 - \omega^2 C_2 L)}.$$

Substituting the value for ω_0^2 from Eq. (11.14) gives

$$AF = \frac{+\beta r_{out}}{r_{out}(C_1/C_2) + R_{in}(C_2/C_1)}.$$

The condition imposed on C_2/C_1 for oscillation is found by equating AF to unity, giving a quadratic equation in C_2/C_1,

$$\left(\frac{C_2}{C_1}\right)^2 \frac{R_{in}}{r_{out}} + 1 = \beta \frac{C_2}{C_1}.$$

Assuming that $\beta(C_2/C_1)$ is much larger than unity, we find that the above equation becomes

$$\frac{C_2}{C_1} = \frac{\beta r_{out}}{R_{in}} = \frac{\beta r_{out}}{r_{bb'} + r_{b'e}}. \qquad (11.15)$$

The ratio C_2/C_1 is seen to have an upper limit, but it can be quite large.

H. The Wien Bridge Oscillator

The circuit of Fig. 11.14 depicts the popular Wien bridge oscillator. This oscillator can operate without the resistors R_1 and R_2, but these legs of the bridge form the basis for the amplitude-stabilized circuit to be considered shortly and will therefore be included in the present discussion. The RC network exhibits zero phase shift at one specific frequency and, if return ratio magnitude conditions are satisfied, the circuit will oscillate at this frequency.

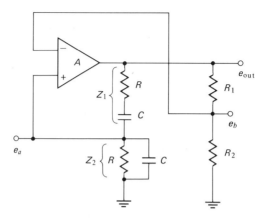

Fig. 11.14 Wien bridge oscillator.

The return ratio or gain around the loop can be found from the expressions

$$e_{out} = A(e_a - e_b),$$

$$e_a = \frac{Z_2}{Z_1 + Z_2} e_{out},$$

and

$$e_b = \frac{R_2}{R_1 + R_2} e_{out}.$$

Combining these expressions gives a value of

$$AF = A\left[\frac{j\omega CR}{1 - \omega^2 C^2 R^2 + 3j\omega CR} - \frac{R_2}{R_1 + R_2}\right]. \quad (11.16)$$

The return ratio will exhibit zero-degree phase shift only at the frequency

$$\omega_0 = 1/CR. \quad (11.17)$$

The magnitude of AF at this frequency becomes

$$|AF| = A\left[\frac{1}{3} - \frac{R_2}{R_1 + R_2}\right]. \quad (11.18)$$

It is obvious that $R_2/(R_1 + R_2)$ must be smaller than $\frac{1}{3}$ or the phase shift will be 180° rather than 0°. Since A is a very large number, $R_2/(R_1 + R_2)$ is selected to be only slightly smaller than $\frac{1}{3}$. In fact, the theoretical value of this ratio to cause unity magnitude of AF is

$$\frac{R_2}{R_1 + R_2} = \frac{1}{3} - \frac{1}{A}. \quad (11.19)$$

A small change in the ratio $R_2/(R_1 + R_2)$ can lead to a large change in amplitude unless A changes very sharply with dc distortion.

Tuning can be accomplished by varying the two capacitors or resistors simultaneously. Commercial oscillators often use capacitor tuning for continuous frequency variation over a given range and resistive tuning to switch to different frequency ranges. Transistor or FET oscillators may be designed to reach the megahertz frequency range.

An FET can be used as a variable resistance to stabilize the amplitude of oscillations, as shown in Fig. 11.15. The FET acts as a variable resistance controlled by the gate-bias voltage. A potentiometer in the comparator circuit establishes the nominal amplitude of oscillations along with the FET gate voltage. If parameter changes now occur causing an increase in amplitude, the peak detection and comparator circuit generates a more negative gate voltage which increases the total resistance R_2. This causes a larger voltage e_b, which tends to offset the increase in e_{out}.

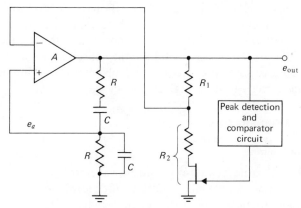

Fig. 11.15 Amplitude-stabilized Wien bridge oscillator.

I. Crystal-Controlled Oscillators

It has been mentioned previously that a high-Q circuit leads to good frequency stability. A piezoelectric quartz crystal represents an element that behaves as a high-Q resonant circuit. The equivalent circuit of the quartz crystal appears in Fig. 11.16. The crystal can exhibit both series and parallel resonance, but is

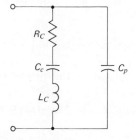

Fig. 11.16 Equivalent circuit for the crystal.

usually applied as a parallel resonant circuit. Slightly below parallel resonance, the crystal appears inductive and can be used to replace an inductance in an oscillator. For example, the crystal could replace L in the Colpitts oscillator of Fig. 11.13(a). The Q of the crystal can be as high as 10^5, whereas the coil Q might typically range from 50 to 100. The overall circuit Q is increased greatly by the use of a crystal. With high circuit Q, a drift in gain requires only a small compensating change in frequency to restore the value of AF to unity as previously discussed.

J. General Discussion of Oscillators

The oscillator circuits considered previously have been analyzed for low and midband frequencies. At higher frequencies, the wiring and parasitic capacitances of the device must be included in the analysis. This complicates the design procedure, but does not change the basic steps. Equations for gain and feedback factor or for loop gain are written to include these additional effects and the same conditions for oscillation apply. It should be noted in both the tuned-circuit and the Hartley oscillators that neutralization of the collector-base feedback path can be affected to remove some high-frequency problems. The techniques of Section 7.4C are applied directly to these high-frequency stages.

11.2 WAVEFORM GENERATION

A. Multivibrator Control

One of the most versatile classes of circuit in waveform generation is the multivibrator. One-shot and astable multivibrators often form the basis for ramp waveforms in addition to pulse and rectangular signals. We have considered the fundamentals of multivibrator operation in Chapter 9; however, we will now discuss certain aspects of design that are pertinent to instrumentation applications.

Possibly the single feature of the one-shot or astable that makes these circuits so versatile is the ease with which the period can be varied. There are several different methods that are used to control the multivibrator period and each applies equally well to the one-shot or astable. For the sake of brevity, we will consider these methods relative only to astable timing.

Variable capacitance. Figure 11.17 shows the basic astable circuit. The period for the astable is given by

$$T = (R_1 C_1 + R_2 C_2) \ln\left(1 + \frac{V_{CC}}{v_t}\right). \tag{9.11}$$

If the timing resistors are returned to the power supply voltage, the equation reduces to

$$T = (R_1 C_1 + R_2 C_2) \ln 2 = 0.69(R_1 C_1 + R_2 C_2). \tag{9.10}$$

The most straightforward means of varying the period is by capacitor variation. Each half-period varies directly with the corresponding capacitor. Furthermore,

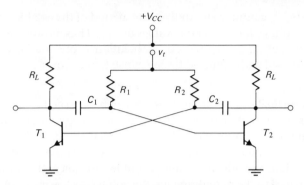

Fig. 11.17 Astable multivibrator.

the only parameter that is affected besides the period is the recovery time, which is insignificant in many applications. The range of periods is limited only by the range of the capacitor values; thus the physical size of the capacitor presents a limitation to the usefulness of this method. It is rather difficult to achieve continuous control of the period over several orders of magnitude, due to the physical problems of constructing the required variable capacitors.

The variable capacitance method is particularly useful when discontinuous period changes over a wide range are necessary. For example, if the period must take on the values 50 μs, 500 μs, 5 ms, and 50 ms, it is a simple matter to arrange switches to present fixed values of capacitance to the multivibrator. Each setting of the switch results in astable capacitances that differ by factors of 10 from values of adjacent switch settings. Continuous period variation at each switch setting can be accomplished by the use of variable timing resistors.

Variable timing resistors. This method offers the advantage of continuous control over a reasonable range of periods. Variable resistors are more readily available, cover a large range of values, and are less expensive than variable capacitors. Unfortunately, the base-drive current is determined by these timing resistors resulting in practical upper and lower limits on these elements. The maximum value of R_1 and R_2 is determined by the requirement that the transistors must saturate when in the on state. The maximum value of these timing resistors is

$$R_{max} = \beta_{0(min)} R_L. \tag{11.20}$$

The minimum value is determined by the allowable overdrive factor. This, in turn, is established by switching time, power dissipation, or current limitations, and may typically range from 10 to 20 times the base saturation current. The minimum resistance is then 1/20 to 1/10 the maximum value. Quite often this minimum resistance is inserted as a fixed resistor in series with a variable resistance. This prevents accidental destruction of the transistor when the variable resistance is set to zero value. Typical continuous period variations with this circuit reach factors of 10 to 20.

Equation (9.10) continues to apply to the period of the astable and is independent of supply voltage for reasonable values of V_{CC}. Thus, power supply variations have a negligible effect on the period. As mentioned previously, variable timing resistors can be used with switched timing capacitors to cover a very wide range of period variation.

Potentiometer method. The timing resistors can be returned to the wiper arm of a potentiometer having outer terminals that connect between ground and V_{CC} or some other value of supply voltage. The potentiometer is normally chosen to have a much smaller value than the timing resistors. This method offers the advantage of control of both half-periods by a single variable resistance. If the timing resistors are large compared to the potentiometer, the voltage appearing at the wiper arm is v_t while the period is

$$T = (R_1 C_1 + R_2 C_2) \ln\left(1 + \frac{V_{CC}}{v_t}\right). \tag{9.11}$$

The base-drive current in this case is

$$I_B = \frac{v_t - V_{BE(on)}}{R},$$

where R is R_1 or R_2. The minimum value of v_t can be selected from

$$v_{t(\min)} = \frac{R V_{CC}}{\beta_{0(\min)} R_L} + V_{BE(on)}, \tag{11.21}$$

where R is the largest value of R_1 or R_2. Often a fixed resistance is inserted in series with the potentiometer to ensure that the minimum required base drive will be present at all wiper arm settings. Figure 11.18 shows one possible arrangement for

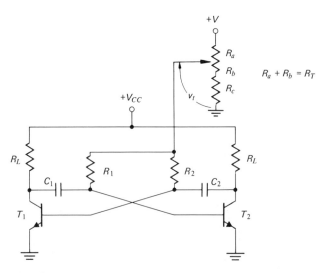

Fig. 11.18 Astable multivibrator with potentiometer control.

this multivibrator. For any particular potentiometer setting the target voltage is

$$v_t = \frac{R_b + R_c}{R_a + R_b + R_c} V. \tag{11.22}$$

When R_b becomes zero this target voltage must equal or exceed that given by Eq. (11.21).

Target voltage control. The target voltage can be controlled by a variable voltage source rather than a potentiometer. The same design considerations again apply if the output resistance of the variable voltage source is small compared to the timing resistors. The period is

$$T = (R_1 C_1 + R_2 C_2) \ln\left(1 + \frac{V_{CC}}{v_t}\right). \tag{9.11}$$

There is one case of particular interest, occurring when the target voltage is much greater than V_{CC}. We use the approximation $\ln(1 + x) \approx x$ for small values of x to write

$$T = (R_1 C_1 + R_2 C_2) \frac{V_{CC}}{v_t} \quad \text{for} \quad v_t \gg V_{CC}. \tag{9.12}$$

For $v_t = 5 V_{CC}$ this approximation leads to a maximum error of 10 percent. The period varies inversely with target voltage. The fundamental frequency and harmonics then vary directly with the target voltage.

It is not unexpected that the period exhibits an inverse relationship with the target voltage when one considers the base voltage waveform of Fig. 11.19. The base voltage follows an exponential path from $-V_{CC}$ toward v_t, but the period ends when this voltage becomes slightly positive. When v_t is large, the base voltage is rising along the linear portion of the exponential curve. If v_t doubles, the initial slope doubles and the period halves.

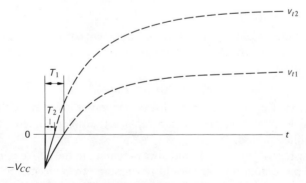

Fig. 11.19 Base voltage waveform for two values of v_t.

Constant current control. The circuit of Fig. 11.20 shows an astable with constant current charging the capacitors. The voltage at the wiper arm of the potentiometer establishes the collector current of both transistors T_3 and T_4. The current I

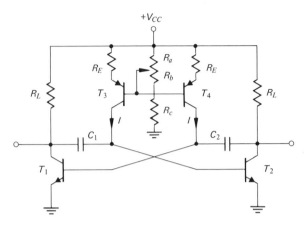

Fig. 11.20 Astable with constant current drive.

charges the capacitor and the base voltage of the off transistor changes from $-V_{CC}$ to approximately zero volts in a linear fashion. The half-periods can be calculated by noting that the capacitor voltage is given by

$$v_c = v_i + \frac{I}{C}t,$$

where v_i is the initial voltage. When a transistor switches off, the initial base voltage is $-V_{CC}$. The half-periods are then

$$T_1 = \frac{V_{CC}C_1}{I} \tag{11.23}$$

and

$$T_2 = \frac{V_{CC}C_2}{I}. \tag{11.24}$$

If the potentiometer is selected so that the base currents of T_3 and T_4 result in negligible loading, the base voltage V will be given by $V_{CC}(R_b + R_c)/(R_a + R_b + R_c)$. The current I is

$$I = \alpha(V_{CC} - V - V_{BE(on)})/R_E. \tag{11.25}$$

To the extent that $V_{BE(on)}$ is negligible, the period varies inversely with the negative of the voltage V. A voltage source can replace the potentiometer to result in a voltage-controlled period.

The transistors T_3 and T_4 should always remain in the active region; thus, the wiper arm of the potentiometer should not assume the extreme positions unless appropriate resistances are inserted on either side of the potentiometer.

Proportional control. There are several applications requiring a period that is proportional to the control voltage. The communication field is one area that

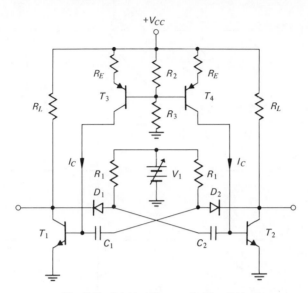

Fig. 11.21 Astable with proportional control.

applies the proportional control circuit in pulse-width modulation. The two-state power amplifier also uses such a circuit. Fig. 11.21 shows an astable multivibrator with half-periods that are proportional to the control voltage V_1 [1].

Transistors T_3 and T_4 provide constant current sources for charging the timing capacitors and supplying base-current drive to transistors T_1 and T_2. We will consider the operation of this multivibrator at the point in time when T_2 has switched on and T_1 switches off. When T_1 shuts off, both sides of the capacitor C_2 are initially at zero volts. The collector of T_1 will rise rapidly to the power supply voltage as T_1 ceases conduction and diode D_1 becomes reverse-biased. The capacitor C_2 charges through R_1 to a target voltage of V_1. The recovery time constant $R_1 C_2$ must be only a fraction of the minimum half-period. The capacitor C_1 charges to a voltage of V_1 during the time T_1 is on. Thus, when T_1 shuts off and T_2 turns on, the base voltage at T_1 drops to a value of $-V_1$. When this voltage charges to zero volts (or slightly positive) T_1 turns on, switching the circuit to the alternate state and dropping the base voltage of T_2 from zero to $-V_1$. The crucial feature of the circuit is that the initial base voltage of either transistor upon switching is equal to the negative of the control voltage V_1. The current I_C will charge either capacitor from this initial voltage to approximately zero volts to end the half-periods. Figure 11.22 shows the pertinent waveforms for two different values of control voltage. This figure pertains to the case of equal timing capacitors. Since the charging current is constant at I_C, the time required to charge C_2 and complete one half-period is

$$T_2 = \frac{C_2}{I_C} V_1.$$

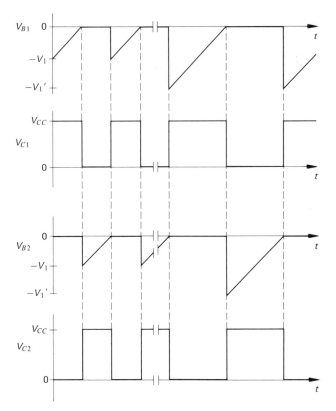

Fig. 11.22 Astable waveforms.

Similarly, the length of the second half-period is

$$T_1 = \frac{C_1}{I_C} V_1.$$

The total period is

$$T = \frac{C_1 + C_2}{I_C} V_1, \tag{11.26}$$

reflecting the proportional control feature of this circuit.

This circuit can be used as a pulse-width modulator by replacing the control voltage with a constant source in series with the modulating signal. Differentiating the output signal and removing the negative-going spikes convert the pulse-width modulator to a pulse-position modulator.

B. The Schmitt Trigger

The Schmitt trigger is a regenerative circuit with feedback that differs depending on which output state is present. This circuit is often used as a trigger-level detector or to convert a sinusoidal input to a rectangular signal with very short rise and fall times. The Schmitt trigger is generally designed to exhibit a hysteresis effect resulting in excellent noise discrimination properties. Hysteresis of the trigger levels means that the input signal required to change the output to state 2 is greater than the input signal required to return the output to state 1. The higher level is called the upper trip point (UTP) and the lower level is the lower trip point (LTP).

The input signal to be detected often contains a noise component. If so, the output response of a single trigger-level circuit reflects the repeated crossing of the trigger level by the input signal. In Fig. 11.23(b) the circuit will trigger only once at the UTP and resume the original state at the LTP corresponding to the desired output signal. Obviously, noise immunity is one advantage of the Schmitt trigger. Another point worth mentioning is that the trigger level is very well defined in terms of the output voltage transition. Once the UTP has been exceeded, regeneration forces the circuit to switch immediately to the opposite state. In a single trigger-level circuit the stage may operate in the linear rather than the saturation region until the trigger level is exceeded by a reasonable voltage.

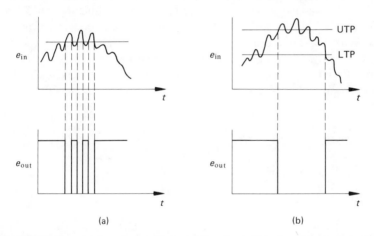

Fig. 11.23 Trigger circuit responses: (a) single trigger level; (b) hysteresis in trigger levels.

Before op amps became so readily available, the Schmitt trigger was constructed from two transistors connected as an emitter-coupled bistable [2]. Figure 11.24 indicates one configuration using an op amp.

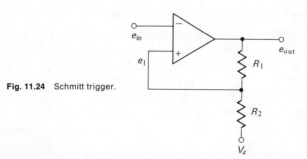

Fig. 11.24 Schmitt trigger.

Let us assume that the limits of output swing are $+V$ and $-V$. When e_{in} is less than e_1 the output signal will equal $+V$ (state 1); when e_{in} is greater than e_1 the output will equal $-V$ (state 2). The voltage e_1 that occurs while the circuit is in state 1 is the UTP and is given by

$$\text{UTP} = \frac{R_1}{R_1 + R_2} V_a + \frac{R_2}{R_1 + R_2} V. \tag{11.27}$$

Once this value is exceed by e_{in}, the output switches to $-V$ dropping e_1 to the LTP. State 2 will prevail until e_{in} drops below the LTP which is

$$\text{LTP} = \frac{R_1}{R + R_2} V_a - \frac{R_2}{R_1 + R_2} V. \tag{11.28}$$

To demonstrate the design procedure, let us take $V = 10$ V and set a UTP $= 2$ V and an LTP $= 1$ V. We might start by choosing $R_1 + R_2 = 10$ kΩ to result in negligible loading of the op amp. Using Eqs. (11.27) and (11.28) to set UTP $= 2$ and LTP $= 1$ gives values of $V_a = 1.58$, $R_1 = 9.5$ kΩ, and $R_2 = 0.5$ kΩ.

It is often necessary to select V_a first, due to some constraint on available voltages. With the configuration of Fig. 11.24, the UTP must always exceed V_a and the LTP must always be less than V_a. It is possible, however, to modify the circuit resulting in a different set of trigger-level equations. Fig. 11.25 shows one such modification. Derivation of the appropriate equations will be left to the student.

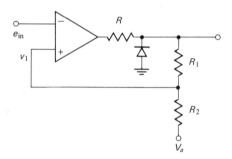

Fig. 11.25 Modified Schmitt trigger.

C. Applications of Multivibrators

The digital computer and related areas probably use multivibrator circuits to a greater extent than any other area. Astable, one-shot, and especially bistable multivibrators are very prominent in timing, delay, and storage operations in logic systems. Since Chapter 9 covers several aspects of multivibrator use in such systems we will not discuss these applications further.

Instrumentation is another important field that relies heavily on multivibrators. The last decade has experienced a growing demand for digital instrumentation and a need for timing circuits and analog-to-digital interfaces. The Schmitt trigger is often used to convert sinusoidal or other analog signals into rectangular signals that are then processed on a digital basis. A simple method of measuring the frequency of a sinusoidal waveform starts by converting the signal to a rectangular waveform by a Schmitt trigger with UTP and LTP near the base line. A timing or count gate opens for a fixed time, allowing the rectangular waveform to enter a digital counter. At the close of the count gate the number contained in the digital counter is displayed. A Schmitt trigger may also be used to produce the timing gate. For lower accuracy systems, the 60 Hz line signal is first attenuated and then shaped by a Schmitt trigger. To produce a 0.1 sec gate, the output of the Schmitt is divided by six to produce a pulse after 6 cycles or 0.1 sec later. A higher accuracy circuit might use a quartz crystal oscillator to supply the Schmitt trigger input. This arrangement would also be used to produce shorter count gates to measure high-frequency signals. A block diagram of a simple frequency counter is shown in Fig. 11.26. This diagram does not include the necessary reset and control section.

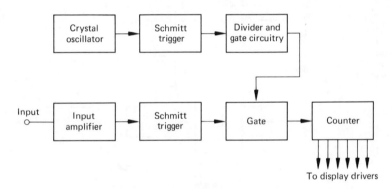

Fig. 11.26 Block diagram of a frequency counter.

Oscilloscope trigger and sweep systems often include Schmitt trigger or tunnel diode trigger circuits. A Schmitt with a variable UTP may control the level which the input signal must reach before the sweep is initiated. The UTP can be controlled by a variable resistance for either R_1 or R_2. A one-shot or astable multivibrator can also be used in the free-running trigger mode of the oscilloscope.

The one-shot or astable multivibrator can be used as a circuit for converting temperature to period length. We discussed the effect of temperature change on one-shot period in Chapter 9, along with methods to minimize this variation. There are cases in which it is desirable to maximize this variation. If a temperature-sensitive resistor is used for the timing resistance of the one-shot, the period exhibits a strong dependence on temperature. It is also possible to generate a current that varies with temperature and then use this current to charge the timing capacitor. Regardless of the method used, it is a simple matter to create a period that is a function of temperature. This relationship may not be a simple one, but to each temperature will correspond a single value of period.

This type of multivibrator can be used in transmitting temperature data from remote locations. In weather prediction applications or water resources projects, the temperatures at several distant locations are monitored continuously and transmitted to a central receiver and data-processing station. The information can be transmitted directly as pulse-width modulated information or it may be converted to a pulse-code modulated signal. A scheme for pulse-code modulation is shown in Fig. 11.27.

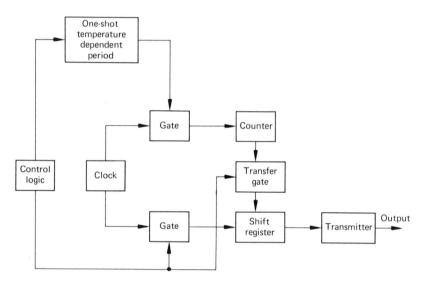

Fig. 11.27 Pulse-code modulation scheme.

The control logic triggers the one-shot which opens a gate from the clock to a binary counter. When the period ends the number in the counter is proportional to the period duration. The control logic then transfers the information from the counter to the shift register in parallel through the transfer gate. The clock gate is then opened to the shift line of the register to transmit the binary code serially. The measurement-transmit operation may be initiated periodically or on command from a central station.

REFERENCES AND SUGGESTED READING

1. D. J. Comer and D. T. Comer, "Directly Linear Period-Controlled Multivibrator." *Electronics* **40**, 115 (August, 1967).
2. D. T. Comer, *Large Signal Transistor Circuits*. Englewood Cliffs, N.J.: Prentice-Hall, 1967.
3. J. Millman and C. Halkias, *Integrated Electronics: Analog and Digital Circuits and Systems*. New York: McGraw-Hill, 1972, Chapter 14.

PROBLEMS

Section 11.1A

* **11.1** If the open-loop gain of the amplifier is

$$A = \frac{-1000}{1 + j\omega/10^7}$$

and has a feedback factor of

$$F = \frac{10^{-2}}{(1 + j\omega/4 \times 10^7)^2},$$

will the circuit oscillate? If so, at what frequency will oscillations occur? If the low-frequency gain is changed to -1250, at what frequency will oscillations occur?

11.2 Repeat Problem 11.1 if the corner frequency of F is changed to 8×10^7 rad/sec.

Sections 11.1B–11.1C

11.3 In the circuit of Fig. 11.1, the gain is given by $A = 3000/r_d$. If $r_d = 26/I_E$ where I_E is in milliamps and $I_E = 2 - 0.1 e_{\text{out}}^2$, find the value of e_{out}.

11.4 The circuit of Fig. 11.1 oscillates at 10 kHz when $L = 0.1$ mH. If L is changed to 0.096 mH, find the new oscillation frequency.

Section 11.1D

11.5 Find the frequency of oscillation and the gain requirement for the phase-shift oscillator of Fig. 11.3, given that $C_3 = 2C_2 = 2C_1$ and $R_1 = R_2 = R_B \|R_3\| (R_F + r_{bb'} + r_{b'e})$.

11.6 The FET of Fig. 11.6 has a $g_m = 4000$ μmhos and $r_{ds} = 30$ kΩ. If $V_{CC} = 12$ V and $R_B = 250$ Ω select all other element values to cause oscillation at a frequency of 24 kHz.

Section 11.1E

11.7 Design an active phase-shift oscillator using op amp stages with $A = 30{,}000$. The oscillation frequency should be adjustable from 4 kHz to 40 kHz. State all assumptions made.

11.8 Repeat Problem 11.7 using FET stages with $g_m = 6 \times 10^{-3}$ mhos and $r_{ds} = 20$ kΩ.

Sections 11.1F–11.1H

* **11.9** Select appropriate values for n, L, C, R_g, and C_g in the tuned-drain oscillator of Fig. 11.11 to result in a 30-kHz oscillation frequency. Assume that the coil $Q = 50$, $g_m = 6 \times 10^{-3}$ mhos, and $r_{ds} = 20$ kΩ.

11.10 A transistor has the values $\beta = 100$, $r_{bb'} = 100$ Ω, and $r_{b'e} = 2$ kΩ. Assuming $r_{\text{out}} = 40$ kΩ, select L, C_1, and C_2 to produce oscillations at 100 kHz with a Colpitts oscillator.

11.11 In the Wien bridge oscillator of Fig. 11.14, $R = 10$ kΩ, $C = 0.01$ μF, and $A = 100$. Select R_1 and R_2 to cause oscillation. At what frequency will the circuit oscillate?

Section 11.2A

*11.12 In the astable multivibrator of Fig. 11.17, $R_1 = R_2 = 40$ kΩ while $C_1 = C_2 = 0.01$ μF. Calculate the total period if $V_{CC} = 12$ V and (a) $v_t = 6$ V; (b) $v_t = 9$ V; and (c) $v_t = 18$ V.

11.13 Using the potentiometer method, design an astable circuit with a period that varies from 1 ms to 2 ms. Assume that $V_{CC} = 10$ V, $\beta = 80$, and $R_L = 2$ kΩ.

11.14 In the astable circuit of Fig. 11.20, $C_1 = C_2 = 0.02$ μF, $R_E = 1$ kΩ, $V_{CC} = +16$ V, and $R_c = 10$ kΩ. If a 20-kΩ potentiometer is used to adjust the period, calculate the period when the wiper arm is set to the midpoint.

11.15 If $\beta = 100$ for the transistors used in the astable of Fig. 11.21 and $V_{CC} = +18$ V, select all element values to cause a period of 1 ms when $V_1 = 18$ V and 0.5 ms when $V_2 = 9$ V.

*__11.16__ Design an inverting Schmitt trigger having a UTP of 0.8 V and an LTP of 0.4 V. The op amp switches between the extremes of $+8$ V and -8 V.

11.17 Choose a configuration using a single op amp to result in a noninverting Schmitt trigger.

12
POWER AMPLIFIERS

Low-power transistor or integrated circuit amplifiers generally produce power ranging from milliwatts to a few watts. There are many requirements today for amplifiers that reach higher output power levels. It is not uncommon to find 50 W stereo systems in large-volume rooms and many public-address systems have considerably higher outputs although more than one power amplifier may be used in such a system. In this audio range of frequencies, amplifiers exceeding 300 W of output power can be constructed. At higher frequencies, the power output capabilities of semiconductor amplifiers decrease. This chapter is concerned primarily with low-frequency or audio amplifier design, although many principles developed here will apply to higher frequency circuits also.

12.1 CLASSIFICATION OF POWER STAGES

There are several types of power output stage that can be used in amplifiers. In general, the first few stages of a power amplifier will make up a voltage amplification section while the last few stages are used primarily to boost the power level. We will consider only the last stage of the amplifier here as a result of the importance of this stage in developing output power.

A class-A stage is one that conducts output current continuously during the complete cycle of a periodic input signal. This stage operates as a linear stage at all times; that is, the output signal developed is always directly proportional to the input signal. We shall later see that this stage is less complex than the class-B or class-C stage, but is less efficient and delivers less power to the load under comparable conditions than do the class-B and class-C stages. Examples of class-A stages are shown in Fig. 12.1.

Class-B operation requires two output stages. One stage conducts output current only when the input signal moves in the positive direction from its zero level. The other stage conducts current only when the input signal moves in the negative direction from the zero level. The currents from both stages are summed and used to create a voltage that is an amplified version of the input signal. For a sinusoid or periodic input signal with half-wave symmetry, one stage conducts

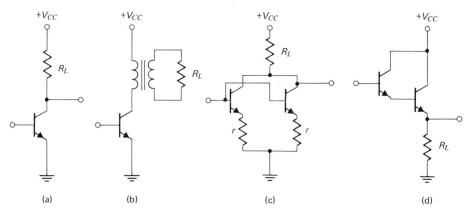

Fig. 12.1 Class-A stages: (a) resistive load; (b) transformer load; (c) parallel output; (d) Darlington connection.

for exactly one-half cycle and the other stage conducts during the alternate half-cycle. It is sometimes advantageous to cause each stage to conduct more than one-half cycle, but less than the full cycle. This mode of operation is called class-AB. Class-B operation is relatively efficient and allows high levels of power output. Figure 12.2 shows two examples of class-B stages.

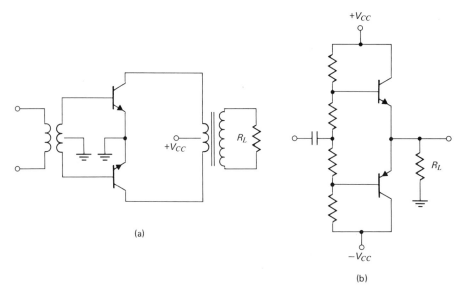

Fig. 12.2 Class-B stages: (a) class-B push-pull; (b) complementary emitter followers.

Class-C operation occurs when a stage conducts during less than a half-cycle of input signal. This type of stage is generally only used to amplify a single frequency. The current conducted by the output stage flows through a resonant circuit and excites an output voltage at the resonant frequency. Although the output current is nonsinusoidal, the output voltage is a sinusoid. Class-C operation can be very efficient, approaching a maximum theoretical efficiency of 100 percent. In narrow-band applications such as radio frequency production and transmission, the class-C stage is quite useful. A class-C amplifier is shown in Fig. 12.3. In this circuit the transformer inductance serves as the resonating inductance.

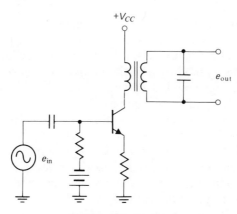

Fig. 12.3 Class-C stage.

In addition to the various types of amplifier previously mentioned, there is a two-state amplifier that achieves amplification by rapidly switching an inductive load between a positive and negative voltage. The inductance acts as a low-pass element and rejects high-frequency components of the rectangular waveform while passing the average value. This amplifier uses a pulse-modulator stage that creates a rectangular signal with a duty cycle that varies in accordance with the input signal to be amplified. The load of the output stage then demodulates this signal, reproducing the input signal at a higher power level. It will later be shown that a transistor dissipates little power if saturated or cut off. The power output stage of the two-state amplifier can deliver a great deal of power to the load while experiencing little power loss by continually operating in either the cutoff or saturation region. A theoretical efficiency of 100 percent can be obtained just as in the case of a class-C amplifier. However, the two-state amplifier can be used for a broader range of audio signals and is not limited to single-frequency operation.

When a transistor is used as a power output stage, it is the external circuit that determines the amount of power dissipated by the transistor. The later sections of this chapter will deal with the determination of the power output of the transistor when employed in various configurations. Before one can choose the proper transistor for a given output, however, one must know the allowable dissipation of the device. The allowable dissipation of a transistor is a function of the construction of the device and the thermal properties of the surrounding medium. The following section will deal with the practical calculation of the allowable device dissipation.

12.2 ALLOWABLE DISSIPATION

The maximum junction temperature is the quantity which ultimately limits the power that a transistor can deliver. The junction temperature will be determined by the power being dissipated by the transistor, the thermal conductivity of the transistor case, and the heat sink that is being used. The collector junction is the point where most power is dissipated; hence it is this junction with which we are concerned.

There are basically two ways in which manufacturers specify the allowable dissipation of a transistor. One way is to specify the maximum junction temperature along with the thermal resistance between the collector junction and the exterior of the case. While this method is very straightforward, it incorrectly implies that the allowable power increases indefinitely as the transistor is cooled to lower temperatures. Actually there is a maximum limit on the allowable dissipation of the transistor which is reflected by the second method of specification. This method shows a plot of allowable power dissipation vs. the temperature of the mounting base. Quite often this plot states that it shows power dissipation vs. ambient temperature, where an infinite heat sink is used. However, if the transistor could be mounted on an infinite heat sink, the ambient temperature would equal the mounting base temperature; thus both plots convey the same information. This second method is used more because the manufacturers want to specify an absolute maximum allowable power dissipation, in addition to the maximum junction temperature.

A. Thermal Resistance

The thermal resistance of a material is defined as the ratio of the temperature difference across the material to the power flow through the material. This assumes that the temperature gradient is linear throughout. The symbol θ is used for thermal resistance, and

$$\theta = \frac{\Delta T}{P} = \frac{\text{Temperature difference across conductor}}{\text{Power flowing through conductor}}. \quad (12.1)$$

An analogy can be made between the thermal circuit and the electrical circuit. If temperature corresponds to voltage, power flow to current flow, and thermal resistance to electrical resistance, Eq. (12.1) is a statement of Ohm's law. The analogy does not end here, since it is obvious that θ will be increased if the material is made longer and will be decreased if the cross-sectional area is made larger. We could then write θ as

$$\theta = \rho \frac{L}{A},$$

where ρ is the thermal resistivity, L is the length of the path, and A is the cross-sectional area perpendicular to the power flow.

Consider the thermal circuit shown in Fig. 12.4 of a transistor surrounded by free air. Here, θ_{JM} is the thermal resistance from collector to mounting base and

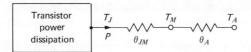

Fig. 12.4 Transistor dissipating power in free air.

θ_A is the thermal resistance of that portion of air in contact with the mounting base; T_A is the temperature of air far away from the transistor; T_M is the mounting base temperature; and T_J is the collector junction temperature. The power P will be determined by the electrical circuit that includes the transistor; P, in turn, will determine the temperatures T_J and T_M. The temperature T_J can be written as

$$T_J = T_A + P(\theta_{JM} + \theta_A). \tag{12.2}$$

Again the analogy to the electrical case is noted. The voltage on one side of a resistance is equal to the voltage on the other side plus the IR-rise across the resistance. Equation (12.2) shows that as more power is dissipated by the transistor, T_J must rise.

For high-power transistors, θ_A is usually many times greater than θ_{JM}. If θ_{JM} were 1°C/W, then θ_A might be 5 to 20°C/W. Of course, θ_A depends on the area of the transistor case.

Next consider the circuit of Fig. 12.5, which shows the transistor mounted on an infinite heat sink.

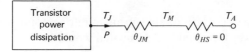

Fig. 12.5 Transistor mounted on infinite heat sink.

The heat-sink temperature will equal the ambient temperature and, since an infinite power can be absorbed by the heat sink without raising its temperature, T_M will also equal T_A. The temperature T_J can now be written as

$$T_J = T_A + P\theta_{JM}. \tag{12.3}$$

For a given power dissipation, T_J will be much less when an infinite heat sink is used than when no heat sink is used. In practice, an infinite heat sink is not obtainable. The values of θ_{HS} presented between the transistor case and free air might range from 0.4°C/W for air-cooled systems to 2°C/W for flat vertical-finned aluminum heat sinks to 8°C/W for cylindrical heat sinks that slide over the transistor. The flat vertical-finned heat sinks are the most commonly used. The manufacturer of the heat sink will specify the thermal resistance of that particular heat sink. Often a long heat sink will be used to accommodate several transistors on the same sink. When this is done, two points should be remembered.

1. Appropriate space should be allotted for each transistor. The manufacturer will usually state the distance to be allowed between adjacent transistors.
2. Since the collector is in contact with the case in power transistors, the case is always at the same electrical potential as the collector. The heat sink is conductive and therefore an insulating washer must be used with transistors mounted on the same heat sink. A mica washer is usually used since it has a high electrical resistance and a low thermal resistance.

B. Method 1. Specification of $T_{J\,max}$ and θ_{JM}.

If the manufacturer specifies a maximum junction temperature and θ_{JM}, the maximum power that can be conducted away from the transistor is

$$P_{max} = \frac{T_{J\,max} - T_A}{\theta_{JM} + \theta_i + \theta_{HS}}, \qquad (12.4)$$

where θ_i is the thermal resistance of the insulating washer. This equation results from solving Eq. (12.2) with θ_A replaced by $\theta_i + \theta_{HS}$.

Example 12.1. Given that $T_{J\,max} = 100°C$, $\theta_{JM} = 1°C/W$, $\theta_i = 0.5°C/W$, and $\theta_{HS} = 2.5°C/W$, find the maximum power that can be dissipated by the transistor when the highest ambient temperature is $T_A = 25°C$.

Solution. Equation (12.4) gives

$$P_{max} = \frac{100 - 25}{1.0 + 0.5 + 2.5} = 18.75 \text{ W}.$$

If the transistor could be mounted on an infinite heat sink so that $\theta_i + \theta_{HS} = 0$, the amount of power that could be dissipated would be

$$P_{max} = \frac{100 - 25}{1.0} = 75 \text{ W}.$$

It is interesting to compare these two figures. For a practical heat sink the transistor can handle 18.75 W. For an infinite heat sink the transistor is capable of 75 W. The manufacturer will specify the transistor as a 75-W device. However, this figure cannot be reached, although if one uses expensive forced air or liquid cooling, the figure might be approached.

C. Method 2. Graphical Determination of Allowable Dissipation

Most manufacturers of power transistors supply a graph of allowable power dissipation vs. transistor mounting-base temperature, as shown in Fig. 12.6. This graph is sometimes called allowable dissipation vs. ambient temperature using an infinite heat sink. If this is the case, $\theta_i + \theta_{HS} = 0$ and $T_A = T_M$. It can be seen from Eq. (12.4) that when the mounting base temperature equals the maximum junction temperature, no power can be allowed to flow. From Fig. 12.6, the intersection of the curve with the $P = 0$ line occurs at $T_{J\,max}$. One would expect the graph to

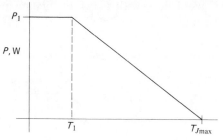

Fig. 12.6 Allowable dissipation vs. T_M.

be continuous, but the manufacturer limits the maximum power of the device by leveling the curve for all values below T_1, which is usually 25°C for germanium and from 25 to 50°C for silicon.

We can evaluate θ_{JM} from the curve, since

$$\theta_{JM} = \frac{T_J - T_M}{P}.$$

The curve gives the *maximum* power that can be dissipated at any T_M; hence, T_J must be the maximum junction temperature. The equation becomes

$$\theta_{JM} = \frac{T_{J\,\text{max}} - T_M}{P},$$

and any P and corresponding T_M falling on the linear portion of the curve can be used. Quite often the point (P_1, T_1) will be used, giving

$$\theta_{JM} = \frac{T_{J\,\text{max}} - T_1}{P_1}. \tag{12.5}$$

Example 12.2. If $\theta_i = 0.5°\text{C/W}$ and $\theta_{HS} = 2.5°\text{C/W}$, and the maximum ambient temperature is to be 25°C, find the allowable dissipation of a transistor whose thermal characteristics are those given in Fig. 12.7.

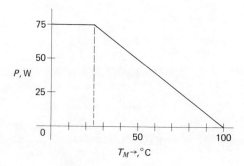

Fig. 12.7 Power dissipation curve.

Solution. If θ_{JM} is evaluated along with $T_{J\,max}$, the first method can again be used. From the curve we see that $T_{J\,max} = 100°C$. From Eq. (12.5) and the curve we can write

$$\theta_{JM} = \frac{100 - 25}{75} = 1°C/W.$$

Equation (12.4) can now be solved to give

$$P_{max} = 18.75 \text{ W}.$$

This same answer can also be obtained graphically. Consider a plot of $(\theta_i + \theta_{HS})$ vs. T_M, as shown in Fig. 12.8. Since one side of the insulating washer is in contact with the transistor case, the temperature at that point will be T_M. The thermal path then travels through the washer and the heat sink to the air at ambient temperature. The slope of the curve in Fig. 12.8 is equal to the reciprocal of $(\theta_i + \theta_{HS})$.

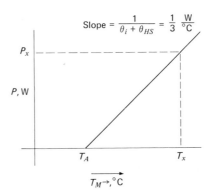

Fig. 12.8 Power conducted through $(\theta_i + \theta_{HS})$ vs. T_M.

When the thermal circuit of the transistor is in equilibrium, the power flowing through θ_{JM} must equal the power flowing through $(\theta_i + \theta_{HS})$. At some particular T_M, the curves of Figs. 12.7 and 12.8 will show the same power flow. This is the simultaneous solution of the equations represented by the two graphs. Figure 12.9 shows this solution, obtained by superimposing the two graphs. The point of inter-

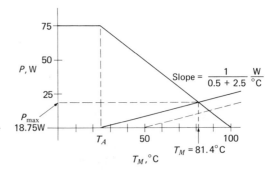

Fig. 12.9 Graphical solution for P_{max}.

section is the only point which will cause the same power to flow through θ_{JM} as that flowing through $(\theta_i + \theta_{HS})$. The graphical technique is a very simple one to apply. The fact that T_A has the same value as T_1 is merely a coincidence. Another problem might give all the same parameters, but this circuit is to work in an ambient temperature of 50°C. This condition is shown by the dotted line on Fig. 12.9, where it can be seen that only 12.5 W could be dissipated in this case.

D. Power Transistors

The maximum junction temperature of a silicon transistor is greater than that of a germanium transistor. For silicon, $T_{J\,max}$ will usually fall in the range of 150°C to 200°C. The corresponding value for germanium is normally near 100°C. In addition, T_1 is often higher for silicon. Figure 12.10 shows a comparison between silicon and germanium for a 200-W transistor. For a given $(\theta_i + \theta_{HS})$, it can be seen that the silicon transistor can dissipate more power than the germanium transistor. It should also be apparent that as T_A is lowered, the allowable dissipation increases. This demonstrates why forced air, liquid cooling, or a refrigeration unit improves the performance.

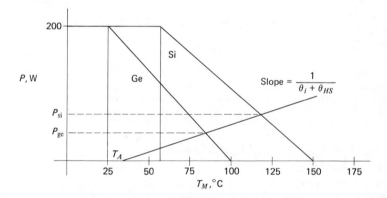

Fig. 12.10 Comparison between silicon and germanium.

Quite often, if more power is needed, many output transistors can be paralleled to deliver the required power. This method is usually the cheapest way to obtain high-power outputs, since power transistor prices are becoming lower. New processes and improvements in old processes continue to extend the performance of power transistors. Many types are available for low-frequency work that are advertised as 200-W transistors. Work is being done at the high-frequency end of the spectrum, and transistors that work up to frequencies of 100 MHz can now deliver 50–100 W of power. However, these units at present are quite expensive.

For switching applications, the capacitance of the thermal circuit must be considered. This subject will not be covered here. In cases where the transistor switches through a region of excessive dissipation, the thermal capacitance is an important consideration.

12.3 CLASS-A STAGES

A. Resistive Stages

According to the maximum-power transfer theorem a source will deliver maximum power to the load if the load resistance equals the source resistance. This theorem applies only to linear operation. In the case of a common-emitter stage as shown in Fig. 12.11, the output circuit can be represented by a current source and an impedance of relatively large value. For a class-A circuit the transistor must operate in the linear region and one might be tempted to apply the maximum-power transfer theorem to select the optimum load impedance. This would result in a load resistance of value $R_L = r_{out}$. If this resistance were used in a power stage that allows the collector current to reach its maximum value, the voltage developed across the load would tend to be very large. Of course, saturation would occur under these conditions and the maximum collector voltage would be greatly exceeded. Thus, the assumption of linear operation would be invalid. We cannot apply linear theory to this case because of the nonlinear constraints imposed on the problem by the transistor.

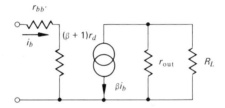

Fig. 12.11 Common-emitter power stage.

In the class-A stage of Fig. 12.1(a) to which the equivalent circuit of Fig. 12.11 applies, the maximum load current that can flow through the load equals the maximum collector-current rating of the transistor, I_{max}. The maximum voltage across the load resistance is limited by the power supply to V_{CC}. If we are interested in delivering maximum dc power to the load we must cause the maximum voltage across the load to occur simultaneously with maximum load current. We can select a load resistance to result in the satisfaction of these conditions if

$$R_{LM} = \frac{V_{CC} - V_{CES}}{I_{max}} \approx \frac{V_{CC}}{I_{max}}, \qquad (12.6)$$

where V_{CES} is the saturation voltage of the transistor corresponding to a collector current of I_{max}. Often V_{CES} will be small enough to be neglected with little loss of accuracy. The maximum dc power dissipated by R_{LM} is

$$P_R = (V_{CC} - V_{CES})I_{max},$$

and it is seen to be independent of the output impedance of the transistor. When the transistor is delivering maximum power to the load, very little power is dissipated by the transistor. When $I_C = I_{max}$, the transistor power is

$$P_T = I_{max} V_{CES} \approx 0.$$

When $I_C = 0$, the transistor voltage is $V_{CE} = V_{CC}$, so

$$P_T = 0 \times V_{CC} = 0.$$

When $I_C = \tfrac{1}{2} I_{max}$, then $V_{CE} = \tfrac{1}{2} V_{CC} + \tfrac{1}{2} V_{CES}$ and

$$P_T = \tfrac{1}{2} I_{max} \left(\frac{V_{CC} + V_{CES}}{2} \right)$$
$$= \tfrac{1}{4}(I_{max} V_{CC} + I_{max} V_{CES}). \qquad (12.7)$$

If V_{CES} is assumed to be zero, maximum power is dissipated by the transistor when $I_C = \tfrac{1}{2} I_{max}$ and $V_{CE} = \tfrac{1}{2} V_{CC}$. The similarity between this result and that predicted by the maximum-power transfer theorem is readily apparent. That is, the dissipation of the transistor is maximum when one-half of the dc source voltage appears across it. We conclude that the transistor delivers maximum power to the load when saturated, and that this power depends on the maximum allowable collector current. Furthermore, we note that the transistor dissipation is very low when the transistor is saturated. As I_C decreases (reducing the power delivered to the load) P_T increases to a maximum value, which occurs at $I_C = \tfrac{1}{2} I_{max}$.

The ac considerations are slightly different from those of the dc case. With a given power supply voltage V_{CC}, the greatest peak-to-peak voltage that can be delivered to a resistive load is equal to V_{CC}, and the quiescent voltage must be chosen as $V_{CEQ} = V_{CC}/2$ for this output voltage to occur. Minimizing the value of R_L will maximize the power delivered to the load. However, the minimum value of R_L is determined by the collector current rating, as shown in Fig. 12.12. The minimum value of R_L that will not result in excessive collector current is

$$R_L = V_{CC}/I_{max}.$$

If the quiescent voltage is $V_{CEQ} = \tfrac{1}{2} V_{CC}$, the collector current will be $I_{CQ} = \tfrac{1}{2} I_{max}$, and the ac power that can be delivered to the load is

$$P_{ac} = \frac{V_{peak}^2}{2R_L} = \frac{V_{CC}^2}{8R_L}. \qquad (12.8)$$

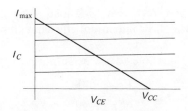

Fig. 12.12 V–I characteristics and load line.

The circuit efficiency of an amplifier is a measure of the effectiveness with which the stage converts the dc input power to ac output power. This efficiency is defined as

$$\eta = \frac{P_{ac}}{P_S} \times 100, \quad (12.9)$$

where P_S is the power delivered by the dc power supply. Using this equation for the resistive load stage gives

$$\eta = \frac{V_{peak}^2/2R_L}{V_{CC}I_{CQ}} \times 100.$$

Both ac power and source power depend on I_{CQ}. If I_{CQ} is less than $\frac{1}{2}I_{max}$, the peak voltage swing that can occur before cutoff is $V_{peak} = I_{CQ}R_L$; thus the circuit efficiency becomes

$$\eta = \frac{I_{CQ}R_L}{2V_{CC}} \times 100 \quad \text{for} \quad I_{CQ} \leq \tfrac{1}{2}I_{max}.$$

The efficiency is seen to increase as I_{CQ} increases up to the point where $I_{CQ} = \frac{1}{2}I_{max}$. Above this value of current, the peak voltage swing is limited by saturation of the transistor. As I_{CQ} increases above $\frac{1}{2}I_{max}$, the maximum possible peak voltage swing decreases and the input source power increases. Therefore we conclude that maximum efficiency occurs when $I_{CQ} = \frac{1}{2}I_{max}$. This value of efficiency is

$$\eta_{max} = \frac{V_{CC}^2/8R_L}{V_{CC} \times \tfrac{1}{2}I_{max}} \times 100.$$

Since $I_{max} = V_{CC}/R_L$,

$$\eta_{max} = 25\%.$$

We note that the quiescent collector current resulting in maximum possible efficiency is the same current that allows maximum output ac power. Unfortunately, this current also results in maximum transistor dissipation, as we have seen earlier. The transistor dissipation with no input signal is

$$P_T = V_{CC}^2/4R_L,$$

which is twice the ac power that can be delivered to the load.

Example 12.3. It is desired to drive a 10-Ω load using a power supply of 20 V. Select a transistor that will allow maximum efficiency to be obtained.

Solution. The maximum possible ac power output is

$$P_{ac} = \frac{V_{CC}^2}{8R_L} = \frac{400}{80} = 5 \text{ W}.$$

The transistor must dissipate twice this power under no signal conditions, so

$P_T = 10$ W. The maximum collector current will be

$$I_{max} = \frac{V_{CC}}{R_L} = \frac{20}{10} = 2 \text{ A}.$$

The transistor selected must handle at least 2 A of collector current and dissipate at least 10 W of power.

The figure of 25 percent efficiency is never obtained in the practical case. The presence of V_{CES} and I_{co}, plus the fact that distortion will be excessive if large voltage swings occur, limit the efficiency of the stage. Values of 15 to 20 percent can be obtained, depending on the amount of allowable distortion.

The power dissipated by the transistor decreases as the ac output voltage increases. This can be seen by considering the product of collector current and collector voltage over a complete cycle. If the current is expressed as $i = I_{CQ} + I \sin \omega t$, the corresponding voltage will be $v = V_{CQ} - V \sin \omega t$. The voltage across the transistor decreases as the collector current increases. The transistor power is

$$P_T = \frac{1}{2\pi} \int_0^{2\pi} (V_{CQ} - V \sin \omega t)(I_{CQ} + I \sin \omega t) \, d(\omega t).$$

If $\omega t = \theta$,

$$P_T = \frac{1}{2\pi} \left[\int_0^{2\pi} V_{CQ} I_{CQ} \, d\theta + \int_0^{2\pi} V_{CQ} I \sin \theta \, d\theta \right.$$
$$\left. - \int_0^{2\pi} I_{CQ} V \sin \theta \, d\theta - \int_0^{2\pi} VI \sin^2 \theta \, d\theta \right]$$
$$= V_{CQ} I_{CQ} - \frac{VI}{2}.$$

This equation shows that no matter where the transistor is biased, the dissipation is maximum when $V = I = 0$, or with no ac output signal. In selecting the transistor, the no-signal power is the important consideration. The allowable dissipation of the device must exceed $V_{CQ} I_{CQ}$, or

$$P_{T \text{ allowable}} > V_{CQ} I_{CQ}.$$

It is of some interest to note that when $V = V_{CQ}$ and $I = I_{CQ}$, maximum power is delivered to the load while the transistor dissipates minimum power. The dissipation in this case is $V_{CQ} I_{CQ}/2$. The source power is constant with output signal and equals the sum of transistor dissipation and load power. It follows that as load power increases from zero to $V_{CQ} I_{CQ}/2$ or $V_{CC}^2/8R_L$, transistor dissipation decreases from $V_{CQ} I_{CQ}$ or $V_{CC}^2/4R_L$ to $V_{CC}^2/8R_L$.

B. Transformer-coupled Stage

Figure 12.13 shows a transformer-coupled stage. The load-line construction of Fig. 12.14 assumes an ideal transformer with zero dc resistance and an ac resistance of $n^2 R_L$.

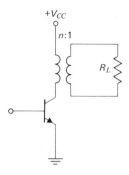

Fig. 12.13 Transformer-coupled stage.

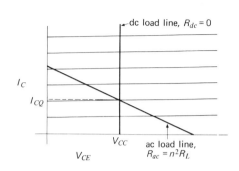

Fig. 12.14 Load lines for circuit of Fig. 12.13.

The equation for the dc load line is

$$V_C = V_{CC} - I_C R_{dc} = V_{CC},$$

which is a vertical line.

The ac load line relates the ac voltage to the ac current, and since the transistor voltage decreases as the current increases, the slope of the ac load line is

$$\text{slope} = -\frac{\Delta I}{\Delta V} = -\frac{1}{R_{ac}}.$$

If the magnitude of ac current is decreased to zero, the output voltage must correspond to the quiescent level. Therefore, the ac load line intersects the dc load line at the quiescent point.

The maximum symmetrical voltage swing that can occur is a value of $2V_{CC}$ peak-to-peak. It seems strange at first thought that the ac voltage can swing more positive than the power supply voltage. However, this is always possible when the ac resistance in the collector circuit is greater than the dc resistance. It can be shown that the maximum efficiency and maximum ac output voltage occur when the quiescent current is

$$I_{CQ} = \frac{V_{CC}}{R_{ac}}. \tag{12.10}$$

Figure 12.15 shows the results of using different values of I_{CQ}.

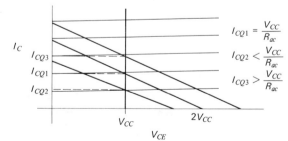

Fig. 12.15 Load lines for various values of I_{CQ}.

For values of I_{CQ} less than V_{CC}/R_{ac}, the peak voltage swing is less than V_{CC}. As I_{CQ} is increased beyond V_{CC}/R_{ac}, the voltage can swing to a peak voltage of V_{CC}, but the efficiency drops off, since the input source power increases with I_{CQ}. The greatest ac output power is

$$P_{ac} = \frac{V_{CC}^2}{2R_{ac}}. \qquad (12.11)$$

The maximum efficiency is then

$$\eta_{max} = \frac{V_{CC}^2/2R_{ac}}{V_{CC}^2/R_{ac}} \times 100 = 50\%.$$

The circuit efficiency is increased over the resistive load case because the load resistor is removed from the dc circuit, and no longer dissipates dc power.

In both cases, the element values and quiescent values can be chosen without referring to the V–I characteristics of the transistor. They have been shown only to demonstrate some pertinent points.

Some of the nonideal effects that will cause the actual design to differ from the theoretical design are:

1. $V_{CES} > 0$;
2. the transformer will have a dc resistance, in addition to some ac losses; and
3. dc power must be supplied to the bias circuit.

The dc resistance of the transformer primary must be known along with the efficiency of the transformer in order to design the output stage. The efficiency of the transformer is defined as the ratio of power output to the power applied to the primary (not including the power dissipated by the dc winding resistance):

$$n_t = P_{out}/P_{in}.$$

Usually, the transformer parameters must be measured, since the manufacturer does not specify these values.

In a stage with little dc resistance in the output circuit, it is possible for thermal runaway to take place. This problem is especially prevalent in germanium power transistors. As the transistor dissipates power, the collector junction temperature increases, causing I_{co} to increase, and this can increase the dissipation, further raising the temperature. This generation can continue until the transistor is destroyed. A bypassed emitter resistance is used to prevent thermal runaway. The additional drop across R_E as I_E increases tends to reduce the forward bias of the emitter-base junction, and limits the increase of current.

Consider the circuit of Fig. 12.16. The load lines are shown in Fig. 12.17.

If maximum-output power is desired, the quiescent point should be set to allow an equal voltage swing on each side of this point. This point can be found by solving the equations for the load lines as follows:

1. $V_{CQ} = R_{ac}I_{CQ}$ (ac condition for maximum output), (12.12)
2. $V_{CQ} = V_{CC} - I_{CQ}R_{dc}$ (dc load line). (12.13)

534 POWER AMPLIFIERS

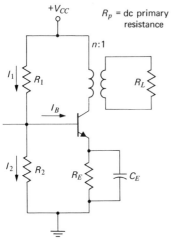

Fig. 12.16 Transformer-coupled stage.

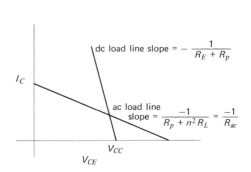

Fig. 12.17 Load lines.

The first of the two equations holds only when biased for maximum output power, while the second equation is simply the dc load line. Solving the equations simultaneously gives

$$I_{CQ} = \frac{V_{CC}}{R_{ac} + R_{dc}} \quad \text{(for maximum output).} \tag{12.14}$$

With a value of quiescent current less than or equal to this maximum value the maximum power that can be delivered to the load is

$$P_{ac} = n_t \frac{I_{CQ}^2 n^2 R_L}{2}. \tag{12.15}$$

Example 12.4. In the circuit of Fig. 12.16, $R_1 = 1 \text{ k}\Omega$, $V_{CC} = 20 \text{ V}$, $R_2 = 100 \text{ }\Omega$, $R_L = 4 \text{ }\Omega$, $R_E = 10 \text{ }\Omega$, $C_E = 100 \text{ }\mu\text{F}$, the turns ratio is 4:1, the transformer efficiency n_t is 0.8, the dc primary resistance is 10 Ω, and the germanium transistor has a $\beta_0 = 25$.

a) Find the greatest output power that can be delivered to the load without saturating or cutting off the transistor; and determine the circuit efficiency.

b) Find the no-signal power dissipated by the transistor.

c) Find the bias point that gives maximum possible circuit efficiency.

Solution. (a) 1. Find I_{CQ}.

i) $I_1 - I_2 = I_{BQ} = \dfrac{I_{CQ}}{25}$

$$\frac{20 - V_B}{1 \text{ k}\Omega} - \frac{V_B}{0.1 \text{ k}\Omega} = \frac{I_{CQ}}{25}$$

ii) $V_B = I_{EQ}R_E + V_{BE} = 1.04I_{CQ}(0.01\text{ k}\Omega) + 0.2$.

Solving simultaneously gives $I_{CQ} = 0.115$ A. The load lines are shown in Fig. 12.18.

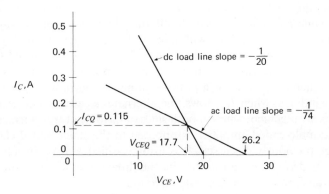

Fig. 12.18 Load lines for Example 12.4.

2. The power that can be delivered to the load is found from Eq. (12.15):

$$P_{ac} = \frac{0.8 \times (0.115)^2 \times 64}{2} = 0.34 \text{ W.}$$

The source power, including the bias power, is

$$P_S = V_{CC}(I_{CQ} + I_1) = 20 \times 0.134 = 2.7 \text{ W.}$$

3. The circuit efficiency is

$$\eta = \frac{P_{ac}}{P_S} \times 100 = \frac{0.34}{2.7} \times 100 = 13\%.$$

b) The no-signal power dissipated by the transistor is

$$P_T = V_{CEQ}I_{CQ} = 17.7 \times 0.115 = 2.04 \text{ W.}$$

We found V_{CEQ} from the dc load line or the corresponding equation.

c) The bias point that gives the highest circuit efficiency is calculated by using Eq. (12.14):

$$I_{CQ} = \tfrac{20}{94} = 0.213 \text{ A.}$$

The circuit efficiency of 13 percent is higher than the actual value that could be obtained in part (b). Distortion would limit the maximum output swing, causing η to be somewhat lower. However, the optimum bias point of part (c) would lead to a higher circuit efficiency. Just as in the case of the resistive-load stage, the maximum transistor dissipation occurs under no-signal conditions.

12.4 CLASS-B POWER OUTPUT STAGES

A class-B stage is shown in Fig. 12.19.

The voltage of the secondary windings of the input transformer will appear between points a and b. For the half-cycle that point a is positive with respect to point b (assuming a sinusoidal input), T_1 will be in its active region and will amplify this signal, while T_2 remains off. On the alternate half-cycle, point b will be positive with respect to point a, and T_2 will amplify this half-cycle, while T_1 is shut off. Since I_{C1} travels in the opposite direction to I_{C2} through the output transformer primary windings, the two currents will cause opposite polarity signals in the secondary. When I_{C1} is flowing, a positive-going waveform will appear across the secondary windings; when I_{C2} flows, the alternate negative-going waveform will appear. This arrangement wherein both sides of the output transformer are driven is called a double-ended amplifier as opposed to the single-ended class-A transformer stage previously discussed. It is also classified as a push-pull stage since one transistor drives the transformer in one direction while the other transistor drives the transformer in the opposite direction.

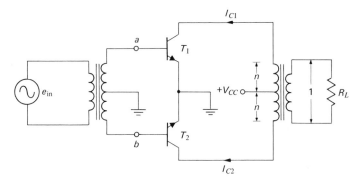

Fig. 12.19 Class-B push-pull.

The class-B stage has certain advantages and disadvantages compared to the class-A stage. These are as follow.

Advantages

1. Very small no-signal power dissipation in T_1 and T_2. The class-A stage causes maximum power dissipation of the transistor under no-signal conditions.
2. Using transistors with the same power rating, the class-B stage will deliver more power to the load than will the class-A stage.
3. Push-pull stages have less distortion than single-ended stages if designed properly.
4. The class-B stage has a higher theoretical maximum for circuit efficiency. This value is $n_{max} = 78.5\%$.

12.4   CLASS-B POWER OUTPUT STAGES

Disadvantages

1. The class-B stage requires two center-tapped transformers.
2. Transistors which have reasonably similar characteristics are required for the class-B stage.
3. The reduction of cross-over distortion in class-B stages requires additional circuitry. Cross-over distortion occurs because of the small base-to-emitter voltage that is required before collector current will flow.

A. Analysis of Class-B Transformer Stages

For a given input signal, the current waveforms I_{C1} and I_{C2} will appear as shown in Fig. 12.20.

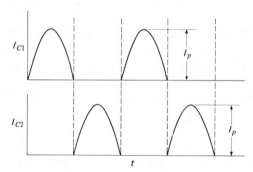

Fig. 12.20 Collector current waveforms.

Each transistor drives collector current through one-half of the total number of windings of the output transformer primary. If the total turns ratio is $2n:1$, the ac load seen by each collector is $n^2 R_L$ plus one-half of the total dc primary resistance. The primary is subjected to a complete sine wave of current of peak value I_P. The total power delivered to the transformer secondary is

$$P_{ac} = \frac{n_t \times I_P^2 n^2 R_L}{2}. \tag{12.16}$$

The maximum amplitude that I_P can reach is that value which causes saturation of the transistor. For the ideal case, this corresponds to a voltage drop of $E_P = V_{CC}$ across the active half of the primary windings, causing zero volts across the transistor. This maximum value of current is then

$$I_{P\,max} = \frac{V_{CC}}{R_{ac}}. \tag{12.17}$$

The maximum power to the load is found by combining Eqs. (12.16) and (12.17) to give

$$P_{ac\,max} = \frac{n_t \times V_{CC}^2 \times n^2 R_L}{2 R_{ac}^2}. \tag{12.18}$$

The input source power is found by taking the product of the power supply voltage and the dc current flowing into the circuit. The current flowing from the source is the sum of I_{C1} and I_{C2}, which is a full-wave rectified waveform. The dc source current is just the average value of this waveform or

$$I_{dc} = 2I_P/\pi.$$

The dc source power is

$$P_S = V_{CC}I_{dc} = \frac{2V_{CC}I_P}{\pi}. \tag{12.19}$$

The greatest power drain will occur for the maximum value of I_P, so

$$P_{S\,max} = \frac{2}{\pi}\frac{V_{CC}^2}{R_{ac}}. \tag{12.20}$$

Equation (12.19) shows that the source power depends on the magnitude of the output signal. In the class-A stage this was not the case. Only the quiescent current affects the source power in the class-A stage and, therefore, this power is independent of the magnitude of the output signal.

The circuit efficiency is calculated from Eqs. (12.16) and (12.19) as

$$\eta = \frac{(n_t \times I_P^2 \times n^2 R_L)/2}{(2V_{CC}I_P)/\pi} \times 100 = \frac{\pi}{4} \times n_t \times \frac{I_P n^2 R_L}{V_{CC}} \times 100. \tag{12.21}$$

The maximum efficiency occurs when maximum output power is present, or when $I_P = I_{P\,max}$, so from Eq. (12.17),

$$\eta_{max} = \frac{\pi}{4} \times n_t \times \frac{n^2 R_L}{R_{ac}} \times 100. \tag{12.22}$$

For the ideal case, the transformer efficiency is unity, and $n^2 R_L = R_{ac}$, giving a maximum circuit efficiency of $\eta_{max} = 78.5\%$.

Transistor dissipation. The input power to the circuit will be accounted for by the dissipation of both transistors, the ac power delivered to the transformer primary, and the power dissipated by the resistances of the collector-emitter circuit. This can be expressed as

$$P_S = \frac{P_{ac}}{n_t} + 2P_T + P_R,$$

where P_R is the resistive losses and P_T is the power dissipated by each transistor. Since we can evaluate P_S, P_R, and P_{ac}, the transistor dissipation can be found from

$$P_T = \frac{P_S - P_R - P_{ac}/n_t}{2}.$$

The ac power delivered to the transformer plus the power dissipated by the dc resistances is

$$\frac{P_{ac}}{n_t} + P_R = \frac{I_P^2}{2}(n^2 R_L + R_{dc}).$$

Since $n^2 R_L + R_{dc} = R_{ac}$ for a class-B stage,

$$P_T = I_P \left(\frac{V_{CC}}{\pi} - \frac{I_P R_{ac}}{4} \right). \tag{12.23}$$

At maximum power output Eq. (12.23) becomes

$$P_T = \frac{0.068 V_{CC}^2}{R_{ac}}, \quad \text{when} \quad I_P = \frac{V_{CC}}{R_{ac}}.$$

The power dissipated by the transistor does not reach its greatest value when maximum output power is being delivered to the load. The power P_T reaches its greatest value when P_{ac} is less than the maximum value. Equation (12.23) can be differentiated with respect to I_P, and the result set equal to zero to find the greatest value of P_T. Setting

$$\frac{dP_T}{dI_P} = \frac{V_{CC}}{\pi} - \frac{I_P R_{ac}}{2} = 0$$

gives a value of $I_P = 2V_{CC}/\pi R_{ac}$. For this value of I_P, P_T will be maximum and will be

$$P_{T\,max} = \frac{0.10 V_{CC}^2}{R_{ac}}, \quad \text{when} \quad I_P = \frac{2}{\pi} \frac{V_{CC}}{R_{ac}}. \tag{12.24}$$

When selecting transistors for this stage, Eq. (12.24) should be used, since this is the maximum possible P_T that can occur.

Distortion in class-B stages. If the stages are perfectly matched, and if T_2 turns on at the exact time that T_1 turns off and conversely, the even harmonic distortion will be very low. Physically, this is easy to visualize. If there is even harmonic distortion in the waveform due to T_1, the same distortion will be present in T_2, but will be out of phase by 180° of the fundamental frequency. This 180° of the fundamental frequency corresponds to phase shifts of integral multiples of 360° for the higher harmonics (360° for second, 720° for fourth, etc.). Since the voltage induced in the transformer by T_2 will be in the opposite direction to that induced by T_1, the even harmonics will cancel in the load. Theoretically, this fact is explained by noting that the waveform will possess half-wave symmetry, and a waveform of this type will have no even harmonics. Since perfectly matched conditions are impossible to achieve in practice, there will be some even harmonic distortion present in the output signal.

A very-high odd harmonic-distortion content will occur if cross-over distortion is allowed to take place. Since the base-emitter junction must be forward-biased a few tenths of a volt before appreciable collector current will flow, the center portion of the input signal will not be amplified unless proper precautions are taken. The effects of this offset voltage which is required for conduction are shown in Fig. 12.21. In order to reduce this distortion, the transistors can be forward-biased enough to overcome this offset voltage. For germanium, the forward bias should be 0.2–0.5 V; for silicon, the voltage should be 0.5–1.0 V. The circuits in Fig. 12.22 can be used to prebias the circuit.

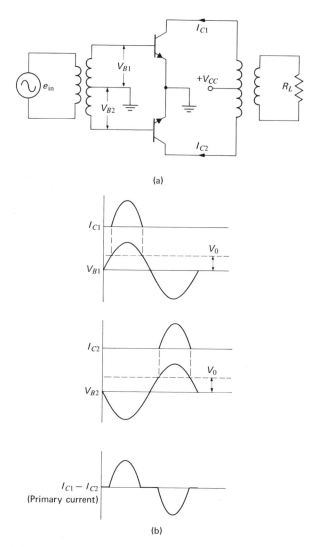

Fig. 12.21 Effect of offset voltage on primary current waveform: (a) actual circuit with no bias; (b) waveforms.

The diode can be biased such that the diode voltage equals or slightly exceeds the offset voltage of the transistors. Usually, a very small collector current is caused to flow to ensure conduction by the transistors for the full half-cycle.

Example 12.5. In the circuit of Fig. 12.23 the total dc primary resistance is 10 Ω and the transformer efficiency is $n_t = 0.9$. Assuming that $V_{CES} = 0$ and $I_{co} = 0$, find

a) the maximum output power before clipping of the waveform occurs,

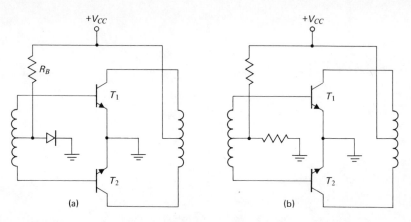

Fig. 12.22 Methods of reducing cross-over distortion: (a) diode bias; (b) resistor bias.

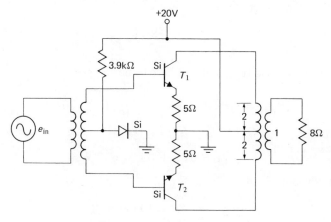

Fig. 12.23 Circuit for Example 12.5.

b) the circuit efficiency for this condition, and

c) the maximum transistor dissipation that could occur in this circuit.

Solution. (a) The reflected ac load between center tap and each side of the primary is $n^2 R_L = 32\,\Omega$. When each transistor conducts, the circuit is as shown in Fig. 12.24. The peak value of current that can occur is

$$I_{P\,\text{max}} = \frac{20\,\text{V}}{42\,\Omega} = 0.476\,\text{A}.$$

The maximum power delivered to the load is found by using Eq. (12.18):

$$P_{\text{ac max}} = \frac{0.9(20)^2 \times 32}{2 \times (42)^2} = 3.27\,\text{W}.$$

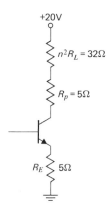

Fig. 12.24 Equivalent circuit of each stage.

(b) The source power when $I_{P\,max}$ is present is

$$P_{S\,max} = \frac{2}{\pi} \times \frac{(20)^2}{42} = 6.07 \text{ W}.$$

The bias power can also be calculated here to be

$$P_{bias} = I_1 \times V_{CC} = \frac{20 - 0.5}{3.9 \times 10^3}(20) = 0.1 \text{ W}.$$

The total source power is

$$P_{S\,total} = 6.07 + 0.1 = 6.17 \text{ W}.$$

The circuit efficiency is then

$$\eta_{max} = \frac{3.27}{6.17} \times 100 = 53\%.$$

(c) The transistor power will be maximum at a current of less than $I_{P\,max}$. Using Eq. (12.24), we find that the greatest transistor dissipation is

$$P_{T\,max} = \frac{(0.10) \times (20)^2}{42} = 0.95 \text{ W}.$$

The allowable distortion will of course limit η_{max} to something less than the value of 53 percent.

B. Class-B, Complementary Emitter Followers

It is possible to design a class-B amplifier without using transformers if complementary power transistors are used. This can reduce the cost and size of the amplifier by eliminating the transformers.

A simple push-pull, complementary emitter-follower stage is shown in Fig. 12.25. The resistors R_1 and R_2 are chosen to set V_Q at a value equal to one-half the power supply voltage. The base of T_1 should be a few tenths of a volt more positive

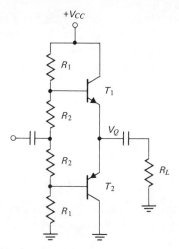

Fig. 12.25 Complementary emitter followers.

than V_Q while the base of T_2 is a few tenths lower. This slight forward bias on each transistor minimizes the cross-over distortion. The positive half-cycle of an input sinusoid leads to conduction by T_1 while the alternate half-cycle leads to conduction by T_2. This power amplifier results in a current gain, but does not exhibit a voltage gain. Any voltage amplification required can be done by earlier stages at a lower power level before reaching the complementary emitter followers.

It should be mentioned that capacitor coupling to the load is often not desirable because a very large capacitance is required to effectively couple the signal to a low resistance load down to reasonable frequencies. If the use of the coupling capacitor is inappropriate in a given application, the circuit of Fig. 12.26 can be

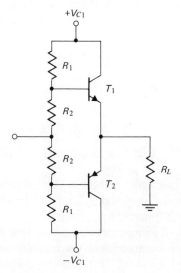

Fig. 12.26 Complementary emitter followers with two sources.

utilized. This circuit requires two voltage sources, but gives performance similar to the previous circuit when each source equals $V_{CC}/2$. The input signal can now be dc-coupled if the average value is zero volts.

The input power from the source for both emitter-follower circuits (if $V_{C1} = V_{CC}/2$) is

$$P_S = \frac{V_{CC}I_P}{\pi} \qquad (12.25)$$

for a peak load current of I_P. The load power is

$$P_{ac} = \frac{I_P^2 R_L}{2}. \qquad (12.26)$$

The power delivered by the source must equal the output power plus the power dissipated by the two transistors. Individual transistor dissipation is then

$$P_T = (P_S - P_{ac})/2 = \frac{V_{CC}I_P}{2\pi} - \frac{I_P^2 R_L}{4}.$$

The maximum transistor dissipation can again be found by differentiating P_T with respect to I_P and equating the result to zero. This peak dissipation occurs for a peak current of

$$I_P = \frac{V_{CC}}{\pi R_L} \qquad (12.27)$$

and is

$$P_{T\,max} = \frac{V_{CC}^2}{4\pi^2 R_L} = \frac{0.025 V_{CC}^2}{R_L}. \qquad (12.28)$$

This value of transistor dissipation is the design value which must be exceeded by the allowable transistor dissipation. Maximum output power is

$$P_{ac\,max} = \frac{V_{CC}^2}{8 R_L} \qquad (12.29)$$

and the corresponding transistor dissipation is

$$P_T = \frac{0.017 V_{CC}^2}{R_L}. \qquad (12.30)$$

Maximum circuit efficiency occurs at maximum output power and again has a theoretical value of $\eta_{max} = 78.5$ percent.

In practice, emitter resistances will often be used to avoid thermal instability. This fact, along with distortion problems, limits the practical maximum efficiency to somewhat lower than 78.5 percent.

There are other types of complementary circuit that can be used to advantage for class-B stages. A quasi-complementary symmetry amplifier is shown in Fig. 12.27. Transistors T_1 and T_3 make up a Darlington pair or double emitter fol-

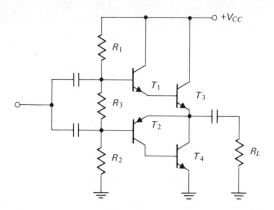

Fig. 12.27 A quasi-complementary symmetry class-B amplifier.

lower. These stages amplify the positive-going portion of the input signal. Transistors T_2 and T_4 form an arrangement that amplifies the negative-going portion of the waveform. When the signal at the base of T_2 swings toward the negative direction, the emitter must follow. Since the emitter is connected to the load, current must be pulled through the load to force the voltage at this point to follow the input signal. The necessary current is conducted primarily by T_4 rather than T_2. As the emitter and collector currents of T_2 increase, the base drive to T_4 increases. Increased base drive to T_4 leads to the desired increase of collector and load current. Hence the appropriate load voltage is established as a result of the emitter-follower action of T_2 while the large amount of current necessary to drive the load is handled by T_4.

One advantage of the circuit of Fig. 12.27 over the complementary symmetry arrangement lies in the ease of finding complementary transistors at lower power levels. Thus, T_1 and T_2 can be matched while T_3 and T_4 can be transistors of the same type.

C. Integrated Amplifier Class-B Stages

While some of the circuits previously considered can be integrated, there are other versions of the class-B stage that are more popular in monolithic amplifiers [5]. Figure 12.28 shows one such stage.

The current source I_1 consists of an integrated constant current stage to provide a small bias current through diodes D_1 and D_2. A very small collector current I_2 is desirable to reduce cross-over distortion at the output. As e_{in} swings positive the input to T_2 goes negative, turning T_2 on and shutting T_1 completely off. A negative-going waveform at the amplifier input drives the bases of T_1 and T_2 positive to shut T_2 off and pass the signal to the output through T_1.

It is possible to make the amplifier "short-circuit proof" by limiting the output current that can flow if the output terminal is accidentally shorted to one of

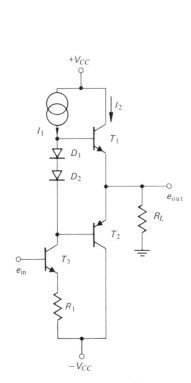

Fig. 12.28 Class-B monolithic stage.

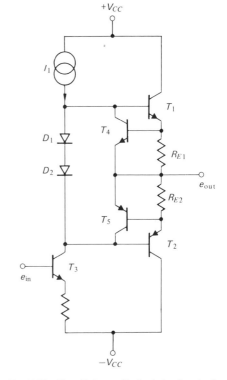

Fig. 12.29 Class-B stage with short-circuit protection.

the supplies. Figure 12.29 indicates the additional circuitry required for this purpose.

The emitter-base junctions of transistors T_4 and T_5 are driven by the voltage drops across the resistances R_{E1} and R_{E2}. Under normal operating conditions, these voltages are too small to turn T_4 and T_5 on; thus, circuit operation is unaffected. If the output is short-circuited to the negative supply voltage, serious damage could result if transistor T_4 were not present. When this occurs, T_4 becomes sufficiently forward-biased to divert the base current drive from T_1. The maximum current that can flow in the output circuit is then limited to V_{BE}/R_{E1}. Typical maximum currents for the short-circuit case range from 10 mA to 50 mA for modern monolithic amplifiers.

The last output stage to be discussed in this section is the totem-pole configuration of Fig. 12.30. This stage is very popular in digital integrated circuits and more recently has been used in monolithic amplifiers.

This configuration has the advantage of using all npn transistors which makes the construction process simpler for monolithic circuits. The diode D and the transistor T_2 replace the pnp transistor that appears in the complementary class-B

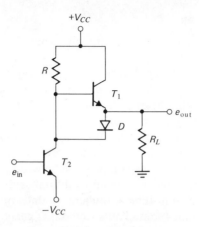

Fig. 12.30 Totem-pole output stage.

stage. When the input signal swings in the negative direction, the collector of T_2 goes positive to cut D off and turn T_1 on. Transistor T_1 functions as an emitter follower for positive output signals. As e_{in} moves in the positive direction, the collector of T_2 goes negative, forward-biasing diode D and pulling the output voltage negatively. Transistor T_1 is cut off for negative output signals.

There is a dead band in this circuit equal to the sum of $V_{BE(on)}$ and V_D, where V_D is the forward voltage drop across the diode. The more complex circuit of Fig. 12.31 can cut this dead band in half [5]. Transistor T_3 is used to set the bias current through R_1 to the proper level.

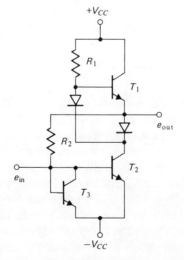

Fig. 12.31 Totem-pole with reduced dead band.

*12.5 CLASS-C AMPLIFIERS

Class-C amplifiers are used for amplification of a single frequency or a very narrow frequency band. In class-B output stages, the output current flows for exactly 180° of the periodic waveform. If the duration of the output current flow is less than one half-cycle, class-C operation is taking place. The periodic output current generates a sinusoidal output voltage by flowing through a resonant circuit tuned to the fundamental frequency or one of the harmonic components.

As the conduction angle of output current approaches zero degrees, the circuit efficiency approaches 100 percent. Unfortunately, the output power also tends toward zero in this instance. Some compromise between good efficiency and high power output is necessary under normal conditions with a resulting typical efficiency of 80 percent. Vacuum-tube amplifiers produce output powers exceeding 50,000 watts in some applications. Transistor stages generally are limited to hundreds of watts of output power although parallel stages can increase this capability considerably. These stages find their greatest use in harmonic generation and radio frequency production.

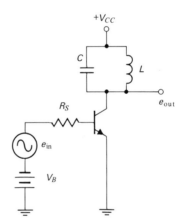

Fig. 12.32 Class-C stage.

A simple class-C amplifier is shown in Fig. 12.32. The resistance R_S is much greater than the input impedance to the transistor. The input voltage, collector current, and collector voltage waveforms are shown in Fig. 12.33.

Since collector current is periodic, it will contain a fundamental component of current and higher harmonics in addition to a dc or average value. The tank circuit or parallel resonant circuit will ideally present zero impedance to dc and all harmonics except the fundamental. A resistance R_P will be presented to the fundamental frequency. This resistance is related to the circuit Q and other tuned-circuit parameters and was derived in Chapter 7. The fundamental frequency component of current will develop an ac voltage across the tuned circuit that is in proportion to this component. The ac output voltage is given by

$$e_{out} = -R_P I \cos \omega t, \tag{12.31}$$

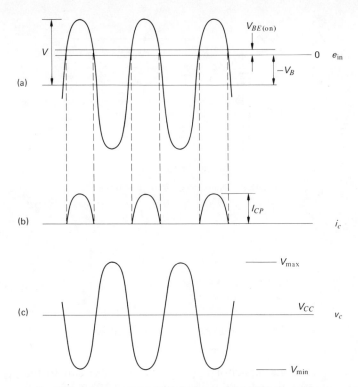

Fig. 12.33 (a) Input voltage. (b) Collector current. (c) Collector voltage.

where I is the magnitude of the fundamental current component. For an ideal tank circuit there will be no distortion of the voltage waveform since load impedance exists only for the fundamental frequency. In practice, the higher components may appear in the voltage waveform due to nonzero impedances presented to harmonics by the tank circuit.

If harmonic generation is required, the tank circuit is tuned to the desired frequency and develops a voltage only at this particular frequency.

To calculate the input power we must find the dc collector current that flows. Output power can be found only if the magnitude of the fundamental frequency component is found. For transistor stages, the collector current waveform closely approximates a portion of a sinusoid. A Fourier analysis of this waveform yields the two quantities of interest.

The equivalent circuit for the class-C stage can be found by using the large signal transistor model of Fig. 4.11(c). Figure 12.34 shows this circuit.

Base current will flow only when the input voltage given by $V \cos \omega t$ exceeds $V_B + V_{BE(\text{on})}$. This current can be written as

$$i_b = \frac{V \cos \omega t - (V_B + V_{BE(\text{on})})}{R_S} \quad \text{(for } i_b > 0\text{)}, \tag{12.32}$$

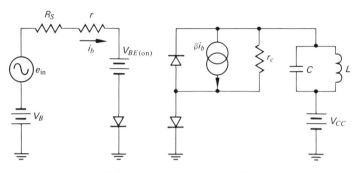

Fig. 12.34 Equivalent circuit of class-C stage.

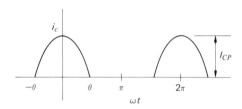

Fig. 12.35 Collector current waveform.

assuming that the input capacitor appears as a short circuit to this frequency. The collector current given by βi_b is shown in Fig. 12.35. The peak value of collector current that flows is

$$I_{CP} = \frac{\beta[V - (V_B + V_{BE(on)})]}{R_S}. \tag{12.33}$$

Collector current can be expressed as

$$i_c = \frac{\beta[V \cos \omega t - (V_B + V_{BE(on)})]}{R_S} \tag{12.34}$$

for values of i_c greater than zero. For values of t which result in a zero or negative value of Eq. (12.34), the collector current is zero. The angle θ can be found by noting that

$$\cos \theta = \frac{V_B + V_{BE(on)}}{V}. \tag{12.35}$$

Equation (12.34) is then valid for times that satisfy

$$-\theta < (\omega t - 2n\pi) < \theta$$

when $n = 0, 1, 2, \ldots$ and

$$\theta = \cos^{-1}\left[\frac{V_B + V_{BE(on)}}{V}\right]. \tag{12.36}$$

A Fourier analysis of the collector current waveform leads to the values of dc and fundamental frequency components. The dc value is

$$I_{dc} = \frac{1}{2\pi} \int_{-\theta}^{\theta} \frac{\beta V}{R_S} [\cos \omega t - \cos \theta] \, d(\omega t)$$

$$= \frac{\beta V}{\pi R_S} [\sin \theta - \theta \cos \theta]. \qquad (12.37)$$

The fundamental component has a magnitude given by

$$I = \frac{1}{\pi} \int_{-\theta}^{\theta} \frac{\beta V}{R_S} [\cos \omega t - \cos \theta] \cos \omega t \, d(\omega t)$$

$$= \frac{\beta V}{\pi R_S} \left[\theta - \frac{\sin 2\theta}{2} \right]. \qquad (12.38)$$

The ac output voltage is then expressible from Eq. (12.31) as

$$e_{out} = \frac{-\beta V R_P}{\pi R_S} \left[\theta - \frac{\sin 2\theta}{2} \right] \cos \omega t. \qquad (12.39)$$

In the preceding equations we have neglected r_{out}, the output resistance of the transistor. In the simple circuit shown this may not be acceptable in view of the fact that R_P, the resonant resistance of the tank circuit, may be quite large. In practical applications transformer mismatching or neutralization of the circuit will generally be used, especially at higher frequencies. Both of these schemes minimize the effect of r_{out}; thus it seems reasonable to neglect this element in the present discussion.

We are now in a position to calculate the various power relationships of the circuit. The dc input power from the source is

$$P_S = V_{CC} I_{dc} = \frac{\beta V_{CC} V}{\pi R_S} [\sin \theta - \theta \cos \theta]. \qquad (12.40)$$

The angle of conduction, which equals 2θ, can be controlled by the bias voltage V_B. As this value approaches the magnitude V of the input signal, the conduction angle approaches zero. The input power in this case also approaches zero as does the output power.

The power dissipated by the transistor equals the source power minus output power. The output power is given by squaring the rms value of output voltage and dividing by R_P, or

$$P_{ac} = \frac{\beta^2 V^2 R_P}{2\pi^2 R_S^2} \left[\theta - \frac{\sin 2\theta}{2} \right]^2. \qquad (12.41)$$

The transistor dissipation is $P_D = P_S - P_{ac}$, or

$$P_D = \frac{\beta V_{CC} V}{\pi R_S} [\sin \theta - \theta \cos \theta] - \frac{\beta^2 V^2 R_P}{2\pi^2 R_S^2} \left[\theta - \frac{\sin 2\theta}{2} \right]^2. \qquad (12.42)$$

It is difficult to apply the power expressions to design without imposing some additional constraints. Thus far we have not discussed methods of selecting θ, R_P, V_{CC}, R_S, or V; consequently, additional information must be considered to specify these values. The power supply voltage is generally chosen from limitations imposed by the transistor or the power supply. The impedance R_P is a function of the load that must be driven and the quality of the tuned circuit. The maximum output voltage and power that can be delivered to the load occur when the magnitude of the output voltage equals V_{CC}. The product of R_P and the fundamental component must then equal V_{CC}, or from Eq. (12.38)

$$V_{CC} = R_P \times \frac{\beta V}{\pi R_S}\left[\theta - \frac{\sin 2\theta}{2}\right].$$

Using this relationship allows the maximum power output to be found from Eq. (12.41). This value is

$$P_{ac\ max} = \frac{\beta V_{CC} V}{2\pi R_S}\left[\theta - \frac{\sin 2\theta}{2}\right] = \frac{V_{CC}^2}{2R_P}. \qquad (12.43)$$

The circuit efficiency for this case is

$$\eta = \frac{\dfrac{\beta V_{CC} V}{2\pi R_S}\left[\theta - \dfrac{\sin 2\theta}{2}\right]}{\dfrac{\beta V_{CC} V}{\pi R_S}[\sin\theta - \theta\cos\theta]}$$

$$= \frac{\theta - \dfrac{\sin 2\theta}{2}}{2\sin\theta - 2\theta\cos\theta}. \qquad (12.44)$$

When θ equals $\pi/2$ the angle of conduction is 180° and the amplifier operates in the class-B mode. The circuit efficiency calculated from Eq. (12.44) is 78.5 percent, which agrees with previous results. As θ approaches zero, the limit of Eq. (12.44) is indeterminate. Applying L'Hospitals' rule indicates that

$$\lim_{\theta \to 0} \frac{\theta - \dfrac{\sin 2\theta}{2}}{2\sin\theta - 2\theta\cos\theta} = 1.$$

An efficiency of 100 percent is approached as θ tends toward zero. However, the output power also tends toward zero resulting in a practical lower limit on θ. Figure 12.36 shows plots of efficiency and output power as a function of θ. The graph of output power is based on the assumption that the amplitude of input voltage V is constant as θ is varied. This can be accomplished by varying the bias voltage V_B. We must note that the development of maximum output power requires the output voltage magnitude to equal V_{CC}. As θ decreases, the fundamental frequency component of collector current decreases in magnitude; thus the resonant

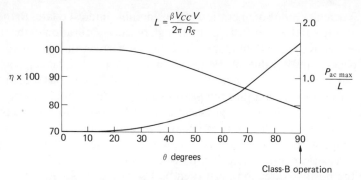

Fig. 12.36 Circuit efficiency and output power as a function of θ.

tank circuit impedance must be increased to develop the maximum output signal. As θ approaches zero, a very small fundamental current component is present. Consequently, an extremely large value of R_P is required to develop the maximum output voltage. It is reasonable to expect little output power under this condition. A reasonable compromise between circuit efficiency and output power often puts θ in the range of 60° to 80°, giving a conduction angle of 120° to 160°.

It should be realized that in applications having a variable input voltage V the output voltage will depend both on V and on θ. Equation (12.36) indicates that θ changes as V is varied. Thus, it is difficult to calculate the maximum dissipation of the transistor. We can calculate this power when maximum output power is being delivered to the load.

The power dissipation of the transistor for maximum output power is found from Eq. (12.42) to be

$$P_{\text{DMO}} = \frac{\beta V_{CC} V}{\pi R_S} \left[1 - \frac{\theta}{2} + \frac{\sin 2\theta}{4} \right]. \tag{12.45}$$

Maximum dissipation by the transistor does not occur simultaneously with maximum output. To calculate the maximum dissipation as θ and V are varied, one could set the partial derivatives of P_D with respect to these variables equal to zero. This would result in values of θ and V that maximize P_D. Following this procedure, however, leads to a value of θ that is greater than 90°. The largest value of P_D for the practical case can be found by setting $(\partial P_D/\partial V) = 0$ and allowing θ to equal 90°. For this case V is found to be

$$V = \frac{4}{\pi} \frac{V_{CC} R_S}{\beta R_P}. \tag{12.46}$$

Using these values of θ and V in Eq. (12.42) leads to a maximum dissipation of

$$P_{D\max} = \frac{2}{\pi^2} \frac{V_{CC}^2}{R_P} = 0.21 \frac{V_{CC}^2}{R_P}. \tag{12.47}$$

This value can be used as a worst-case design value on which to base the allowable transistor dissipation.

One particular mode of operation of considerable interest relates to the class-C modulator stage. If the load R_P is great enough to cause saturation of the transistor during the positive peaks of fundamental collector current, the output voltage will reach a peak-to-peak value of $2V_{CC}$. An increase in collector current will not increase this upper limit of output voltage if the tank circuit has a high Q value. This can be seen by noting that the peak swing of output voltage in the negative direction is V_{CC} volts, from the quiescent level of V_{CC} down to zero volts at saturation. The tank circuit offers zero impedance to all harmonics of the fundamental; thus a voltage across this network can only be developed at the fundamental frequency. A perfect sinusoid at this frequency must be the resultant output. The swing in the positive direction must then be V_{CC} volts. In practice, a small impedance at the higher harmonics may exist and a modest amount of distortion will appear in the waveform.

As the base drive increases beyond that necessary for maximum output voltage, the collector current waveform distorts considerably. While current distortion is severe, the output variable of interest is voltage which is entirely free of distortion as explained.

We will see in Chapter 14 that a class-C collector modulator operates under saturation-limited conditions. The output voltage is then determined by the value of the power supply voltage. To change the amplitude of the ac output voltage or carrier signal the power supply voltage is modulated, resulting in a proportional change in magnitude in the carrier.

*12.6 PULSE-WIDTH MODULATED AUDIO POWER AMPLIFIERS

The integrated circuit offers many advantages over the conventional, lumped-element circuit in a number of applications. Chief among these are digital networks, low-power amplifiers, and low-power, subminiature communication systems. On the other hand, applications requiring large capacitors, inductors, transformers, or high-power capability rarely utilize the integrated circuit. A direct, integrable replacement for an inductor or transformer has yet to be found while large capacitors cannot be integrated due to volume requirements.

A power amplifier requires that the output stage or stages dissipate a certain fraction of the power delivered to the circuit from the source. For class-A resistive stages, the maximum circuit efficiency is 25 percent. The device in this case dissipates twice the power delivered to the load or one-half of the total power delivered by the source. For a 5-W output signal the output transistor must be capable of dissipating 10 W. In a class-B stage the situation is improved somewhat as a result of obtainable efficiencies of 50–60 percent. A 5-W output signal now results in a 4-W to 5-W device dissipation. The heat generated by the output stages must be conducted away from the device by means of a high thermal conductance path. The minimum size of an integrated circuit is now limited by thermal resistance requirements. Not only does the small integrated circuit experience difficulty in conducting heat from the device, but the reliability of the unit is adversely affected when the chip temperature is elevated. It is then desirable to develop more efficient amplifiers that do not require transformers or large capacitors.

There is a pulse-width modulation scheme known for many years that appears to be quite appropriate for integrated circuit power amplifiers. This type of amplification has been used successfully in high-power transistor amplifiers to deliver several hundred watts of audio frequency power to a load. Integrated circuit amplifiers have been reported with efficiencies of approximately 90 percent at 1-W output and a frequency response from dc to 15 kHz [4]. Five watts of audio power at an efficiency of 70 percent have been obtained by integrated circuits occupying less than half the chip area of a comparable class-B integrated circuit [3]. We will consider only the qualitative aspects of the "two-state amplifier" since the theoretical calculations involved are quite complex.

Any periodic waveform can be represented by a Fourier series consisting of a dc component (if present), a fundamental frequency component, and higher harmonics of the fundamental frequency. A rectangular wave with 50% duty cycle contains no dc or average value, but a change of duty cycle will give the waveform an average value as shown in Fig. 12.37. If T remains constant as the duty cycle is varied, the average value of a voltage waveform of amplitude V is

$$V_{av} = V(2x - 1), \qquad (12.48)$$

where x is that fraction of the total period over which the waveform is positive (duty cycle). The average value varies directly with x. The Fourier coefficients of the ac components also vary as x is changed and new frequencies centered about the ac components may be introduced, but in general these components are of no interest to us as they can be easily eliminated. Let us assume that the repetition frequency is 100 kHz. All ac components of the waveform will be greater than this

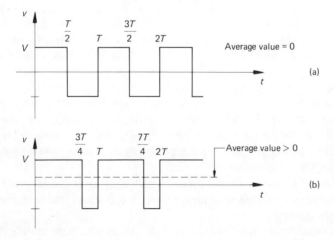

Fig. 12.37 (a) Rectangular wave with 50% duty cycle. (b) Rectangular wave with 75% duty cycle.

value and far out of the audio range. Now if we vary the duty cycle sinusoidally at some low frequency, the average value will also vary sinusoidally. Mathematically, we can express this by saying that if $x = 0.5 + k \sin \omega t$, then the average value is

$$V_{av} = 2 \text{ kV} \sin \omega t. \tag{12.49}$$

The waveform with variable duty cycle can be filtered by a low-pass filter to eliminate all frequencies above ω. The result is a low-frequency output sinusoid whose amplitude varies in the same manner as the duty cycle. Thus, we can propose a system of amplification as shown in Fig. 12.38.

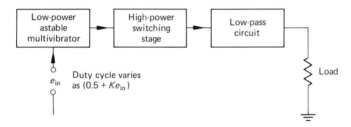

Fig. 12.38 Pulse-width modulated amplifier.

The astable multivibrator has a 50% duty cycle when no input signal is present. As e_{in} becomes nonzero, the duty cycle varies proportionally. The high-power switching stage amplifies this rectangular wave and applies the output to a low-pass circuit which allows only the changes in average value to pass to the load. The output signal is then proportional to the input signal, but can be at a much higher power level than the input signal. Obviously, there are some details of the system that require further consideration. However, we might stop at this point to consider the advantages of the pulse-width modulated amplifier.

The major advantage is that the output transistors need operate in only two states to produce the rectangular waveform: either fully on or fully off. In saturation we know that the very small voltage drop across a transistor leads to very low power dissipation. A very small dissipation is also present when the transistor is cut off. If switching times were negligible, no device power loss would occur during the transition between states. Actually, there is a finite switching time and this leads to an increased total dissipation of the output stages. Still, the efficiency figures for the two-state amplifier are very high, as reported earlier. This leads to higher possible power outputs for lumped-element circuits and to smaller chip areas for integrated amplifiers.

The stages can also be direct-coupled which eliminates the necessity of capacitors. Nonlinear distortion is less than that of class-B stages and matching of transistors is unnecessary.

On the other hand, the disadvantages of this amplifier ultimately dictate the limits of usefulness of the pulse-width modulation scheme. The upper frequency response is limited to a small fraction of the switching frequency. The operating frequency of power transistors generally decreases with higher power ratings. It follows that the upper corner frequency of the amplifier may be lower for higher power transistors. Furthermore, a low-pass filter is required to eliminate the unwanted frequency components of the waveform. An inductance or capacitance will often be required for the filter circuit, although active filtering methods can be applied to reduce the values of any reactive elements used. The generation of radio frequencies by the switching circuits can also present problems in certain applications.

In addition to compound emitter followers, the power output stages can be designed in several arrangements. Figure 12.39 shows two possible configurations [3].

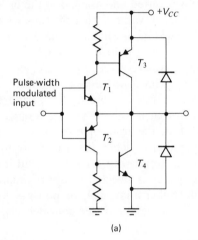

(a)

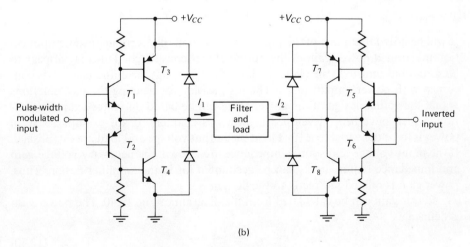

(b)

Fig. 12.39 Push-pull output stage. (b) Bridge output.

The diodes appearing across the output transistors are present to protect the transistors against inductive voltage surges. If the filter is inductive, the current reversals that occur over short switching times generate very high voltage spikes unless the protective diodes are used.

In the push-pull circuit, the low-power, pulse-width modulated input turns T_1 and T_3 on when the signal is at its maximum value. Transistors T_2 and T_4 are off at this time and current is pushed through the load. When the signal switches to the minimum value, T_1 and T_3 go off while T_2 and T_4 turn on to pull current through the load. This circuit allows the load to have a common ground with the amplifier.

Figure 12.39(b) shows a bridge circuit that requires a floating load. When the input signal reaches its maximum value, T_1 and T_3 are on while T_2 and T_4 are held off. The input signal is inverted and applied to the bases of T_5 and T_6. This inverted signal is at its minimum value during the time when the normal input is maximum; thus, T_5 and T_7 will be off while T_6 and T_8 are on. Current will leave the collector of T_3, flow through the load, and enter the collector of T_8. When the input assumes the most negative value, T_1, T_3, T_6, and T_8 turn off while T_2, T_4, T_5, and T_7 turn on. Current now leaves the collector of T_7, flows through the load, and enters the collector of T_4. During this period, the load current flows in the opposite direction to that flowing when the input is maximum. The load current then reverses each time the input signal makes a transition.

We will not discuss the filter in detail. It is worth noting that for many applications a large inductance in series with a resistive load is sufficient for filtering purposes. The corner frequency is selected to approximately equal the highest frequency to be amplified. Higher frequency components, for example those near the fundamental frequency, will be attenuated considerably.

12.7 MISCELLANEOUS ASPECTS OF POWER AMPLIFICATION

A. Power Gain

It will be noted that none of the previous six sections concerning power amplification mentioned power gain of the output stages considered. This may appear to be somewhat unusual, especially in view of the fact that the entire chapter is concerned with power amplification. The justification for deleting these calculations is that the actual design of the power stage can be based on voltage gain calculations which have been developed in previous chapters. If, for example, 5 W of power is to be delivered to a 10-Ω load the output voltage and load are established. If the input voltage and source impedance are known, the necessary voltage gain and impedance requirements can be accounted for in the amplifier design. Thus, power gain is related indirectly to the design.

Power gain can be calculated from the diagram of Fig. 12.40. The power gain is defined as

$$G_P = \frac{P_{ac}}{P_{in}}. \tag{12.50}$$

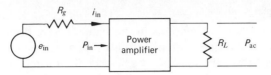

Fig. 12.40 Power amplifier.

This gain can be related to voltage gain for sinusoidal waveforms since

$$P_{ac} = [e_{out}]^2/R_L$$

and

$$P_{in} = (i_{in})^2 R_{in} = \left[\frac{e_{in}}{R_{in} + R_g}\right]^2 R_{in}.$$

Equation (12.50) can now be written as

$$G_P = \left(\frac{e_{out}}{e_{in}}\right)^2 \frac{(R_{in} + R_g)^2}{R_L R_{in}} = (A_v)^2 \frac{(R_{in} + R_g)^2}{R_L R_{in}}. \quad (12.51)$$

Thus, power gain can easily be related to voltage gain and impedance levels. For the amplifier in which R_{in} is much larger than R_g, Eq. (12.51) can be written as

$$G_P = \left(\frac{e_{out}}{e_{in}}\right)^2 \frac{R_{in}}{R_L} = (A_v)^2 \frac{R_{in}}{R_L}. \quad (12.52)$$

Let us assume that an ideal source drives the class-B amplifier of Example 12.5. The turns ratio of the input transformer is 1:1 between primary and each side of the secondary to center tap. Losses in this transformer are negligible. If we assume that the secondary voltage of the output transformer can be related to the primary voltage by

$$e_s = \frac{e_p \sqrt{n_t}}{n},$$

where n is the turns ratio as defined by Fig. 12.19 and n_t is the transformer efficiency, we can easily calculate the voltage gain. It is given by

$$A_v = \frac{\beta n^2 R_L}{r_{bb'} + r_{b'e} + (\beta + 1)R_E} \frac{\sqrt{n_t}}{n}.$$

If $\beta = 50$, $r_{bb'}$ is 20 Ω, and $r_{b'e}$ is taken to be 10 Ω, the value of voltage gain is $A_v = -2.7$. The input impedance is 280 Ω, giving a power gain of

$$G_P = (2.7)^2 \frac{280}{8} = 266.$$

While the voltage gain is rather low in power output stages, the power gain is high as a result of the difference between input and load impedance levels. In general, a voltage amplifier will precede the output stages.

B. Sinusoidal Signal Analysis

All efficiency and power calculations in this chapter have been based on the assumption of a sinusoidal signal. Actually, it is a rare occasion when a pure sinusoid is to be amplified; thus, the maximum efficiency figures may be somewhat misleading. A square or triangular waveform would result in a different theoretical efficiency figure. It should be realized that all periodic waveforms, which incidentally include many speech sounds, are composed of several sinusoidal components. To simplify the analysis to the point of tractability by assuming sinusoidal signals would appear to be justifiable on this basis.

C. Distortion in Power Stages

There are two major causes of distortion in transistor stages. The first is the change of current gain β with changes in I_C or V_{CE}. If a nondistorted base current enters the device, the output current is distorted as a result of the change in current gain as the output quantities vary. This effect can also be explained in terms of the output characteristics which show unequal spacing as I_C or V_{CE} is changed. Obviously if the output signal amplitude is limited, less distortion occurs. In power stages, very large output current changes are prevalent and higher distortion levels are to be expected.

The second important cause of distortion arises from the nonlinearity of the base-emitter characteristics of the transistor. When a voltage source drives a common-emitter stage with little series resistance, the base current can be quite distorted. It is possible to use this input distortion to partially offset the output distortion; however, feedback techniques are most often used in distortion reduction.

REFERENCES AND SUGGESTED READING

1. C. L. Alley and K. W. Atwood, *Electronic Engineering*, 3rd ed. New York: Wiley, 1973, Chapters 11 and 18.
2. L. Blaser and H. Franco, "Push-pull Class AB Transformerless Power Amplifiers." *Fairchild Semiconductor Application Data*, APP 51, 1963.
3. H. R. Camenzind, *Circuit Design for Integrated Circuits*, Reading, Mass.: Addison Wesley, 1968, Chapter 6.
4. H. R. Camenzind, "Modulated Pulse Audio Power Amplifiers Integrated Circuits." *IEEE Transactions on Audio and Electroacoustics* **14**, 3: 136 (September, 1966).
5. A. B. Grebene, *Analog Integrated Circuit Design*. Cincinnati, Ohio: Van Nostrand–Reinhold, 1972, Chapter 5.

PROBLEMS

Section 12.2

12.1 The thermal characteristics of a power transistor are shown in the figure. Find the maximum allowable power dissipation for a maximum ambient temperature of 40°C given that the heat sink and insulating washer present a total thermal resistance of (a) 2°C/W;

(b) 4°C/W. Calculate the allowable dissipation when no heat sink is used given that the thermal resistance between transistor case and air is 10°C/W.

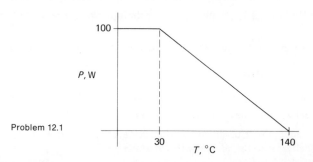

Problem 12.1

* **12.2** Repeat Problem 12.1 assuming that refrigeration is used to maintain an ambient temperature of $-20°C$.

12.3 Calculate the ambient temperature required in Problem 12.1(a) to allow a transistor dissipation of 100 W.

12.4 A power stage must dissipate 20 W. If a transistor with thermal characteristics, as shown in the figure for Problem 12.1, is to be used and the maximum ambient temperature is limited to 30°C, what is the maximum usable value of thermal resistance of the heat sink and insulating washer?

12.5 Repeat Problem 12.4 given that the transistor must dissipate 10 W.

12.6 A low-power transistor is rated as a 750-mW transistor (infinite heat sink). The specifications state that above 25°C the rating should be reduced 10 mW per °C. If the thermal resistance between the case and free air is 200°C/W, what is the maximum allowable dissipation for an ambient temperature of 30°C? What is the maximum allowable junction temperature?

* **12.7** Repeat Problem 12.6, given that a heat sink and insulating washer having a total thermal resistance of 100°C/W are used.

12.8 Repeat Problem 12.6 for an ambient temperature of 0°C.

12.9 Given that $\beta_0 = 50$ and $V_{BE} = 0.6$ V find the power dissipated by the transistor shown in the figure. Assume that the switching time of the transistor is zero. If the transistor

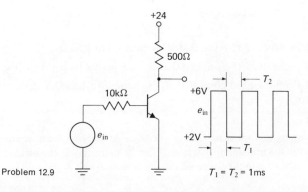

Problem 12.9

thermal characteristics correspond to those given in Problem 12.6, what is the maximum thermal resistance that the heat sink and insulating washer can have to prevent the destruction of the transistor? Assume a maximum ambient temperature of 30°C.

*12.10 Repeat Problem 12.9, assuming that T_1 of the input waveform equals 4 ms and T_2 equals 1 ms.

12.11 Repeat Problem 12.9 given that T_1 of the input waveform equals 1 ms and T_2 equals 4 ms.

Section 12.3A

12.12 Calculate the power dissipated by the 500-Ω resistor in Problem 12.9. Calculate the power delivered to the circuit by the dc source.

12.13 Calculate the power dissipated by the 500-Ω resistor in Problem 12.10. Calculate the power delivered to the circuit by the dc source.

*12.14 Calculate the power dissipated by the 500-Ω resistor in Problem 12.11. Calculate the power delivered to the circuit by the dc source.

12.15 The transistor in the circuit shown in the figure has a current gain of $\beta_0 = 60$, and $V_{BE} = 0.6$ V. Calculate
(a) the no-signal power dissipated by the transistor,
(b) the maximum ac power that can be delivered to the load before cutoff or saturation occurs,
(c) the power dissipated by the transistor when maximum ac power output is present (assume a nondistorted output signal), and
(d) the power delivered by the dc source.

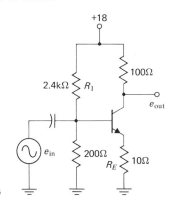

Problem 12.15

12.16 Repeat Problem 12.15 given that R_1 is changed to 1.5 kΩ.

12.17 Repeat Problem 12.15 assuming that R_E is bypassed with a capacitor.

*12.18 Repeat Problem 12.15 given that R_1 is changed to 1.5 kΩ and R_E is bypassed with a capacitor.

Section 12.3B

12.19 (a) Assuming that the bias point can be adjusted and that the saturation voltage is zero, what is the maximum power that can be delivered to the ideal transformer in the circuit shown in the figure on the following page?

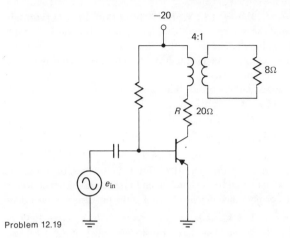

Problem 12.19

(b) In order to deliver the maximum power to the load, what value of quiescent collector current is required?

(c) What is the no-signal transistor dissipation at the bias point calculated in part (b)?

*12.20 Repeat Problem 12.19 given that a 12-V power supply is used and that R is 10 Ω.

12.21 If the transistor dissipation curve corresponds to that shown in the figure, and $\theta_i + \theta_{HS} = 15°C/W$, what is the maximum power that can be delivered to the load in the circuit of Problem 12.19 without exceeding the allowable transistor dissipation? Assume a maximum ambient temperature of 25°C.

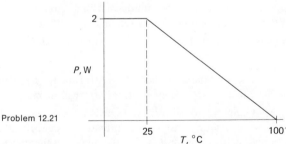

Problem 12.21

12.22 Repeat Problem 12.21 for $\theta_i + \theta_{HS} = 10°C/W$ and a maximum ambient temperature of 40°C.

12.23 Work parts (a) and (b) of Example 12.4 for $\beta_0 = 100$.

12.24 In Example 12.4, assume that the transistor has a $T_{J\,max} = 100°C$ and $\theta_{JM} = 7.5°C/W$. What is the maximum value of $\theta_i + \theta_{HS}$ that will allow maximum power to be delivered to the load without destroying the transistor? The maximum ambient temperature will be limited to 20°C by forced-air cooling.

*12.25 A transistor is rated as a 10-W transistor. The maximum junction temperature is 150°C and $\theta_{JM} = 12°C/W$. The transistor is used in a class-A stage, as shown in Fig. 12.16,

with $R_E = 5\,\Omega$, $V_{CC} = +12$ V, and a 4:1 transformer turns ratio. Given that $\theta_i + \theta_{HS} = 10°\text{C/W}$, and the maximum ambient temperature is to be 50°C, find:
(a) the allowable dissipation of the transistor,
(b) the collector-base junction temperature when the transistor is dissipating the maximum allowable power,
(c) the value of load resistance that causes maximum power to be delivered to the load (assuming an ideal transformer), and
(d) the maximum power delivered to the load.

12.26 Repeat Problem 12.25 given that the transformer turns ratio is 2:1.

12.27 In the circuit of Fig. 12.16, $R_L = 8\,\Omega$, $V_{CC} = +20$ V, $R_E = 10\,\Omega$, and the dc resistance of the transformer primary is $10\,\Omega$. Find the turns ratio n that will cause maximum power to be delivered to the load, given that the maximum collector current is 200 mA. If the transformer efficiency is 0.9, what is the maximum power delivered to R_L?

12.28 In the circuit of Fig. 12.16, $R_L = 8\,\Omega$, $V_{CC} = +40$ V, $R_E = 10\,\Omega$, $R_p = 20\,\Omega$, the turns ratio is 4:1, $R_2 = 240\,\Omega$, and the transformer efficiency is 0.9. The transistor has $\beta_0 = 40$, $V_{BE} = 0.7$ V, $\theta_{JM} = 4.5°\text{C/W}$, $T_{J\,max} = 130°\text{C}$, and a maximum power dissipation of 20 W; $\theta_i + \theta_{HS} = 10°\text{C/W}$, and the ambient temperature will never exceed 50°C. Find:
(a) the maximum power that can be delivered to the load without exceeding transistor dissipation,
(b) the quiescent collector current to deliver maximum power,
(c) R_1 to give this value of collector current,
(d) the circuit efficiency at this point,
(e) the no-signal transistor dissipation,
(f) the required value of $\theta_i + \theta_{HS}$ such that maximum circuit efficiency can be obtained,
(g) the maximum circuit efficiency,
(h) the no-signal transistor dissipation for the maximum circuit efficiency condition, and
(i) the quiescent collector current for maximum circuit efficiency.

*****12.29** Repeat Problem 12.28 given that the transformer turns ratio is 6:1.

Section 12.4A

12.30 Rework Example 12.5 for the case where $R_E = 10\,\Omega$ and the transformer turns ratio is doubled. Assume that $R_p = 20\,\Omega$ for the transformer primary.

*****12.31** Design a class-B amplifier to deliver 2 W to an 8-Ω load. Assume that a 24-V power supply exists and allow for a transformer dc primary resistance of 20 Ω (total) and $n_t = 0.9$. Specify a reasonable value for $\theta_i + \theta_{HS}$ assuming that the transistor thermal characteristics are as given in the figure for Problem 12.21 and that the maximum ambient temperature is 20°C.

12.32 Repeat Problem 12.31 given that the amplifier is to deliver 3 W to the load.

12.33 Plot the transistor dissipation as a function of peak collector current I_P for a class-B stage with a sinusoidal input. Assume $V_{CC} = 20$ V and $R_{ac} = 100\,\Omega$.

12.34 Rework Example 12.5, given that the reflected ac load between center tap and each side of the primary is $n^2 R_L = 64\,\Omega$.

Section 12.4B

12.35 Compare Eq. (12.24) to Eq. (12.28). Both equations correspond to maximum dissipation of a transistor in a class-B circuit. Explain why the two equations differ.

*12.36 It is desired to deliver a 10-W signal to a 16-Ω resistive load using the circuit of Fig. 12.26. Select an appropriate value of V_{C1} and find the minimum allowable dissipation of the transistors.

Section 12.5

12.37 A transistor can dissipate 10 W with the heat sink being used. If the power supply voltage for the stage is 30 V, $V_B = 1$ V, $\beta = 100$, $V_{BE(on)} = 0.5$ V and $V_{CE(sat)} = 0$, find:

(a) the minimum value of R_P that can be used,
(b) the maximum output power that can be delivered to R_P,
(c) the value of θ, V, and R_S to deliver this power, and
(d) the circuit efficiency of the stage.

Section 12.7

*12.38 Calculate the voltage gain and power gain of the class-A stage of Example 12.4. Assume that $r_{bb'} = 10\ \Omega$, $R_g = 10\ \Omega$, $\beta = 40$, and the stage operates at a quiescent collector current of 0.115 A.

13

APPLICATIONS OF THE OP AMP

The operational amplifier, or op amp, has become a basic element in the design of electronic systems. This amplifier is used in linear amplification schemes, analog computer circuits, digital computers, sweep generation methods, and waveshaping applications. The use of the op amp is now so widespread in the field of electrical engineering that it is infeasible to consider all important applications. We shall then restrict our considerations to a representative sampling of op amp circuits.

13.1 DIGITAL-TO-ANALOG CONVERSION

The op amp can accept any number of signal inputs forming the negative of the sum of these signals at the output. The gain factor can be controlled by the selection of the ratio of feedback resistance to the summing resistance. The circuit of Fig. 13.1 has an output given by

$$e_{out} = -\left(\frac{R_F}{R_a}e_a + \frac{R_F}{R_b}e_b + \frac{R_F}{R_c}e_c + \frac{R_F}{R_d}e_d\right). \tag{13.1}$$

Since the impedance between the negative terminal and ground is very low, this point can be considered a virtual ground. The resistance presented to each input source is equal to the corresponding summing resistor.

One use of the summer is the conversion of a digital signal to an analog signal. Let us assume that we have a four-bit binary code that must be converted to an analog voltage with amplitude proportional to the binary number present. Let us further assume that the digital one bit is represented by a positive voltage of V_1 volts while the zero bit is represented by zero volts. The four voltages e_a, e_b, e_c, and e_d in Fig. 13.1 can carry the binary information, with e_a being the least significant bit, e_b the next least significant bit and so forth. The number 13 would result, for example, if $e_a = V_1$, $e_b = 0$, $e_c = V_1$, and $e_d = V_1$. The conversion factor is determined by the selection of the ratio of R_F to R_a. The output voltage per unit is given by

$$\text{CF} = \frac{R_F}{R_a} V_1 \text{ volts/unit.} \tag{13.2}$$

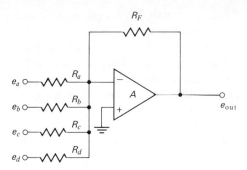

Fig. 13.1 Summing amplifier.

This value can be related to the desired maximum output voltage of the converter. For an n-bit converter, if V_0 is the output desired when all n inputs are binary ones, then

$$\text{CF} = V_0/(2^n - 1). \tag{13.3}$$

The binary number 0001 will produce an output voltage of $R_F V_1/R_a$. The binary number 0010 should produce two times this voltage; thus R_b is selected to equal $R_a/2$. The binary numbers 0100 and 1000 should produce outputs equal to four and eight times the voltage $R_F V_1/R_a$. The resistance R_c is selected to be $R_a/4$, while R_d is $R_a/8$. Each input voltage will now contribute to the output in proportion to the significance of the bit represented by that input voltage.

Example 13.1. Select values of R_F, R_a, R_b, R_c, and R_d in Fig. 13.1 such that a four-bit binary number can be converted to proportional voltage. The binary number 1111 should result in an output voltage of -5 V. The one bit is represented by a $+12$ V level while 0 V represents the zero bit.

Solution. Since the binary number 1111 which is equivalent to decimal 15 is to produce an output voltage of -5 V, the conversion factor is

$$\text{CF} = 5 \text{ V}/15 \text{ units} = \tfrac{1}{3} \text{ volt/unit}.$$

We will arbitrarily select R_F to be 100 kΩ; thus we can use Eq. (13.2) to calculate the resistance R_a; $R_a = R_F V_1/\text{CF} = 10^5 \times 12/(\tfrac{1}{3}) = 360$ kΩ. The resistances R_b, R_c, and R_d are then found to be 180 kΩ, 90 kΩ, and 45 kΩ, respectively.

There are several factors to be considered when larger numbers must be converted. The active region of any op amp is limited to some finite value, typically 20 V or less for integrated circuits. In calculating the largest conversion factor that can be used, we find that

$$\text{CF}_{\max} = V_{OM}/N_M, \tag{13.4}$$

where V_{OM} is the magnitude of the maximum allowed output voltage of the amplifier and N_M is the largest binary number to be converted. Usually $N_M = 2^n - 1$, but this may not always be true. As N_M increases to larger numbers, the output voltage per unit becomes very small. Errors can now be introduced by variations of input voltage (especially the voltage of the most significant bit since gain is high for this bit), variations in power supply voltage, resistance changes in summing or feedback resistors, and deviations from linearity in amplifier gain including drift effects. As these changes lead to voltage errors that are comparable to the conversion factor, accuracy of the D/A converter decreases.

To demonstrate the significance of the error-producing variables let us consider an eight-bit D/A converter with input signals that change from 0 V to 12 V. The maximum output signal is 10 V. For this case, the maximum conversion factor is

$$CF_{max} = 10 \text{ V}/255 = 39.2 \text{ mV/unit}.$$

The gain from the least significant bit input to output is then

$$0.0392/12 = 3.27 \times 10^{-3}.$$

The gain from the most significant bit input to output must be 2^7 or 128 times this value, giving a gain of 0.4186. We note that if the input signal of the most significant bit changes 1 part in 128 (0.7%) the output will reflect a voltage change equivalent to a one-unit change in the binary number. Changes in resistance or gain may also contribute to this uncertainty in output voltage. For conversion of large binary numbers, very precise components and input voltages are required.

There are two specifications of major interest to the user of a D/A converter: resolution and accuracy [3]. Resolution is specified by the number of possible output voltage levels and is often expressed in number of bits of binary information that the converter will handle. For example, a 9-bit resolution means there are $2^9 - 1 = 511$ distinguishable output voltage levels.

Accuracy relates to the measure of deviation of the analog output signal from the predicted signal value. It may be expressed as a percentage of full-scale output voltage or as a fraction of the least significant bit. Typically the accuracy is $\pm\frac{1}{2}$ LSB, meaning that the error voltage will never exceed one-half of the conversion factor; that is,

$$V_E \leqq \frac{V_O}{2N_M} \approx \frac{V_O}{2^{n+1}}. \tag{13.5}$$

Resolution and accuracy are not necessarily equal. A converter with 12-bit resolution might have only 10-bit accuracy.

One of the more serious problems of the binary-weighted D/A converter is the large spread in resistor values. For integrated circuit fabrication this spread is very inconvenient. To overcome this problem, the configuration of Fig. 13.2 is often used for monolithic D/A conversion, especially if six or more bits are to be converted. At each node in the network the input current divides equally since the equivalent impedance to the right of the node is always $2R$ as is the impedance to ground. A current switch is used to control the current path. If a bit equals one, the corresponding current switch diverts current to the amplifier input. If a bit equals zero the current is diverted to ground. The output voltage will be propor-

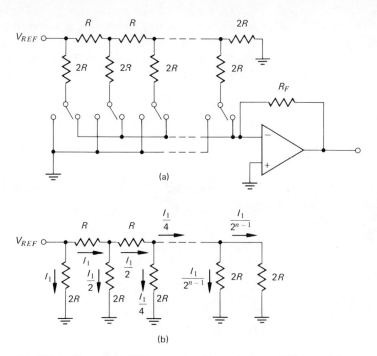

Fig. 13.2 Ladder method of D/A conversion: (a) actual circuit; (b) current relations.

tional to the sum of the current entering the input node. The conversion factor for this circuit is

$$\mathrm{CF} = \frac{V_{REF} R_F}{2^n R}. \qquad (13.6)$$

The summing amplifier is also conveniently used in control systems to sum input, error, and damping signals to create a drive signal for the mechanism to be controlled. The correct proportions of these various signals are determined simply by the selection of the summing resistances.

In the last few years, the use of special-purpose computers in process control has become quite widespread. The input is in the form of a digital address which must be converted to an analog drive signal. The conversion of this address and the summing of the analog error and damping signals can all be accomplished with the summing amplifier, creating the appropriate control signal to apply to the power amplifier stage.

13.2 INTEGRATORS AND SWEEP CIRCUITS

The op amp can be used to construct an integrator with the analog computer serving as a prime example of this application. Very accurate sweep circuits can also be constructed with op amps. The sweep circuit is basically an integrator that accepts a step function input and forms a linear ramp as the output.

The circuit of Fig. 13.3 demonstrates an integrator. If it is assumed that the amplifier has a very high input impedance and that e_a is very small compared to e_{in} and e_{out}, the output voltage can be calculated quite simply. The current charging the capacitor equals the input current and is

$$i_c = e_{in}/R.$$

The voltage across the capacitor equals the negative of the output voltage; thus

$$e_{out} = -v_c = -\frac{1}{C}\int i\, dt = -\frac{1}{RC}\int e_{in}\, dt. \tag{13.7}$$

The output voltage is equal to the negative of the integral of the input voltage multiplied by the scale factor $1/RC$.

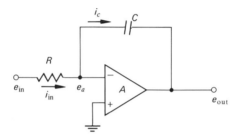

Fig. 13.3 An integrator.

In practice, the input impedance to the op amp and noninfinite gain A introduce errors into the integration. These effects will be considered later in conjunction with the discussion on Miller sweep circuits.

Sweep circuits, which generate a linear ramp voltage or current, are used in many systems, two notable examples being oscilloscopes and television receivers. While a perfectly linear sweep signal is often desirable, most schemes result in an exponential output waveform to approximate a linear signal. A measure of the difference between the ideal and actual sweep waveforms is the deviation from linearity d_t [2]. Referring to Fig. 13.4, the deviation from linearity is defined by

$$d_t = \frac{Te'(0) - e(T)}{Te'(0)}, \tag{13.8}$$

where T is the sweep time, $e(T)$ is the actual voltage at $t = T$, and $e'(0)$ is the slope of the voltage waveform at the start of the sweep ($t = 0$).

For an exponential signal the early portion of the waveform can be approximated closely by

$$\exp(-t/\tau) = 1 - t/\tau + t^2/2\tau^2, \tag{13.9}$$

where τ is the time constant. A charging curve can then be represented by

$$e(t) = E[1 - \exp(-t/\tau)] = E[t/\tau - t^2/2\tau^2] \tag{13.10}$$

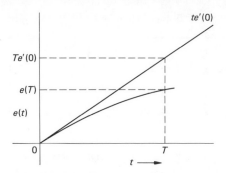

Fig. 13.4 Quantities used in defining the deviation from linearity.

over the near-linear portion of the curve. In this case E is the target voltage. A perfectly linear waveform would reach a voltage V_s at the end of the sweep time T. The slope of the curve would be V_s/T. The approximate waveform given by Eq. (13.10) can be made to have the same initial slope as the ideal waveform by setting

$$V_s/T = E/\tau. \tag{13.11}$$

The deviation from linearity of the exponential waveform will then be

$$d_t = \frac{V_s - E[T/\tau - T^2/2\tau^2]}{V_s} = T/2\tau. \tag{13.12}$$

In terms of the target voltage E and the sweep voltage V_s,

$$d_t = V_s/2E. \tag{13.13}$$

To minimize the deviation from linearity, the time constant τ must be much larger than the sweep time and the target voltage E must be much larger than the required sweep voltage.

The voltage on the capacitor of a simple RC network departs from linearity because charging current decreases as the capacitor charges. If the charging current could be maintained at a constant value, the voltage build-up would be linear, with slope given by

$$dv_c/dt = I/C, \tag{13.14}$$

where I is the charging current and C is the value of capacitance. As v_c increases in Fig. 13.5(a) the charging current decreases; however, in Fig. 13.5(b) the charging current will be maintained constant due to the dependent source voltage. The network of Fig. 13.5(b) would create a sweep voltage that is perfectly linear since the total voltage across R is constant, resulting in a constant current charge.

A. Bootstrap Sweep Circuit

Figure 13.6 shows one alternative form in the practical realization of the sweep circuit of Fig. 13.5(b). This circuit is called a bootstrap sweep circuit.

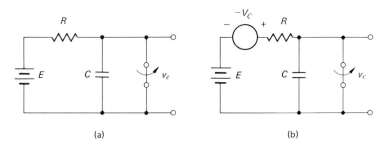

Fig. 13.5 Charging networks: (a) RC network; (b) RC network with dependent source.

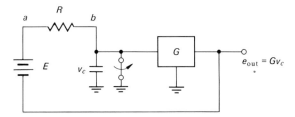

Fig. 13.6 Bootstrap sweep circuit.

Assuming an infinite input impedance to the amplifier, the voltage appearing at point a will be $(E + Gv_c)$ while the voltage at point b is v_c. The drop across the resistance is then $(E + Gv_c - v_c)$. If $G = 1$, this resistive voltage is constant leading to a constant current charge and a perfectly linear sweep. An op amp with unity gain can be used as the amplifying element. Deviations from linearity in the practical circuit result from values of gain differing from unity and finite input impedance to the amplifier. Generally the voltage source is replaced by a large storage capacitor to eliminate the necessity of a floating source. The loss of charge on this capacitor during the sweep time can also lead to errors. Let us consider these effects leading to deviations from linearity in more detail.

If gain G were unity and the amplifier input impedance were infinite, the circuit of Fig. 13.6 would generate a linear sweep. If G were slightly less than unity with infinite input impedance an exponential waveform would result. The charging current can be written as

$$i_c = \frac{E - (1 - G)v_c}{R}.$$

It follows that the first derivative of capacitor voltage is

$$\frac{dv_c}{dt} = \frac{1}{C}i_c = \frac{E - (1 - G)v_c}{RC}.$$

Solving the first-order differential equation

$$\frac{dv_c}{dt} + \frac{(1 - G)v_c}{RC} = \frac{E}{RC} \qquad (13.15)$$

results in

$$v_c = \frac{E}{1-G}\left[1 - \exp\left(-\frac{t[1-G]}{RC}\right)\right]. \quad (13.16)$$

The voltage v_c has a target value of $E/(1-G)$ which may be several orders of magnitude larger than E. The time constant is increased by the same factor to $RC/(1-G)$. The initial slope of the voltage is the same as the simple RC network, but the target voltage is much larger. The required sweep voltage is generally much smaller than $E/(1-G)$; thus only the very linear portion of the charging curve is used in generating the sweep voltage. Figure 13.7 shows the waveform of the simple RC sweep and the bootstrap sweep circuit with nonunity gain.

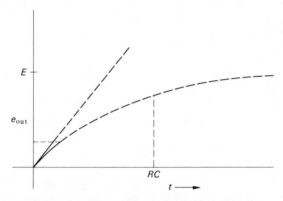

Fig. 13.7 Bootstrap sweep compared to simple RC sweep.

Expanding the exponential of Eq. (13.16) in a power series leads to a deviation from linearity of

$$d_t = (1-G)T/2RC \quad (13.17)$$

which is much smaller than that for the simple RC case.

We next consider the effect of finite amplifier input impedance on the bootstrap circuit. Figure 13.8 shows an equivalent circuit which can be used to account for the amplifier input impedance R_i.

The Thévenin equivalent voltage presented to the capacitor is now $(E + Gv_c)R_i/(R + R_i)$ and the equivalent resistance is $R_{th} = RR_i/(R + R_i)$. The equation

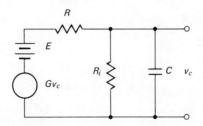

Fig. 13.8 Equivalent circuit for bootstrap sweep with nonunity gain and finite amplifier input impedance.

describing capacitor voltage is

$$\frac{dv_c}{dt} + \frac{\left(1 - \dfrac{GR_i}{R + R_i}\right)}{R_{th}C} v_c = \frac{\dfrac{ER_i}{R + R_i}}{R_{th}C}. \qquad (13.18)$$

Equation (13.18) is solved to give

$$v_c = \frac{ER_i}{R + (1 - G)R_i}\left[1 - \exp\left(-\frac{t}{RC}[1 - G + R/R_i]\right)\right]. \qquad (13.19)$$

The target voltage is lowered from the infinite impedance case to $ER_i/[R + (1 - G)R_i]$. The time constant is also modified. The deviation from linearity can be found to be

$$d_t = \frac{T}{2RC}(1 - G + R/R_i). \qquad (13.20)$$

As input impedance decreases to the point that it is not negligible compared to R, the deviation from linearity increases.

If the floating source E is replaced by a large storage capacitor C_b, the charge lost by this capacitance over the sweep period leads to a further deviation from linearity. The charge on C is now supplied by C_b. Current through the input impedance R_i during sweep time also leads to a loss of charge on C_b; however, this component of charge is negligible compared to that absorbed by C. Figure 13.9 shows a circuit that can be used to calculate the sweep parameters. Assuming all charge lost by C_b is absorbed by C we can write the change in voltage on C_b as

$$\Delta E = -\frac{C}{C_b}v_c. \qquad (13.21)$$

The charging current of C can be written as

$$i_c = C\frac{dv_c}{dt} = \frac{E - \dfrac{C}{C_b}v_c - (1 - G)v_c}{R} - \frac{v_c}{R_i}. \qquad (13.22)$$

Solving Eq. 13.22 gives

$$v_c = \frac{E}{1 - G + \dfrac{C}{C_b} + \dfrac{R}{R_i}}\left[1 - \exp\left(-\frac{t}{RC}\left[1 - G + \frac{C}{C_b} + \frac{R}{R_i}\right]\right)\right]. \qquad (13.23)$$

The deviation from linearity is

$$d_t = \frac{T}{2RC}\left(1 - G + \frac{C}{C_b} + \frac{R}{R_i}\right), \qquad (13.24)$$

which represents a further degradation of linearity over the circuit that uses a floating source.

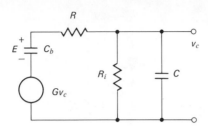

Fig. 13.9 Bootstrap circuit including several nonideal effects.

We now return to the question of selecting an amplifier that offers a stable gain of approximately unity and a high input impedance. Figure 6.22 shows a noninverting amplifier with midband gain given by

$$G_{MB} = 1 + R_F/R_2 \tag{6.41}$$

and an input impedance of

$$R_i = 2r_{cm} \| rA_{MB}F = 2r_{cm} \left\| \frac{rA_{MB}R_2}{R_2 + R_F} \right. . \tag{6.45}$$

If R_F is selected to be zero by providing a short circuit between output and the inverting terminal, the voltage gain is exactly unity and the input resistance is quite large. Thus, the op amp can be used to construct a bootstrap sweep circuit.

Returning to Eq. 13.24 it can be seen that d_t will be minimized if G is selected to be slightly larger than unity, so that

$$G = 1 + C/C_b + R/R_i. \tag{13.25}$$

The feedback resistance R_F can be selected to obtain the necessary gain to compensate for the effects of C_b and R_i. It should be noted that G can become too large, causing the sweep voltage to exceed the desired linear sweep. An example will demonstrate the design of the bootstrap sweep circuit using an op amp.

Example 13.2. A sweep circuit with a 6-V, 100-ms sweep is to be designed having a deviation from linearity of 0.01 or one percent. The op amp to be used has the following characteristics: $A_{MB} = 50{,}000$, $r = 200$ kΩ, $r_{cm} = 10$ MΩ, with a bandwidth of 20 kHz.

Solution. The circuit of Fig. 13.10 is chosen for this design. When the input signal is near zero volts the transistor is saturated and the emitter potentiometer is adjusted to cause exactly zero volts to appear at the collector.

Two possible methods of solution are (1) to select an amplifier gain of unity and divide the sweep error equally between effects due to C_b and R_i, or (2) to choose reasonable values of circuit elements C, C_b, and R using a gain greater than unity to achieve the appropriate error.

METHOD 1. The input impedance to the op amp with $R_F = 0$ (unity gain) is $2r_{cm} \| rA_{MB} = 20$ MΩ. If the contribution to the deviation from linearity due to

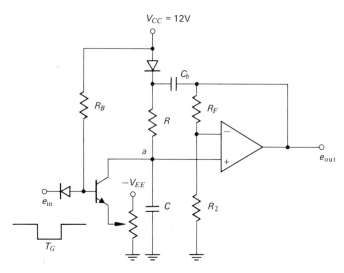

Fig. 13.10 Practical bootstrap sweep circuit.

noninfinite input impedance is to equal that due to discharge of C_b, then

$$\frac{R}{R_i} = \frac{C}{C_b}.$$

The charging current will determine the voltage at the end of the sweep since

$$v_c = \frac{i_c}{C} t.$$

This current is given by

$$i_c = (V_{CC} - V_d)/R = 11.4/R.$$

We can now write

$$V_s = \frac{11.4}{RC} T \quad \text{or} \quad RC = \frac{11.4 \times 0.1}{6} = 0.190$$

and

$$d_t = 0.01 \times 6 = 0.06 = \frac{T}{2RC}\left(\frac{C}{C_b} + \frac{R}{R_i}\right) = \frac{T}{CR_i} = \frac{5 \times 10^{-9}}{C}.$$

Combining these two equations gives $R = 2.28$ MΩ and $C = 0.0833$ μF. The value of $C_b = CR_i/R = 0.731$ μF. For unity gain $R_F = 0$ and R_2 is infinite.

METHOD 2. Let us assume in this case that 100 μF is the largest usable capacitance. This restriction could be imposed due to economics, size limitations, or availability of this value of capacitance. If C is selected to be 1 μF, R must be calculated to give the appropriate sweep time. The charging current in this case must be

$$i_c = \frac{CV_s}{T} = \frac{10^{-6} \times 6}{10^{-1}} = 0.06 \text{ mA}.$$

The value of R is then

$$R = (V_{CC} - V_d)/i_c = 11.4/0.06 = 190 \text{ k}\Omega.$$

The contribution to the deviation from linearity equation due to R and C is

$$\frac{C}{C_b} + \frac{R}{R_i} = \frac{1}{100} + \frac{190}{20000} = 0.0195.$$

In order to achieve minimum deviation from linearity, the gain of the amplifier must satisfy Eq. (13.25),

$$1 - G + C/C_b + R/R_i = 0.$$

The gain must be $G = 1.0195$. It should be noted that the precision with which G can be determined depends on the precision of the resistors R_F and R_2. Thus, the overall deviation from linearity will never be zero as one might expect from the formulas. Values of $R_F = 1.95$ kΩ and $R_2 = 100$ kΩ are reasonable to use for this circuit.

We must now consider the discharge circuit to calculate a suitable value for R_B. When the input gate returns to its positive voltage level the transistor turns on to end the sweep and discharge C. The discharge current will equal the collector current less the current still flowing through R. If transistor current gain can be considered constant as collector voltage changes, the discharge current is

$$I_d = \beta_0 I_B - V_{CC}/R.$$

The current I_B is approximated by V_{CC}/R_B; thus,

$$I_d = V_{CC}(\beta_0/R_B - 1/R).$$

Since discharge current is constant, the discharge of C will be linear. The required change in voltage equals the sweep voltage V_s; therefore, the discharge time is

$$T_d = \frac{CV_s}{I_d}.$$

For a discharge time of 1 ms, the required value of I_d for the case of $C = 1$ μF is

$$I_d = \frac{10^{-6} \times 6}{1 \times 10^{-3}} = 6 \text{ mA}.$$

If $\beta_0 = 50$, a value of $R_B = 99$ kΩ will result in the appropriate current. After each cycle, charge lost by C_b must be replaced. Charge will be supplied by C_b during both sweep time and discharge time and must be replaced after discharge time is completed through the diode and the output impedance of the op amp. The output impedance of the op amp is generally low enough to lead to a very short recovery time before the input gate can again be applied.

B. Miller Sweep Circuit

Returning to Fig. 13.5(b) we see that a dependent voltage generator is used to ensure constant charging current. The position of this generator can be moved within the loop, maintaining the same current conditions. Fig. 13.11 shows a second possible configuration that leads to the Miller sweep circuit. As voltage builds up on the capacitor the generator creates an equal voltage of opposite polarity. The voltage across the resistance continues to equal E, leading to constant charging current as v_c increases. In this case the output voltage is of opposite polarity to the capacitor voltage.

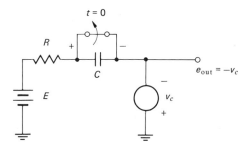

Fig. 13.11 Constant current charging network.

The ideal op amp of Fig. 6.21 used in the inverting mode provides the dependent voltage source required by the Miller sweep circuit. Figure 13.12 indicates the appropriate configuration. The voltage across the capacitance is

$$v_c = e_1(1 + A),$$

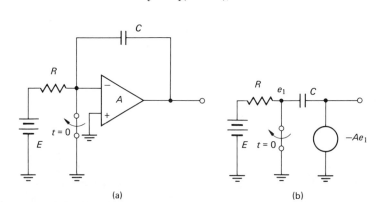

Fig. 13.12 Miller sweep circuit: (a) actual circuit; (b) equivalent circuit using ideal op amp.

while the output voltage is
$$e_{out} = -Ae_1 \approx -v_c;$$
thus the situation of Fig. 13.11 is approximated to the extent that $A \gg 1$. Summing current at the amplifier input gives
$$\frac{E - e_1}{R} = C\frac{dv_c}{dt} = C(1 + A)\frac{de_1}{dt}.$$
Solving this differential equation for e_1 results in
$$e_1 = E(1 - e^{-t/RC(1+A)}). \tag{13.26}$$
Output voltage is given by
$$e_{out} = -AE(1 - e^{-t/RC(1+A)}). \tag{13.27}$$
The time constant of this circuit is
$$\tau = RC(1 + A), \tag{13.28}$$
while the initial slope is
$$AE/RC(1 + A) \approx E/RC. \tag{13.29}$$
Approximating the exponential by the first three terms of the series expansion leads to
$$e_{out} = \frac{-AE}{RC(1 + A)}t + \frac{AE}{2[RC(1 + A)]^2}t^2. \tag{13.30}$$
For a given sweep time T, the deviation from linearity is
$$d_t = \frac{T}{2RC(1 + A)}. \tag{13.31}$$
As A approaches infinity the sweep becomes perfectly linear and equal to $-(E/RC)t$.

The same results could have been achieved by the reflected impedance method. Reflecting the effects of C to the input side of the amplifier as done in Chapter 7 gives the circuit of Fig. 13.13. The time constant of e_1 is now obviously equal to $RC(1 + A)$ and the output voltage is
$$e_{out} = -AE(1 - e^{-t/RC(1+A)}) \tag{13.27}$$
as before.

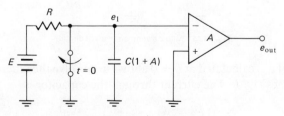

Fig. 13.13 Reflected impedance method.

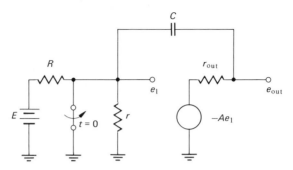

Fig. 13.14 Miller sweep circuit with finite amplifier input and output impedances.

If input and output impedances are significant the circuit of Fig. 13.14 applies. The equivalent circuit of Fig. 6.24 is used for the op amp in this case.

A Thévenin equivalent taken at the amplifier input results in the circuit of Fig. 13.15. When the switch is initially opened the capacitor appears as an instantaneous short. The output voltage immediately jumps to e_1 which can be written as

$$e_1 = E_{th} - R_{th} \frac{E_{th} + Ae_1}{R_{th} + r_{out}}.$$

Solving for e_1 gives

$$e_1(0) = e_{out}(0) = E_{th} \frac{r_{out}}{r_{out} + (1 + A)R_{th}} \approx \frac{E_{th} r_{out}}{(1 + A)R_{th}}. \quad (13.32)$$

The output voltage jumps to this positive value at the start of the sweep. For reasonably high values of gain this initial output voltage is negligible.

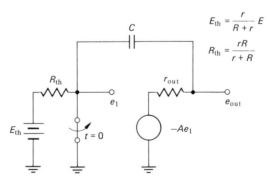

Fig. 13.15 Miller sweep equivalent circuit.

The value of e_{out} calculated at $t = 0$ allows us to solve the differential equation for e_{out} at all times $t \geq 0$. The current through the capacitor is

$$i_c = \frac{E_{th} - e_1}{R_{th}} = C \frac{d(e_1 - e_{out})}{dt}. \quad (13.33)$$

The output voltage is

$$e_{out} = -Ae_1 + r_{out}i_c = -Ae_1 + (E_{th} - e_1)\frac{r_{out}}{R_{th}}. \quad (13.34)$$

Equation (13.34) is used to eliminate e_1 from Eq. (13.33). The differential equation is then

$$\frac{de_{out}}{dt} + \frac{e_{out}}{C[r_{out} + (1+A)R_{th}]} = \frac{-AE_{th}}{C[r_{out} + (1+A)R_{th}]}. \quad (13.35)$$

Solving this equation and evaluating the arbitrary constant using Eq. (13.32) give

$$e_{out} = -AE_{th} + \frac{E_{th}(1+A)[r_{out} + AR_{th}]}{r_{out} + (1+A)R_{th}} e^{-t/C[r_{out} + (1+A)R_{th}]}. \quad (13.36)$$

If $A \approx A + 1$, the output voltage becomes

$$e_{out} \approx -AE_{th}(1 - e^{-t/C[r_{out} + (1+A)R_{th}]}). \quad (13.37)$$

The deviation from linearity is then

$$d_t = \frac{t}{2C[r_{out} + (1+A)R_{th}]}. \quad (13.38)$$

If a transistor switch is used to start and stop the sweep, the finite switch resistance will lead to a capacitor discharge time. Letting the total resistance between the amplifier input node and ground equal R_S when the transistor saturates, the discharge time constant will be

$$\tau_d = [R_S(1+A) + r_{out}]C. \quad (13.39)$$

This is calculated by reflecting the effects of the series combination of C and r_{out} to the amplifier input.

13.3 ACTIVE FILTERS

Over the years a great deal of effort has been devoted to the development of LC or RLC filter theory. While RL or RC filters are used for low or moderate falloff rates it is necessary to use both inductors and capacitors to achieve sharp cutoff, passive filters. Highly selective filters have presented a rather formidable problem for the integrated circuit industry since inductors are not readily integrable. The last decade has seen many significant advancements concerning highly selective integrated filters.

When one considers filter theory from the standpoint of pole locations in the Laplace plane, the transfer function of a highly selective filter has complex poles near the $j\omega$-axis. Several other filters, not requiring high selectivity, also require complex poles in the transfer function. While these transfer functions may not include poles near the $j\omega$-axis, both inductors and capacitors are necessary to synthesize passive filters of this type. Active circuits can be used to eliminate the inductors while achieving the desired pole locations. Examples of this type of filter are the maximally flat magnitude response or Butterworth filter, the maximally flat phase response filter, and the Chebyshev response filter.

Another area that has been important for several years is that of low-frequency filters. Large inductors or capacitors are required in passive circuits to achieve low corner frequencies. Active filters have been quite useful in reducing the required size of reactive elements for low-frequency filters. Capacitors with values that are integrable can then be used to synthesize these filters in relatively small volume circuits.

The field of active filters is so extensive that only a small number of selected topics can be treated in this section. Inclusion of a particular topic does not imply that this subject is more important than others that are not included.

A. Low-pass Transfer Function

Figure 13.16 indicates a network with a transfer function $T(s) = e_0/e_i$, where s is the Laplace variable. For sinusoidal excitation, this variable reduces to $s = j\omega$. There may be other transfer functions of interest such as e_0/i_i or i_0/e_i, but we will restrict the discussion here to the voltage transfer function. This function is useful in active filter design since impedance levels are conveniently controlled by op amps.

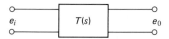

Fig. 13.16 Filter network.

The basic problem in filter synthesis can be subdivided into two parts. First an appropriate transfer function must be selected; then a network must be designed to yield this transfer function. The selection of an appropriate transfer function is relatively easy for a number of filter applications. Transfer functions for maximally flat magnitude response filters and Chebyshev filters are tabulated in many books. These types of response are useful for the bulk of filter circuits used and are therefore of great practical significance. The transfer functions for maximally flat magnitude response up to fourth order (four pole) are listed. These are also called the denormalized Butterworth polynomials.

$$T(s) = \frac{K_1}{s + B} \quad (13.40a)$$

$$T(s) = \frac{K_2}{s^2 + \sqrt{2}Bs + B^2} \quad (13.40b)$$

$$T(s) = \frac{K_3}{s^3 + 2Bs^2 + 2B^2s + B^3} \quad (13.40c)$$

$$T(s) = \frac{K_4}{s^4 + 2.613Bs^3 + 3.414B^2s^2 + 2.613B^3s + B^4}. \quad (13.40d)$$

The 3-db bandwidth is designated by B. Due to the availability of tabulated transfer functions it is generally necessary only to refer to a table to select the desired function. We must of course assume that the desired characteristics can be related to the tabulated functions by the designer. The first part of the filter synthesis problem is therefore solved and we can now proceed to the second part.

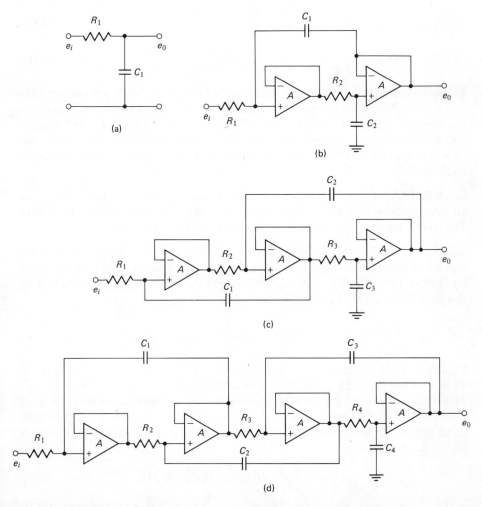

Fig. 13.17 Low-pass filters: (a) one-pole; (b) two-pole; (c) three-pole; (d) four-pole.

There are always several possible solutions to a given synthesis problem. One method of synthesis is based on the circuits of Fig. 13.17 [5]. An n-pole low-pass filter can be constructed by simply extending the pattern of Fig. 13.17 to n stages. If we define the parameters $\omega_1 = 1/R_1 C_1$, $\omega_2 = 1/R_2 C_2$, $\omega_3 = 1/R_3 C_3$, and

$\omega_4 = 1/R_4C_4$ the transfer functions for the four circuits become

$$T(s) = \frac{\omega_1}{s + \omega_1}, \qquad (13.41a)$$

$$T(s) = \frac{\omega_1\omega_2}{s^2 + \omega_1 s + \omega_1\omega_2}, \qquad (13.41b)$$

$$T(s) = \frac{\omega_1\omega_2\omega_3}{s^3 + \omega_1 s^2 + \omega_1\omega_2 s + \omega_1\omega_2\omega_3}, \qquad (13.41c)$$

$$T(s) = \frac{\omega_1\omega_2\omega_3\omega_4}{s^4 + \omega_1 s^3 + \omega_1\omega_2 s^2 + \omega_1\omega_2\omega_3 s + \omega_1\omega_2\omega_3\omega_4}. \qquad (13.41d)$$

The transfer function for an n-pole filter can be found by extending these formulas.

If a maximally flat magnitude response is desired, the transfer function corresponding to the appropriate number of poles is selected from Eq. (13.40). The corresponding equation from Eq. (13.41) is used and circuit element values are chosen to make Eq. (13.41) equal to Eq. (13.40). An example will demonstrate this method.

Example 13.3. Design a low-pass, maximally flat magnitude response filter with an asymptotic rolloff of 18 db/octave. The corner frequency should be 2 kHz. The source resistance is 100 Ω while the load resistance is 10 kΩ.

Solution. For Butterworth filters the rolloff rate is $6n$ db/octave where n is the number of poles of the transfer function. This design requires three stages to reach an 18 db/octave rolloff. Equation (13.40c) corresponds to the transfer function for this case. For a 2-kHz bandwidth the equation becomes

$$T(s) = \frac{K_3}{s^3 + 8\pi \times 10^3 s^2 + 32\pi^2 \times 10^6 s + 64\pi^3 \times 10^9}.$$

Equation (13.41c) can be made to equal this equation if $\omega_1 = 8\pi \times 10^3$, $\omega_2 = 4\pi \times 10^3$, and $\omega_3 = 2\pi \times 10^3$ and we assume that $K_3 = \omega_1\omega_2\omega_3 = 64\pi^3 \times 10^9$. The circuit of Fig. 13.17(c) has the appropriate transfer function if the element values are properly selected. Let us choose $R_1 = 10$ kΩ to create a relatively high input impedance compared to the 100 Ω source resistance. The required value of C_1 is

$$C_1 = \frac{1}{\omega_1 R_1} = 0.0398 \ \mu F \approx 0.04 \ \mu F.$$

Selecting R_2 and R_3 as 10-kΩ resistances leads to values of $C_2 = 0.08 \ \mu F$ and $C_3 = 0.16 \ \mu F$. The transfer function of this circuit is

$$T(s) = \frac{64\pi^3 \times 10^9}{s^3 + 8\pi \times 10^3 s^2 + 32\pi^2 \times 10^6 s + 64\pi^3 \times 10^9}.$$

Figure 13.18 shows the frequency response of this filter.

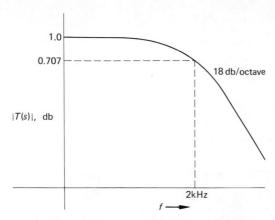

Fig. 13.18 Frequency response of three-pole Butterworth filter.

A buffer stage can be used to obtain more favorable loading conditions if necessary. However, there is a great deal of flexibility in impedance level with a circuit of this type. The input impedance can be made to have a high value by selecting R_1 to be large. The output impedance is determined by the output impedance of the last stage and this value can be quite small. Low-frequency filters down to corner frequencies of a few hertz can be constructed with reasonable values of capacitance by this method.

Starting with Bessel or Chebyshev polynomials results in the realization of maximally flat phase or equal-ripple response filters by this same method.

B. High-pass Transfer Function

Given the low-pass transfer function for a desired response it is a simple matter to obtain the high-pass transfer function yielding the same type of response. A frequency transformation that preserves the same bandwidth can be used to convert from the low-pass function to the high-pass function. One such transformation replaces the variable s with B^2/s, where B is the bandwidth applying to both low- and high-pass functions. For example, the two-pole Butterworth response of Eq. (13.40b) will transform as shown below upon applying the preceding frequency transformation.

$$\frac{K_2}{s^2 + \sqrt{2}Bs + B^2} \to \frac{K_2}{\dfrac{B^4}{s^2} + \dfrac{\sqrt{2}B^3}{s} + B^2} = \frac{(K_2/B^2)s^2}{s^2 + \sqrt{2}Bs + B^2} = \frac{K'_2 s^2}{s^2 + \sqrt{2}Bs + B^2}.$$

The Butterworth high-pass transfer functions are maximally flat about $s = \infty$ whereas the low-pass functions are maximally flat about $s = 0$. These high-pass functions up to the fourth order are listed.

$$T(s) = \frac{sK'_1}{s + B} \qquad (13.42a)$$

$$T(s) = \frac{s^2 K_2'}{s^2 + \sqrt{2}Bs + B^2} \qquad (13.42b)$$

$$T(s) = \frac{s^3 K_3'}{s^3 + 2Bs^2 + 2B^2 s + B^3} \qquad (13.42c)$$

$$T(s) = \frac{s^4 K_4'}{s^4 + 2.613Bs^3 + 3.414B^2 s^2 + 2.613B^3 s + B^4}. \qquad (13.42d)$$

Again the 3-db frequency is given by B.

The circuits of Fig. 13.17 can be converted to high-pass filters simply by interchanging the positions of the resistors and capacitors. The transfer functions then become

$$T(s) = \frac{s}{s + \omega_1}, \qquad (13.43a)$$

$$T(s) = \frac{s^2}{s^2 + \omega_2 s + \omega_2 \omega_1}, \qquad (13.43b)$$

$$T(s) = \frac{s^3}{s^3 + \omega_3 s^2 + \omega_3 \omega_2 s + \omega_3 \omega_2 \omega_1}, \qquad (13.43c)$$

$$T(s) = \frac{s^4}{s^4 + \omega_4 s^3 + \omega_4 \omega_3 s^2 + \omega_4 \omega_3 \omega_2 s + \omega_4 \omega_3 \omega_2 \omega_1}. \qquad (13.43d)$$

The high-pass filters can be constructed by choosing the parameters $\omega_1, \omega_2, \ldots, \omega_n$ such that the applicable equation from Eq. (13.43) equals the appropriate equation from Eq. (13.42).

Bessel, Chebyshev, and other responses can be realized for the high-pass filter also, requiring the appropriate starting polynomial.

There are many other methods that can be used to synthesize active filters. Some methods require fewer op amps than the method discussed and these approaches can be found in the literature [4, 5].

C. Bandpass Filters

High-Q bandpass filters have traditionally used resonant circuits containing both capacitance and inductance. Pressure to develop inductorless bandpass filters has come not only from the integrated circuit area, but also from the low-frequency communication field. Sharp, low-frequency filters are very difficult to synthesize with inductors due to the low values of coil Q at low frequencies. Research on active bandpass filters predates the development of the integrated circuit, but this development has certainly led to intensified efforts in the active filter area.

There are several theoretical methods of synthesizing general transfer functions containing complex poles near the $j\omega$-axis. So much information is available in the literature that even a brief review of this area cannot be attempted here. Thus, the ensuing discussion will be limited to a consideration of second-order, highly selective bandpass filters that include the op amp as the basic component.

Gyrators. The gyrator is a two-port device that can be used to simulate an inductance. Synthesis of the gyrator can be accomplished with op amps. Figure 13.19 shows the symbol for this device. The expressions for terminal voltages in terms of currents for the ideal gyrator are

$$v_1 = -Ki_2$$

and

$$v_2 = Ki_1.$$

In matrix form these expressions become

$$\begin{bmatrix} v_1 \\ v_2 \end{bmatrix} = \begin{bmatrix} 0 & -K \\ K & 0 \end{bmatrix} \begin{bmatrix} i_1 \\ i_2 \end{bmatrix}. \quad (13.44)$$

The parameter K is a positive constant called the gyrostatic coefficient or gyration impedance.

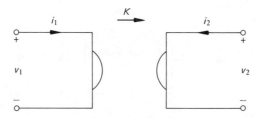

Fig. 13.19 Gyrator symbol.

When a load Z_L is placed across the output terminals, the input impedance is

$$Z_{in} = \frac{v_1}{i_1} = \frac{v_1}{v_2/K} = \frac{v_1}{-Z_L i_2/K} = \frac{v_1}{Z_L v_1/K^2} = \frac{K^2}{Z_L}. \quad (13.45)$$

The ideal gyrator defined by Eq. (13.44) is a positive impedance inverter. More specifically, if a capacitance of impedance $1/j\omega C$ is used as the load, the input impedance becomes

$$Z_{in} = j\omega K^2 C. \quad (13.46)$$

This is an inductive impedance equivalent to that presented by an inductor of value

$$L = K^2 C. \quad (13.47)$$

A second capacitor placed across the input terminals will resonate with this simulated inductance forming a tuned filter. Thus, a tank circuit can be constructed without using an inductance. Generally the resonating capacitance also has a value equal to C.

In passive tank circuits inductive losses are quite significant in determining the Q of the circuit and hence, the selectivity. To this point we have neglected all nonideal effects of the gyrator and the capacitor losses represented by a finite value of Q for the capacitor. In fact, the gyrator will exhibit finite rather than zero values of open-circuit input and output impedances. Furthermore, the gyration impedance will be frequency-dependent and further complicate the simple picture previously considered.

Figure 13.20 shows one method of gyrator implementation. The transconductance amplifiers can be based on the differential op amp, but will have finite input and output admittances. In terms of admittance or y parameters, the practical gyrator obeys the equation

$$\begin{bmatrix} i_1 \\ i_2 \end{bmatrix} = \begin{bmatrix} G_1 & 1/K \\ -1/K & G_2 \end{bmatrix} \begin{bmatrix} v_1 \\ v_2 \end{bmatrix}, \qquad (13.48)$$

where G_1 is the input admittance to the circuit measured with the output terminals shorted while G_2 is the output admittance measured with the input terminals shorted. The practical gyrator equivalent circuit is shown in Fig. 13.21 including input and output admittances.

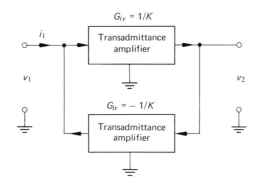

Fig. 13.20 Gyrator realization.

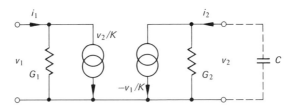

Fig. 13.21 Practical gyrator equivalent circuit.

With a capacitor placed across the output terminals the input impedance is found to be

$$Z_{in} = \frac{v_1}{i_1} = \frac{G_2 + j\omega C}{(1/K^2) + G_1 G_2 + j\omega C G_1}. \qquad (13.49)$$

This impedance can be expressed in terms of a real part and an imaginary part as

$$Z_{in} = \frac{G_2 + G_1 G_2^2 K^2 + \omega^2 C^2 G_1 K^2 + j\omega C}{(1/K^2) + 2G_1 G_2 + G_1^2 G_2^2 K^2 + \omega^2 C^2 G_1^2 K^2} \tag{13.50}$$
$$= R(\omega) + j\omega L(\omega).$$

The real part of the input impedance is

$$R(\omega) = \frac{G_2 + G_1 G_2^2 K^2 + \omega^2 C^2 G_1 K^2}{1/K^2 + 2G_1 G_2 + G_1^2 G_2^2 K^2 + \omega^2 C^2 G_1^2 K^2}, \tag{13.51}$$

while the reactive part equals that caused by an inductance of value

$$L(\omega) = \frac{C}{1/K^2 + 2G_1 G_2 + G_1^2 G_2^2 K^2 + \omega^2 C^2 G_1^2 K^2}. \tag{13.52}$$

The input circuit can be represented by a series combination of $R(\omega)$ and $L(\omega)$. The Q of the equivalent inductance is given by

$$Q = \frac{\omega L(\omega)}{R(\omega)} = \frac{\omega C}{G_2 + G_1 G_2^2 K^2 + \omega^2 C^2 G_1 K^2}. \tag{13.53}$$

The maximum value of Q occurs at a frequency of

$$\omega_m = \frac{G_2}{C}\sqrt{1 + \frac{1}{G_1 G_2 K^2}} \tag{13.54}$$

and is given by

$$Q_m = \frac{\sqrt{1 + \frac{1}{G_1 G_2 K^2}}}{2(1 + G_1 G_2 K^2)}. \tag{13.55}$$

In a properly designed gyrator the conductances G_1 and G_2 are approximately equal and much smaller than $1/K$. If we let $G_1 = G_2 = G$, the maximum value of Q becomes

$$Q_m = \frac{1}{2GK} \tag{13.56}$$

and occurs at a frequency of

$$\omega_m = \frac{1}{KC}. \tag{13.57}$$

Figure 13.22 indicates one version of the gyrator [4] that results in reasonable values of Q. Simulated inductances with values of Q approaching 1000 at frequencies of 50 kHz have been obtained with this configuration.

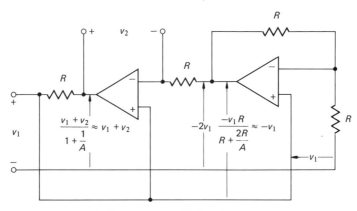

Fig. 13.22 Gyrator realization with op amps.

The admittance matrix can be found to be

$$Y = \begin{bmatrix} \dfrac{1}{(1+A)R} & \dfrac{-A}{(1+A)R} \\ \dfrac{A}{(1+A)R} & \dfrac{1}{(1+A)R} \end{bmatrix}. \tag{13.58}$$

The capacitor losses can be represented by a shunt resistance R_C which can be included in the diagonal terms, giving

$$G'_1 = \frac{1}{(1+A)R} + \frac{1}{R_C} = G'_2, \tag{13.59}$$

assuming equal capacitors across input and output terminals. The Q value can be expressed as

$$Q = \frac{1}{\dfrac{2}{A} + \dfrac{2R(A+1)}{R_C A}} \approx \frac{R_C}{2R}. \tag{13.60}$$

The lower limit on R is imposed by the constraint that, for the gain approximations used in this circuit, R must be much greater than the output impedance of the op amps. When the operating frequency is low compared to the amplifier bandwidth, the gyrator is a useful tool in active filter design. The sensitivity of Q to parameter changes is often lower than that of the passive tank circuit.

Sallen and Key resonator. The circuit shown in Fig. 13.23 was originally described by Sallen and Key [6]. The transfer function of this circuit is found to be

$$\frac{e_{out}}{e_{in}} = \frac{-\dfrac{A}{(A+1)R_1 C_1} s}{s^2 + \left(\dfrac{R_1 C_1 + R_1 C_2 + R_2 C_2}{(A+1)R_1 C_1 R_2 C_2}\right) s + \dfrac{1}{(A+1)R_1 C_1 R_2 C_2}}, \tag{13.61}$$

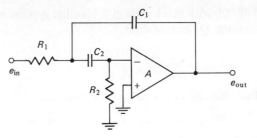

Fig. 13.23 Sallen and Key resonator.

where $s = j\omega$. For a second-order system of reasonably high Q, the transfer function can be expressed as

$$\frac{e_{out}}{e_{in}} = \frac{N(s)}{s^2 + s(\omega_n/Q) + \omega_n^2}, \tag{13.62}$$

where ω_n is the resonant frequency. This parameter is actually the undamped natural frequency, but differs little from the resonant frequency for sharply tuned circuits.

Equating coefficients of like powers of s in the denominator polynomials of Eqs. (13.61) and (13.62) results in

$$\omega_n = \frac{1}{\sqrt{(A+1)R_1C_1R_2C_2}} \tag{13.63}$$

and

$$Q = \frac{\sqrt{(A+1)R_1C_1R_2C_2}}{R_1C_1 + R_1C_2 + R_2C_2}. \tag{13.64}$$

The gain of the circuit at resonance is

$$A_R = \frac{-AR_2C_2}{R_1C_1 + R_1C_2 + R_2C_2}. \tag{13.65}$$

If $R_1 = R_2 = R$ and $C_1 = C_2 = C$, the preceding three equations simplify to

$$\omega_n = \frac{1}{\sqrt{A+1}\,RC}, \tag{13.66}$$

$$Q = \frac{\sqrt{A+1}}{3}, \tag{13.67}$$

and

$$A_R = \frac{A}{3}. \tag{13.68}$$

While it can be shown that the sensitivity of Q to any passive component is quite low, this circuit suffers from frequency limitations. If a resonant frequency of 1 kHz is required, good design practice dictates an amplifier bandwidth of at least

10 kHz. A gain of 40,000 with this bandwidth is rather typical of a high-quality op amp; thus the maximum Q obtainable is

$$Q = \frac{\sqrt{40,001}}{3} = 67.$$

It can be shown that this circuit is limited to frequencies such that

$$100\omega_n Q^2 < \text{Gain bandwidth of op amp.} \tag{13.69}$$

Phase-shift filters. The last second-order system to be discussed is based on the all-pass network shown in Fig. 13.24. If loading effects of the op amp and capacitive losses are neglected, the output voltage is

$$e_{\text{out}} = A(e_b - e_a),$$

where

$$e_a = e_{\text{in}} \frac{R_A}{R_A + R_B} + e_{\text{out}} \frac{R_B}{R_A + R_B}$$

and

$$e_b = \frac{e_{\text{in}}}{1 + sCR}.$$

The overall gain is

$$\frac{e_{\text{out}}}{e_{\text{in}}} = \frac{1 - \dfrac{sCRR_A}{R_B}}{1 + sCR} \frac{AR_B}{R_A + (A+1)R_B}.$$

If R_A and R_B are selected to be equal, the gain expression reduces to

$$\frac{e_{\text{out}}}{e_{\text{in}}} = \frac{A}{A+2} \frac{1 - sCR}{1 + sCR}. \tag{13.70}$$

The magnitude of the gain is $A/(A+2)$, which approximates unity for typical values of A, while the phase difference between the output and input signals is

$$\theta = -2 \tan^{-1} \omega CR. \tag{13.71}$$

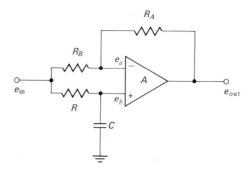

Fig. 13.24 Op amp all-pass network.

The phase shift varies from zero to $-180°$ as frequency increases. All magnitudes are passed by this network with unity gain, but phase shift is a function of frequency.

The all-pass network is an important element in several applications requiring phase control of signals. At this point, however, we are primarily interested in those features of this network that relate to active filters. Figure 13.25 shows a narrow-band filter that utilizes this all-pass element [1]. The transfer function for the filter is

$$\frac{e_{out}}{e_{in}} = \frac{s^2 + \dfrac{2}{RC}s + \dfrac{1}{R^2C^2}}{s^2 + \dfrac{2}{RC}\left(\dfrac{R_1 - R_2}{R_1 + R_2}\right) + \dfrac{1}{R^2C^2}}, \qquad (13.72)$$

where we have assumed that $A \gg 2$. Comparing this transfer function to Eq. (13.62) results in

$$\omega_n = \frac{1}{RC}, \qquad (13.73)$$

$$Q = \frac{R_1 + R_2}{2(R_1 - R_2)}, \qquad (13.74)$$

and

$$A_R = \frac{R_1 + R_2}{R_1 - R_2} = 2Q, \qquad (13.75)$$

where A_R is the overall resonant gain.

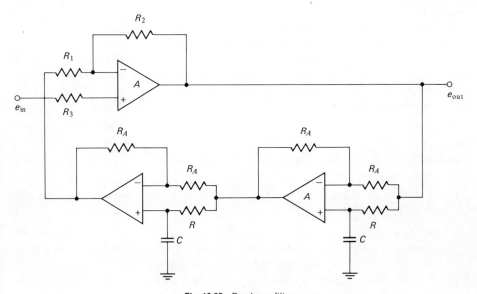

Fig. 13.25 Bandpass filter.

The selectivity and resonant gain of this filter depend strongly on the difference $R_1 - R_2$. For a Q of 100, the difference must be one percent. Resistances can be controlled closely enough to establish practical values of Q in the range of 100 to 200 for lumped or integrated circuits. If R_1 and R_2 are physically constructed from the same material, the temperature coefficients will be equal minimizing temperature-sensitivity problems.

In the preceding analysis we have neglected loading of the input impedances of the op amp, the quality factor of the capacitor, and temperature variations of A. While it is necessary to select high-quality components, this circuit leads to a reasonably sharp bandpass filter.

REFERENCES AND SUGGESTED READING

1. D. J. Comer and J. E. McDermid, "Inductorless bandpass characteristics using all-pass networks." *IEEE Transactions on Circuit Theory*, CT-15, December 1968.
2. D. T. Comer, *Large Signal Transistor Circuits*. Englewood Cliffs, N.J.: Prentice-Hall, 1967.
2. A. B. Grebene, *Analog Integrated Circuit Design*. Cincinnati, Ohio: Van-Nostrand Reinhold, 1972, Chapter 9.
4. P. L. Huelsman, *Active Filters*. New York: McGraw-Hill, 1970.
5. S. J. Mitra, *Analysis and Synthesis of Linear Active Networks*. New York: Wiley, 1969, Chapter 8.
6. R. P. Sallen and E. L. Key, "A practical method of designing RC active filters." *IRE Transactions on Circuit Theory*, CT-2, March 1955.

PROBLEMS

Section 13.1

13.1 Design a five-bit D/A converter. The full-scale output should equal -10 V. The input levels for the 1 and 0 bits are $+6$ V and 0 V, respectively.

* **13.2** A D/A converter has a maximum output of -8 V. If the conversion factor is 0.058 volts/unit, what is the maximum number of binary input lines allowed? Using binary-weighted resistors and an ideal op amp, design this D/A converter. Assume input levels of $+6$ and 0 V.

13.3 Redesign the converter of Problem 13.2 using a ladder network and an op amp.

Section 13.2

* **13.4** Rework Example 13.2 to create an 8-V, 500-ms sweep voltage. The discharge time should not exceed 0.5 ms.

13.5 Construct the sweep circuit of Example 13.2 using a Miller sweep circuit. State any assumptions made.

Sections 13.3A–13.3B

13.6 Design a 4-pole, low-pass, Butterworth filter having a 3-db point at 14 kHz using unity gain stages. The source resistance is 1 kΩ while the load resistance is 10 kΩ.

* **13.7** Design a 3-pole, high-pass, Butterworth filter with a 3-db point of 20 kHz. The source resistance is 1 kΩ while the load resistance is 10 kΩ.

Section 13.3C

* **13.8** Assume op amps are available with zero output and infinite input impedance. Design a resonant circuit using a gyrator to result in a $Q = 100$ at $f = 10$ kHz. Assume the capacitors used have $Q_{CAP} = 240$ at $f = 10$ kHz. Specify capacitor values, resistance values R, and amplifier gains.

13.9 Repeat Problem 13.8 for $Q = 50$ using a Sallen and Key resonator.

13.10 Repeat Problem 13.8 using the phase-shift filter of Fig. 13.25.

MODULATORS, RECEIVERS, AND COMMUNICATION SYSTEMS

It is almost superfluous to emphasize the significance of the field of communications. Yet most people have become so familiar with everyday applications of communications that they fail to appreciate the absolute necessity of this field to our society. Not only has the importance of the telephone, radio, and television continued to grow, but also new areas not so well known to the public have been developing. Digital data transmission and satellite communications are examples of two rather recent areas of great activity and importance.

It is quite difficult to keep abreast of the new developments in the communications field. Newer techniques have improved the capabilities of many communication-related devices by several orders of magnitude in the last few decades. Improvement can take place in many ways. For example, the decrease in physical size of radio receivers represents one type of advancement while the increase in the number of separate conversations that can be transmitted over telephone lines simultaneously represents improvement in a different direction. Still another important advancement is the satellite communication system now in operation. These satellite links allowing live television transmission among most continents of the world represent a quantum jump in broadcast capability.

It is obviously impossible to cover all topics of importance to such a dynamic field in a single chapter. We will, however, consider fundamental ideas that will lay a foundation for further study at a more advanced level.

14.1 COMMUNICATIONS

Table 14.1 lists the various frequency ranges of interest [1]. Although modulation has not yet been discussed it is useful to consider the transmission characteristics and usage of each band. For transmission of electromagnetic waves through the atmosphere antennas must be used to radiate the desired signal. Antennas tend to operate efficiently only when the antenna dimensions are of the same order of magnitude as the wavelength of the transmitted signal. An electromagnetic wave travelling in free space has a wavelength of

$$\lambda = c/f,$$

Table 14.1 Classification of frequency spectrum.

Range	Name	Approximate transmission distance
10 to 30 kHz	Very low frequency (VLF)	Over 1000 miles
30 to 300 kHz	Low frequency (LF)	1000 miles
300 to 3000 kHz	Medium frequency (MF)	0–1000 miles
3 to 30 MHz	High frequency (HF)	Horizon
30 to 300 MHz	Very high frequency (VHF)	Horizon
300 to 3000 MHz	Ultrahigh frequency (UHF)	Horizon
3 to 30 GHz	Superhigh frequency (SHF)	Horizon
30 to 300 GHz	Extra high frequency (EHF)	Horizon

where c is the velocity of light and f is the frequency of the signal. A very low frequency signal wavelength may be several kilometers long while a VHF signal wavelength may be only a few meters. The transmission of low frequencies requires very large antenna systems sometimes occupying acres of land, whereas high-frequency antennas are far more compact. A second disadvantage of lower frequency signal transmission is the falloff in antenna radiation efficiency with frequency.

There are some rather important advantages of lower frequency transmission that lead to practical transmission systems in the VLF and LF bands. Lower frequency signals are attenuated less by the atmosphere than higher frequency signals. Furthermore, the index of refraction of the atmosphere varies with distance from the surface of the earth in such a way as to bend the path of the lower-frequency electromagnetic wave toward the earth. These signals can then follow the curvature of the earth for considerable distances. The range of VLF signals is over 1000 miles. Under certain weather conditions VLF signals travel far around the earth. Unfortunately, the available bandwidth is too low for speech, limiting the traffic to telegraphy. This band is used for very few applications, although some military installations still use VLF transmission.

The LF band is very dependable and is less affected by ionospheric storms than are higher bands. In the field of navigation the LF band is used as a radio beacon or to determine long-range position by both civil and military organizations. This band is also used for backup communications in conjunction with systems in higher bands that may be adversely affected by weather conditions.

The MF band contains the familiar commercial radio band of 535 kHz to 1605 kHz. This band is reserved in the United States and most parts of the world for transmission of amplitude-modulated information to general audiences. Several smaller frequency bands within the MF band are allocated to aeronautical services. The range from 1605 kHz to 2850 kHz is used primarily for short-distance fixed circuits. Aeronautical frequencies extend into all higher bands with short distance point-to-point and ground-air-ground communications falling in the 2850 kHz to 4063 kHz range. Aircraft flying scheduled routes are allocated specific channels within this band while nonscheduled routes are allocated separate channels. The MF band antenna is of reasonable size and radiation efficiency.

In the United States the band from 3.5 MHz to 4 MHz is reserved for amateur radio operation, although most other countries use this band for fixed and mobile services.

The HF band is used for long-distance point-to-point and transoceanic ground-air-ground communications. Maritime ship-to-coast frequencies also use several channels in this band for telephone and telegraph communications. Land mobile frequencies extend from the MF band through the HF band and on up through the higher bands. In the United States specific frequency allocations for several services have been assigned by the Federal Communications Commission. For example, the petroleum industry is assigned the frequency ranges 25.02 to 25.3 MHz, 30.66 to 30.82 MHz, and several others. Assignments have been made to industries related to business, forest products, manufacturers, motion pictures, power, telephone maintenance, and others.

Land transportation services such as taxicab, railroad, bus, or auto emergency have been assigned specific channels in the HF, VHF, UHF, and higher bands. Citizens' bands have been reserved within the HF and UHF bands.

An important user of the VHF band is the television industry. Television channel assignments range from 54 MHz to 88 MHz, 174 MHz to 216 MHz, and 470 MHz to 890 MHz. The first two ranges fall within the VHF band while the last range is in the UHF band. Commercial FM radio stations transmit carrier frequencies ranging from 88 MHz to 108 MHz.

Applications involving higher bands include satellite communication, relay links for radio or television, radar, and special-purpose communications.

14.2 MODULATION

At the outset of our discussion on modulators it should be noted that modern high-power transmitters use special-purpose vacuum tubes for the final amplifier. High-power triodes and tetrodes have been used for many years in class-B operation as transmitter output stages. Carrier power outputs of 10 kW are typical of these devices in the popular MF band. Class-C amplification is generally used in the MF and HF bands for higher output power levels. Tetrodes are now used almost universally as the final stage of high-power broadcast transmitters [4]. A typical carrier power level for these transmitters is 100 kW, although transmitters reaching 1000 kW are also used in special cases. The output power levels of commercial transmitters are so great that special cooling systems are required to keep the temperature of the output stage at an acceptable value.

Solid-state devices are not often used in these high-power transmitters, but are important in other communication applications. Cable transmission systems require considerably lower power signals, thus allowing the use of transistor circuits. Frequency-division multiplexing is now used extensively in telephone systems to simultaneously transmit several conversations over a single pair of lines.

Cable television systems offer a possible solution to the overcrowded broadcast spectrum and are expected to become significantly more important in the next decade. These and other cable systems use transistor circuits extensively.

Low-power transistor transmitters are also useful in public-safety radio services (police, fire, local government), industrial and land transportation services (taxicab, railroad), and amateur radio services.

Although communication satellites often operate at higher microwave frequencies with special devices such as travelling-wave tubes, lower frequency bands are also used. Relatively small amounts of power are transmitted from these satellites to minimize source power drain. Transistor and other solid-state devices are used in the VHF and UHF bands for some satellite systems and may become more important in the near future.

From the preceding discussion it is obvious that semiconductor devices are important to the communications industry. While these devices are not so universal in this field as they are in most other fields of electronics, they play a significant role, and thus warrant a detailed study of communication applications.

This section will be primarily concerned with modulation techniques for frequencies in and above the MF band.

A. Amplitude Modulation

A carrier frequency scheme is always used to transmit audio frequency information. The required broad bandwidth cannot be effectively transmitted directly due primarily to antenna limitations. The carrier frequency is much greater than that of the highest frequency to be transmitted. For example, the commercial AM broadcast band extends from 535 kHz to 1605 kHz while the bandwidth of the information to be transmitted is typically 5 kHz. The carrier signal is combined in such a way with the lower frequency signal to produce a high frequency signal. The transmitted signal frequency varies little from the carrier frequency and thus, the antenna design problem is considerably simplified.

It is not only the audio spectrum that is transmitted by amplitude modulation. Carrier frequency applications are important because of the great demand on the frequency spectrum for transmission of information. The use of prespecified carrier frequencies along with modulated information allows information to be transmitted over appropriate channels while other information is simultaneously transmitted on another channel. Video information is transmitted by AM waveforms with each channel corresponding to a different carrier frequency. Several telephone conversations are transmitted over the same line by using amplitude modulation techniques. Thus, an important feature of modulation is the use of carrier frequencies to identify specific information channels.

In order to transmit information using a carrier signal, the amplitude of the carrier signal is caused to vary in accordance with the instantaneous value of the signal to be transmitted. Figure 14.1 demonstrates this amplitude modulation, using a sine wave as the information signal. The amplitude-modulated signal is easily transmitted, since it is a high frequency signal. In Fig. 14.1 the modulating

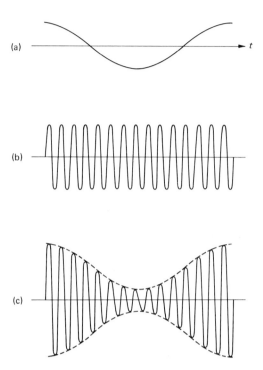

Fig. 14.1 Amplitude modulation: (a) information to be transmitted (modulating signal); (b) carrier signal; (c) carrier signal modulated by information to be transmitted.

signal is shown with a frequency of $\frac{1}{20}$ to $\frac{1}{10}$ the carrier frequency. In truth, the carrier frequency might be 100 to 10,000 times the frequency of the modulating signal. The AM signal is received and demodulated to obtain the desired information. That is, a signal is generated by the demodulator which varies in accordance with the peaks of the AM waveform, giving the original modulating signal.

It is convenient to base the theoretical discussion of amplitude modulation on the assumption of a sinusoidal modulating signal. This assumption is useful in practice because most audio sounds to be transmitted are made up of harmonic components. The extension of the single component input to an audio input modulating signal is thus rather straightforward.

A specific method of achieving amplitude modulation that is conceptually simple is shown in Fig. 14.2. In Fig. 14.2, the gain of the amplifier depends linearly on the modulating signal. If the modulating signal is zero, the carrier signal is amplified by a constant value. As e_m becomes positive, the gain increases as does the output signal. For negative values of e_m, the output signal decreases.

Let us assume that e_{in} is a sinusoidal signal expressed by $C \cos \omega_c t$, while e_m can be represented by $M \cos \omega_m t$. The output signal is given by

$$e_{out} = A_0 C \cos \omega_c t + A_1 MC \cos \omega_m t \cos \omega_c t, \tag{14.1}$$

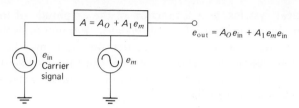

Fig. 14.2 Simple amplitude modulator.

which can also be expressed as

$$e_{out} = A_0 C \left[1 + \frac{A_1 M}{A_0} \cos \omega_m t \right] \cos \omega_c t. \tag{14.2}$$

Equation (14.2) can be considered to be a sinusoid of radian frequency ω_c with an amplitude that varies slowly with time, as given by

$$A_0 C \left[1 + \frac{A_1 M}{A_0} \cos \omega_m t \right].$$

It is desirable that the amplitude of the sinusoid never be less than zero; thus the term $A_1 M / A_0$ should never exceed a value of unity. This quantity controls the degree of modulation. If $A_1 M / A_0$ approaches zero, the output waveform approaches a constant signal of amplitude $A_0 C$. As $A_1 M / A_0$ becomes larger, the modulation or change in amplitude of the output signal increases. If $A_1 M / A_0 = 1$, the amplitude of the carrier signal becomes zero as the modulating signal reaches its negative peak. As the modulating signal reaches its positive peak, the amplitude of the carrier signal approaches a value of $2 A_0 C$. This condition is referred to as 100 percent modulation.

Regardless of how the AM waveform is obtained we can express it in general terms as

$$e = E_c \cos \omega_c t + E \cos \omega_m t \cos \omega_c t$$
$$= E_c \left[1 + \frac{E}{E_c} \cos \omega_m t \right] \cos \omega_c t. \tag{14.3}$$

The amplitude of the unmodulated carrier is E_c while the ratio E/E_c expresses the maximum fractional deviation of amplitude with respect to the unmodulated value. This quantity is called the index of modulation and is designated by m, that is

$$m = E/E_c. \tag{14.4}$$

The index of modulation is then defined as the ratio of the maximum amplitude deviation from the unmodulated value to the amplitude of the unmodulated signal. Equation (14.3) can now be written

$$e = E_c (1 + m \cos \omega_m t) \cos \omega_c t. \tag{14.5}$$

For practical systems, m is a function of the amplitude of the modulating signal and certain parameters of the modulator. For nonsinusoidal signals or nonlinear modulators, the deviation in carrier amplitude may be greater in the positive direction than in the negative direction. For this case two different values of m can be defined; however, we will limit our discussion to symmetrical modulation with a single applicable value of m.

The percent modulation of an AM waveform is obtained by multiplying m by 100. When $E = E_c$, 100-percent modulation is present. For $E < E_c$ as shown in Fig. 14.1, something less than 100-percent modulation is present. If $E > E_c$, the waveform is said to be overmodulated and the envelope of the AM waveform may not follow the value of the modulating signal. Figure 14.3 shows a 100-percent modulated waveform and an overmodulated signal. While Eq. (14.5) is valid for values of m greater than unity, many physical modulators will remain cutoff or saturated when the amplitude of the carrier tends to be negative. The waveform of Fig. 14.3(b) corresponds to a typical output rather than the equation for an overmodulated signal.

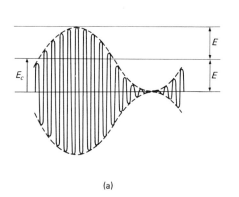
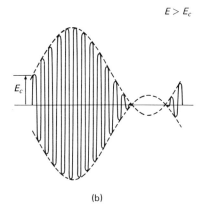

Fig. 14.3 Amplitude-modulated waveforms: (a) 100-percent modulation; (b) overmodulation.

The AM waveform is not a pure sinusoid as is the unmodulated carrier signal. For sinusoidal modulation the AM waveform is repetitive, however, and therefore must be composed of sinusoidal components. Returning to Eq. (14.5) and applying the trigonometric identity

$$\cos \omega_c t \cos \omega_m t = \tfrac{1}{2}[\cos (\omega_c - \omega_m)t + \cos (\omega_c + \omega_m)t],$$

we can write

$$e = E_c \cos \omega_c t + \frac{mE_c}{2} \cos (\omega_c - \omega_m)t + \frac{mE_c}{2} \cos (\omega_c + \omega_m)t. \qquad (14.6)$$

The output voltage is expressible as three sinusoidal components: one of radian frequency ω_c, one of frequency $\omega_c - \omega_m$, and one of frequency $\omega_c + \omega_m$. Since ω_m

is generally much smaller than ω_c, the frequency content of the AM waveform is limited to a small range of frequency centered about ω_c. Figure 14.4 shows the components of a 100-percent modulated waveform. For this case $m = 1$ and the signal becomes

$$e = E_c \cos \omega_c t + \frac{E_c}{2} \cos (\omega_c - \omega_m)t + \frac{E_c}{2} \cos (\omega_c + \omega_m)t. \qquad (14.7)$$

The component of frequency $\omega_c - \omega_m$ is called the lower sideband frequency while the component of frequency $\omega_c + \omega_m$ is the upper sideband frequency. The three constant amplitude signals add together to give an AM waveform.

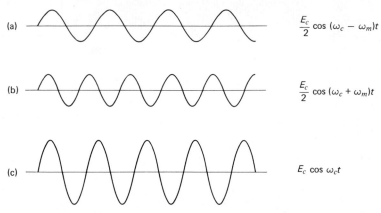

Fig. 14.4 The three components of a sinusoidally modulated waveform (100-percent modulation): (a) lower sideband frequency; (b) upper sideband frequency; (c) carrier.

It must be remembered that audio signals contain several components rather than a single component. Furthermore, in the dynamic speech process, the components vary with time. A typical speech sound might appear as shown in Fig. 14.5(a). If this sound serves as the modulating signal the AM waveform of Fig. 14.5(b) results. The spectral representation of part (c) is valid only while the speech sound generates the repetitive waveform shown in part (a). The upper sideband contains all components of frequency greater than ω_c, while the lower sideband contains all components of frequency less than ω_c.

The transistor modulator. The dynamic resistance of the base-emitter junction can be modulated to create a variable voltage gain in a single-stage transistor amplifier. Under proper conditions the amplification of a carrier signal is made to vary directly with modulating signal producing an AM waveform at the output. This modulator corresponds closely to the basic scheme suggested in Fig. 14.2. The transistor modulator produces rather small indexes of modulation, but has some practical importance in low-power, low-fidelity communication systems. Figure 14.6 illustrates the basic circuit. In this circuit we assume that the capacitors C_1 and C_2 offer low impedance to the carrier signal and high impedance to the

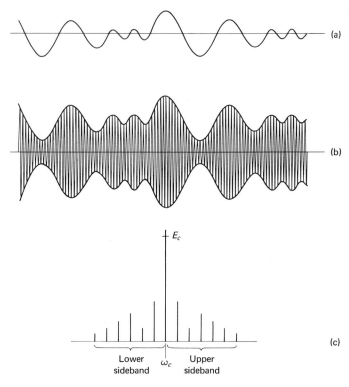

Fig. 14.5 Speech sound as the modulating signal: (a) modulating signal (speech sound); (b) resulting AM waveform; (c) spectral representation at a given instant.

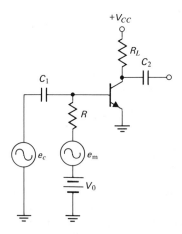

Fig. 14.6 Transistor modulator.

modulating signal. This is easy to achieve if there is a large difference between carrier and modulating frequency. If not, the capacitors can be replaced by more effective filters. It will also be useful to require that $R \gg r_{bb'} + (\beta + 1)r_d$. If e_m is zero the gain of the circuit is

$$A = \frac{e_{out}}{e_c} = \frac{-\alpha R_L}{(1 - \alpha)r_{bb'} + r_d} \approx \frac{-\alpha R_L}{r_d}.$$

The dc emitter current will be determined by V_0 and R and will fix the value of r_d. To modulate the gain and hence the output signal, a finite value of e_m is imposed to control the emitter current and r_d. The emitter current is now given by

$$I_E = (\beta + 1)I_B = (\beta + 1)\frac{[V + e_m]}{R}, \qquad (14.8)$$

where $V = V_0 - V_{BE(on)}$. The output signal is then

$$e_{out} = \frac{-\alpha R_L e_c}{r_d} = -\frac{\beta R_L e_c V}{0.026R}\left[1 + \frac{e_m}{V}\right]. \qquad (14.9)$$

Expressing the carrier signal as $e_c = C \cos \omega_c t$ and the modulating signal as $e_m = M \cos \omega_m t$ gives

$$e_{out} = -\frac{\beta R_L V}{0.026R} C \cos \omega_c t \left[1 + \frac{M}{V} \cos \omega_m t\right]$$
$$= -E_c(1 + m \cos \omega_m t) \cos \omega_c t, \qquad (14.10)$$

where

$$E_c = \frac{\beta R_L V C}{0.026R} \qquad (14.11)$$

and

$$m = \frac{M}{V}. \qquad (14.12)$$

We must now consider the relative amplitudes of the applied signals. The carrier signal must be small to have a negligible effect on the emitter current; that is, small-signal conditions must apply to the carrier signal. The component of emitter current caused by e_c is

$$i_e = \frac{(\beta + 1)C \cos \omega_c t}{r_{bb'} + (\beta + 1)r_d}$$

with a maximum value of

$$i_{e\,max} = \frac{(\beta + 1)C}{r_{bb'} + (\beta + 1)r_d}.$$

As e_m varies, r_d will change; thus $i_{e\,\max}$ will not be constant. We can, however, calculate $i_{e\,\max}$ at the quiescent value of dc emitter current. From Eq. (14.8) with $e_m = 0$,

$$I_E = \frac{(\beta + 1)V}{R}.$$

The maximum component of emitter current caused by the carrier signal should be much smaller than this quiescent level or

$$\frac{(\beta + 1)C}{r_{bb'} + (\beta + 1)r_d} \ll \frac{(\beta + 1)V}{R}.$$

This inequality can also be written

$$\frac{C}{r_{bb'} + (\beta + 1)r_d} \ll \frac{V_0 - V_{BE(\text{on})}}{R}.$$

If we allow the maximum component of emitter current to equal one-tenth of the quiescent level, the carrier signal magnitude is limited to

$$C \leq 0.1(V_0 - V_{BE(\text{on})}) \frac{r_{bb'} + (\beta + 1)r_d}{R}. \tag{14.13}$$

The equation for dynamic diode resistance applies only when the dc bias current is much greater than the ac signal current and when the transistor remains in the active region. We can consider the bias current for the modulator to result from the voltage $V_0 + e_m$. As e_m swings positive to result in a maximum bias voltage of $V_0 + M$, the transistor must not saturate. As e_m swings negative to a value of $V_0 - M$, the transistor must not cut off. From Eq. (14.12) 100-percent modulation cannot be obtained unless $M = V_0 - V_{BE(\text{on})}$; thus the percent modulation must be limited to some value less than 100 percent in order to avoid cutoff. Further limiting the index of modulation is the fact that the approximation of $r_d + (1 - \alpha)r_{bb'} \approx r_d$ used in developing Eq. (14.9) will not hold at high currents. The net result of these nonideal effects is that unwanted nonlinearities are introduced into the modulator. The results of these nonlinearities can be reduced by filtering in some applications, but in general cause a limitation to be placed on the index of modulation. An index of modulation ranging from 0.1 to 0.2 is rather typical for a transistor modulator of this type.

The transistor modulator is used in low fidelity applications wherein distortion is rather unimportant. Low-power wireless intercom or two-way radio applications might use this modulation scheme.

Effect of distortion in the output signal. If a signal with a narrow bandwidth is to be transmitted, for example a half-octave bandwidth, it is sometimes possible to achieve a higher index of modulation and remove the resulting nonlinear effects by filtering. Let us consider the circuit of Fig. 14.7. We will now analyze this circuit to examine the effects of large magnitudes of the carrier and modulating signals.

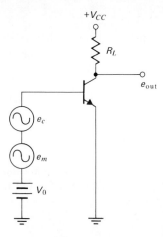

Fig. 14.7 Transistor modulator with series input sources.

The input current (base current) as a function of input voltage (base-emitter voltage) is the familiar diode relationship,

$$I_B = I_{BO}\left[\exp\left(\frac{qV_{EB}}{kT}\right) - 1\right] = I_{BO}\left[\exp\left(\frac{V_{EB}}{\delta}\right) - 1\right], \quad (14.14)$$

where I_{BO} is a constant and $\delta = kT/q$. The input voltage V_{EB} consists of the dc bias voltage, the modulating signal, and the carrier signal; thus $V_{EB} = V_0 + e_m + e_c$. Equation (14.14) can be written as

$$I_B = I_{BO}\left[\exp\left(\frac{V_0}{\delta}\right)\exp\left(\frac{e_m + e_c}{\delta}\right) - 1\right] \approx D\exp\left(\frac{e_m + e_c}{\delta}\right), \quad (14.15)$$

where $D = I_{BO}\exp(V_0/\delta)$. A power-series expansion of the exponential allows I_B to be written as

$$I_B = D\left[1 + \frac{(e_m + e_c)}{\delta} + \frac{(e_m + e_c)^2}{2!\,\delta^2} + \frac{(e_m + e_c)^3}{3!\,\delta^3} + \cdots\right]. \quad (14.16)$$

Assuming that the current gain from base to collector remains constant with output voltage, we see that the collector voltage is

$$V_C = V_{CC} - I_C R_L = V_{CC} - \beta I_B R_L.$$

Therefore the output voltage is

$$V_C = V_{CC} - \beta D R_L\left[1 + \frac{(e_m + e_c)}{\delta} + \frac{(e_m + e_c)^2}{2!\,\delta^2} + \frac{(e_m + e_c)^3}{3!\,\delta^3} + \cdots\right]. \quad (14.17)$$

We can now consider the above equation, term by term, to find those terms that are important to the modulation process.

Constant terms

The quantity $V_{CC} - \beta D R_L$ has a constant value and is simply the dc bias term. The quiescent collector voltage is given by this quantity.

First-order term

The quantity

$$\beta DR_L \left(\frac{e_m + e_c}{\delta}\right)$$

expresses the fact that the amplifier exhibits gain. That is, this term is directly proportional to the input signals. This can be written as $\beta DRL(M \cos \omega_m t + C \cos \omega_c t)/\delta$ for sinusoidal inputs.

Second-order term

The second-order term is

$$\frac{\beta DR_L}{2\delta^2}(e_m^2 + 2e_m e_c + e_c^2)$$

$$= \frac{\beta DR_L}{2\delta^2}[M^2 \cos^2 \omega_m t + 2MC(\cos \omega_m t)(\cos \omega_c t) + C^2 \cos^2 \omega_c t].$$

Applying the proper trigonometric identities allows this quantity to be expressed as

$$\frac{\beta DR_L}{2\delta^2}\left[\frac{M^2 \cos 2\omega_m t + M^2}{2} + \frac{C^2 \cos 2\omega_c t + C^2}{2} + MC\{\cos(\omega_c + \omega_m)t + \cos(\omega_c - \omega_m)t\}\right].$$

The second-order term has caused frequency components of $(\omega_c + \omega_m)$ and $(\omega_c - \omega_m)$ which are the sideband frequencies of the AM waveform. If the first-order term and second-order term are combined, we get a voltage equal to

$$\frac{\beta DR_L}{\delta}\left[C \cos \omega_c t + \frac{MC}{2\delta}\cos(\omega_c - \omega_m)t + \frac{MC}{2\delta}\cos(\omega_c + \omega_m)t\right]$$
$$+ \frac{\beta DR_L}{\delta}\left[M \cos \omega_m t + \frac{M^2 \cos 2\omega_m t + M^2}{4\delta} + \frac{C^2 \cos 2\omega_c t + C^2}{4\delta}\right].$$

The first quantity in parentheses is recognized as an AM waveform given by Eq. (14.6). In addition to the desired AM signal, however, we also have a signal which is made up of frequency components of dc, ω_m, $2\omega_m$, and $2\omega_c$. These signals can easily be removed from the overall signal by filtering, since they lie outside of the frequency range of the AM signal.

Third-order term

The third-order term is

$$\frac{\beta DR_L}{6\delta^3}[e_m^3 + 3e_m^2 e_c + 3e_m e_c^2 + e_c^3]$$

$$= \frac{\beta DR_L}{6\delta^3}[M^3 \cos^3 \omega_m t + 3M^2 C(\cos^2 \omega_m t)(\cos \omega_c t)$$

$$+ 3MC^2(\cos \omega_m t)(\cos^2 \omega_c t) + C^3 \cos \omega_c t]$$

$$= \frac{\beta DR_L}{6\delta^3}\left[(\tfrac{3}{4}M^3 + \tfrac{3}{2}MC^2) \cos \omega_m t + \frac{M^3}{4}\cos 3\omega_m t + \frac{3M^2 C}{4}\cos(\omega_c - 2\omega_m)t\right.$$

$$+ (\tfrac{3}{4}C^3 + \tfrac{3}{2}M^2 C) \cos \omega_c t + \frac{3M^2 C}{4}\cos(\omega_c + 2\omega_m)t + \frac{3MC^2}{4}\cos(2\omega_c - \omega_m)t$$

$$\left. + \frac{3MC^2}{4}\cos(2\omega_c + \omega_m)t + \frac{C^3}{4}\cos 3\omega_c t\right].$$

The third-order term results in components at frequencies of ω_m, $3\omega_m$, $\omega_c - 2\omega_m$, ω_c, $\omega_c + 2\omega_m$, $2\omega_c - \omega_m$, $2\omega_c + \omega_m$, and $3\omega_c$. Most of these components can be removed by filtering since they are generally far removed from the carrier frequency. However, the components ω_c, $\omega_c - 2\omega_m$, and $\omega_c + 2\omega_m$ are near the carrier frequency and must be considered in more detail. The component of frequency ω_c is relatively unimportant since the AM waveform will contain the carrier signal. The components $(\omega_c - 2\omega_m)$ and $(\omega_c + 2\omega_m)$ cannot be easily filtered unless ω_m is a fixed frequency or extends over a narrow bandwidth. Even then it is difficult to attenuate these components without attenuating the desired sidebands.

If f_m varies from say 100 Hz to 2000 Hz, the bandwidth of each sideband is 1900 Hz. The component $f_c - 2f_m$ then varies from $f_c - 200$ Hz to $f_c - 4000$ Hz and a portion of the spectrum of these unwanted signals falls within the lower sideband. Thus, the third-order term leads to distortion in the AM waveform. If both M and C are limited to very small values, the magnitude of the undesired terms compared to the desired sideband magnitudes decreases.

The second-order term in the expression is responsible for amplitude modulation. All higher-order terms contribute components that must be removed from the output signal to obtain the AM waveform. In many cases it is impossible to filter the components near the sideband frequency and, therefore, the index of modulation must be limited for the transistor modulator.

Returning to the transistor modulator discussed in the last section, we can now see why a tuned circuit was not necessary to obtain the AM signal. When the amplitude of e_c is small compared to the bias voltage, the gain of the circuit is

$$A = \frac{e_{\text{out}}}{e_c} = \frac{-\alpha R_L}{r_d} = \frac{-\alpha R_L}{0.026} I_E.$$

The modulating signal determines I_E; hence

$$e_{out} = \frac{-\beta R_L}{0.026} \frac{[V + e_m]}{R} e_c.$$

The output voltage is then a function of the product of e_m and e_c, which is a second-order term. When the amplitudes of e_c and e_m are limited so that the above equation is true, no higher-order terms appear in e_{out}. However, the modulation will be small as discussed previously. If 100-percent modulation is desired, a different type of modulator must be used.

We can conclude from this section that any signal that can be expanded as a power series in terms of $(e_m + e_c)$ and that possesses a second-order term, will contain an AM signal. Hence any electrical element with nonlinear V–I characteristics can be used as a modulator. The filtering problem is simplified if the nonlinear characteristics give rise to a large second-order term, compared to higher-order terms.

Class-B and class-C base-modulation circuit. The class-A base-modulation scheme discussed previously suffers from several problems. It is possible to achieve far greater circuit efficiencies and higher indexes of modulation by using class-B or class-C stages using either base or collector modulation.

The basic idea of the base-modulation circuit is to control the collector current magnitude by the modulating voltage applied at the transistor base. Although the collector current waveform is highly distorted, a tuned circuit allows only the fundamental frequency component to appear at the output. As the magnitude of collector current varies with modulating signal, the output voltage also varies. We will start our analysis by examining the circuit of Fig. 14.8. This circuit is a class-C stage for values of V_0 greater than zero and becomes a class-B stage when V_0 is zero. Base and collector currents will flow only when the applied carrier voltage exceeds $V_0 + V_{BE(on)}$. The collector current can be expressed as

$$i_c = \beta i_b = \frac{\beta(C \cos \omega_c t - V_0)}{R_S} \qquad (14.18)$$

if we assume that $V_{BE(on)}$ is negligible and that R_S is much larger than $r_{bb'} + r_{b'e}$. The peak value of collector current is

$$I_{CP} = \frac{\beta[C - V_0]}{R_S}. \qquad (14.19)$$

Since the collector current is repetitive it contains a fundamental component of frequency ω_c. If the tuned circuit of the collector resonates at this frequency, a sinusoidal voltage will appear at the output. This voltage will be approximately proportional to the peak value of collector current. Although higher harmonics are present in the collector current waveform no voltage will develop across the tuned circuit for these components. A modulating signal of value $e_m = M \cos \omega_m t$ can be added in series with V_0 to modulate the value of I_{CP}. As I_{CP} changes with

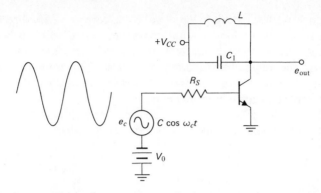

Fig. 14.8 Class-C amplifier.

e_m, so also does the magnitude of the output signal. The peak collector current becomes

$$I_{CP} = \frac{\beta[C - V_0 + M \cos \omega_m t]}{R_S}. \qquad (14.20)$$

The input and output waveforms are shown in Fig. 14.9. If 100-percent modulation is desired the peak collector current must equal zero on the negative crest of the modulating signal. This condition occurs when $M + V_0 = C$.

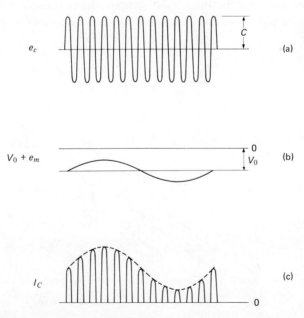

Fig. 14.9 Waveforms of a class-C base-modulation circuit: (a) carrier signal; (b) modulating signal and dc bias; (c) collector current.

From Eq. (14.20) the peak value of collector current varies directly with the modulating signal. If the fundamental component varied directly with I_{CP}, the modulation would be perfectly linear. Unfortunately, this is not the case. As I_{CP} varies with e_m, the angle of conduction varies also. This leads to a variation of fundamental component with e_m that is not perfectly linear. We will now examine this deviation from linearity in more detail.

For a given value of e_m and V_0 the collector current is shown in Fig. 12.35. Note that the total angle of conduction is given by 2θ, where

$$\theta(t) = \cos^{-1}\left[\frac{V_0 - e_m}{C}\right] = \cos^{-1}\left[\frac{V_0 - M\cos\omega_m t}{C}\right]. \tag{14.21}$$

The amplitude of the fundamental component was evaluated in Chapter 12 and is given by

$$I = \frac{\beta C}{\pi R_S}\left[\theta(t) - \frac{\sin 2\theta(t)}{2}\right]. \tag{12.38}$$

If R_P is the resonant collector impedance the output voltage is

$$e_{\text{out}} = \frac{\beta C R_P}{\pi R_S}\left[\theta(t) - \frac{\sin 2\theta(t)}{2}\right]\cos\omega_c t. \tag{12.39}$$

Figure 14.10 plots the variation of the amplitude of e_{out} with e_m for the three conditions $V_0 = 0.5C$, $V_0 = 0.25C$, and $V_0 = 0$.

It is observed that the linearity of output signal with modulating voltage improves as class-B operation is approached. As V_0 becomes smaller, the angle

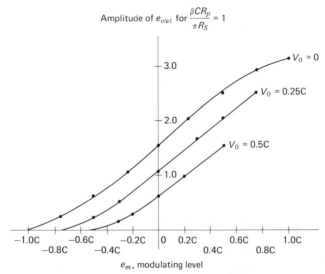

Fig. 14.10 Amplitude of e_{out} as a function of e_m.

of conduction increases and efficiency decreases. In applications requiring higher output power levels it is important to minimize device dissipation by operating with high circuit efficiency. In this case, the value of V_0 may be relatively high sacrificing linearity for circuit efficiency. Figure 14.11 shows two practical realizations of base-modulation circuits.

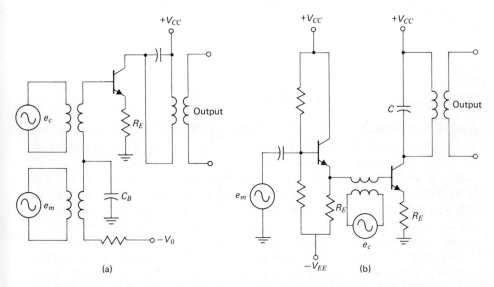

Fig. 14.11 Class-C base-modulation circuits: (a) transformer-coupled; (b) emitter-coupled.

In Fig. 14.11(a), the capacitor C_B is used to prevent carrier frequency current from passing through the modulation transformer. However, C_B should present a high impedance to the modulating frequency. Figure 14.11(b) shows an alternate method of applying the modulating signal to the base of the output transistor. The emitter follower can be biased to have a zero or negative quiescent voltage at the emitter (class-B or class-C). The resonant impedance of the tank circuit should be such that the transistor never saturates. This limits the maximum value of I_{CP} that can occur. If the collector current is a perfect half-wave rectified signal, the amplitude of the collector voltage is found from Eq. (12.39) to be

$$|e_{out}| = \frac{\beta C R_P}{2R_S}.$$

The collector voltage will always be greater than or equal to zero if

$$\frac{\beta C R_P}{2R_S} \leq V_{CC},$$

assuming the angle of conduction never exceeds 180°.

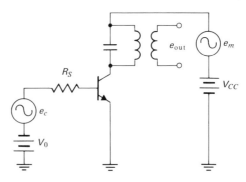

Fig. 14.12 Collector modulation.

Collector-modulation circuits. The collector-modulation circuit, shown in Fig. 14.12, can be operated as a class-B or class-C stage. The fundamental component of current must develop enough voltage across the tank circuit to saturate the transistor at the positive peaks of e_c. Again the collector current waveform is high in harmonic content, but the tank circuit develops an output voltage only for the signal of frequency ω_c. The higher-order current harmonics develop negligible voltage across the tuned circuit. As discussed in Section 12.5 the amplitude of the output signal under saturation-limited conditions equals the power supply voltage. A change in the power supply voltage then changes the output signal proportionately. The modulating signal is applied in series with the dc voltage supply giving a total power supply voltage of $e_m + V_{CC} = M \cos \omega_m t + V_{CC}$. For 100-percent modulation the modulating voltage amplitude M must equal V_{CC}. The power supply voltage is then $V_{CC}(1 + \cos \omega_m t)$ and the AM waveform is given by

$$e_{out} = V_{CC}(1 + \cos \omega_m t) \cos \omega_c t. \tag{14.22}$$

The pertinent waveforms for the collector-modulation stage are shown in Fig. 14.13.

When the instantaneous collector-supply voltage reaches $2V_{CC}$, the collector current must be sufficient to saturate the transistor. The fundamental component has a magnitude of

$$I = \frac{\beta C}{\pi R_S} \left[\theta - \frac{\sin 2\theta}{2} \right]. \tag{12.38}$$

In this case θ is constant with time and is

$$\theta = \cos^{-1}\left(\frac{V_0}{C}\right), \tag{14.23}$$

neglecting $V_{BE(on)}$. The parameters and element values must be chosen such that

$$IR_P \geq 2V_{CC}.$$

When the collector-supply voltage approaches zero, approximately the same current drive is present; hence the transistor will be considerably overdriven. Since

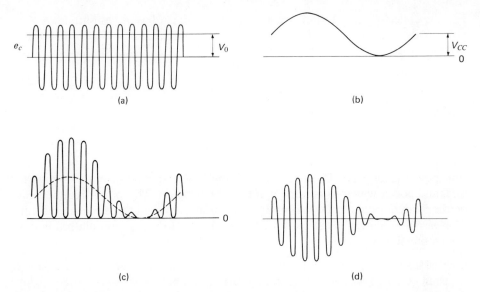

Fig. 14.13 Waveforms for a collector-modulation circuit: (a) input signal, V_{be}; (b) collector-supply voltage; (c) total voltage developed at collector [the amplitude of output is $2V_{CC}(1 + \cos \omega_m t)$ peak to peak]; (d) output voltage—100-percent modulated carrier.

the tank circuit is not ideal some distortion will be introduced by this overdriving and the linearity of modulation may be affected.

The collector-modulation circuit generally exhibits more linearity than does the base-modulation stage, especially as high indexes of modulation are approached. One disadvantage of this stage is the large amount of modulator power required. The base modulator requires less modulation power since the active device provides modulator power gain. It is possible, however, to transmit a given power level with less device dissipation using the collector modulator. This is important since transistor transmitter applications are often limited by the allowable device dissipation.

Power relations in AM waveforms. The equation for an AM waveform modulated by a sinusoid is

$$e = E_c \cos \omega_c t + \frac{mE_c}{2} \cos (\omega_c - \omega_m)t + \frac{mE_c}{2} \cos (\omega_c + \omega_m)t. \quad (14.6)$$

For 100-percent modulation $m = 1$ and each sideband amplitude equals one-half the carrier amplitude. Since the impedance across which these components appear is approximately constant for these three frequencies, the power in each component is proportional to the square of the voltage amplitudes. The sidebands each contain one-fourth the power of the carrier, giving a total power of

$$P_T = P_C + P_{LS} + P_{US} = P_C(1 + \tfrac{1}{4} + \tfrac{1}{4}) = \tfrac{3}{2} P_C. \quad (14.24)$$

For the collector-modulation circuit, the carrier signal has an amplitude of V_{CC} volts while that of the sidebands (100-percent modulation) is $V_{CC}/2$. The total output is then

$$P_T = \left(\frac{3}{2}\right)\frac{V_{CC}^2}{2R_P}. \tag{14.25}$$

In this case the unmodulated carrier signal contains the power

$$P_C = \frac{V_{CC}^2}{2R_P}.$$

This power is furnished by the dc power supply. The modulator must supply the sideband power which equals one-half the carrier power. This power requirement on the modulator of a collector-modulation circuit can present a serious problem. The base-modulation stage requires far less power from the modulator, as discussed previously.

Carrier suppression. The commercial radio band extending from 535 to 1605 kHz transmits the AM waveform containing carrier and both sidebands. Receivers for this band contain rather simple detectors that reproduce the modulating signal quite faithfully. We note that the carrier signal conveys little or no information and yet requires the bulk of the power used in transmission. It is possible to suppress the carrier and transmit only the two sidebands at a greatly increased power level. It is further possible to eliminate the carrier and one sideband, concentrating all available transmission power into a single sideband which theoretically conveys all necessary information. These possibilities result in carrier-suppressed transmission and single-sideband transmission, respectively. While the complexity of the transmitter may not be greatly increased for these systems, the receiver is far more complex than the simple AM waveform receiver. Furthermore, the quality of the demodulated signal is generally poorer. Certain civil and military organizations use carrier-suppressed or single-sideband systems to capitalize on the lower bandwidth and lower power requirements.

Carrier suppression is conveniently accomplished by a balanced modulator such as that shown in Fig. 14.14. Assuming perfect modulators with equal gains results in an output signal of $2A_1 e_m e_c$. If $e_m = M \cos \omega_m t$ and $e_c = C \cos \omega_c t$, the

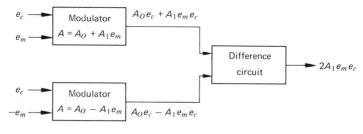

Fig. 14.14 Block diagram of balanced modulator.

output signal consists only of two sidebands, giving

$$e_{out} = A_1 MC[\cos(\omega_c - \omega_m)t + \cos(\omega_c + \omega_m)t]. \tag{14.26}$$

The balanced modulator is actually a multiplier circuit forming the product of the carrier and modulating signals. Figure 14.15 shows a possible circuit for the balanced modulator. The carrier signal is applied in phase to both bases simultaneously while the modulating signals applied to the bases are 180° out of phase. The nonlinearity of the base-emitter junctions modulates the carrier signal. The resulting collector currents flow through the transformer in such a way as to generate a difference voltage across the secondary winding. This circuit implements the block diagram of Fig. 14.14.

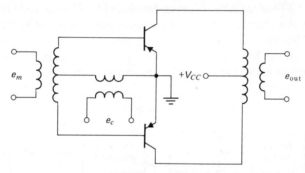

Fig. 14.15 Balanced modulator circuit.

It must be noted that the modulators are imperfect, as discussed earlier, and small unwanted components result. Furthermore, the two modulators must be perfectly balanced to lead to complete suppression of the carrier. While it is possible to use base or collector modulators to decrease distortion, the balanced modulator is more complex than the simple modulators.

Single-sideband systems. The most widely used modulation technique in frequency-division multiplexing results in single-sideband transmission [5]. In telephone transmission of the human voice, quality is not of prime importance and the single-sideband method is applicable. In this instance the voice frequencies transmitted range from 300 to 3400 Hz. The spacing in such a system is typically 4 kHz for each voice channel. Up to 12 channels, for example, are assembled into a group band with a frequency range of 60 to 108 kHz. As many as five group bands can be used to allow a 60-channel system to transmit simultaneous conversations over a single transmission line.

The simplest method of single-sideband generation uses filtering to eliminate one sideband from a suppressed carrier signal. Let us assume that we wish to transmit the upper sideband of a signal resulting from modulating a 60 kHz carrier by a signal varying from 300 to 3400 Hz. Using a balanced modulator results in a lower sideband extending from 56.6 to 59.7 kHz. A bandpass filter with a pass band extending from 60.3 to 63.4 kHz follows the modulator. The filter is designed to fall sharply as frequency decreases from 60.3 to 59.7 kHz. Typically a 40- to 60-db difference in response at these frequencies must result at the filter output. A rather sharp filter is required as the frequency change is only one percent in this case.

With higher carrier frequencies it becomes very difficult to use the simple method just described due to severe filter requirements. For example, if the same modulating signal is to be transmitted with a 10-MHz carrier and 40-db attenuation between sidebands, a very complex filter would be required. This filter would have to provide 40-db attenuation within 600 Hz at a frequency of 10 MHz, a 0.006-percent frequency change. It is possible to use two stages of modulation as shown in Fig. 14.16 to ease the filtering problem [5]. The second filter now must provide 40-db attenuation between the frequencies 9.8997 and 10.1003 MHz, a 200.6-kHz or two-percent difference. Use of this two-step method requires more bandwidth per channel, but some applications are more concerned with obtaining the single-sideband signal than with the channel spacing.

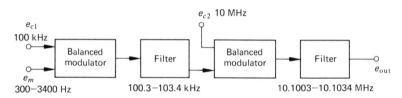

Fig. 14.16 Single-sideband generation.

In general, the receiver is designed to reinsert the carrier signal to obtain the desired information. Modulating or mixing the single-sideband signal with the carrier results in a lower sideband of frequency equal to that of the modulating signal. Unfortunately, the reinserted carrier frequency may differ slightly from the transmitter carrier frequency, causing some harmonic distortion.

There are other methods of generating single-sideband signals using wideband phase-shift networks [5], but these methods are usually not as useful as the preceding method.

Vestigial-sideband. This system transmits the carrier and one sideband with little or no attenuation along with the other sideband which has been heavily attenuated. This method is used to transmit the video information for black-and-white television. Some distortion can be tolerated, but if the lower sideband is completely eliminated the distortion is unacceptable. Only a portion of the lower sideband is transmitted to reduce distortion while conserving bandwidth. Filtering is again applied to achieve the vestigial-sideband signal and rolloff rates are less critical than for single-sideband transmission.

B. AM Detection Circuits

The diode detector. The information transmitted by a simple AM wave is carried by the peak amplitude of the waveform. Thus a peak-detecting circuit should recover this information. The simple circuit of Fig. 14.17 will detect peaks so long as the peak amplitude does not decrease. This detector will not follow the peaks of decreasing amplitude unless a resistor is used to slightly discharge the capacitor between cycles, as shown in Fig. 14.18. The output voltage of the detector contains three components. It has a high-frequency ripple voltage, a much lower frequency voltage which is proportional to the amplitude of the peaks of the AM waveform, and a dc voltage equal to the amplitude of the unmodulated carrier signal. The signal that contains the transmitted information is the one that is proportional to the peaks of the waveform. The dc signal and the high frequency signal are undesirable and must be removed. The detector voltage might ultimately be used to activate a loudspeaker; hence a coupling capacitor should be used to remove the dc component. The high frequency component can be filtered or even ignored in some applications, since it is far above the audio range. The signal shown in Fig. 14.18 does not demonstrate the true frequency difference between the carrier and the modulating signal. We must bear in mind that carrier frequency is several hundred times greater than the modulating frequency. The time constant of the RC-combination is chosen to be much larger than the period of the carrier and much smaller than the period of the modulating signal. A reasonable value, although by no means the only value, is

$$RC = \sqrt{1/\omega_m \omega_c}. \qquad (14.27)$$

Equation (14.27) gives a time constant which is the geometric mean of the two periods (divided by 2π). The discharge between cycles of the carrier signal should

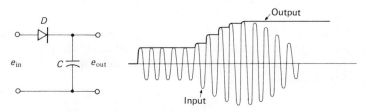

Fig. 14.17 Peak detector.

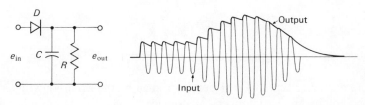

Fig. 14.18 Simple diode detector.

be negligible, but the discharge rate must be high enough to follow accurately the decrease in amplitude of the peaks. Two practical demodulators are shown in Fig. 14.19.

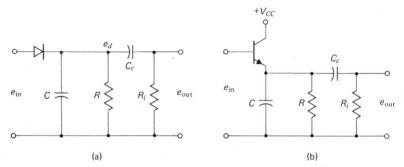

Fig. 14.19 Practical detector: (a) diode detector; (b) emitter diode detector.

The coupling capacitor C_c removes the dc component of the demodulated signal from the output voltage. In the figure, R_i might represent the input impedance to an amplifying stage. If an unmodulated carrier is applied to the input, the voltage e_d will be a dc signal equal to the amplitude of the carrier. The voltage e_{out} will equal zero. The average current drawn from the source is then

$$I_{in} = C_0/R,$$

where C_0 is the amplitude of the unmodulated carrier signal. If the signal were 100-percent modulated, the input current could be expressed as

$$I_{in} = \frac{C_0}{R} + \frac{C_0}{R_{eq}} \cos \omega_m t, \qquad (14.28)$$

where $R_{eq} = RR_i/(R + R_i)$. Unless $R_i \gg R$, the expression for input current will become negative as $\cos \omega_m t$ approaches -1. Since the diode cannot pass negative current, the output waveform distorts. To avoid this distortion the index of modulation must be limited such that

$$\frac{C_0}{R} = \frac{mC_0}{R_{eq}}$$

or

$$m = \frac{R_i}{R + R_i}. \qquad (14.29)$$

Only when R_i is much greater than R can m approach unity with no resulting distortion.

There are other types of AM detectors such as the square-law detector that utilize the nonlinear characteristics of a device to recover the modulating signal. The square-law detector is usually used only in applications where distortion is unimportant since it adds some harmonic distortion to the recovered signal. Most applications apply the rectifying junction in the form of a diode or transistor to construct the detector.

Heterodyning. When a signal is multiplied by an auxiliary sinusoidal signal to form sidebands the operation is called mixing, or heterodyning. This process is basically a modulation process, but is usually performed on an AM waveform without destroying the information. In effect, heterodyning translates the AM waveform in such a way that the sidebands are centered about a new carrier frequency. One very useful application of mixing occurs in all standard commercial radio receivers. Information is transmitted with a carrier frequency falling in the 535 to 1605 kHz range. The desired signal is translated from the incoming carrier frequency to a carrier of 455 kHz carrying the same modulation information. A fixed-frequency tuned amplifier is then used to amplify this signal. Other signals can be amplified with this same amplifier simply by translating each AM waveform to this 455-kHz carrier frequency. This idea of a fixed intermediate-frequency amplifier is useful in any receiver that is to receive any one of several different signals within a specified frequency range. Heterodyning can also be useful in single-sideband detectors either to reinsert a carrier frequency or to recover the original modulating frequency. It has been mentioned that single-sideband is useful in transmitting several telephone channels over a single transmission line. A pilot carrier frequency having a specified time relationship to each channel carrier frequency is also transmitted to be used at the receiver in generating a new carrier frequency. This new carrier frequency can then be mixed with the single-sideband signal of interest creating an AM waveform that is easily detected by conventional detectors.

Heterodyning can be performed by means of a balanced modulator although a simple nonlinear element such as a transistor or FET is generally used in practice. Let us assume that we apply a constant amplitude component $e_1 = D \cos \omega_1 t$ to the modulating input of a balanced modulator. To the other input is applied an AM waveform

$$e = E_c \cos \omega_c t + \frac{mE_c}{2} \cos (\omega_c - \omega_m)t + \frac{mE_c}{2} \cos (\omega_c + \omega_m)t.$$

The resulting signal is

$$e_h = \frac{D}{2}\left[E_c \cos (\omega_c - \omega_1)t + \frac{mE_c}{2} \cos (\omega_c - \omega_1 - \omega_m)t \right.$$
$$\left. + \frac{mE_c}{2} \cos (\omega_c - \omega_1 + \omega_m)t \right] + \frac{D}{2}\left[E_c \cos (\omega_c + \omega_1)t \right.$$
$$\left. + \frac{mE_c}{2} \cos (\omega_c + \omega_1 - \omega_m)t + \frac{mE_c}{2} \cos (\omega_c + \omega_1 + \omega_m)t \right]. \quad (14.30)$$

The voltage of Eq. (14.30) represents two AM waveforms, one centered about a carrier of frequency $(\omega_c - \omega_1)$ and one centered about $(\omega_c + \omega_1)$. Under usual conditions both ω_c and ω_1 are much larger than the maximum modulation frequency; thus the sidebands associated with this waveform do not overlap. A tuned amplifier with a bandwidth of $2\omega_m$ can be used to separate either of the AM waveforms from the other. In commercial applications the signal with the frequency $(\omega_c - \omega_1)$ is generally amplified as the intermediate frequency.

We note that if $\omega_c = \omega_1$, the lower frequency AM signal contains a dc term and a term equal to $mE_c \cos \omega_m t$. This signal is proportional to the original modulating signal. Although it is difficult to synchronize the frequencies ω_c and ω_1 some special-purpose receivers use this basic idea to recover the transmitted information.

The mixer in most commercial receivers uses a simple transistor mixer rather than a perfect multiplier, but additional output frequencies are far out of the intermediate-frequency band and present no problems to the heterodyne process.

The block diagram of a typical superheterodyne receiver is shown in Fig. 14.20. The AM waveform is received by the antenna and amplified by the *RF* amplifier. Some filtering can be done in this amplifier, but the selectivity is determined primarily by the *IF* section. The local oscillator is tuned such that the oscillation frequency minus the carrier frequency of the desired signal equals the intermediate frequency. In a superheterodyne system the oscillator frequency is always greater than the *RF* carrier frequency. The *IF* section passes the signal centered at the intermediate frequency while discriminating heavily against signals of other frequencies. The signal is then demodulated, amplified, and filtered if necessary to produce the output signal. This signal is then used to drive a speaker or other output device.

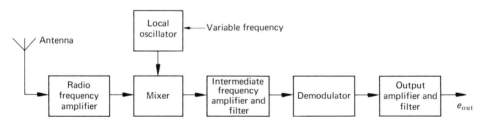

Fig. 14.20 Superheterodyne receiver.

C. Phase Modulation

Phase modulation occurs when the phase of a carrier signal is changed from its quiescent value by the application of a modulating signal. Generally, the phase deviation varies directly with the amplitude of the modulating signal. Phase modulation, as such, is not very important in communication systems. Phase modulators can form the basis for frequency-modulating systems and thus become important in this respect.

The system of Fig. 14.21 shows a small-deviation phase modulator [5]. The phase of the carrier is shifted 90° and added to a much smaller AM waveform having the same carrier frequency. The resultant waveform is the vector sum of the two components, as shown in Fig. 14.21(b). As the AM waveform varies in amplitude, the output voltage E_C varies in phase and slightly in amplitude. If the phase deviation is limited to approximately 30° ($\pm 15°$), a reasonably linear variation of phase with modulating voltage results. The limiter is used to keep the amplitude variations of the output waveform at a minimum level. The expression for the output voltage is then

$$e_{\text{out}} = C \cos(\omega_c t + \phi_0 + m_p \sin \omega_m t),$$

which corresponds to a phase-modulated wave. The quantity m_p is called the phase modulation index.

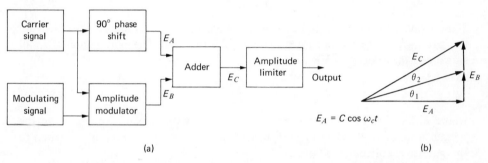

Fig. 14.21 Phase-modulating system: (a) block diagram; (b) vector diagram.

The circuit of Fig. 14.22 shows a large-deviation phase modulator [2] using a varistor (silicon-carbide voltage-variable resistor). This modulator allows linear phase shifts of $\pm 50°$ or greater and does not require an amplitude limiter. When

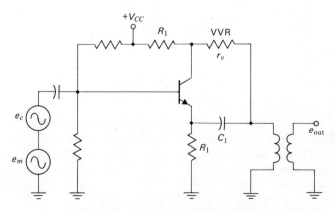

Fig. 14.22 Large-deviation phase modulator.

the modulating voltage is zero, the circuit reduces to a phase-shift circuit similar to that of Section 13.3. If the amplitude of the carrier is fairly small, it will pass through the phase-shift circuit with constant amplitude and a phase shift of

$$\phi = 2 \tan^{-1} \frac{1}{\omega_c C_1 r_v}. \tag{14.31}$$

For phase shifts up to 100°, Eq. (14.31) can be approximated by

$$\phi = \frac{2}{\omega_c C_1 r_v}. \tag{14.32}$$

The phase shift can be varied by changing r_v, but the variation is an inverse one. The voltage variable resistor (VVR) [2] is a device which has a V–I relationship as expressed by

$$I = kV^2. \tag{14.33}$$

The exponent can be made to lie anywhere from 2 to 7, depending on the construction process, but for our purposes a value of 2 is required. The incremental or small signal resistance of the VVR is

$$r_v = \frac{dV}{dI} = \frac{K}{V} \tag{14.34}$$

where $K = 1/2k$. The dynamic resistance of the VVR depends inversely on the voltage applied to the device. The phase shift of the carrier can then be related to the quiescent collector voltage as

$$\phi_{CQ} = \frac{2}{\omega_c C_1 r_v} = \frac{2}{\omega_c C_1 K} V_{CQ}.$$

If the modulating voltage has a much larger amplitude than the carrier signal, the resistance r_v will be

$$r_v = \frac{K}{V_{CQ} + M \sin \omega_m t},$$

and the phase shift of the carrier is

$$\phi = \frac{2}{\omega_c C_1 K} [V_{CQ} + M \sin \omega_m t]. \tag{14.35}$$

The phase deviates linearly with the amplitude of e_m. The transformer must be so chosen as to offer a low impedance to the modulating signal and a high impedance to the carrier. The transformer can be replaced with an inductor.

The output signal is then given by

$$e_{out} = C \cos \left(\omega_c t + \frac{2 V_{CQ}}{\omega_c C_1 K} + \frac{2M}{\omega_c C_1 K} \sin \omega_m t \right). \tag{14.36}$$

Comparing Eq. (14.36) to that of a phase-modulated waveform results in the equations

$$\phi_0 = \frac{2V_{CQ}}{\omega_c C_1 K}$$

and

$$m_p = \frac{2M}{\omega_c C_1 K}.$$

D. Frequency Modulation

Frequency modulation occurs when the frequency of a carrier signal is caused to change in accordance with a modulating signal. For most communication systems the frequency deviation is related linearly to the amplitude of the modulating signal. The expression for frequency of an FM signal with sinusoidal modulation is

$$f = f_c + k_f M \cos \omega_m t, \tag{14.37}$$

where k_f is a constant.

In visualizing the process of phase modulation we recognize that during a given cycle of the carrier signal, the frequency must change as the phase is changing. Thus, there is obviously a relationship between phase and frequency modulation. This relationship can be derived mathematically starting with the expression for a phase-modulated waveform

$$e_{out} = C \cos (\omega_c t + \phi_0 + m_p \sin \omega_m t).$$

The frequency of this signal is not constant since the argument

$$\theta = (\omega_c t + \phi_0 + m_p \sin \omega_m t)$$

does not vary linearly with time. The instantaneous radian frequency is found from

$$\omega = d\theta/dt = \omega_c + m_p \omega_m \cos \omega_m t. \tag{14.38}$$

As the phase of a sinusoid is modulated the frequency is also modulated. The instantaneous frequency change is proportional to the derivative of the phase-modulating signal. Thus, one method of obtaining an FM signal with a frequency of

$$\omega = \omega_c + A(t)$$

is to use $\int A(t) \, dt$ as the modulating signal of a phase modulator. For example, if the desired FM signal frequency is

$$\omega = \omega_c + 2\pi k_f M \cos \omega_m t,$$

then the phase-modulating signal should be

$$\frac{2\pi k_f M}{\omega_m} \sin \omega_m t.$$

There are other methods of expressing the instantaneous frequency of the FM waveform, but perhaps the most useful form is given by Eq. (14.38). The maximum frequency deviation is given by

$$\Delta \omega_{max} = \omega_m m_p \tag{14.39}$$

or

$$\Delta f_{max} = f_m m_p. \tag{14.40}$$

The modulation index or deviation ratio is given by the ratio of the maximum frequency deviation to the modulating frequency or

$$m_p = \frac{\Delta f_{max}}{f_m}. \tag{14.41}$$

In commercial FM broadcasting, a maximum frequency deviation of 75 kHz is allowed by the Federal Communications Commission. This deviation may be typically imposed on a carrier frequency of 100 MHz. The maximum frequency deviation is only 0.075 percent requiring a very stable carrier frequency. A small drift in frequency of the carrier will lead to a false signal being detected at the receiver. There are other FM applications in communications and control systems requiring less precise control over the transmitted carrier frequency.

The tuned-collector modulator. Figure 14.23 shows an FM modulator based on the tuned-collector oscillator. If the collector depletion-region capacitance is neglected, the reverse-biased diode determines the tank circuit capacitance. This capacitance is a function of the reverse-bias voltage across the diode, as shown by Eq. (2.24):

$$C_{dl} = \frac{k}{\sqrt{\phi + V}} \approx \frac{k}{\sqrt{V}}.$$

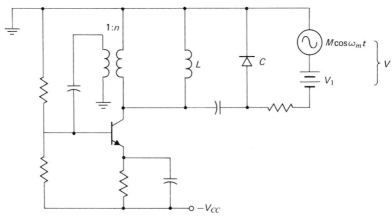

Fig. 14.23 Frequency modulator.

When V is much larger than the barrier voltage, the above approximation is quite valid. The resonant frequency of the tuned circuit is (neglecting stray capacitance)

$$\omega_0 = \frac{1}{\sqrt{LC_{dl}}} = \frac{V^{1/4}}{\sqrt{Lk}} \qquad (14.42)$$

The voltage V is the total reverse-bias voltage across the diode, which is equal to $V_1 + M \cos \omega_m t$, where V_1 is the dc bias voltage. Equation (14.42) can then be expressed as

$$\omega_0 = \frac{V_1^{1/4}[1 + (M/V_1) \cos \omega_m t]^{1/4}}{\sqrt{Lk}}.$$

Using the binomial expansion for the numerator term, we obtain

$$\omega_0 = \frac{V_1^{1/4}}{\sqrt{Lk}} \left(1 + \frac{1}{4} \frac{M}{V_1} \cos \omega_m t - \frac{3}{32} \frac{M^2}{V_1^2} \cos^2 (\omega_m t) + \cdots \right). \qquad (14.43)$$

Applying the identity $\cos^2 \omega_m t = \frac{1}{2} + \frac{1}{2} \cos 2\omega_m t$ allows Eq. (14.43) to be written as

$$\omega_0 = \frac{V_1^{1/4}}{\sqrt{Lk}} \left(1 - \frac{3}{64} \frac{M^2}{V_1^2} + \frac{1}{4} \frac{M}{V_1} \cos \omega_m t - \frac{3}{64} \frac{M^2}{V_1^2} \cos 2\omega_m t \right).$$

If

$$\frac{V_1^{1/4}}{\sqrt{Lk}} \left[1 - \frac{3}{64} \frac{M^2}{V_1^2} \right] \approx \frac{V_1^{1/4}}{\sqrt{Lk}}$$

is equated to the carrier frequency and

$$\frac{V_1^{1/4}}{\sqrt{Lk}} \times \frac{1}{4V_1} = 2\pi k_f,$$

then the first three terms in Eq. (14.43) describe the frequency of an FM wave as given by Eq. (14.37). Comparing the amplitude of the second harmonic term to that of the fundamental shows that the percent distortion is

$$\frac{3M}{16V_1} \times 100. \qquad (14.44)$$

In order to keep this distortion at a negligible level, M must be somewhat smaller than V_1. This requirement imposed on M limits the linear frequency deviation that can be obtained with this circuit. This type of circuit can allow frequency deviations of two to five percent of the carrier frequency without serious distortion.

In practice it is possible to use the depletion layer capacitance of the collector-base junction to resonate the tank circuit. The modulating voltage can then be applied to the transistor base and the diode is eliminated from the tank circuit.

While this modulator can be used for noncritical applications such as wireless microphones, a crystal oscillator must be used to accurately control the carrier frequency for most communication circuits.

It should be noted that, unlike AM transmitters that generally use high-power modulators near the transmitter output, the FM waveform is usually produced at a low power level and then amplified before transmission. Often the modulation is applied to a frequency much lower than the desired carrier frequency. Frequency doublers or triplers are then used to produce the appropriate carrier frequency. The varactor diode or class-C amplifier can be used to translate the input frequency to the transmission frequency. Limiting of amplitude variations can also be accomplished with the class-C stage operating into the saturation region.

Frequency modulation with an AM modulator. The system of Fig. 14.21 can be converted from a phase modulator to a frequency modulator simply by inserting an integrator between the modulating signal generator and the amplitude modulator. This system can be very stable if the carrier signal is produced by a crystal-controlled oscillator. Instead of applying the carrier signal to the system input, a subharmonic might be applied. The modulation can then be performed to produce a low-level, comparatively low-frequency FM signal. A harmonic of this signal is then amplified and limited by the class-C output amplifier. Greater indexes of modulation can be obtained by cascading phase modulators.

Large-deviation frequency modulator. A frequency modulator [2] that gives linear frequency deviations of 50 percent of the carrier frequency is shown in Fig. 14.24. Frequency deviations of this magnitude are not required in conventional FM transmitters, but certain automatic-control applications require large-deviation frequency modulation. The quiescent collector voltage controls the resistance r_v. If r_v and C_1 are selected to give a phase shift of $90°$ at the carrier frequency, the circuit becomes an oscillator at this frequency. This frequency is found from Eqs. (14.32)

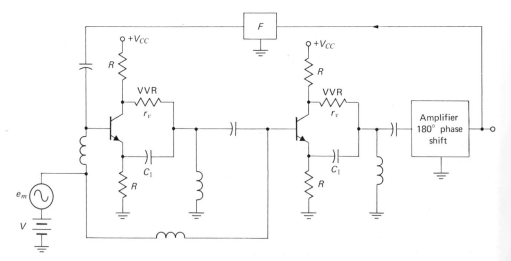

Fig. 14.24 Large-deviation frequency modulator.

and (14.34) to be

$$f_c = \frac{2}{2\pi\phi C_1 r_v} = \frac{0.2}{C_1 r_v} = \frac{0.2 V_{CQ}}{C_1 K}.$$

If the modulating voltage now causes the collector voltage to vary as

$$V_C = V_{CQ} + M \cos \omega_m t,$$

the carrier frequency will change as

$$f = \frac{0.2}{C_1 K}(V_{CQ} + M \cos \omega_m t)$$

$$= f_c + \frac{0.2M}{C_1 K} \cos \omega_m t. \tag{14.45}$$

This frequency change is necessary to satisfy the phase-shift requirements of the oscillator as V_C is changed. Very large frequency deviations, varying linearly with the amplitude of the modulating voltage, can be obtained with this circuit.

One final method of obtaining an FM waveform involves heterodyning. In this method the frequency modulation is performed at a lower frequency and then heterodyned up to the desired carrier frequency. This method requires a large-deviation frequency modulator since the deviation ratio achieved at the low frequency decreases as the frequency is translated upward. For example, the audio information for television is transmitted by an FM waveform at perhaps 200 MHz. This signal is limited to a frequency deviation of 25 kHz. It is customary to generate an FM signal at a center frequency of 2 MHz and then heterodyne this signal to 200 MHz. The frequency deviation at 2 MHz must be 25 kHz also as the deviation will be unaffected by heterodyning. Thus, the frequency deviation is 1.25 percent at the lower frequency and is 0.0125 percent at the transmission frequency. It is to be noted that this percentage remains constant when a chain of harmonic multipliers is used.

It is possible to use heterodyning in connection with multipliers to control the index of modulation. Let us assume that an FM waveform is produced at a center frequency below that of the required carrier frequency. Let us further assume that if multiplers are used to reach the final carrier frequency the resulting frequency deviation is too high. If so, fewer multipliers are used and the signal is then heterodyned up to reach the carrier frequency. If the multipliers lead to less frequency deviation than required, more multipliers can be used and the signal is heterodyned down in frequency somewhere along the multiplication chain.

By way of example, let us assume that we desire to generate an FM signal having a carrier frequency of 200 MHz with a maximum frequency deviation of 25 kHz. The modulating signal frequency varies from 50 Hz to 5000 Hz. The system is based on a phase modulator that has a modulation index of 0.25 and operates at a frequency of 200 kHz.

At the lowest modulating frequency of 50 Hz the phase modulator will produce a frequency deviation of

$$\Delta f = m_p f_m = 0.25 \times 50 \text{ Hz} = 12.5 \text{ Hz}.$$

To lead to a deviation of 25 kHz requires a frequency multiplication of

$$\frac{25 \times 10^3}{12.5} = 2000.$$

Multiplying the frequency of the modulator by 2000 leads to the required 25 kHz deviation, but the carrier frequency would then be

$$2000 \times 200 \text{ kHz} = 400 \text{ MHz}.$$

It would be possible to heterodyne this signal down to 200 MHz preserving the correct deviation; however, it is more convenient to heterodyne the signal earlier in the multiplication chain. Figure 14.25 indicates one method of producing the required output.

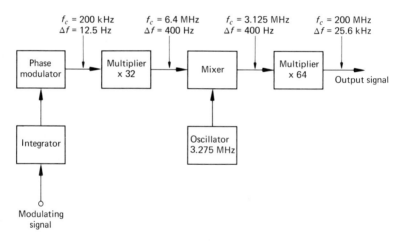

Fig. 14.25 An FM waveform generator using multipliers.

A significant feature of this system is that the actual modulation takes place at a low frequency. This allows transistor circuits to find application in most FM communication systems.

Bandwidth requirements of FM waveforms. The instantaneous frequency of an FM waveform varies from a low value of $f_c - \Delta f$ to a high value of $f_c + \Delta f$. It might appear that a bandwidth of $2 \Delta f$ is sufficient to transmit the FM signal. This conclusion is incorrect due to the fact that additional sidebands are generated as the instantaneous frequency varies. Carson's rule [5] states that the bandwidth required to transmit the entire FM waveform is twice the sum of the maximum frequency deviation and the modulating frequency; that is,

$$BW = 2(\Delta f + f_m). \tag{14.46}$$

We will not derive this result here, but will outline the major steps of the derivation which can be found in more detail elsewhere [5].

Starting with the expression

$$e_{out} = \cos\left(\omega_c t + \frac{\Delta\omega}{\omega_m} \sin \omega_m t\right),$$

where $\Delta\omega$ is constant with ω_m, we use a trigonometric identity to write

$$e_{out} = \cos \omega_c t \cos\left(\frac{\Delta\omega}{\omega_m} \sin \omega_m t\right) - \sin \omega_c t \sin\left(\frac{\Delta\omega}{\omega_m} \sin \omega_m t\right).$$

All terms in this expression are periodic in time and can be expanded in terms of a Fourier series. After evaluation of the Fourier coefficients and application of some trigonometric identities, the voltage becomes

$$\begin{aligned} e_{out} &= J_0\left(\frac{\Delta\omega}{\omega_m}\right) \cos \omega_c t - J_1\left(\frac{\Delta\omega}{\omega_m}\right) [\cos(\omega_c - \omega_m)t - \cos(\omega_c + \omega_m)t] \\ &+ J_2\left(\frac{\Delta\omega}{\omega_m}\right) [\cos(\omega_c - 2\omega_m)t + \cos(\omega_c + 2\omega_m)t] \\ &- J_3\left(\frac{\Delta\omega}{\omega_m}\right) [\cos(\omega_c - 3\omega_m)t - \cos(\omega_c + 3\omega_m)t] \\ &+ \cdots \end{aligned}$$

The functions $J_n(\Delta\omega/\omega_m)$ are known as the Bessel functions of the first kind of order n. As numerical values of these functions are evaluated, it is found that the only terms of significance are those of order $n = (\Delta\omega/\omega_m + 1)$ or less where $\Delta\omega/\omega_m$ is taken as an integer. The sidebands corresponding to this value of n occur at frequencies $\omega_c - (\Delta\omega/\omega_m + 1)\omega_m$ and $\omega_c + (\Delta\omega/\omega_m + 1)\omega_m$. All lower-order terms have sidebands occurring between these two extremes. The required radian bandwidth is then

$$BW_{rad} = 2\left(\frac{\Delta\omega}{\omega_m} + 1\right)\omega_m = 2(\Delta\omega + \omega_m).$$

Any frequency-selective network through which the FM signal must pass with negligible distortion must have a bandwidth given by Eq. (14.46).

E. FM Detectors

An FM detector consists of two sections. The first converts the constant amplitude signal into a signal whose amplitude varies with frequency. This section converts the FM signal into an AM signal. The second section is an AM detector which demodulates the newly created AM waveform.

A resonant circuit tuned to a slightly different frequency than the carrier of the FM signal can be used for the first section. If a current proportional to the FM signal enters the tank circuit, the output-voltage amplitude can be found by considering the impedance curve of the tank circuit shown in Fig. 14.26. As the frequency varies about the carrier frequency, the output voltage varies from the unmodulated level. One problem with this circuit is that the output is linear with frequency deviation only for small frequency deviations.

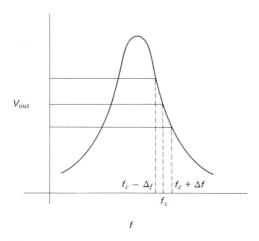

Fig. 14.26 Response of a tuned circuit to an FM waveform.

Other types of detectors can be used to improve the linearity. The Foster-Seeley discriminator and the ratio detector are quite linear FM detectors. The Foster-Seeley discriminator is shown in Fig. 14.27. It can be shown that the voltage e_1 leads the primary voltage e_p by 90° and e_2 lags e_p by 90° when the secondary circuit is tuned to resonance. The voltages e_{1a} and e_{2a} applied to the AM detector circuits are $e_{1a} = e_1 + e_p$ and $e_{2a} = e_2 + e_p$, as shown in Fig. 14.28. As the frequency varies from resonance, the phases of e_1 and e_2 change. The voltage e_{1a} increases

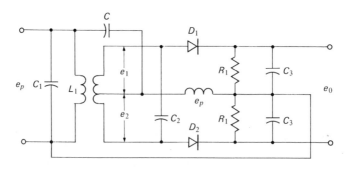

Fig. 14.27 Foster-Seeley discriminator.

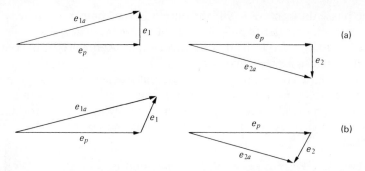

Fig. 14.28 Voltage relations in the FM discriminator: (a) at resonance; (b) off resonance.

and e_{2a} decreases as the frequency varies in one direction from resonance. As the input frequency varies in the opposite direction from resonance, e_{1a} decreases and e_{2a} increases. The output voltage equals the difference between e_{1a} and e_{2a}, and this voltage varies linearly with frequency deviation over a sizable range of frequency.

The phase-locked loop or PLL is often used as an FM demodulator. This circuit consists of a phase comparator and a voltage-controlled oscillator (VCO) as shown in Fig. 14.29.

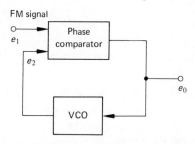

Fig. 14.29 Phase-locked loop used as FM demodulator.

There are two types of phase comparator that can be used in the PLL. One is a multiplier circuit or balanced modulator that forms the product of the two input signals. If the inputs are given by $e_1 = C \cos(\omega_c t + \phi_1)$ and $e_2 = D \cos(\omega_c t + \phi_2)$ the product is

$$CD \cos(\omega_c t + \phi_1) \cos(\omega_c t + \phi_2) = \frac{CD}{2}[\cos(2\omega_c t + \phi_1 + \phi_2) + \cos(\phi_1 - \phi_2)].$$

Passing the output of the multiplier through a low-pass filter results in an output signal of

$$e_0 = \frac{CD}{2}[\cos(\phi_1 - \phi_2)]. \tag{14.47}$$

When the phase difference is $-90°$ the output of the comparator is zero. As this difference becomes greater than $-90°$ the output becomes negative while a phase difference less than $-90°$ leads to a positive output. Another type of comparator converts the input signals to square waves and then compares the phase. Regardless of the type of phase comparator used the output reaches zero at $-90°$ phase difference. We will assume a linear variation of comparator output with phase shift near the $-90°$ phase difference point. For the multiplying comparator this is only true for small departures from the $-90°$ phase shift since $\cos(\theta - 90°) = \sin\theta \approx \theta$ for small values of θ. The comparator output is then

$$e_0 = K(\phi_1 - \phi_2 + 90°). \tag{14.48}$$

Figure 14.30 shows the comparator output as a function of phase difference. Equation (14.48) is valid only for values of $\phi_1 - \phi_2$ between $-180°$ and $0°$.

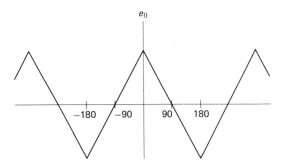

Fig. 14.30 Comparator output as a function of phase difference.

The VCO is an oscillator with an output frequency linearly controlled by an applied voltage. This circuit is basically an FM modulator with a frequency deviation at least as large as that of the input signal.

For use as an FM demodulator the PLL is so designed that the VCO oscillates at the frequency of the incoming carrier signal when no control voltage is applied. For nonzero values of e_0 the frequency is expressed as $f_0 = f_c + Ge_0$. Equilibrium is reached when the phase difference between the FM input and the VCO signal is $-90°$. To verify this, let us assume that the system initially exhibits a phase difference of only $-60°$; that is, the VCO output signal leads the FM carrier by $60°$. A positive voltage will be generated by the comparator according to Eq. (14.48). This voltage will increase the frequency of the VCO. With a slightly higher VCO frequency the absolute phase difference will increase over each cycle. As this difference approaches $-90°$, the comparator output approaches zero and the VCO frequency approaches the carrier frequency. Other values of initial phase difference also lead to the same equilibrium point. Once the two frequencies are "locked in" a variation in the FM signal frequency generates a signal e_0 that is proportional to the frequency difference. An abrupt variation of input frequency from f_c to $f_c + \Delta f_1$ leads to a phase change in ϕ_1 of

$$\Delta \phi_1 = \Delta \omega_1 t$$

since

$$d\phi_1/dt = \Delta \omega_1.$$

This phase change generates a positive voltage at the comparator output which increases the frequency of the VCO until a new equilibrium point has been established. Obviously this new VCO frequency must equal the FM signal frequency for equilibrium and e_0 is proportional to the difference between the incoming frequency and f_c. If the incoming signal has a frequency given by

$$f = f_c + m(t),$$

the waveform can be expressed as

$$e = C \cos(\omega_c t + \int m(t)\, dt).$$

The change in ϕ_1 is given by $\int m(t)\, dt$. Since the oscillator frequency is $f_0 = f_c + Ge_0$ and $\phi_2 = 90°$ for $e_0 = 0$, the change in ϕ_2 is $\int Ge_0\, dt$. From Eq. (14.48) we can write

$$e_0 = K[\phi_1 - \phi_2 + 90°] = K[\int m(t)\, dt - \int Ge_0\, dt].$$

Differentiating both sides and transposing gives

$$\frac{de_0}{dt} + KGe_0 = Km(t). \tag{14.49}$$

Let us first consider the case of an abrupt change in modulating voltage from zero to m_1. Solving Eq. (14.49) results in a solution

$$e_0 = \frac{m_1}{G}[1 - e^{-t/\tau}], \tag{14.50}$$

where

$$\tau = 1/GK. \tag{14.51}$$

The output voltage reaches the desired voltage exponentially. If the time constant τ is small, the output voltage follows a variation that is proportional to the modulating signal.

Now let us assume that the input frequency is given by

$$f = f_c + m_1 \sin \omega_m t.$$

The differential equation for e_0 becomes

$$\frac{de_0}{dt} + KGe_0 = Km_1 \sin \omega_m t. \tag{14.52}$$

Solution of this differential equation gives

$$e_0 = \frac{K\tau m_1}{1 + \omega_m^2\tau^2} \sin \omega_m t + \frac{K\tau^2\omega_m m_1}{1 + \omega_m^2\tau^2}(e^{-t/\tau} - \cos \omega_m t). \qquad (14.53)$$

If we require the time constant τ to be much smaller than the period of the modulating signal so that $\omega_m^2\tau^2 \ll 1$, Eq. (14.53) reduces to

$$e_0 = K\tau m_1 \sin \omega_m t = \frac{m_1 \sin \omega_m t}{G}. \qquad (14.54)$$

The output voltage follows the sinusoidal modulating signal.

The PLL is important in applications that require the recovery of a noisy signal. More complex PLLs can be designed to perform even better than the simple PLL described.

F. Pulse Modulation

Pulse modulation is a very important area due primarily to the data transmission field. It has also become prominent in telephony and time-division multiplexing and is now used to carry more conversations over a single transmission line than the single-sideband, frequency-division multiplexing method. There are many types of pulse modulation including pulse-amplitude modulation (PAM), pulse-width modulation (PWM), pulse-position modulation (PPM), and pulse-code modulation (PCM). Examples of the first three types of modulation are shown in Fig. 14.31.

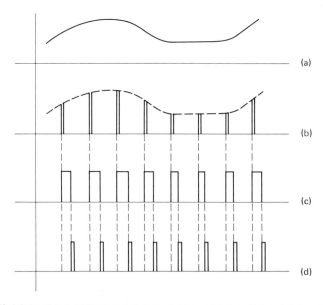

Fig. 14.31 (a) Modulating signal. (b) Pulse-amplitude modulation. (c) Pulse-width modulation. (d) Pulse-position modulation.

In PAM the period and duty cycle remain constant as the amplitude of the pulses follows the envelope of modulating voltage. The amplitude and period remain constant in PWM while the width of the signal varies in accordance with the modulation envelope. In PPM both amplitude and pulse width are held constant while the position of the pulse within the period carries the envelope information. It is possible to send information pertaining to several different channels by reserving specific time slots in each period for each channel using PAM. This process is called time-division multiplexing and can be very efficient in transmitting several channels of information simultaneously over a transmission line.

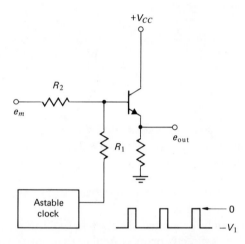

Fig. 14.32 A simple PAM signal generator.

A simple method of generating a PAM signal is shown in Fig. 14.32. The astable clock determines the period and the pulse width of the signal. When the clock output equals $-V_1$ the emitter follower must be reverse-biased, resulting in a zero-volt output signal. When the clock signal swings to the zero-volt level, the voltage at the output is

$$e_{out} = e_m \frac{R_1}{R_1 + R_2} \tag{14.55}$$

neglecting the emitter-base diode drop. During the short half-period of the clock, the output signal is proportional to e_m while the output is zero during the long half-period of the clock. One problem with this circuit is the slight error introduced by the emitter-base diode drop. This problem is eliminated in the circuit of Fig. 14.33. When both sampling control signals are at zero volts, the output signal also is at the zero-volt level, regardless of the level of modulating voltage. During the short period of time that the sampling signals depart from the zero-volt level, the output signal equals the modulating voltage. All diodes will be forward-biased with a voltage of $e_m + V_d$ appearing at point a and a voltage of $e_m - V_d$ appearing at

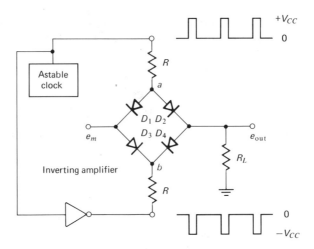

Fig. 14.33 Sampling circuit used for PAM signal generator.

point b. The voltage V_d is the forward-biased diode drop. A voltage drop of V_d appears across D_2 and a rise of V_d appears across D_4; thus the output voltage equals e_m during this time period.

In both PAM signal generating schemes the modulating voltage must always be greater than or equal to zero. It may be necessary to add a bias voltage to ensure that this is the case.

The PWM signal can be generated as shown in Fig. 14.34. The one-shot circuit must have a period that is directly proportional to the control voltage. This circuit is based on the scheme indicated in Figure 11.21. The period is constant, determined by the astable clock, while the pulse width is directly proportional to the modulating signal. Again the modulating voltage must always be positive.

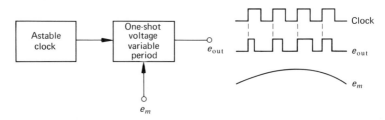

Fig. 14.34 A PWM signal generator.

A constant period one-shot, triggered on the trailing edge of the PWM output signal, converts the circuit to a PPM signal generator. The period of this one-shot plus the maximum period of the voltage-variable one-shot must be less than the total period of the astable clock to prevent overlaps of the output pulse into the next time slot.

In transmitting signals with amplitude proportional to a modulating signal a certain amount of uncertainty is introduced by noise. One can never be certain that a given pulse height is h or $h + \Delta h$, where Δh is a measure of the noise of the channel. Because of this uncertainty it is often inefficient to go to great lengths in accurately reproducing the modulating signal. Often the signal is quantized so that if the modulating signal falls anywhere within a specified range, a single, fixed-amplitude signal will be transmitted. It is found that a voice signal can be transmitted intelligibly for as few as eight levels of transmitted signal. Rather than transmit an amplitude-modulated signal, a binary code representing the amplitude of the modulating signal can be transmitted. This method of information transmission is called pulse-code modulation or PCM. In PCM systems the signal is sampled at fixed intervals with a binary code generated at each sampling point as shown in Fig. 14.35. This code modulation scheme reduces the error rate considerably because constant-amplitude logic-level signals are transmitted for all values of modulating signal level.

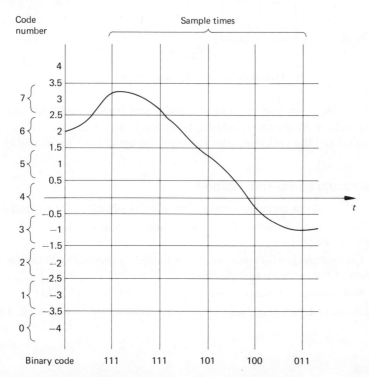

Fig. 14.35 Conversion of modulating signal to PCM signal.

While much could be written on the subject of PCM signal generation and detection, the basic elements of a PCM communication system are shown in Fig. 14.36. A sampled signal is converted to a digital signal by the A/D converter.

640 MODULATORS, RECEIVERS, AND COMMUNICATION SYSTEMS

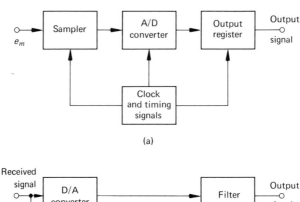

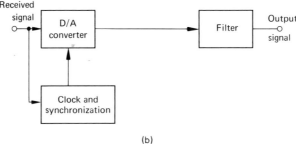

Fig. 14.36 A PCM system: (a) transmitter; (b) receiver.

This digital code is set into the output register and transmitted under control of the timing section. At the receiving end, synchronization is developed from the incoming signal and the D/A converter converts the signal back to an analog signal.

REFERENCES AND SUGGESTED READING

1. R. Brown, *Telecommunications*. Garden City, N.Y.: Doubleday, 1970, Chapter 3.
2. D. J. Comer, "Large-deviation phase and frequency modulators." *Electronic Engineering* **39,** 495, August 1967.
3. M. E. Cookson and E. M. Thompson, "Multiplexing," *Communication System Engineering Handbook*. London: McGraw-Hill, 1967.
4. V. O. Stokes, *Radio transmitters*. London: Van Nostrand-Reinhold, 1970, Chapter 2.
5. H. Taub and D. L. Schilling, *Principles of Communication Systems*. New York: McGraw-Hill, 1971.

PROBLEMS

Section 14.2A

14.1 A signal varying from 100 Hz to 5 kHz is to be transmitted by means of an AM system with a 600-kHz carrier frequency. What frequency range is required to transmit the AM signal?

* **14.2** A square wave, varying from -1 V to $+1$ V, is transmitted by an AM system. The

square wave can be approximated by the first three terms of the Fourier series:

$$f(t) = (4/\pi)[\cos \omega t - (1/3) \cos 3\omega t + (1/5) \cos 5\omega t].$$

The period of the square wave is 0.5 ms. If the carrier frequency is 600 kHz with a 10 V amplitude, what frequency components will be present in the AM waveform? Assuming that 100-percent modulation is to be used, calculate the amplitudes of all components of the transmitted signal.

14.3 Repeat Problem 14.2 for the transmission of a sawtooth waveform that is represented by

$$f(t) = (2/\pi)[\sin \omega t - (1/2)\sin 2\omega t + (1/3)\sin 3\omega t - (1/4)\sin 4\omega t].$$

The period of the sawtooth waveform is 1 ms and the waveform varies from -1 V to $+1$ V.

14.4 Design a transistor modulator such as that shown in Fig. 14.6 to give 15-percent modulation. The unmodulated carrier output should have a 0.5-V amplitude. Assume $\beta = 100$, $V_{CC} = 18$ V, $f_c = 600$ kHz, and $f_m = 5$ kHz.

* **14.5** A class-C, base-modulation circuit is to be used to produce an AM waveform. The carrier frequency is 1 MHz and the highest modulation frequency is 5 kHz. A 10-μH coil with $Q_0 = 120$ is used in the resonant circuit. Calculate the required value of C_1 and the shunt resistance that must be added to the tank circuit.

14.6 If the carrier amplitude applied to the circuit of Problem 14.5 is 10 V and $V_0 = 2.5$ V what is the magnitude of modulating signal required for 80-percent modulation?

* **14.7** A carrier signal of 5-V amplitude and 0.71-MHz frequency is applied to the circuit shown in the figure. The modulating frequency is 5 kHz. Assuming that the output resistance of the transistor is 100 kΩ and that $L = 40$ μH with $Q_0 = 120$, calculate all other element values required. Specify the amplitude of e_m to give 100-percent modulation if $V_0 = 2$ V and $\beta_0 = 80$.

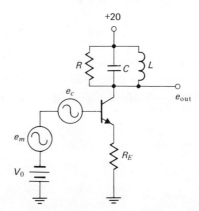

Problem 14.7

14.8 Design a collector-modulated stage using the same coil as in Problem 14.7. The modulating voltage now appears in series with the power supply and 80-percent modulation is desired. The carrier amplitude is now 12 V while $V_0 = 2$ V.

* **14.9** Evaluate the carrier power, the upper sideband power, and the lower sideband power in Problems 14.7 and 14.8.

14.10 Evaluate the power supplied by the modulator in Problems 14.7 and 14.8.

Section 14.2B

14.11 Select values for the diode detector of Fig. 14.19(a) if $R_i = 10$ kΩ. The maximum modulation frequency is 5 kHz and the carrier is 800 kHz. The index of modulation will never exceed 0.8.

*__14.12__ If the signal of Problem 14.11 is to be detected by a superheterodyne receiver with 455-kHz intermediate frequency, what must be the frequency of the local oscillator? What must be the Q of the IF stages if the 3-db points are to correspond to the maximum and minimum frequencies present?

Section 14.2C

14.13 Design a phase modulator using the circuit of Fig. 14.22. The carrier signal is $0.1 \cos \omega_c t$, where $f_c = 100$ kHz, and the modulating signal is $e_m = 3 \sin \omega_m t$, where $f_m = 200$ Hz. Assume that $r_v = 90$ kΩ with 1 V appearing across the VVR. The quiescent phase shift should be 70°. Calculate m_p for your circuit.

Section 14.2D

14.14 Design a frequency modulator using the tuned-collector modulator of Fig. 14.23. The carrier frequency is to be 1 MHz and a 5-percent second-harmonic distortion is allowed. What is the maximum frequency deviation that can be obtained with your circuit? State any assumptions made.

*__14.15__ Assume that a frequency modulator is used with a 1-MHz carrier and a 0.2-percent frequency deviation. If a 96-MHz carrier with a 75-kHz deviation is required for transmission, show how to mix and multiply this waveform to obtain the desired signal.

Section 14.2E

14.16 The maximum frequency of a modulating signal is 10 kHz. Select an appropriate product for GK of the phase-locked loop used to detect this signal.

Section 14.2F

14.17 Design a PAM signal generator to transmit a 3-kHz sine wave of amplitude 2 V. The pulse repetition frequency should be 100 kHz. Use the configuration of Fig. 14.33 and state all assumptions made. Show element values of all circuits.

14.18 Design a PWM to transmit the signal of Problem 14.17.

14.19 Design a PPM to transmit the signal of Problem 14.17.

15

SEMICONDUCTOR POWER SUPPLIES

The operation of active electronic circuits depends on the presence of an energy source. In general, the primary energy source for the circuit will be a dc voltage source. The importance of the power supply is emphasized by noting that all electronic circuits discussed in the first fourteen chapters of this book require a dc source of power. Amplifiers, digital circuits, oscillators, and modulators all depend on the presence of a fixed dc voltage for proper operation.

Low current and power requirements of some transistor circuits (for example, the transistor radio) have made the dry cell battery a particularly effective source of power. Modern-day capabilities of the dry cell have further extended the usefulness of this power source in the electronics field. Solid-state automotive ignition systems can derive power from the wet cell battery already present in the automobile; thus batteries are seen to be an important power source.

There are many circuit applications that cannot use the battery as a result of one or more disadvantages of this type of energy source. Among the shortcomings of the battery that may preclude its use are (1) the relatively short lifetime, (2) the relatively large volume requirement, and (3) the relatively low currents that can be continuously supplied without recharging. A given battery may not exhibit all three shortcomings simultaneously; for example, electric fork lifts use very large dc wet cell batteries that provide currents in the hundreds of amps, but require quite sizable volumes.

Probably the major source of power to electronic circuits is the dc supply that converts an ac input signal to a dc output voltage. This type of supply is popular because power to homes and industry has traditionally been transmitted and distributed in the form of an ac voltage. The ease of transforming ac voltages to higher or lower values has made this form of power transmission very popular. Large ac generators produce the voltage, and then it is stepped up to a high voltage by a transformer. The high voltage allows efficient transmission of power with fewer losses than does low voltage transmission. In the 1960s it became feasible to transmit large dc voltages even more efficiently than ac voltages. Development of high-power, solid-state control circuitry allowed the construction of a limited number of dc power transmission systems. At the present time, however, ac transmission is by far the most popular form of power transmission and will probably remain so for many years.

15.1 RECTIFIERS

A. Half-wave Rectification

Since a single frequency ac signal has no dc component, it must be distorted in some way to obtain this component. The circuit of Fig. 15.1 shows a half-wave rectifier which allows current to pass to the resistor in only one direction. The output voltage is shown in Fig. 15.1(b). The average forward diode resistance is represented by R_f, and the reverse-biased resistance is usually large enough to be represented by an open circuit. The piecewise linear diode model is shown in Fig. 15.1(c). It is obvious that the waveform possesses an average or dc value. This average value is found by integrating the voltage over one period and dividing by the period. The result is the average voltage over one period, but since the waveform repeats, the average values over all periods are equal. The dc voltage is then

$$V_{dc} = \frac{1}{2\pi} \int_0^\pi \frac{V_m R_L}{R_L + R_S + R_f} \sin(\omega t)\, d(\omega t) + \frac{1}{2\pi} \int_\pi^{2\pi} 0 \cdot d(\omega t)$$
$$= \frac{1}{\pi} \frac{V_m R_L}{R_L + R_S + R_f}. \tag{15.1}$$

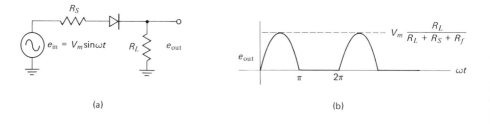

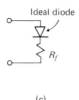

Fig. 15.1 Half-wave rectifier: (a) rectifier circuit; (b) output waveform; (c) diode model.

If R_L is much larger than $R_S + R_f$, then

$$V_{dc} = \frac{V_m}{\pi}. \tag{15.2}$$

In addition to this dc component of output voltage, there are also several ac components with frequencies harmonically related to the input frequency. Fourier's Theorem states that any periodic, single-valued function possessing only a finite number of discontinuities within the period is made up of harmonically related ac components plus a dc component equal to the average value of the waveform [3]. Furthermore, it tells us that the frequency of the lowest frequency ac component will equal the reciprocal of the period of the repetitive waveform and an infinite number of higher-order harmonics will be present. For most engineering applications, the waveform can be approximated quite accurately by the dc component plus a few of the lower frequency components. The Fourier representation of the half-wave rectified signal is

$$e_{out} = V\left[\frac{1}{\pi} + \tfrac{1}{2}\cos \omega t + \frac{2}{3\pi}\cos 2\omega t \right.$$
$$\left. - \frac{2}{15\pi}\cos 4\omega t + \frac{2}{35\pi}\cos 6\omega t - \cdots\right], \tag{15.3}$$

where V is the peak voltage of the waveform. It is characteristic of waveforms encountered in electrical engineering to show a decreasing coefficient with the higher-order harmonics. This is the characteristic that makes Fourier analysis so useful to the engineer. Rather than requiring an infinite number of terms to represent the waveform, only a few need be included in the analysis. When the waveform is applied to a linear circuit, the overall effect of the circuit can be found by applying the superposition theorem. The effect of the circuit on each component can be calculated separately and the overall effect is found by summing these separate signals.

B. Full-wave Rectification

The full-wave rectifier is more efficient and more expensive than the half-wave circuit. This circuit is shown in Fig. 15.2. A secondary voltage between points c

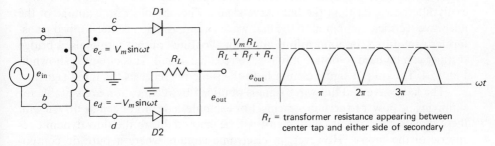

Fig. 15.2 Full-wave rectifier.

and d is induced by e_{in}. During the positive portion of the secondary waveform, point c is positive with respect to point d. During the alternate half-cycle, point d is positive with respect to c. This means D1 will be forward-biased and D2 reverse-biased when point c is more positive than d. When point d is more positive than c, D1 will be reverse-biased and D2 forward-biased. The equivalent circuits of Fig. 15.3 pertain to these cases.

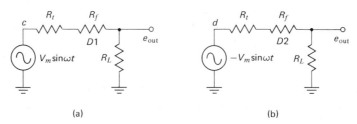

Fig. 15.3 Equivalent circuits for full-wave rectifier: (a) during first half-cycle; (b) during second half-cycle.

Current is always conducted through the load resistor in one direction only and it occurs during each half-cycle, rather than on alternate half-cycles as in the half-wave rectifier. The dc value of the waveform is obviously twice that of the half-wave case or

$$V_{dc} = \frac{2V_m R_L}{R_L + R_f + R_t} \cdot \frac{1}{\pi}. \tag{15.4}$$

If R_L is much larger than $R_t + R_f$, then

$$V_{dc} = \frac{2V_m}{\pi}. \tag{15.5}$$

The Fourier expansion of the full-wave rectified voltage is

$$e_{out} = V\left[\frac{2}{\pi} + \frac{4}{3\pi}\cos 2\omega t - \frac{4}{15\pi}\cos 4\omega t + \frac{4}{35\pi}\cos 6\omega t \cdots\right], \tag{15.6}$$

where V is the peak voltage of the waveform. Note that, since the waveform now has a period equal to π rather than 2π, the lowest frequency component has a frequency equal to twice the fundamental of the input signal. This results in a more easily filtered signal than the half-wave signal. The obvious disadvantage of the full-wave rectifier, compared to the half-wave circuit, is that a center-tapped transformer is required. In applications allowing "floating" output terminals, a bridge rectifier can be used to obtain a full-wave rectified signal. This circuit is shown in Fig. 15.4. The output signal cannot have the same reference as the input signal.

The half-wave and full-wave signals obtained by rectification are suitable for supplying dc power to certain systems. For example, a dc motor can be driven with these signals, since the ac components will be ignored due to the slow dynamic response of the motor. However, in electronic circuits where a pure dc compo-

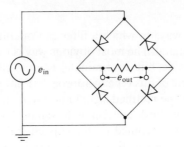

Fig. 15.4 Bridge rectifier.

nent is required, additional steps are taken to rid the rectified signals of their ac components.

C. Voltage Doublers

It is possible using capacitive circuits to develop a peak output voltage that equals twice the peak value of the input signal. Figure 15.5 shows two schemes for voltage doubling.

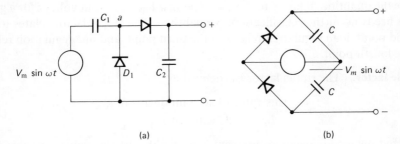

Fig. 15.5 Voltage doublers: (a) clamping circuit; (b) additive circuit.

The first circuit has the advantage of a single reference for both input and output signals. Diode D_1 will not allow the voltage at point a to become negative; thus, capacitor C_1 will charge to the peak value of the input waveform on the negative portion of the input signal. As the input signal now increases, diode D_1 becomes reverse-biased and diode D_2 will allow C_2 to be charged positively by the input source and C_1. At the positive peak of the input signal, point a can reach a value of $2V_m$ and eventually charge C_2 to this voltage. If current is drawn from the circuit, the output voltage will drop. This effect relates to the regulation of the circuit, which will be considered shortly.

The second circuit of Fig. 15.5 simply charges one capacitor to the positive peak value while the other capacitor is charged to the negative peak value. A voltage of $2V_m$ appears across the two capacitors, but cannot share the reference line with the input signal.

15.2 FILTERS

There are basically two ways in which a filter can operate to reduce the ac components of the rectified signal. The most obvious way is to filter the rectified signal with a circuit that passes dc, but discriminates against all ac components. The *L*-section filter is an example of this type of approach. For appreciable values of load current the filter accepts the rectified signal and attenuates the ac components to a large extent while passing the dc content of the signal. A second mode of operation of the power supply filter is typified by the capacitor filter. In this type of filter, the presence of the filter modifies the rectifier output such that the dc content is increased while the ac components are decreased. Further filtering reduces the ac components still further. We will consider these approaches in more detail in succeeding sections.

Once the output voltage waveform of the filter is found, there are four quantities of direct interest to aid in the specification of power supply performance. These are (1) the dc voltage contained in the output signal for a specified ac input signal, (2) the ac content of the output signal relative to the dc content, (3) the variation of dc output voltage as the load current varies, and (4) the dc output voltage variation with change in ac input signal.

In order to make meaningful comparisons between different supplies, we must define the quantities relating to the above points. The dc output voltage for a given input needs no further clarification. A quantity called ripple factor relates to the second point, load regulation relates to the third point, and line regulation relates to the fourth point.

Ripple factor. The ripple factor r is defined as

$$r = \frac{\text{rms value of ac components of } e_{\text{out}}}{\text{dc value of } e_{\text{out}}}. \tag{15.7}$$

If all but one ac frequency component is negligible, the rms value is easily found as the amplitude of this component divided by $\sqrt{2}$. In many power supplies there will be several ac components contained in the output signal. In this case we can find the ac portion of the waveform by subtracting the dc signal from the output voltage, that is,

$$e_{\text{out(ac)}} = e_{\text{out}} - V_{\text{dc}}.$$

The usual case will find known values for e_{out} and V_{dc}, allowing $e_{\text{out(ac)}}$ to be calculated. The rms value of $e_{\text{out(ac)}}$ can then be found from

$$\begin{aligned}
\text{rms of } e_{\text{out(ac)}} &= \sqrt{(1/2\pi) \int_0^{2\pi} [e_{\text{out(ac)}}]^2 \, d(\omega t)} \\
&= \sqrt{(1/2\pi) \int_0^{2\pi} [e_{\text{out}} - V_{\text{dc}}]^2 \, d(\omega t)} \\
&= \sqrt{(1/2\pi) \left[\int_0^{2\pi} e_{\text{out}}^2 \, d(\omega t) - 2 \int_0^{2\pi} e_{\text{out}} V_{\text{dc}} \, d(\omega t) + \int_0^{2\pi} V_{\text{dc}}^2 \, d(\omega t) \right]}.
\end{aligned}$$

The second integral divided by 2π reduces to

$$\frac{V_{dc}}{2\pi} \int_0^{2\pi} e_{out}\, d(\omega t) = V_{dc}^2$$

since $(1/2\pi) \int_0^{2\pi} e_{out}\, d(\omega t)$ is the average value of the waveform. Combining the second and third integrals gives a final result of

$$\text{rms of } e_{out(ac)} = \sqrt{(\text{rms of } e_{out})^2 - V_{dc}^2}. \tag{15.8}$$

Example 15.1. Calculate the ripple factor of an unfiltered full-wave rectified signal with amplitude V_m.

Solution. The rms value of the full-wave rectified signal is

$$\text{rms of } e_{out} = \frac{V_m}{\sqrt{2}}.$$

The average or dc value of e_{out} has been found previously to be

$$V_{dc} = \frac{2V_m}{\pi}.$$

Using these values in Eq. (15.8) gives

$$\text{rms of } e_{out(ac)} = \sqrt{\frac{V_m^2}{2} - \frac{4V_m^2}{\pi^2}} = 0.307 V_m.$$

Equation (15.7) gives a ripple factor of

$$r = \frac{0.307 V_m}{2V_m/\pi} = 0.48.$$

This ripple factor is very large compared to the values that apply to the output signal of filtered power supplies.

Load regulation. Regulation is usually expressed in percent and is defined by

$$\text{Reg} = \frac{V_{dc(no\ load)} - V_{dc(full\ load)}}{V_{dc(no\ load)}} \times 100. \tag{15.9a}$$

An alternate definition in terms of load current is

$$\text{Reg} = \frac{V_{dc}(I_{dc} = 0) - V_{dc}(I_{dc} = I_{rated})}{V_{dc}(I_{dc} = 0)} \times 100, \tag{15.9b}$$

where I_{dc} is the current supplied to the load. In general, as load current is increased in unregulated supplies the output voltage decreases. If this effect is pronounced the operation of the circuit to which the supply furnishes power can be adversely affected.

Line regulation. The line regulation can be defined in several ways. Often it will be specified as a percentage variation in output voltage for a given variation in input voltage. For example, the line regulation may be specified to be $+0.2$ percent of the nominal dc output voltage for a variation of $+10$ percent of the nominal ac input voltage.

A. Capacitor Filter

The circuit of Fig. 15.6 shows the capacitor filter. This circuit distorts the rectifier output such that the dc content is increased over the value of that of a simple half-wave rectified signal. When no load is presented to the capacitor, this element charges to V_m on the first cycle of the waveform. The capacitor voltage remains indefinitely at V_m so long as no load current is drawn. Thus the input sinusoid is converted into a pure dc output signal. To be useful, however, load current must be drawn from the supply resulting in a discharge of the capacitor. The power supply will be designed to deliver some maximum current at a specified voltage. The value of the load resistance can be chosen to represent these specifications if

$$R_L = \frac{V_{dc}}{I_{dc(max)}}.$$

The capacitor discharges between successive peaks of the input voltage and charges as the input signal reaches its maximum value. The waveform shown in Fig. 15.6(c) indicates that the steady-state output voltage consists of a large dc component plus a small repetitive component with period equal to that of the input voltage.

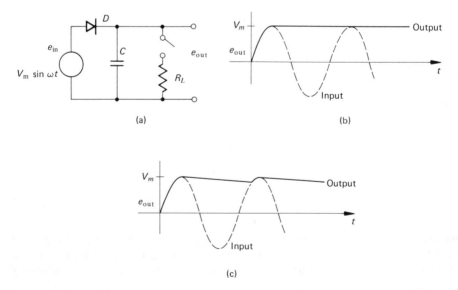

Fig. 15.6 The capacitor filter: (a) actual circuit; (b) output waveform for no load case; (c) output waveform for loaded condition.

This means that in addition to the dc component of the waveform, ac components that are harmonically related to the input frequency must be present.

The problem of analyzing the output voltage waveform of Fig. 15.6(c) could be very difficult. We would have to find the exact points at which the capacitor starts discharging and charging, and express the voltage between these points as an exponentially decaying function of time. In practice, it is found that a simple approximation allows an analysis which gives very accurate results. We note that for a capacitor filter to be practical, the discharge between successive peaks must be quite small compared to V_m. This means the $R_L C$ time constant governing the decay must be very large compared to T, the period of the input signal. If this is the case, the decay will be linear, since the exponential function is linear over a small portion of the time constant. Furthermore, the uncertainty in the position of the starting points of charge and discharge is small compared to the total period T. These considerations allow us to represent the steady-state output waveform as a triangular voltage, as shown in Fig. 15.7. The three quantities of interest, that is V_{dc}, r, and Reg can be evaluated from this figure.

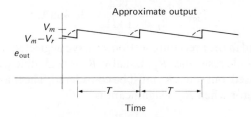

Fig. 15.7 Approximate output voltage of capacitor filter.

Output voltage. We see that V_{dc} is obviously equal to $V_m - \frac{1}{2} V_r$, where V_r is the amplitude of the triangular portion of the output waveform. We can evaluate V_r by considering the discharge rate of the capacitor. If the dc current drain on the capacitor is constant at I_{dc}, the loss of voltage can be found from

$$I_{dc} = \frac{\Delta Q}{T} = C \frac{\Delta V}{T}.$$

Solving for V_r gives

$$V_r = \Delta V = \frac{I_{dc} T}{C} = \frac{I_{dc}}{fC},$$

where f is the frequency of the input signal. This value can be written in terms of V_{dc}, since $I_{dc} = \dfrac{V_{dc}}{R_L}$, giving

$$V_r = \frac{V_{dc}}{R_L f C}. \tag{15.10}$$

The output voltage is then written as

$$V_{dc} = V_m - \frac{V_{dc}}{2R_L f C},$$

which is solved to give

$$V_{dc} = \frac{V_m}{1 + 1/2R_L f C}. \tag{15.11}$$

When R_L or C approaches infinity, V_{dc} approaches V_m, as noted previously.

Example 15.2. For the half-wave rectifying capacitor filter of Fig. 15.6(a), $R_L = 1\ \mathrm{k\Omega}$, $C = 200\ \mu\mathrm{F}$, and $e_{in} = 40 \sin(2\pi \times 60)t$. Find V_{dc} and I_{dc}.

Solution. Using Eq. (15.11),

$$V_{dc} = \frac{40}{1 + 1/(2 \times 10^3 \times 2 \times 10^{-4} \times 60)} = 38.4\ \mathrm{V}.$$

The dc current through R_L is

$$I_{dc} = \frac{V_{dc}}{R_L} = \frac{38.4}{1\ \mathrm{k\Omega}} = 38.4\ \mathrm{mA}.$$

In this example and in the preceding section we have neglected the source resistance and the forward diode resistance R_f. Usually, R_f is negligible when semiconductor diodes are being used, but often R_S must be considered. The maximum voltage that reaches the capacitor when R_S is significant is

$$V'_m = \frac{R_L}{R_L + R_S} V_m.$$

The dc voltage can then be written as

$$V_{dc} = \frac{V'_m}{1 + 1/2R_L f C} = \frac{V_m}{1 + 1/2R_L f C} \frac{R_L}{R_L + R_S}. \tag{15.12}$$

Equation (15.12) can be interpreted physically in the following way: The dc output voltage equals V_m when no load is present. When R_L is placed across the capacitor, it has two effects. One is to discharge the capacitor between cycles, and this effect is accounted for by the $1 + 1/2R_L f C$ term. The second effect is the drop occurring across the source resistance when load current is drawn. The term $R_L/(R_L + R_S)$ accounts for this voltage drop. Of course R_S must be much smaller than R_L for the above analysis to apply.

Ripple factor. The ripple factor can be found from Eq. (15.8). The term $(e_{out} - V_{dc})$ is as shown in Fig. 15.8. The waveform for the first period can be written as $V_r/2 - V_r t/T$.

The ripple factor is then

$$r = \frac{\sqrt{(1/T)\int_0^T (V_r/2 - V_r t/T)^2\, dt}}{V_{dc}} = \frac{V_r/\sqrt{12}}{V_{dc}} = \frac{1}{\sqrt{12R_L f C}}. \tag{15.13}$$

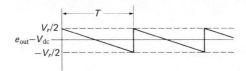

Fig. 15.8 The sum of the ac components of the waveform of Fig. 15.7.

For Example 15.2, the ripple factor is calculated to be

$$r = \frac{1}{\sqrt{12} \times 10^3 \times 60 \times 2 \times 10^{-4}} = 0.024.$$

Regulation. The load regulation can be found by noting that when $R_L = \infty$, $V_{dc} = V_m$. Equation (15.12) can be used to calculate V_{dc} at rated current using

$$R_L = \frac{V_{dc}}{I_{dc}}.$$

At rated current

$$V_{dc(\text{full load})} = \frac{V_m}{1 + 1/2R_L fC}.$$

The load regulation is found to be

$$\text{Reg} = \frac{V_m - \frac{V_m}{1 + 1/2R_L fC}}{V_m} \times 100 = \frac{1}{1 + 2R_L fC} \times 100. \qquad (15.14)$$

For Example 15.2 where R_S is assumed negligible the regulation is

$$\text{Reg} = \frac{1}{1 + 2 \times 10^3 \times 60 \times 2 \times 10^{-4}} \times 100 = 4\%.$$

The voltage changes from 40 V at no load to $40 - 0.04 \times 40 = 38.4$ V at full load.

Practical considerations. We must now consider the specifications of the diode in terms of the current and voltage. When the input voltage swings negative to a value of $-V_m$, the capacitor voltage will remain equal to $+V_m$ for the case of no output current. A total voltage of $2V_m$ appears across the diode; therefore the minimum breakdown voltage or peak inverse voltage rating of the diode should equal $2V_m$; that is,

$$PIV = 2V_m \qquad \text{(capacitor filter)}. \qquad (15.15)$$

The maximum forward current of the diode will occur when the power supply is initially turned on and the capacitor is uncharged. Since the circuit appears as shown in Fig. 15.9 as the input initially swings positive, the maximum current can be found by assuming C is a short circuit. The current can then be written as

$$i = \frac{V_m \sin \omega t}{R_f + R_S}.$$

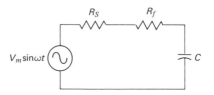

Fig. 15.9 Equivalent circuit on first positive quarter-cycle.

The maximum magnitude of this value is

$$i_{max} = \frac{V_m}{R_f + R_S}. \tag{15.16}$$

The surge-current rating of the semiconductor diode should equal or exceed this value of i_{max}. The continuous-current rating of the diode should be equal to or greater than the maximum dc current that is supplied to the load. The surge-current rating of the diode is usually 5 to 20 times greater than the continuous-current rating.

When a full-wave rectifier is used in conjunction with the capacitor filter, the period of the triangular output is

$$T = \frac{1}{2f}.$$

The equations derived previously are modified by changing f to $2f$, resulting in the following.

Output voltage:

$$V_{dc} = \frac{V_m}{1 + 1/4R_L f C} \frac{R_L}{R_L + R_S}. \tag{15.17}$$

Ripple factor:

$$r = \frac{1}{2\sqrt{12} R_L f C}. \tag{15.18}$$

The small size of semiconductor diodes allows several to be placed in series to increase the *PIV* rating. Parallel diodes can be used to increase the current capabilities. The following data* demonstrate that semiconductor diodes cover a wide range of current and voltage requirements.

	Continuous forward current	Surge current	PIV
MR 1297	1000 A	18,000 A	400 V
IN 4725	3 A	25 A	1200 V
IN 3211	15 A	250 A	400 V

---

* Motorola Semiconductor Data Manual.

The dc output voltage of the full-wave rectifier with capacitor filter can approach the peak value of input voltage. However, as current is drawn by the load, the dc component of voltage contained by the input waveform decreases. If the diodes were ideal, the output waveform would approach that of the full-wave rectifier without filtering, as load current becomes excessive. The capacitor charges to follow the input signal on the positive-going portion of the cycle. Due to the high current drain, the capacitor would also discharge to follow approximately the decreasing portion of the input. Of course, the theory developed in the previous section no longer holds because the discharge of the capacitor is no longer linear. The dc voltage now is

$$V_{dc} = \frac{2V_m}{\pi}. \tag{15.19}$$

As the current is further increased, V_{dc} no longer decreases for zero source and diode resistance. The voltage regulation is now quite good, but one must remember that the ripple factor is very poor due to the presence of the large ac components. Hence we see that when the current is increased to the point where each diode conducts for exactly one half-cycle, the regulation becomes very good and the ripple factor very high.

B. The *L*-Section Filter

The circuit of Fig. 15.10 shows an *L*-section filter. This circuit is designed to take advantage of the fact that when each diode conducts for a full half-cycle, the regulation is improved. The inductor reduces the ac voltage that reaches the load, resulting in a reasonable ripple factor. When there is no load current being drawn, the capacitor charges to the peak voltage of the filter input V_m, just as in the case of the capacitor filter. The diodes would not conduct at all once the capacitor is fully charged. As the load current increases, the diodes begin to conduct over a portion of each half-cycle. The circuit behaves much like the capacitor filter, in terms of regulation, until enough load current is drawn to cause the diodes to conduct for the full half-cycle. The minimum dc load current required to cause half-cycle conduction by the diodes is called the critical value of current I_{cr}. This value can be found by writing the inductor current i_L in terms of the applied voltage and solving for the load current required to guarantee that i_L never becomes

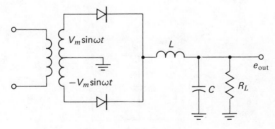

Fig. 15.10 *L*-section filter.

negative. If each diode conducts for one-half cycle, the input voltage to the filter is as given by Eq. (15.6)

$$e_{in} = V_m \left[\frac{2}{\pi} + \frac{4}{3\pi} \cos 2\omega t - \frac{4}{15\pi} \cos 4\omega t + \frac{4}{35\pi} \cos 6\omega t - \cdots \right].$$

The circuit of Fig. 15.11 applies to this case. The capacitor will offer a negligible impedance to the ac components of e_{in}, while L offers no impedance to the dc component. The inductor current can then be expressed as

$$i_L = I_{dc} + \frac{(4V_m/3\pi) \cos 2\omega t}{2j\omega L} - \frac{(4V_m/15\pi) \cos 4\omega t}{4j\omega L} + \frac{(4V_m/35\pi) \cos 6\omega t}{6j\omega L} - \cdots$$

$$= I_{dc} - j \frac{2V_m}{3\pi\omega L} \cos 2\omega t + j \frac{V_m}{15\pi\omega L} \cos 4\omega t - j \frac{2}{105\pi\omega L} V_m \cos 6\omega t - \cdots$$

The last term is negligible, compared to the first ac component in the above equations; thus for reasonable accuracy the inductor current can be expressed as

$$i_L = I_{dc} - j \frac{2V_m}{3\pi\omega L} \cos 2\omega t + j \frac{V_m}{15\pi\omega L} \cos 4\omega t. \tag{15.20}$$

In order for i_L to be greater than zero, the following inequality must hold at all times:

$$I_{dc} > -\left(\frac{2V_m}{3\pi\omega L} \cos 2\omega t - \frac{V_m}{15\pi\omega L} \cos 4\omega t \right) \tag{15.21}$$

The right-hand side of the above inequality is shown in Fig. 15.12. The maximum negative value is reached at $\omega t = \pi/2$. At this point the quantity becomes

$$-\left[\frac{2V_m}{3\pi\omega L} + \frac{V_m}{15\pi\omega L} \right] = -\frac{11}{15} \frac{V_m}{\pi\omega L}.$$

The critical value of dc current to the load is found from inequality (15.21) to be

$$I_{cr} = \frac{11}{15} \frac{V_m}{\pi\omega L}. \tag{15.22}$$

When I_{dc} exceeds I_{cr}, the input signal to the filter is a perfect full-wave signal. The dc component of this signal is passed to the capacitor, meaning that the output dc

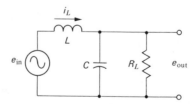

Fig. 15.11 Equivalent circuit for L-section filter.

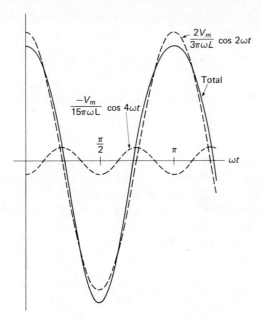

Fig. 15.12 Plot of $(2V_m/3\pi\omega L)\cos 2\omega t - (V_m/5\pi\omega L)\cos 4\omega t$.

voltage is

$$V_{dc} = \frac{2V_m}{\pi}.$$

Assuming no transformer resistance and no diode resistance, we find that the variation of output dc voltage with output current will be as shown in Fig. 15.13. Below values of I_{cr}, the regulation is similar to that of the capacitor filter. Above this value, the total dc voltage contained in the full-wave rectified voltage is passed to the output. If the drops across the transformer, diode, and inductor are considered, the output voltage will be

$$V_{dc} = \frac{2V_m}{\pi} - (R_t + R_f + R_l)I_{dc}, \tag{15.23}$$

where R_l is the dc resistance of the choke.

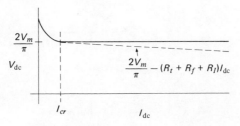

Fig. 15.13 Regulation curve for L-section filter.

In Eq. (15.22) we solved for the critical value of current in terms of the inductance. We could express I_{cr} in terms of a critical load resistance also. That is,

$$I_{cr} = \frac{V_{dc}}{R_{cr}} = \frac{2V_m}{\pi R_{cr}}. \tag{15.24}$$

The critical value of resistance is found by solving Eqs. (15.22) and (15.24) to give

$$R_{cr} = \tfrac{30}{11}\omega L. \tag{15.25}$$

In practice, the current required by the load is often variable. In order to ensure that I_{dc} never becomes less than I_{cr}, a resistor of value R_{cr} is shunted across the output terminals of the supply. The load current can then vary from zero to the maximum value, and the output voltage variation is very low.

The ripple factor can be found by assuming that the load offers a high impedance to the ac components of the waveform compared to that offered by the capacitor. The ac circuit then reduces to that shown in Fig. 15.14 (neglecting resistive drops). The magnitude of the output ac voltage is

$$e_{ac} = e_{in} \left| \frac{1/j\omega C}{j\omega L + 1/j\omega C} \right| \approx \frac{e_{in}}{\omega^2 LC}. \tag{15.26}$$

Equation (15.26) shows that the ac components of the input voltage are greatly attenuated, depending on the frequency of the component. Using Eq. (15.6), we can find the attenuation of the first two components:

1. $|e_{in}(2\omega)| = \dfrac{4V_m}{3\pi},\quad |e_{ac}(2\omega)| = \dfrac{V_m}{3\pi\omega^2 LC}.$

2. $|e_{in}(4\omega)| = \dfrac{4V_m}{15\pi},\quad |e_{ac}(4\omega)| = \dfrac{V_m}{60\pi\omega^2 LC}.$

Since the component of frequency 4ω is 20 times less than that of frequency 2ω, only the second harmonic component of the input frequency need be considered. In other words, the output signal of the L-section filter can be approximated quite accurately by the dc output voltage plus a second harmonic signal of magnitude

$$|e_{ac}(2\omega)| = \frac{V_m}{3\pi\omega^2 LC}. \tag{15.27}$$

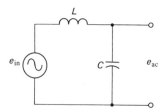

Fig. 15.14 The equivalent ac circuit of an L-section filter.

The rms-value of the ac signal can be found directly, and the ripple factor is

$$r = \frac{1}{6\sqrt{2}\omega^2 LC}. \tag{15.28}$$

The surge-current requirement of the diodes is reduced considerably over the capacitor filter, due to the presence of the inductor. The maximum surge current that can flow, assuming that C is initially uncharged, is

$$i_{max} = \frac{V_m}{\omega L}.$$

Quite often this value is less than the continuous dc current that is drawn by the load.

Example 15.3. Design an L-section filter giving a V_{dc} of 28 V at $I_{dc} = 100$ mA. The ripple factor should be less than 0.001 and the value of inductance available is 4 H. The input signal is 110 V (rms) at 60 Hz. Assume that $R_f = 10\,\Omega$ and $R_t = 20\,\Omega$, and that the resistance of the inductor is 20 Ω. Calculate I_{cr} and the output voltage at I_{cr}. If the critical value of resistance is used as a bleeder, what load current will flow when $V_{dc} = 28$ V?

Solution. The circuit to be used is shown in Fig. 15.10. The capacitor is calculated from Eq. (15.28) to be

$$C = \frac{1}{6\sqrt{2}\omega^2 Lr} = \frac{1}{6 \times \sqrt{2} \times (2\pi \times 60)^2 \times 4 \times 0.001} = 208\ \mu\text{F}.$$

A 220-μF capacitor will be selected to guarantee that the ripple factor does not exceed 0.001.

The maximum value of voltage that must be applied to the filter is found from Eq. (15.23). Since

$$28\text{ V} = \frac{2V_m}{\pi} - (20 + 10 + 20) \times 0.1\text{ A},$$

then

$$V_m = (28 + 5)\frac{\pi}{2} = 51.8\text{ V}.$$

The turns ratio of the transformer from primary to secondary (total windings) is

$$n = \frac{110 \times \sqrt{2}}{103.6} = 1.5.$$

It should be noted that the resistance of the transformer windings has already been assumed in previous calculations. This value could now be checked making any necessary modifications in calculating V_m.

The critical current is

$$I_{cr} = \frac{11}{15} \times \frac{51.8}{\pi \times 2\pi \times 60 \times 4} = 8.02 \text{ mA}.$$

The output voltage at I_{cr} is

$$V_{dc}(I_{cr}) = \frac{2V_m}{\pi} - 50 \times 0.008 = 32.6 \text{ V}.$$

If a resistor is shunted across the terminals to draw I_{cr}, the value would be

$$R_{cr} = \tfrac{30}{11} \times 2\pi \times 60 \times 4 = 4.1 \text{ k}\Omega.$$

If R_{cr} is used, this resistance will draw 6.8 mA at $V_{dc} = 28$ V. The load current at $V_{dc} = 28$ V is $100 - 6.8 = 93.2$ mA.

The regulation is calculated to be (assuming a full load current of 93.2 mA when R_{cr} is present)

$$\text{Reg} = \frac{32.6 - 28.0}{32.6} \times 100 = 14.1\%.$$

C. Multiple L-Section Filters

The analysis of multiple L-section filters is simply an extension of the single-section filter theory. The output impedance of the L-section is very low for both dc and ac signals. The input impedance to the L-section is very high for both dc and ac signals; hence succeeding sections will not load the previous stages. The output voltage of the first stage consists of a dc component of magnitude

$$V_{dc} = \frac{2V_m}{\pi} - I_{dc}(R_t + R_f + R_{l1})$$

and an ac component of frequency 2ω, with an rms value of

$$e_{ac} = \frac{V_m}{3\sqrt{2\pi\omega^2 L_1 C_1}}.$$

The output voltage and ripple factor can easily be found for the two-section filter of Fig. 15.15 to be

$$V_{dc} = \frac{2V_m}{\pi} - I_{dc}(R_t + R_f + R_{l1} + R_{l2}) \tag{15.29}$$

and

$$r = \frac{1}{24\sqrt{2\omega^4 L_1 C_1 L_2 C_2}}.$$

There are several other possible configurations for power supply filters. Often these configurations amount to a combination of those configurations already discussed, and the theory is easily extended to cover these filters.

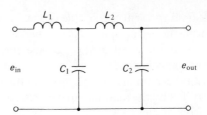

Fig. 15.15 Two L-section filters.

Throughout the previous discussions we have assumed a constant input ac voltage. This is often an invalid assumption. If the magnitude of input or line voltage changes, the dc output voltage will also change. This variation can be expressed in terms of the line regulation of the power supply. In the case of L-section filters, the dc output voltage varies directly with V_m, the magnitude of the input voltage. The resulting line regulation is quite poor. We will see that regulating circuits can be used to improve both line and load regulation.

15.3 REGULATORS

Regulators are used to improve both the load and line regulation of a filter; however, they generally lead to a decreased ripple factor also. Complexity of a regulator varies from that of the simple Zener diode to that of complex temperature-compensated, transistorized systems that measure output voltage, compare this value to an established reference, and exert control over the output signal to maintain the desired value. We will first consider Zener-diode regulating circuits before proceeding to more sophisticated regulators.

A. Zener Diode Regulator

This regulating element is one of the simplest forms of voltage regulators. The V–I characteristics of the Zener diode are shown in Fig. 15.16. The forward

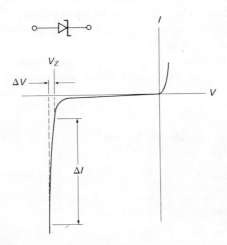

Fig. 15.16 Zener diode characteristics.

characteristics represent the simple diode. When an increasing negative voltage is applied to the diode, a point V_Z is reached when the diode "breaks down," that is, begins to conduct a great deal of reverse current. While this breakdown is often due to an avalanche effect, it was first thought to be due to Zener breakdown, and this name has been applied to the diode to the present day. The breakdown voltage V_Z is controlled by the manufacturing process and can be made to fall anywhere in the range from about 2.6 V to 200 V. At reverse voltages greater than V_Z, the curve exhibits a fairly constant slope. The reciprocal of this slope is equal to the dynamic resistance of the Zener diode. This value is then

$$r_Z = \frac{\Delta V}{\Delta I}, \tag{15.30}$$

where ΔV and ΔI are taken from that portion of the curve beyond V_Z.

The circuit of Fig. 15.17 is used to explain the basic mechanism underlying the use of the Zener diode as a regulator. If V_{in} is smaller than the breakdown voltage of the Zener diode, the diode will behave as a normal, reverse-biased diode. Very little current flows and the drop across the resistance R is negligible. The total input voltage appears across the diode. Obviously the Zener diode will not regulate in this mode since the output voltage varies directly with input voltage. As V_{in} increases to higher values than V_Z, the diode breaks down and conducts current, maintaining a nearly constant output voltage of V_Z. The current flowing through R can be found from Ohm's law as

$$I = \frac{V_{in} - V_Z}{R}.$$

Increasing V_{in} does not change the output voltage significantly, but the current increase will lead to a slight voltage change. From Fig. 15.16 we can see that the current increase causes a diode voltage increase that is dependent on the slope of the curve or the dynamic resistance. A zero resistance or infinite slope would indicate perfect regulation as input voltage varies. This discussion suggests that the equivalent circuit of Fig. 15.18 is applicable to the Zener diode in the regulating mode.

To demonstrate the effectiveness of the Zener diode in regulating output voltage as input varies, we can consider Example 15.4.

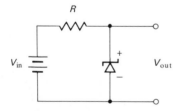

Fig. 15.17 Zener regulator.

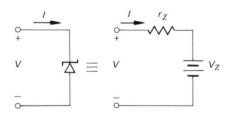

Fig. 15.18 Equivalent circuit for Zener diode.

Example 15.4. In the circuit of Fig. 15.17 the resistance is $R = 1$ kΩ, $V_Z = 10$ V at 1 mA, and $r_Z = 30$ Ω. Given that V_{in} changes from 11 V to 20 V, calculate the Zener current change and the output voltage change.

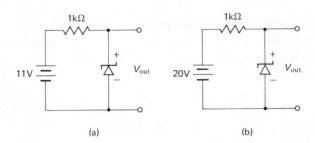

Fig. 15.19 Circuits for Example 15.4: (a) 11-V input; (b) 20-V input.

Solution. The circuits of Fig. 15.19 can be used to work this example. In the first circuit the output voltage is equal to V_Z. The current is

$$I_Z = \frac{V_{in} - V_{out}}{1 \text{ k}\Omega} = \frac{11 - 10}{1 \text{ k}\Omega} = 1 \text{ mA}.$$

When V_{in} increases to 20 V, the Zener current increases, causing the output voltage to be slightly higher than V_Z. The output voltage can now be written as

$$V_{out} = V_Z + \Delta I r_Z, \tag{15.31}$$

where ΔI is the increase in current due to the increase in V_{in}. The new current is then

$$I_Z = \frac{V_{in} - (V_Z + \Delta I r_Z)}{R} = \frac{V_{in} - (V_Z + (I_Z - 0.001)r_Z)}{R}.$$

From this equation, I_Z is found to be

$$I_Z = 9.76 \text{ mA}.$$

From a practical standpoint, the Zener current can be found by assuming that $V_{out} \approx V_Z$ for all values of input voltage. This value of current can then be used in Eq. (15.31) to find the actual output voltage. Thus

$$I_Z = \frac{V_{in} - V_Z}{R}, \tag{15.32}$$

which, in this case, becomes

$$I_Z = \frac{20 - 10}{1 \text{ k}\Omega} = 10 \text{ mA}.$$

The change in Zener current is $\Delta I = 10 - 1 = 9$ mA, giving an output voltage of

$$V_{out} = 10 + 0.009 \times 30 = 10.27 \text{ V},$$

when $V_{in} = 20$ V. The input voltage has almost doubled while the output voltage changes only

$$\frac{\Delta V}{V} \times 100 = \frac{0.27}{10} \times 100 = 2.7\%.$$

It appears then that the line regulation properties of the Zener diode are excellent. If we assume that there is an ac component of rms-value 1 V present in addition to the 20-V dc input, the ripple reduction property can also be demonstrated. The output ac component will be

$$e_{out} = e_{in} \times \frac{r_Z}{R + r_Z}$$
$$= 1 \times \frac{30}{1030} = 0.029 \text{ V}. \qquad (15.33)$$

The ripple factor of the input signal is

$$r = \tfrac{1}{20} = 0.05,$$

while the output ripple factor is

$$r = \frac{0.029}{10} = 0.0029.$$

The ripple factor has been decreased by a factor of more than 17.

We have seen that the Zener diode can greatly improve regulation and the ripple factor, but we have yet to examine the limitations on the device and the load regulation. The maximum current that can be allowed to flow through the Zener diode is limited by the allowable device dissipation. The power dissipated by the Zener is

$$P_Z = V_Z I_Z. \qquad (15.34)$$

The maximum allowable power, which is always specified by the manufacturer, limits the maximum Zener current to

$$I_{Z\,max} = \frac{P_{Z\,max}}{V_Z}. \qquad (15.35)$$

The minimum current that can flow and still cause the diode to be in the regulating region is usually taken to be

$$I_{Z\,min} = \tfrac{1}{10} I_{Z\,max}. \qquad (15.36)$$

The characteristic curve of the Zener is shown again in Fig. 15.20 with these key points noted.

To demonstrate the effect of the current limitations on circuit performance, we shall consider Fig. 15.21. If we assume that V_{in} is constant and that the current required by the load can vary from 0 to $I_{L\,max}$, we can calculate the two extremes of I_Z. Since $I_{in} = (V_{in} - V_Z)/R$, the input current will be approximately constant, independent of the value of load current. When $I_L = I_{L\,max}$, minimum current will

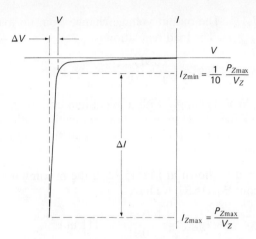

Fig. 15.20 Zener diode characteristics showing key values.

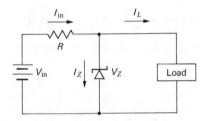

Fig. 15.21 Zener-diode regulating circuit.

flow through the Zener. This value must equal or exceed $I_{Z\,min}$, thus

$$I_{in} = I_{L\,max} + I_{Z\,min}.$$

On the other hand, when $I_L = 0$, maximum Zener current flows and

$$I_{in} = I_{Z\,max}.$$

Combining these equations to find $I_{L\,max}$ gives

$$I_{L\,max} = I_{Z\,max} - I_{Z\,min} = 0.9 I_{Z\,max}. \tag{15.37}$$

The maximum current that can be supplied to the load is then equal to nine-tenths of the maximum diode current.

We can now consider the load regulation of the Zener diode. If a power supply is followed by the resistor-diode circuit, the input current to this regulating circuit will be constant. Therefore, as the load on the regulator varies, the output current of the power supply remains constant. Voltage variations at the input of the regulator are eliminated due to this constant current draw on the power supply. There will be a load voltage variation due to the change in Zener current as the load is varied. As load current varies from zero to $I_{L\,max}$, the Zener current decreases from $I_{Z\,max}$ to $0.1 I_{Z\,max}$. The output voltage change from no load to full load is then given by $0.9 I_{Z\,max} r_Z$. The load regulation is

$$\text{Reg} = \frac{0.9 I_{Z\,max} r_Z}{V_Z + 0.9 I_{Z\,max} r_Z} \times 100. \tag{15.38}$$

Example 15.5. A 20-V dc supply with a maximum current rating of 100 mA is required. Choose a Zener diode to be used in conjunction with the L-section filter of Example 15.3 to satisfy the specifications. Calculate the resulting load regulation and the ripple factor for the circuit.

Solution. The circuit is shown in Fig. 15.22. If the maximum load current is to be 100 mA, then from Eq. (15.37) we have

$$I_{Z\,max} = \frac{100 \text{ mA}}{0.9} = 111 \text{ mA}.$$

Since the Zener voltage must be 20 V, the allowable dissipation must equal or exceed a value of

$$P_{Z\,max} \geq 20 \times 0.111 = 2.22 \text{ W}.$$

A 2.5-W, 20-V Zener is chosen to be used for this example. The minimum current that must flow is then

$$I_{Z\,min} = \frac{1}{10} \times \frac{2.5 \text{ W}}{20 \text{ V}} = 12.5 \text{ mA}.$$

The input current must then be

$$I_{in} = I_{L\,max} + I_{Z\,min} = 112.5 \text{ mA}.$$

To calculate the resistor R, the output voltage of the L-section filter when the output current is 112.5 mA must be found. Using Eq. (15.23), we have

$$V_{in} = \frac{2 \times 51.8}{\pi} - (50)0.1125 = 27.3 \text{ V}.$$

The resistance R is then

$$R = \frac{V_{in} - V_Z}{I_{in}} = \frac{27.3 - 20}{0.1125} = 65 \text{ }\Omega.$$

In selecting the Zener, the dynamic resistance would be given. A typical value for this Zener is $r_Z = 8\,\Omega$. The regulation can then be calculated by noting that when no current is drawn by the load, 112.5 mA flows through the Zener giving

$$V_{no\,load} = V_Z + (I_{in} - I_{Z\,min}) r_Z = 20 + (0.1125 - 0.0125) = 20.8 \text{ V}.$$

The full load voltage will be 20 V; since when $I_L = I_{L\,max}$, the Zener current is $I_{Z\,min}$, which corresponds to the rated Zener voltage, the regulation is then

$$\text{Reg} = \frac{20.8 - 20}{20.8} \times 100 = 3.85\%.$$

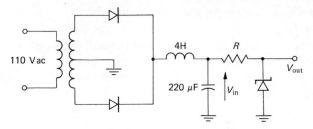

Fig. 15.22 Power supply circuit for Example 15.5.

The rms-value of the ac component at the output is found by

$$e_{out} = e_{in} \frac{r_Z}{R + r_Z},$$

where e_{in} is the ac voltage at the output of the L-section. The rms-value of e_{in} is found from Eq. (15.27) to be

$$e_{in} = \frac{V_m}{3\sqrt{2}\omega^2 LC\pi} = \frac{51.8}{3 \times \sqrt{2} \times (2\pi \times 60)^2 \times 4 \times 220 \times 10^{-6}\pi} = 0.031 \text{ V}.$$

Therefore e_{out} is

$$e_{out} = 0.031 \times \frac{8}{65 + 8} = 0.0034.$$

The ripple factor is

$$r = \frac{0.0034}{20} = 0.00017.$$

Emitter follower-Zener diode circuits. In some applications the relatively small output current range of the Zener diode regulator restricts the utility of the circuit. An emitter follower or multiple emitter followers can overcome this limitation with current amplification. Figure 15.23 shows the Zener diode circuit followed by a transistor stage. The voltage presented to the load is $V_{out} = V_Z - V_{BE(on)}$. If R_E is large compared to the effective resistance of the load, the load current will equal the emitter current of the transistor. The current leaving the Zener

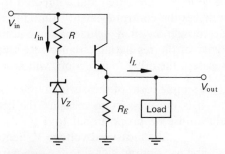

Fig. 15.23 Zener regulator with current gain.

diode regulator can swing from some low value to $0.9I_{Z\,\text{max}}$ as I_Z changes from approximately $I_{Z\,\text{max}}$ to $0.1I_{Z\,\text{max}}$. This current corresponds to transistor base current; thus emitter or load current can vary from a small value up to $0.9(\beta + 1)I_{Z\,\text{max}}$. The current output capability is increased by the current gain of the transistor.

The transistor dissipation must not exceed the allowable dissipation of the transistor $P_{T\,\text{max}}$; that is,

$$P_{T\,\text{max}} \geqq V_{CE}I_E = (V_{\text{in}} - V_{\text{out}})I_E.$$

Neglecting the base-emitter voltage drop and noting that maximum transistor dissipation occurs at the point of maximum load current, we can write

$$P_{T\,\text{max}} \geqq (V_{\text{in}} - V_Z)0.9(\beta + 1)I_{Z\,\text{max}}. \tag{15.39}$$

For relatively high current supplies, the transistor dissipation can become excessive. This dissipation can be reduced without affecting load current by reducing the collector voltage of the transistor. A collector resistance can be chosen to reduce the collector voltage to a value that exceeds the emitter voltage by only a few volts, thereby reducing transistor dissipation.

In higher current applications more than one emitter follower can be used. Figure 15.24 shows two such possibilities. In tandem emitter-follower stages the last stage will experience maximum power dissipation since this stage conducts more emitter current. If current requirements on the output transistor become excessive, parallel output stages can be used as shown in Fig. 15.24(b). A very small resistance is inserted at the emitter of each output stage to equalize the input characteristics of these stages. If one stage conducts more current than the other two for a given base-emitter voltage, the added drop across R_e will tend to reduce the base-emitter voltage for this stage. Thus, the effect of "current-hogging" is eliminated by the small emitter resistors.

There are some obvious disadvantages of regulators of this type. For example, the base-emitter drops vary slightly with load current and with temperature, leading to poorer regulation and temperature stability than more sophisticated regulated supplies.

B. Regulating Circuits

A regulated power supply can be divided into five basic sections, as shown in Fig. 15.25. The output voltage is measured and compared to a reference voltage. The comparator then drives the control element to modify the output voltage if necessary. The comparator section often includes a stage or more of amplification to increase the sensitivity of the regulating circuitry. Rectifier and filter circuits have already been discussed; the other four sections will now be covered.

Measurement section. A simple form of measuring circuit is the voltage divider shown in Fig. 15.26(a). It is unnecessary to measure the actual output voltage, but the measured value must be proportional to V_{out}. As the output current increases tending to drop V_{out}, the measured voltage V_m tends to drop also. This voltage is sufficient to serve as an input signal to the comparator. The adjustable

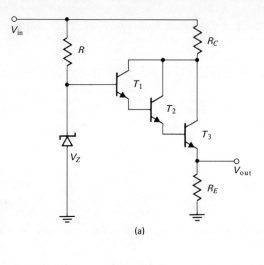

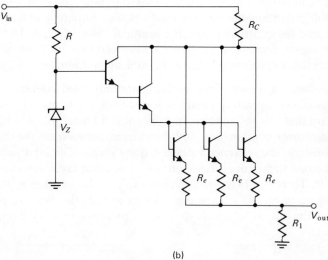

Fig. 15.24 High-current, Zener regulator stages: (a) tandem emitter followers; (b) tandem emitter followers with parallel output stages.

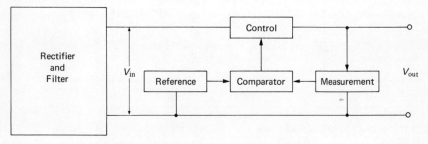

Fig. 15.25 Block diagram of a regulated dc supply.

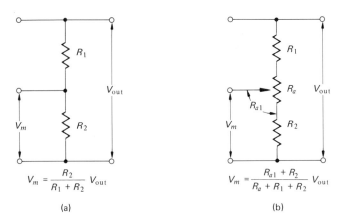

Fig. 15.26 Measuring circuits: (a) fixed measuring circuit; (b) adjustable measuring circuit.

measuring circuit of Fig. 15.26(b) is often used to obtain a variable voltage supply. Changing the potentiometer setting modifies V_m, initiating a reaction of the comparator and the control section. If a nominal value of $V_m = 10$ V leads to a 20-V output signal, decreasing V_m to 9.5 V may lead to a 30-V output. The comparator interprets a decreasing value of V_m as a signal to increase V_{out}.

Reference section. A Zener diode is the commonly used reference element. A simple Zener-diode regulating circuit will establish V_Z as the reference voltage. It is important that this voltage remain constant at all times. If the voltage applied to the diode-resistor combination and the current drawn from the diode remain relatively constant, the reference voltage is quite stable. The input current to the comparator must then remain constant and it is advantageous to minimize this input current. There are two possible choices as to the location of the reference circuit as shown in Fig. 15.27. Some regulators include the diode-resistor combination across the output terminals to ensure a more stable voltage applied to the

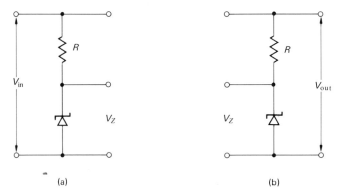

Fig. 15.27 (a) One possible location of reference circuit. (b) Improved location for reference circuit for fixed output voltage.

combination. Since this voltage is highly regulated the reference voltage generated will be very stable. Of course, if the output voltage is to be variable over a large range, this arrangement is inappropriate. Figure 15.27(a) shows the reference circuit appearing across the input voltage (nonregulated voltage). This arrangement leads to a less stable reference voltage since load and line regulation of the voltage V_{in} is generally poor. A second Zener diode can improve the stability of the reference voltage as indicated in Fig. 15.28.

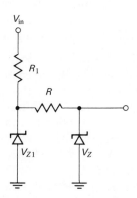

Fig. 15.28 A two-stage reference circuit.

Comparison section. The comparison element can be a simple common-emitter circuit, a difference amplifier, or a more elaborate circuit. Figure 15.29 shows two possible choices of comparison elements. The comparator shown in part (a) compares the Zener diode voltage appearing at the emitter to the measurement circuit voltage appearing at the base. When the output voltage is at the desired value, a particular value of quiescent collector current will flow to the control element. If

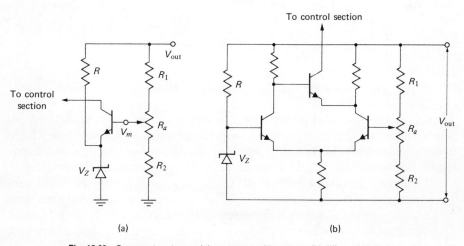

Fig. 15.29 Comparator stages: (a) common-emitter stage; (b) differential stage.

V_{out} drops, less current will flow to the control element. The control element must be so designed that a lower current flow from the comparator causes the output voltage to increase. It is not necessary that the comparator stage have extremely good dc stability, since it is part of a feedback loop. If drift does occur, causing V_{out} to change, the change in V_{out} will be fed back to the comparator to decrease the total drift.

The differential amplifier stage is a popular circuit to be used as a comparator section. The voltage between collectors will be proportional to the difference in voltages presented at the two inputs. A scheme such as that shown in Fig. 15.29(b) will generate a control signal with little loading of the reference or measuring circuits.

Control section. The control element is driven by the comparator in such a way as to compensate for any changes in V_{out} that occur. There are two types of control elements: series controllers and shunt controllers. Both types are shown in Fig. 15.30. The shunt controller requires that the transistor withstand the full output voltage, but it does not have to pass the load current. The series controller only requires a voltage rating V_{CE} of $V_{in} - V_{out}$, which can be much smaller than V_{out}. However, the total load current must be conducted by the emitter of the transistor. The logical conclusion is that the series element is used in high-voltage, low-current applications, while the shunt element is used in low-voltage, high-current applications. The emitter follower in the regulator of Fig. 15.23 is a series element. A compound connection can be used to increase the current drive to the control element. One such arrangement is shown in Fig. 15.31.

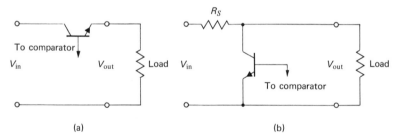

Fig. 15.30 Controllers for regulated supplies: (a) series controller; (b) shunt controller.

A complete regulated power supply is shown in Fig. 15.32. Since there are several stages included in the feedback loop, there is often a tendency toward high-frequency instability. Fortunately, since only dc signals need be amplified, the ac signals can be shunted to ground by capacitors to prevent oscillations.

It is often desirable to stabilize the output voltage of the supply as the temperature varies. The same type of compensating circuits discussed in connection with biasing schemes (Chapter 3) and differential stages (Chapter 5) can be applied. Low temperature-coefficient resistors are also useful to temperature-compensate the regulated power supply.

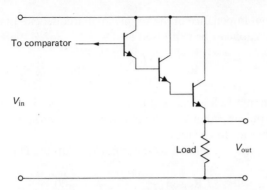

Fig. 15.31 Compound-series control element.

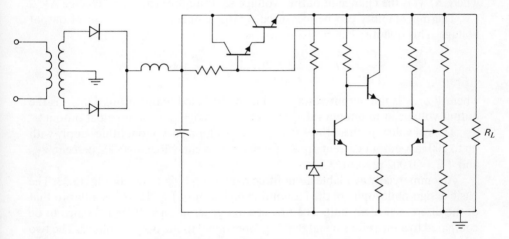

Fig. 15.32 Regulated supply with series controller.

C. Monolithic Voltage Regulators [1]

Integrated circuit or monolithic voltage regulators have some advantages over discrete-element regulators that make them attractive in many applications. The monolithic regulator can be temperature-compensated very accurately since all temperature-sensitive elements are included in a monolithic chip that will tend to have a uniform temperature. It is rather inexpensive to add devices to an integrated circuit so that more complex circuitry can be employed. A disadvantage that can be overcome by using an integrated-discrete circuit approach is that of relatively low power capability of monolithic circuits. The controller can be constructed with power transistors while the remaining circuitry of the regulator can be included in the monolithic chip.

To this point we have considered load regulation from a dc or slowly varying current load. In practice the output current can change very abruptly especially when supplied to switching or high frequency circuits. It is more meaningful to consider the output impedance of the supply at all frequencies of interest. The load regulation in this case can be expressed as

$$\text{Reg} = \frac{\Delta I \times Z_0}{V_{dc}} \times 100,$$

where ΔI is the change in load current from no load to full load and Z_0 is the magnitude of the output impedance at the frequency of interest. Obviously, the regulation will be better for low values of output impedance.

The line regulation is related to the input regulation. This figure is given by

$$I \text{ Reg} = \frac{\Delta V_{dc}}{\Delta V_{in} \times V_{dc}} \times 100,$$

where ΔV_{dc} is the change in output voltage resulting from an input change ΔV_{in}.

Thermal stability can be measured by the temperature coefficient of output voltage. This quantity can be expressed as

$$TC = \frac{V_{dc\,max} - V_{dc\,min}}{V_{dc}(T_{max} - T_{min})} \times 100,$$

where $V_{dc\,max}$ is the output voltage at maximum rated temperature, $V_{dc\,min}$ is the output voltage at minimum rated temperature, and V_{dc} is the nominal output.

Typical values of these three figures of merit for a 36-V monolithic supply with up to 200 mA output current are $\text{Reg} = 0.005$ percent, $I \text{ Reg} = 0.002$ percent/V_{in}, and $TC = 0.002$ percent/°C.

A commercially available monolithic regulator [1] is shown in Fig. 15.33. The basic design philosophy of the monolithic regulator of Fig. 15.33 is similar to that of the discrete element unit. The first section generates a voltage reference to be compared to a measured signal that is proportional to the output voltage. The two signals are compared in the level-shifting amplifier by means of a differential amplifier. The error signal is then amplified and applied to the series controller circuit.

15.4 HIGH-CURRENT, VARIABLE VOLTAGE SUPPLY

The silicon-controlled rectifier (SCR), which was discussed in Chapter 8, is often used in high-current applications. The SCR is a three-terminal semiconductor device having properties very similar to the semiconductor diode. The important difference between the diode and the SCR is the fact that the voltage at the additional terminal of the SCR controls the conduction of the diode. The symbol for the SCR is shown in Fig. 15.34(a). If a positive voltage is applied to the anode of the SCR, as shown in part (b) of the figure, the SCR will not conduct when V_g is zero or negative. If V_g is increased to some small positive value (specified by the manufacturer) the SCR will "break down" and conduct. Once the SCR has fired or broken down, the gate loses control over the conduction process. Causing V_g to decrease will not turn off the SCR. Current will continue to flow until the anode

15.4 HIGH-CURRENT, VARIABLE VOLTAGE SUPPLY

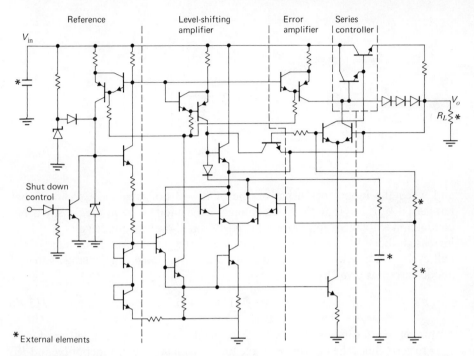

Fig. 15.33 Monolithic voltage regulator, Motorola MC1560. (Courtesy of Motorola Semiconductor Products, Inc.)

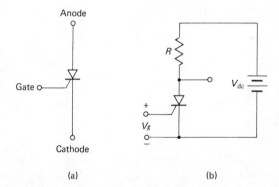

Fig. 15.34 Silicon-controlled rectifier: (a) symbol for SCR; (b) SCR circuit.

current is interrupted. When this occurs, and if V_g is below the firing level, the circuit shuts off and resumes the original nonconducting state. The characteristics of the SCR allow this device to be used as a controllable rectifier. Consider the circuit of Fig. 15.35.

If V_g is always greater than the firing voltage, the SCR will act as a half-wave rectifier. When the input signal to the anode goes positive, the SCR will conduct;

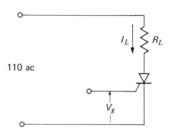

Fig. 15.35 The SCR as a half-wave rectifier.

and when the input goes negative, the SCR will not conduct. Consider what happens if V_g is less than the firing voltage as the anode voltage goes positive. No conduction will take place. During the positive half-cycle of anode voltage, if V_g is increased above the firing level, the SCR will conduct for the remainder of the half-cycle. The resulting waveform is shown in Fig. 15.36. The average value of anode current is given by

$$I_{dc} = \frac{I_m}{2\pi}(1 + \cos\theta). \tag{15.40}$$

Assuming that θ can vary from 0 to 180°, the dc current can then be varied from I_m/π to zero. Since the dc voltage to the load is equal to the product of I_{dc} and R_L, the dc voltage can be varied from $R_L I_m/\pi$ to zero.

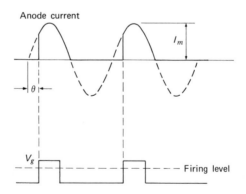

Fig. 15.36 Anode current for the circuit of Fig. 15.35.

There are two fairly simple methods of controlling the angle θ or the point where V_g exceeds the firing voltage. The first is shown in Fig. 15.37. The comparator puts out a trigger pulse, initiating the one-shot period, when the ac voltage starts its positive swing. The period of the one-shot can be adjusted from some very small value (say 100 μs) to some larger value (say 10 ms). The SCR will fire at the end of the period as the gate swings positive. The firing angle θ can vary from some small

15.4 HIGH-CURRENT, VARIABLE VOLTAGE SUPPLY

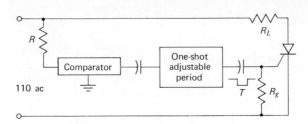

Fig. 15.37 Variable-output voltage circuit.

value to greater than 180°. The output dc voltage is then controlled by the period of the one-shot, which is adjustable.

A second method of control uses a phase-shifter circuit to shift the phase of V_g with respect to the input voltage. If the voltage of the gate is of the same frequency as the anode voltage, but lagging in phase, the waveforms of Fig. 15.38 can be applied to the SCR-circuit.

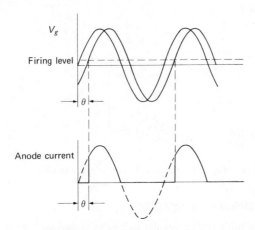

Fig. 15.38 Phase-shift control.

The circuits of Figs. 15.39 are similar, but the second requires a dc voltage to be present. In many cases this can easily be done, since the collector supply need be only a low current source. Both circuits reduce to the equivalent circuit of Fig. 15.40. The output voltage is equal in amplitude to the input voltage, but shifted in phase by

$$\phi = -2 \tan^{-1} \omega CR \tag{15.41}$$

As R varies from zero to infinity, ϕ varies from 180° to zero. If the output of the phase shifter is applied to the gate, the output dc voltage can then be varied by changing R.

The SCR is ideally suited for high current operation and can supply several hundreds of amps to a load.

678 SEMICONDUCTOR POWER SUPPLIES

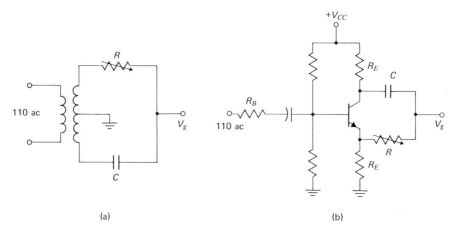

Fig. 15.39 Phase-shift circuits: (a) passive; (b) active.

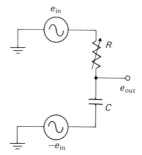

Fig. 15.40 Equivalent circuit of phase shifters.

REFERENCES AND SUGGESTED READING

1. D. Kesner, "Monolithic voltage regulators." *IEEE Spectrum* 7, 4: 24 (April, 1970).
2. Motorola Staff, *Motorola Semiconductor Data Manual*, 4th ed. Phoenix, Ariz.: Motorola, Inc., 1969.
3. H. H. Skilling, *Electrical Networks*. New York: Wiley, 1974, Chapter 13.
4. Texas Instrument Staff, *Transistor Circuit Design*. New York: McGraw-Hill, 1963, Chapter 9.

PROBLEMS

Section 15.1

15.1 A diode with an average forward resistance of $R_f = 20\ \Omega$ is used in a half-wave rectifier circuit. Given that the source resistance is $10\ \Omega$, calculate the dc output voltage for a 110-V(rms) input signal when (a) $R_L = 500\ \Omega$; (b) $R_L = 80\ \Omega$.

* **15.2** Calculate the dc voltage of a full-wave rectifier and the magnitude of the fundamental frequency if forward resistance of each diode can be taken as $10\ \Omega$ and the secondary resistance to center tap is 10 Ω. The secondary voltage (referenced to center tap) is 220 V(rms). The load resistance is 100 Ω.

15.3 The input signal to the bridge rectifier of Fig. 15.4 is 100 cos ωt. If the load resistance is 50 Ω and the forward resistance of the diode is 5 Ω, calculate the peak value of the load voltage and the dc output voltage.

Section 15.2A

* **15.4** Calculate the ripple factor for a half-wave rectified signal of peak amplitude V_m.

15.5 Calculate the ripple factor for the waveform shown in the figure.

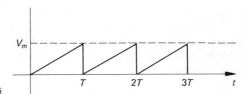

Problem 15.5

* **15.6** In the capacitor filter of Fig. 15.6(a), $R_L = 500$ Ω, $C = 1000$ μF, and $e_{in} = 30 \sin(2\pi \times 60)t$. Assuming that the source resistance is 20 Ω, find (a) V_{dc}, (b) I_{dc}, and (c) the ripple factor.

15.7 Repeat Problem 15.6 if $R_L = 4$ kΩ and $C = 40$ μF.

15.8 Repeat Problem 15.6 if a full-wave rectifier replaces the half-wave rectifier.

15.9 Repeat Problem 15.6 if a full-wave rectifier replaces the half-wave rectifier, $R_L = 4$ kΩ, and $C = 40$ μF.

15.10 Calculate the load regulation for the filter of Problem 15.6 if the 500-Ω load resistance results in the rated current.

***15.11** Calculate the load regulation for the filter of Problem 15.9 if the 4-kΩ load resistance results in the rated current.

15.12 Design a capacitor filter with a half-wave rectifier input to supply a 12-V dc signal to a load of 1 kΩ. The ripple factor should not exceed 0.02. Calculate the required turns ratio of the transformer, given that the input signal is 110 V rms at 60 Hz. Neglect transformer and forward diode resistances. What is the load regulation of the supply? Specify reverse-voltage rating of the diode.

15.13 Repeat Problem 15.12 if a full-wave rectifier is used.

Sections 15.2B–15.2C

***15.14** The L-section filter of Fig. 15.10 has the following values: $R_l = 2$ Ω, $R_t = 10$ Ω (for one-half of secondary windings), $R_f = 5$ Ω, $R_L = 400$ Ω, $L = 2$ H, $C = 40$ μF, and $V_m = 40$ V at 60 Hz. Find (a) V_{dc}; (b) I_{dc}; (c) I_{cr}; (d) the ripple factor; and (e) the load regulation as R_L varies from R_{cr} down to 400 Ω.

15.15 Rework Problem 15.14 if L is changed to 4 H and C to 100 μF.

15.16 Design an L-section filter that supplies 1 A and 40 V to a load. The ripple factor should be less than 0.001. A 20-H choke is available with a 4-Ω resistance. The input signal is 110 V rms at 60 Hz, $R_f = 0$, and $R_t = 4$ Ω. Shunt a resistance across the output terminal to guarantee that I_{cr} always flows in the circuit. Calculate the regulation as the load current varies from 0 to 1 A.

Section 15.3A

15.17 A Zener diode regulator, such as that shown in Fig. 15.17, is used to supply 10 V to a load having a variable current requirement from 0 to 20 mA. Given that the input dc voltage is 15 V, calculate the required value of R and the maximum power dissipated by the Zener diode.

*15.18 Repeat Problem 15.17 assuming that the load current must vary from 0 to 1 mA.

15.19 Repeat Problem 15.17 assuming that the load current must vary from 0 to 100 mA.

15.20 Repeat Problem 15.17 if $V_{in} = 18$ V. If the ripple factor at the input is 0.1 calculate the output ripple factor with $r_Z = 25 \, \Omega$.

15.21 Use a 400-mW Zener diode and an emitter follower to supply 20 V to a load. The current requirements of the load vary from 5 to 500 mA. The current gain of the transistor is 80 and the dc input signal V_{in} is 28 V under no load conditions. V_{in} drops linearly with current to 24 V when the current drain is 500 mA. If $r_Z = 20 \, \Omega$ and $V_{BE(on)} = 0$ calculate the element values required and the load regulation.

*15.22 For the circuit of Fig. 15.24(b) the load current is 20 A, $V_{in} = 36$ V, $V_Z = 20$ V, and $V_{BE(on)} = 0.5$ V. Select reasonable values of R_C and R_e to limit dissipation to 20 W per transistor. What is the required power rating of R_C?

Section 15.3B

15.23 In Fig. 15.27(a) the input voltage varies from 24 to 28 V. Calculate the reference voltage change if $R = 1 \, k\Omega$, $V_Z = 12$ V, and $r_Z = 50 \, \Omega$. Compare this to the change in reference voltage for the circuit of Fig. 15.28 for the same input voltage variation. Use values of $R_1 = 0.5 \, k\Omega$, $R_2 = 1 \, k\Omega$, $V_{Z1} = 18$ V, $V_{Z2} = 12$ V, $r_{Z1} = 50 \, \Omega$, and $r_{Z2} = 50 \, \Omega$.

Section 15.4

15.24 Derive Eq. 15.40.

15.25 Derive Eq. 15.41.

*15.26 The input frequency to the circuit of Fig. 15.37 is 60 Hz. If the minimum period of the one-shot is 2 ms, what is the maximum dc load current that can occur?

15.27 If the maximum period of the one-shot in Problem 15.26 is 7 ms, what is the minimum dc load current that can occur?

15.28 If the capacitor in the circuit of Fig. 15.39(a) is 1 μF, what range of values must R have to result in phase shifts varying from $-10°$ to $-170°$? The input frequency is 60 Hz.

*15.29 Repeat Problem 15.28 given that the input frequency is 1 kHz and $C = 0.01 \, \mu$F.

ANSWERS TO SELECTED PROBLEMS

CHAPTER 1

1. $n_i = 2.05 \times 10^{14}/\text{cm}^3$
4. $n_i = 1.329 \times 10^9/\text{cm}^3$
6. $n = 1 \times 10^{17}/\text{cm}^3$; $p = 9.01 \times 10^{10}/\text{cm}^3$
7. $p = 2 \times 10^{14}/\text{cm}^3$; $n = 9.8 \times 10^5/\text{cm}^3$
10. $\sigma_i = 4.08 \times 10^{-7}\,\mho/\text{cm}$
12. $N_A = 1.042 \times 10^{17}/\text{cm}^3$
15. $I_p = 3.33$ mA
16. $p(0) = 1.359 \times 10^{13}/\text{cm}^3$
19. $p_e = p_e(t=0)e^{-t/\tau_p}$
22. $n(0) = 1.838 \times 10^{10}/\text{cm}^3$

CHAPTER 2

1. $\phi = -0.759$ V
4. $I = 0.32$ mA
6. (a) $I_S = 6.77\,\mu\text{A}$; (b) $T = 310°\text{K}$
11. $i_{\text{in}} = 107 \sin 1000t\,\mu\text{A}$
13. (a) $C = 30$ pF; (b) $C = 14.6$ pF
16. $P = 67.5$ mW
19. (a) $E_{\text{out}} = -7 + 1.5 \sin \omega t$; (b) $E_{\text{out}} = -4.0 + 1.5 \sin \omega t$
21.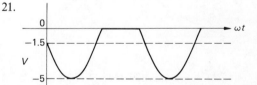

24. $V_{CEQ} = -8$ V; $V_{CE\,\text{max}} = -3$ V; $V_{CE\,\text{min}} = -14$ V
26. $A = -58$. Gain increases with load resistance.
29. (a) $\beta = 70$; (b) $\beta = 85$; (c) $\beta = 90$; (d) $\beta = 100$ 31. $A = 4$
34. This expression applies strictly to drain current at the edge of the pinch-off region. Above this value of V_{DS}, however, the current increases only slightly; therefore, this equation approximates I_{DS} for values of V_{DS} above pinch-off.
36. $g_m = 1$ m$\mho$ from curve; $g_m = 0.93$ m$\mho$ from Eq. (2.36)

CHAPTER 3

1. $R_B = 800$ kΩ; 3% error

4. (a) $R_B = 1.95$ MΩ; (b) $R_B = 4.88$ MΩ; (c) $R_B = 1.30$ MΩ

9. (a) $V_{CQ} = 6.67$ V; (b) $I_{co} = 112$ μA; (c) $V_{CQ} \approx 0$ V (saturation)

12. $R_C = 18.5$ kΩ

14. $R_1 = 168$ kΩ

18. $R_1 = 197$ kΩ; $R_1 = 212$ kΩ; 7.6% change

20. (a) $R_1 = 143$ kΩ; (b) $V_{CQ} = 14.5$ V; (c) for (a) 12 V peak-peak; for (b) 11 V peak-peak

22. $V_{CQ} = 9$ V upper limit; $V_{CQ} = 4.2$ V lower limit. If $R_2 = 15$ kΩ, $R_1 = 96$ kΩ.

24. $R_1 = 42.5$ kΩ; $R_2 = 2.92$ kΩ

28. $R_F = 78$ kΩ; $V_{CQ} = 6.05$ V

30. $R_1 = 156$ kΩ; $R_2 = 16$ kΩ

32. $\beta_{min} = 94.4$; $\beta_{max} = 105.6$

35. $V_{CQ\,max} = 4.25$ V; $V_{CQ\,min} \approx 0$ V

37. $V_{EB} = 0.39$ V from Eq. (3.37); $V_{EB} = 0.40$ V from Eq. (3.41)

38. $R_B = 238$ Ω

42. $R_3 = 405$ kΩ

CHAPTER 4

5. Since 10 kΩ is much greater than the forward resistances of the diode model, the drop across the diode resistance is negligible when forward-biased. However, the 0.4-V offset voltage will cause the ideal diode model to be inaccurate for small voltages. With this 50-V input signal, the ideal model will be very accurate.

7. For e_{in} positive, $e_{out} = 10$ V; for e_{in} negative, $e_{out} = 0$ V

10. $I = 0.59$ mA

13. $\tau = 0.013$ μsec for both cases

16. When $V_1 = 1$ V, $e_{out} = 9.34$ V; when $V_1 = 6$ V, $e_{out} = 0$ V

19.

20. $A_v = -132$; $R_{in} = 1500$ Ω; $V = 8$ V peak-peak

23. $A(5\,μA) = -24.8$; $A(30\,μA) = -34.0$

25. $A_v = -14.7$; $R_{out} = 2$ kΩ; $R_{in} = 13.5$ kΩ

28. $R_{in} = 2.4$ kΩ; $A_v = -187$

31. $R_B = 2.08$ MΩ; $V_{CQ} = 14.6$ V

35. $R_E = 179$; $R_2 = 3.6$ kΩ; $R_1 = 48.4$ kΩ; $V_{max} = 11$ V peak-peak; $S_V = -35.8$ mV/°C

39. $S_V = +8.4$ mV/°C; $V_{CQ} = -10.4$ V; $A_v = -18.6, -18.8, -19.8$

41. $V_{EQ} = 4.98$ V; $R_{in} = 36.4$ kΩ; $A_v = 0.978$; $R_{out} = 16.8$ Ω

45. $A_v = -26.8$; $A_I = g_m R_{GS}$, which is very high since R_{GS} may be 10 MΩ.

47. $A_v = -6.16$; $R_{out} = 5.85$ kΩ

51. [graph showing current-voltage characteristic: horizontal line at 2A for V > 2V]

53. [circuit diagram with 4mA current source, diode, 2kΩ resistor, two diodes with 3V and −2V sources, 625Ω and 2.5kΩ resistors]

CHAPTER 5

2. $A_o = -3.6$
9. $A_o = -6.67$
13. $C_{c1} = 0.29$ μF; $C_{c2} = 0.16$ μF;
$$A_o = \frac{-16}{\left(1 + j\frac{50}{f}\right)\left(1 + j\frac{20}{f}\right)}, 55.1 \text{ Hz}$$
20. $A_{MB} = -5.63$; $f_1 = 0.16$ Hz; $f_{1S} = 26.5$ Hz; $f_{2S} = 49$ Hz
27. 3 to 4
41. $f_{2E} = 4110$ Hz

6. $R_C = 21.4$ kΩ
11. $A_o = -242$
18. $f_{2E} = 58.1$ Hz

22. $A_{MB} = -35.7$; $f_{2E} = 19.3$ Hz
37. $A_o = 71.5$
49. $R_{C1} = R_{C2} = 5.5$ kΩ; $R_{E2} = 290$ Ω; $R_E = 400$ Ω; $R_1 = 54.2$ kΩ; $R_2 = 5.43$ kΩ

CHAPTER 6

1. (a) $G = 10$; (b) $G = 100$; (c) $G = 1000$
7. SNR = 1401
12. $R_{in} = 19.6$ Ω; $R_{out} = 39.2$ Ω
21. Voltage-shunt $R_{ia} = 1.58$ kΩ; $G = -7.65$; $R_{in} = 15.7$ Ω

4. $S_{G,T} = 0.0164$
8. $G_{MB} = 19.6$; $\omega_{2F} = 51 \times 10^7$ rad/sec
15. $G = -1.89$; $R_{in} = 67$ Ω; $R_{out} = 154$ Ω
24. $G = -18$; $R_{in} = 38.4$ Ω

684 ANSWERS TO SELECTED PROBLEMS

28. $G = 3116$; $R_{in} = 0.083\ \Omega$

34. No. $\omega = 1.1 \times 10^6$ rad/sec

37.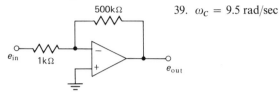

39. $\omega_C = 9.5$ rad/sec

$f_2 = 200$ kHz; $G = -500$

CHAPTER 7

2. $C = 8.08$ pF
11. $A_{MB} = -57.8$; $f_2 = 152$ kHz
20. $f_2 = 121$ kHz
26. $C_1 = 173$ pF; $A_{MB} = -5.98$; $f_2 = 1.47$ MHz
37. $f = 0.832$ MHz
44. $R = 60.6\ \Omega$; $L = 0.264\ \mu H$; $f_2 = 47.7$ MHz
53. $C_n = 50$ pF; $r_n = 100$ kΩ; $R = 162$ kΩ; $C = 246$ pF; $A_o = -211$

6. (a) $C_D = 37.7$ pF; (b) $C_D = 188$ pF
18. $A_{MB} = -36.3$; $f_2 = 540$ kHz
24. $|A_{MB}| = 3.31$
35. $f_2 = 1.155$ MHz
41. $A_o = 250$; $\omega_{2o} = 1.10$ MHz
50. $C = 633$ pF; $Q_{eff} = 46.9$; $BW = 21.3$ kHz
56. 48 Ω upper limit; 20 Ω lower limit; $L_T = 2.4 \times 10^{-9}$ H

CHAPTER 8

2. $V_{in} = 2.25$
8. When $e_{in} = 0$, $e_{out} = 10.9$ V; when $e_{in} = 6$ V, $e_{out} = 0$ V
14. $t'_d = 0.97\ \mu sec$
20. $T = 1.28\ \mu sec$
26. $t'_s = 0.65\ \mu sec$
33. Switches from 8.64 V to 0 V and back to 8.64 V; $t_{on} = 0.74\ \mu sec$; $t'_s = 3.22\ \mu sec$; $t_f = 7.30\ \mu sec$

5. $R_C = 1.64$ kΩ
11.
18. $C_B = 17.3$ pF; $\tau_c = 0.036\ \mu sec$
23. $t_{off} = 0.59\ \mu sec$
30. $T = 1.85\ \mu sec$
38. $R_2 = 8.34\ \Omega$

CHAPTER 9

2. (a) $e_{out} = 2.23$ V; (b) $e_{out} = 2.08$ V; (c) $e_{out} = 0$ V
7. $e_{in} = 1.0$ V
15. $t_1 = t_2 = 49.8\ \mu sec$

5. $e_{out} = 2.45$ V
9. $C = 9662$ pF; $R_1 = R_2 = 100$ kΩ
21. $R_{L1} = R_{L2} = 3$ kΩ; $R_1 = 196$ kΩ; $C = 5428$ pF

CHAPTER 10

2. (a) 11111111 = 255;
   (b) 10010110 = 150;
   (c) 11100011 = 227;
   (d) 1101010101 = 853

5. (a) 0.11110 = 0.9375;
   (b) 0.1010101 = 0.664;
   (c) 0.000001 = 0.016;
   (d) 0.100101 = 0.5625

8. False

11. (a), (b)

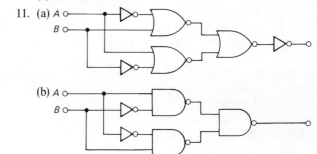

14. $f = AB\bar{C} + BCD$

17. Gates = 64; inputs = 384

20. Gates = 16,640; inputs = 34,560

25. The one-shot must start with the first transition of a word and last 4-3/4 bit times. If the circuit were retriggerable, the period of the one-shot would be extended beyond this time since each positive transition of input data would restart the period.

28.

40. 32 inhibitors

CHAPTER 11

1. If $A_{MB} = -1250$, $f_{osc} = 7.80$ MHz

5. $\omega_o = \dfrac{1}{\sqrt{11}\,CR}$; $A_{MB} = 37.5$

9. $L = 1$ mH; $C = 0.0281\ \mu$F;
   $n = 38.4$; $R_g = 1$ MΩ; $C_g = 10\ \mu$F

12. (a) $T = 0.879$ ms; (b) $T = 0.678$ ms;
    (c) $T = 0.409$ ms

16. $R_2 = 1$ kΩ; $R_1 = 39$ kΩ;
    $V_a = 0.615$ V; see Fig. 11.24

CHAPTER 12

2. (a) $P = 51.6$ W; (b) $P = 31.4$ W;
   (c) $P = 14.4$ W

7. $P = 350$ mW

10. $P_D = 171.5$ mW; $\theta_i + \theta_{HS} = 308°$C/W

14. $P_R = 296.5$ mW; $P_S = 552$ mW

18. (a) $P_{TQ} = 608$ mW; (b) $P_{ac} = 89$ mW;
    (c) $P_T = 519$ mW; (d) $P_S = 2124$ mW

20. (a) $P_{ac} = 0.421$ W; (b) $I_{CQ} = 0.0811$ A;
    (c) $P_{TQ} = 0.907$ W

25. (a) $P_{max} = 4.55$ W; (b) $T_J = 150°$C;
    (c) $R_L = 0.963\ \Omega$ (d) 1.72 W

29. (a) $P_{ac} = 1.82$ W; (b) $I_{CQ} = 0.118$ A; (c) $R_1 = 3492\ \Omega$; (d) $\eta = 35.5\%$;
    (e) $P_{TQ} = 4.30$ W; (f) $\theta_i + \theta_{HS} = 14.1°$C/W; (g) 35.5%; (h) 4.30 W; (i) 0.118 A

31. $R_E = 10\ \Omega$; $n = 3.26$;
 $P_{T_{max}} = 0.55$ W; $\theta_i + \theta_{HS} = 108°$C/W
38. $A_v = -87.5$; $G_p = 3.89 \times 10^4$

36. $V_{C1} = 18$ V; $P_{T_{max}} = 0.253$ W

CHAPTER 13

2. $n = 7$; $R_a = 640$ kΩ; $R_b = 320$ kΩ;
 $R_c = 160$ kΩ; $R_d = 80$ kΩ;
 $R_e = 40$ kΩ; $R_f = 20$ kΩ;
 $R_g = 10$ kΩ; $R_F = 227.72$ kΩ

7. Let $R_1 = R_2 = R_3 = R_4 = 10$ kΩ;
 $C_1 = 2080$ pF; $C_2 = 1040$ pF;
 $C_3 = 609$ pF; $C_4 = 305$ pF

4. $G = 0$; $R_F = 0$; $R_2 = \infty$;
 $C = 0.3125\ \mu$F; $R = 2.28$ MΩ;
 $R_B = 120$ kΩ

8. $A = 1205$; $R = 10$ kΩ; $C = 1590$ pF

CHAPTER 14

2. 10 V at 600 kHz; $\dfrac{20}{\pi}$ V at 598 and 602 kHz; $\dfrac{20}{3\pi}$ V at 594 and 606 kHz; $\dfrac{4}{\pi}$ V at 590 and 610 kHz

5. $C = 2533$ pF; $R = 37.5$ kΩ

7. $C = 126$ pF; $R = 45.4$ kΩ;
 $M = 3$ V; $R_E = 2.64$ kΩ

9. $P_C = 1.42$ mW; $P_{US} = P_{LS} = 0.352$ mW;
 $P_C = 15.7$ mW; $P_{US} = P_{LS} = 2.51$ mW

12. $f_{osc} = 345$ kHz; $Q = 45.5$

15.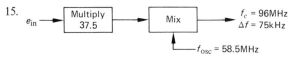

CHAPTER 15

2. $V_{dc} = 165$ V

6. (a) $V_{dc} = 28.4$ V; (b) $I_{dc} = 56.8$ mA;
 (c) $r = 0.0096$

14. $I_{cr} = 12.4$ mA; $V_{dc} = 24.4$ V;
 $I_{dc} = 61.1$ mA; Reg $= 0.041$;
 $r = 0.0104$

22. $R_e = 0.075\ \Omega$; $R_C = 0.725\ \Omega$;
 $P_R = 1160$ W $= 4$ times dissipation

29. $R_{max} = 182$ kΩ; $R_{min} = 1.40$ kΩ

4. $r = 1.21$

11. Reg $= 2.51\%$

18. $R = 4.5$ kΩ; $P_Z = 11.1$ mW

26. $I_{dc} = 0.275\ I_m$

INDEX

Acceptor atoms, 7, 8, 11, 27
Acceptor impurities, 7
Active filters, 581
   Bessel, 585
   Butterworth, 582
   Chebychev, 582
   maximally flat phase, 582
Active loads, 213
Adder, 208, 464, 467
   binary, 464
   full, 466
   half, 465
   parallel, 466
   serial, 466
α, 51, 146–148
   inverted, 147
   normal, 146
Aluminum, 74
Amplification factor, 144
Amplifier, 57, 161–214, 220–263, 270–318, 330–335
   common-base, 57
   common-emitter, 57
   current, 161, 230, 244
   dc, 203
   differential, 205
   feedback, 220–263
   high frequency, 270, 318
   integrated circuit, 315
   multistage, 166–169
   operational, 239
   power, 161
   transadmittance, 229, 237
   transimpedance, 230, 238
   tuned, 271, 319–335
   tunnel diode, 330–335
   video, 271
   voltage, 161–214
   wideband, 271
Amplitude limiting, 489
Amplitude modulation (AM), 599–618
   index, 601
   100 percent, 601
   overmodulation, 602
AND gate, 379, 381–383, 430
Antennas, 596
   radiation efficiency of, 597
Antimony, 8, 72
Arithmetic circuits, 464–469
Arithmetic unit, 424
Arsenic, 8, 72
Astable clock, 450
Astable multivibrator, 395, 407–411, 506
Asymptotic frequency response, 172, 174, 176
Available states, 5
Avalanche breakdown, 36, 45
Avalanche multiplication, 46
Average velocity of electron, 11

Bandpass filters, 586–594
Bandwidth, 226–227
   improvement due to feedback, 226
Bandwidth shrinkage, 295
   formula, 296
Barrier voltage, 28, 30
Base, 49

Base compensation, 286–290
Base current bias, 85–87
Base width of transistor, 49
  effective, 49
  modulation of, 49, 53
Base-width modulation, 130
Beta cutoff frequency, 271
B-factor, 352
Biasing, 82–95
  base current bias, 85–87
  collector-base bias, 94–95
  emitter bias, 90–93
  FET stage, 105–107
  integrated circuit, 109–112
  procedure for, 85
Binary-coded decimal code, 427
Binary counter, 452–457
  down, 454
  synchronous, 454
  up-down, 457
Binary numbers, 423, 426–428
Binary pulse counter, 452
Bipolar transistor, 26
Bistable multivibrator, 395, 411
Boltzmann's constant, 6
Bonding, 6
  covalent, 5
  ionic, 5
  metal, 5
  molecular, 5
Boolean algebra, 428–439
Bootstrap circuit, 188–189
  for high impedance, 188–189
  for sweep generation, 571–578
Boron, 7, 72
Broadband factor, 278
Broadbanding, 282
Butterworth polynomials, 582

Cable television, 599
Capacitor filter, 650–655
Carrier signal, 599
Carrier suppression, 616
Cascode connection, 315
Central processing unit, 424
Channel, 63
  of FET, 63

Characteristics, 33, 48, 52, 55, 65, 69, 70,
    105, 123, 138
  common-base, 52
  common-emitter, 55, 123
  diode, 33
  FET, 65, 69, 105, 138
  MOSFET, 69, 70
  tunnel diode, 48
Charge carrier, 4, 6
Circuit efficiency, 530
  for class-A stage, 530, 533
  for class-B stage, 538
  for class-C stage, 552
Clock gate, 457
Clock signal generation, 449
CMOS, 393–395
Code conversion, 447
Coincident-current memory, 475
Collector, 49
Collector-base bias, 94–95
Collector leakage current, 343
Common base amplifier, 186
Common base configuration, 52–54
Common-collector/common-base pair,
    318
Common-collector/common-emitter pair,
    317
Common-collector configuration, 56
Common-emitter configuration, 54–56
Common-emitter switch, 348
Common mode, 206, 252, 256
  impedance, 252
  rejection, 256
Communication satellites, 599
Communication systems, 596–640
Comparison section of power supply, 671
Compensation, 14, 99–101
  diode, 99
  impurity, 14
  temperature, 99–101
Compensation of op amp, 257–261
  lag, 260
Complementary emitter follower, 190
Component tolerance, 135
Computer organization, 423–428
Computer-aided design models, 144–152
Concentration gradient, 17
Conditional stability, 247

Conduction, 5, 11
  by electrons, 5, 11
  by holes, 5, 12
Conduction band, 3
Conductivity, 1, 12
  conductor, 1
  germanium, 2
  insulator, 1
  silicon, 2
Constant current source, 210
Construction processes, 71–77
Control section of power supply, 672
Core plane, 476
Core storage, 473–480
Corner frequency, 171, 175
Counter, 452
  down, 454
  synchronous, 454
Cross-over distortion, 537
Crystal, 5
Current components in the diode, 35
Current density, 12
Current flow by diffusion, 14
Current mode switching, 392
Current select lines, 475
Current sink, 110
Cutoff, 82, 91, 341
Cycle time of core memory, 480

Data entry gate, 479
Data register, 479
DC amplification, 202–205
Decibels, 172
Decoding, 439–445
  minimization of gates, 440
  minimization of inputs, 440
Degenerate doping, 47
Delay time, 348–352
Delayed clock, 451
DeMorgan's laws, 432
Depletion/enhancement MOSFET, 70
Depletion mode, 69
Depletion region, 28, 47, 49
  diode, 28
  transistor, 49
  tunnel diode, 47
Depletion-region capacitance, 41, 121, 271, 347, 350

average value, 351
of diode, 41–43
Deposition, 72
Destructive readout (DRO) memory, 475–479
Detection circuits, 619–622, 631–636
  diode, 619
  FM, 631
  heterodyne, 621
  phase-locked loop (PLL), 633
  superheterodyne, 622
D-factor, 277–352
Diamond structure, 5
Dielectric constant, 41
Difference circuit, 221
Differential amplifiers, 205
  common mode gain, 206
  monolithic, 213–214
  stability, 209
  voltage gain, 207
Diffusion capacitance, 43, 121, 272, 347
  of diode, 43–45
  of transistor, 272, 347
Diffusion constant, 17
  for electrons in germanium, 17
  for electrons in silicon, 17
  for holes in germanium, 17
  for holes in silicon, 17
Diffusion current, 14, 19, 31, 51
  in diode, 31
  in presence of recombination, 19
  in transistor base, 51
Diffusion equation, 17
Diffusion of impurities, 72
Diffusion windows, 76
Digital circuits, 341
Digital computer, 423
Digital-to-analog (D/A) converter, 566
  accuracy of, 568
  binary weighted, 568
  conversion factors of, 566
  ladder type, 569
  resolution of, 568
Diode, 26, 29, 35
  germanium, 29
  silicon, 35
Diode-biased stage, 110
Diode equation, 32

Diode-steering network, 416
Discrete energy levels, 3
Dissipation, 522–527
   allowable, 522
Distortion, 82, 124, 537
   cross-over, 537
   even harmonic, 539
   in class-B stage, 539
   in power stages, 560
Donor atoms or impurities, 8, 9, 10, 27
Doping, 5, 7, 47
   degenerate, 47
Double-ended output, 206
Drain, 61
   of FET, 61
Drift, 205, 253–256
   referred to input, 205
Drift current, 14, 20
Drift velocity, 12
DTL, 388
Dynamic resistance, 40, 120

Ebers-Moll equations, 147
Ebers-Moll model of transistor, 145–148,
   342, 349
ECAP, 149
ECAP II, 148, 349
ECL, 392
Edge-triggered JK, 419
Effective mass, 11
   of electron, 11
   of hole, 11, 12
Electron suppression, 11
Electrons, 2, 28
Emitter, 49
Emitter bias, 90–93
Emitter compensation, 290–294
Emitter efficiency, 146
Emitter follower, 56, 187–190
   complementary, 190
Empty states, 4
Encoding, 445–447
   decimal-to-BCD, 445
Energy, 2
   bands, 3
   diagram, 3, 47
   gap, 3, 6
   levels, 3

   states, 3
   thermal, 2, 6
Energy diagram of semiconductor, 7, 9
   with acceptor impurities, 7
   with donor impurities, 9
   tunnel diode, 47
Energy gap, 7
   germanium, 7
   silicon, 7
Enhancement mode, 69
Epitaxial region, 72, 74
Epitaxy, 72
Equivalent circuits, 48, 116–152
   diode, 117–122
   FET, 138–144
   transistor, 122–137
   tunnel diode, 48
Etching, 73
Excess base charge, 361
Exclusive OR gate, 433, 465
Expanders for logic gates, 440
Extraneous noise, 225

Fall time, 348
Fan-in, 386
Fan-out, 385, 428
Feedback, 220–263, 487
   current, 227
   factor, 221–234
   negative, 220, 246
   positive, 246, 487
   series, 227
   shunt, 227
   voltage, 227
Field-effect transistor (FET), 60–71, 105,
   168
   amplifier, 168
   characteristics, 105
   junction, 61–68
   MOS, 68–71
File memory, 424
Filters, 581, 648–661
   active, 581
   Bessel, 585
   Butterworth, 582
   capacitor, 650
   Chebychev, 582

L-section, 665
   maximally flat phase, 582
Flip-flop, 411–419, 447, 451, 456, 471
   D, 447
   JK, 417, 451
   master-slave, 456
   RS, 471
   T, 416
Forbidden gap, 47
   of tunnel diode, 47
Forbidden region, 3, 6
Forward-biased junction, 30
Foster-Seeley discriminator, 632
Free charge carrier, 4, 6
Free electrons, 5
Free-running multivibrator, 395
Frequency modulation, 625–636
   bandwidth requirements of, 630
   index, 626
   large deviation, 628
Frequency response, 172–184
Frequency spectrum, 270, 597
   classification of, 597

Gain bandwidth product, 250, 283, 334
Gain stability, 184–186, 222
Gallium, 7
Gallium-arsenide, 46
Gate, 61
   of FET, 61
Gated clock, 450
Gate-to-drain capacitance, 281
Gate-to-source capacitance, 281
Gate trigger voltage of SCR, 368
Gating of bistable, 416
Generation of carriers, 9, 17
   rate, 9, 18
Gradient of carriers, 17
Graphical analysis, 58
   of transistor, 58–60
Gyrator, 587–590
   equivalent circuit, 588
   impedance, 588

Harmonic distortion, 224
Heat sink, 523
Heterodyning, 621
High-frequency amplifiers, 270–318

High-frequency models, 120, 273, 281
   of diode, 120
   of FET, 281
   of transistor, 273
High-pass transfer function, 585
Holding current of SCR, 369
Hole suppression, 11
Holes, 5, 6, 28
HTL, 388
Hybrid-π circuit, 129, 271
Hybrid stages, 169
   FET-bipolar, 169
Hysteresis curve of ferrite core, 474

I_{co} variation with temperature, 102–103
Impurities, 5, 8
Incandescent lamp driver, 373
Index of modulation, 601, 623, 626
   AM, 601
   FM, 626
   PM, 623
Indium, 7
Inductance, 48
   lead inductance of tunnel diode, 48
Injection of carriers, 19, 22
Inhibit line, 476
Input characteristics of transistor, 60, 86
Input impedance of feedback amplifier, 231
Instability, 246, 487
Integrated circuits, 26, 109–112
   biasing, 109–112
Integrated circuit memory, 480
Integration, 71
   large-scale, 71
   medium-scale, 71
   small-scale, 71
Integrator, 569
Intermediate-frequency (IF) amplifier, 621
Intrinsic concentration, 7
   germanium, 7
   silicon, 7
Intrinsic conductivity, 14
   for germanium, 14
   for silicon, 14
Intrinsic material, 10
Inverter, 431

Ionization, 3, 45
Ionizing agent, 3
  electrical energy, 3
  optical energy, 3
  thermal energy, 3
Isolation diffusion, 75
Iterative stages, 295–309

Junction, 30, 33
  forward-biased, 30
  reverse-biased, 33
Junction formation, 27

Karnaugh map, 435–439
Kronig-Penny model, 11

Large-scale integration (LSI), 71, 379
Large-signal circuits, 341–375
Large-signal model, 117–119, 122–124, 138–141
  diode, 117–119
  FET, 138–141
  transistor, 122–124
Large-signal operation, 36
  of diode, 36–39
Lattice, 2, 3, 8, 28
Leakage current, 53
  collector to base, 53
Level shift stage, 204
Line printer, 425
Load-line, 36, 58, 106
  construction, 36–37
Logic circuits, 423–483
Logic families, 379, 385
Logic gates, 380
Low frequency falloff, 169–173, 178–182
  due to coupling capacitor, 169–173
  due to emitter-bypass capacitor, 178–181
  in FET, 182–184
Low-pass transfer function, 582–585
Lower sideband, 603
Lower trip point (LTP), 513
L-section filter, 655–661
  critical current, 658
  critical resistance, 658
  multiple, 660

Magnetic core storage, 473
Magnetic head, 469
Magnetic recording, 469–473
  NRZ, 471
  NRZI, 471
  RZ, 470
Majority carriers, 23, 27
Masking, 74
Master-slave JK, 419
Maxwell-Boltzmann statistics, 29
Measurement section of power supply, 668
Medium-scale integration (MSI), 71, 379
Memory, 480
  address, 477
  core, 473
  cycle, 480
  integrated circuit, 480
  ROM, 482
Metal-semiconductor junction, 26
Mica washer, 524
Miller effect, 274, 282, 315, 347
Minimization of logic functions, 434
Minority carrier distribution in a transistor, 49–50
Minority carriers, 23, 27
Mismatching of stages, 326
Mobility of electrons, 12
  in germanium, 12
  in silicon, 12
Mobility of holes, 13
  in germanium, 13
  in silicon, 13
Modeling, 116–152
Models, 117–152
  computer-aided design, 144–152
  diode, 117–122
  Ebers-Moll, 145
  FET, 138–144
  piecewise linear, 149–152
  transistor, 122–137
  tunnel diode, 151
Modulating signal, 600
Modulation, 598–618, 622–640
  amplitude, 599–618
  frequency, 625–636
  phase, 622–625
  pulse, 636–640

Modulator, 512, 603
 balanced, 616
 base, 610–613
 class-B, 610
 class-C, 610
 collector, 614–615
 pulse-position, 512
 pulse-width, 512
 transistor, 603–606
 tuned-collector, 626
Momentum, 3
 electron spin, 3
 orbital angular, 3
Monostable multivibrator, 395–407, 506
MOS field-effect transistor (MOSFET), 26, 68–71
Multiplexing, 598
 frequency-division, 598
Multistage amplifiers, 166–169
Multivibrator, 506
 applications, 514
 astable, 506
 one-shot, 506
Multivibrator period control, 506–512
 capacitor variation, 506
 current, 509
 proportional, 510
 resistor variation, 507
 voltage, 509

NAND gate, 384
Narrowband amplifier, 271, 322–335
Negative logic, 379, 428
Negative resistance, 46, 330
 of tunnel diode, 46–48
NET-1, 349
Neutralization of $C_{b'c}$, 324
Neutralizing network, 325
Nickel-chromium, 74
n-MOS, 393
Noise, 225
 extraneous, 225
 margin, 385
 $1/f$, 225
 white, 225
Nondestructive readout (NDRO) memory, 479
Nonlinear distortion, 224

Nonretriggerable monostable, 407
Nonreturn to zero (NRZ) code, 448
NOR gate, 383
Nucleus, 2

Octal code, 428
Offset, 253–256
 current, 254
 voltage, 254
Offset voltage of diode, 117
Ohmic resistance, 36, 120
 of diode, 36, 120
 of transistor base, 125
One-shot multivibrator, 395–407, 449, 506
 nonretriggerable, 449
Operational amplifier or op amp, 239, 248–263, 310, 566–594
 applications, 566–594
 differential, 248
 ideal, 248
 nonideal, 252
OR gate, 379–381, 428
Orbit, 2
Oscillator, 492
 Colpitts, 501–503
 crystal-controlled, 492, 505
 Hartley, 501
 phase shift, 492–498
 RC, 492–496
 tuned-circuit, 498–501
 Wien bridge, 503–505
Output impedance of feedback amplifier, 232
Overdrive factor, 357

Parent atoms, 2, 5
Passivation, 72
Passive delay time, 349–352
Pauli exclusion principle, 3
Peak inverse voltage (PIV), 653
Piecewise linear approximation, 38
Piecewise linear circuit, 116, 149
Phase-locked loop (PLL), 633
Phase modulation, 622–625
 index, 623
 large deviation, 623
Phase-shift filters, 592–594

Phosphorus, 8, 72
Photolithography, 72
Photoresist, 73
Pinchoff, 65
　region, 65
　voltage, 66
Planar process, 71
p-MOS, 393
pn-junction, 26
Positive feedback, 487
Positive impedance inverter, 587
Positive logic, 379, 428
Power amplifier, 519–560
　class-A, 519, 528–535
　class-AB, 520
　class-B, 519, 536–547
　class-C, 519, 548–554
　complementary emitter followers, 542
　integrated circuit, 545
　pulse-width modulated, 554–558
　resistive load, 528
　short-circuit protection, 546
　transformer-coupled, 531
　two-state, 521, 554–558
Power gain, 558
Power in AM waveform, 615
Power supplies, 643–678
　filters, 648–661
　high-current, 674–678
　regulated, 668
Power transistor, 527
Preamp, 162, 226
Propagation delay time, 386
Protons, 2
p-type material, 8
Pulse-amplitude modulation (PAM), 636
Pulse-code modulation (PCM), 636
Pulse-position modulation (PPM), 512, 636
Pulse repetition rate of bistable, 411
Pulse-width modulation (PWM), 512, 636

Q of a coil, 320

Race condition in logic systems, 445
Rate effect, 370
Ratio detector, 632

Read gate, 480
Read-only memory, 482
Read-write cycle of core memory, 480
Recombination of carriers, 9, 17, 20, 51, 361
　average time, 18
　current, 20, 361
　rate, 9, 18
　in transistor base, 51
Recovery time of multivibrator, 398, 408
Rectifiers, 644–647
　bridge, 647
　full-wave, 645
　half-wave, 644
Reference section of power supply, 670
Reflected impedance, 242
Regeneration, 397
Registers, 448, 460–464
　dynamic, 460
　input/output, 463
　shift, 460
　static, 460
Regulated power supply, 668–674
　monolithic, 673
Regulation, 648, 661
　line, 648
　load, 648
Regulators, 661
　emitter-follower, 667
　Zener diode, 661
Relaxation oscillator, 488
Resistance, 48
　bulk resistance of tunnel diode, 48
Resonant frequency, 321
Retriggerable monostable, 407
Return ratio, 246, 488
Return to zero (RZ) code, 448
Reverse saturation current, 32
Reversible current driver, 373
Ring counter using SCR, 374
Ripple factor, 648
Rise time, 348
RTL, 387

Sallen and Key resonator, 590–592
Saturation, 59, 82, 91, 341, 348, 361
　current, 32
　region, 54

storage time, 348, 361
  voltage, 343
SCEPTRE, 148, 349
Schmitt trigger, 512
Schottky-clamped TTL, 391
Schottky diode, 391
SCR, 366–375
Self-bias resistance, 105, 143
Sensitivity, 222, 590
  of circuit Q, 590
  function, 222
Serial transmission, 449
Series controller, 672
Shift pulse, 448
Shockley diode, 366
Short-circuit current gain, 276
Shunt controller, 672
Shunt peaking, 304–307
Shunt-series cascade, 315
Signal-to-noise ratio, 224–226
  improvement with feedback, 225
Silicon-controlled rectifier (SCR),
    366–375, 674–678
Silicon-controlled switch, 366
Silicon nitride, 69
Silicon resistor, 210
Single-ended output, 206
Single-sideband, 616–618
Slew rate of op amp, 261–262
Small-scale integration (SSI), 71, 379
Small-signal model, 120, 124–137
  common-base, 126, 131
  common-emitter, 128, 131
  diode, 120
  emitter follower, 136
  FET, 141–144
  transistor, 124–137
Small-signal operation, 39
  of diode, 39–41
Small-signal resistance, 40
  of diode, 40
SNR, 224
Source, 61
  of FET, 61
Speed-up capacitor, 286, 356
Split-phase clock, 451
Stability, 88–99, 107, 246, 489
  amplitude, 489

factor, 88
of feedback amplifier, 246
of FET, 107
frequency, 489
parameter, 96
with respect to changes in β, 95
Start bit, 449
Storage time constant, 347
Substrate, 72
Subtractor, 468
  binary, 468
Superheterodyne receiver, 622
Suppression, 11
  of electrons, 11
  of holes, 11
Sure-starting astable, 408
Sweep circuits, 569–581
  bootstrap, 571–578
  deviation from linearity of, 570, 581
  Miller, 578–581
Switching, 341, 347
  circuits, 341
  high-speed, 347
  low-speed, 341
Synchronization, 448, 458
Synchronous counter, 454

Tank circuit, 321
Tantalum, 74
Teletype terminal, 424
Temperature coefficient, 223
Temperature drift, 205
  in dc amplifiers, 205
Temperature stability, 89, 107
  of FET, 107
Thermal energy, 2, 6
Thermal resistance of transistor, 522
Thermal runaway, 533
Three-db frequency, 171
Thyristor, 366
Timing circuits, 448
Tin oxide, 74
Totem-pole output stage, 547
Transconductance, 68, 139
  of FET, 68
Transformer, 498
Transient response of transistor switch,
    348–365

Transistor, 49
  bipolar, 49–60
Transit time, 281
Transition region, 28
Trigger recovery time, 417
Truth table, 381, 429–439
TTL, 388–391
Tuned amplifier, 271, 319–335
Tunnel current, 47
Tunnel diode, 46–48, 151, 330–335, 365
  amplifier, 330–335
  model, 151
  switch, 365
Turn-off time, 358–360, 370
  of SCR, 370
Turn-on time, 353–358, 370
  of SCR, 370

Unilateral equivalent circuit, 275
Unloaded gain, 163
Upper sideband, 603
Upper trip point (UTP), 513

V_{BE} variation with temperature, 104
Valence band, 3, 6, 7, 9

Valence electrons, 5, 8, 10, 45
Vapor diffusion, 27
Velocity of electron, 11
Vestigial-sideband, 618
Video amplifier, 271
Virtual ground, 250
Voltage-controlled resistance, 61
Voltage doubler, 647
Voltage gain, 59
Voltage stability factor, 88–95
Voltage transfer, 162
Voltage variable resistor (VVR), 624
Volume control, 192

Wave shaping, 487
White noise, 224
Wideband amplifier, 271
Worst-case design, 98
Write gate, 479

Zener breakdown, 36
Zener-coupled dc amplifier, 203
Zener diode, 45–46, 661
  equivalent circuit, 662
Zone refining, 72